Penetration of light into aquatic ecosystems is greatly affected by the absorption and scattering processes that take place within the water. Thus within any water body, the intensity and colour of the light field changes greatly with depth. This has a marked influence on both the total productivity of, and the kinds of plant that predominate in, the ecosystem.

This study presents an integrated and coherent treatment of the key role of light in aquatic ecosystems. It ranges from the physics of light transmission within water, through the biochemistry and physiology of aquatic photosynthesis to the ecological relationships which depend on the underwater light climate.

It will be required reading for all those working on aquatic photosynthesis and productivity, and will be an invaluable source of information for students and researchers in plant physiology, limnology and oceanography.

Light and photosynthesis in aquatic ecosystems

Light and photosynthesis in aquatic ecosystems

Second edition

JOHN T. O. KIRK

Division of Plant Industry, Commonwealth Scientific and Industrial Research Organization, Canberra, Australia

Published by the Press Syndicate of the University of Cambridge
The Pitt Building, Trumpington Street, Cambridge CB2 1RP
40 West 20th Street, New York, NY 10011-4211, USA
10 Stamford Road, Oakleigh, Melbourne 3166, Australia

First published 1983
Second edition 1994
Reprinted 1996

Printed in Great Britain by J. W. Arrowsmith Ltd, Bristol

A catalogue record for this book is available from the British Library

Library of Congress cataloguing in publication data
Kirk, John T. O. (John Thomas Osmond), 1935–
Light and photosynthesis in aquatic ecosystems / John T. O. Kirk. – 2nd ed.
p. cm.
Includes bibliographical references (p.) and indexes.
ISBN 0 521 45353 4 (hc). – ISBN 0 521 45966 4 (pb)
1. Photosynthesis. 2. Plants, Effect of underwater light on. 3. Aquatic plants – Ecophysiology. 4. Underwater light. I. Title.
QK882.K53 1994
574.5′263–dc20 93–37395 CIP

ISBN 0 521 45353 4 hardback
ISBN 0 521 45966 4 paperback

SE

Contents

Preface to the first edition

Four things are required for plant growth: energy in the form of solar radiation; inorganic carbon in the form of carbon dioxide or bicarbonate ions; mineral nutrients; and water. Those plants which, in the course of evolution, have remained in, or have returned to, the aquatic environment have one major advantage over their terrestrial counterparts: namely, that water – lack of which so often limits productivity in the terrestrial biosphere – is for them present in abundance; but for this a price must be paid. The medium – air – in which terrestrial plants carry on photosynthesis offers, within the sort of depth that plant canopies occupy, no significant obstacle to the penetration of light. The medium – water – in which aquatic plants occur, in contrast, both absorbs and scatters light. For the phytoplankton and the macrophytes in lakes and rivers, coastal and oceanic waters, both the intensity and spectral quality of the light vary markedly with depth. In all but the shallowest waters, light availability is a limiting factor for primary production by the aquatic ecosystem. The aquatic plants must compete for solar radiation not only with each other (as terrestrial plants must also do), but also with all the other light-absorbing components of the aquatic medium. This has led, in the course of evolution, to the acquisition by each of the major groups of algae of characteristic arrays of light-harvesting pigments which are of great biochemical interest, and also of major significance for an understanding both of the adaptation of the algae to their ecological niche and of the phylogeny and taxonomy of the different algal groups. Nevertheless, in spite of the evolution of specialized light-harvesting systems, the aquatic medium removes so much of the incident light that aquatic ecosystems are, broadly speaking, less productive than terrestrial ones.

Thus, the nature of the light climate is a major difference between the terrestrial and the aquatic regions of the biosphere. Within the underwater environment, light availability is of major importance in determining how much plant growth there is, which kinds of plant predominate and, indeed, which kinds of plants have evolved. It is not the whole story – biotic factors, availability of inorganic carbon and mineral nutrients, temperature, all make their contribution – but it is a large part of that story. This book is a study of light in the underwater environment from the point of view of photosynthesis. It sets out to bring together the physics of light transmission through the medium and capture by the plants, the biochemistry of photosynthetic light-harvesting systems, the physiology of the photosynthetic response of aquatic plants to different kinds of light field, and the ecological relationships which depend on the light climate. The book does not attempt to provide as complete an account of the physical aspects of underwater light as the major works by Jerlov (1976) and Preisendorfer (1976); it is aimed at the limnologist and marine biologist rather than the physicist, although physical oceanographers should find it of interest. Its intention is to communicate a broad understanding of the significance of light as a major factor determining the operation and biological composition of aquatic ecosystems. It is hoped that it will be of value to practising aquatic scientists, to university teachers who give courses in limnology or marine science, and to postgraduate and honours students in these and related disciplines.

Certain features of the organization of the book merit comment. Although in some cases authors and dates are referred to explicitly, to minimize interruptions to the text, references to published work are in most cases indicated by the corresponding numbers in the complete alphabetical reference list at the end of the book. Accompanying each entry in the reference list is (are) the page number(s) where that paper or book is referred to in the text. Although coverage of the field is, I believe, representative, it is not intended to be encyclopaedic. The papers referred to have been selected, not only on the grounds of their scientific importance, but in large part on the basis of their usefulness as illustrative examples for particular points that need to be made. Inevitably, therefore, many equally important and relevant papers have had to be omitted from consideration, especially in the very broad field of aquatic ecology. I have therefore, where necessary, referred the reader to more specialized works in which more comprehensive treatments of particular topics can be found. Because its contribution to total aquatic primary production is usually small I have not attempted to deal with bacterial photosynthesis, complex and fascinating though it is.

I would like to thank Professor D. Branton, Dr M. Bristow, Mr S. Craig,

Dr W. A. Hovis, Dr S. Jeffrey, Dr D. Kiefer, Professor V. Klemas, Professor L. Legendre, Dr Y. Lipkin, Professor W. Nultsch, Mr D. Price, Professor R. C. Smith, Dr M. Vesk, and Biospherical Instruments Inc., who have provided original copies of figures for reproduction in this work, and Mr F. X. Dunin and Dr P. A. Tyler for unpublished data.

John Kirk

Canberra
November 1982

Preface to the second edition

The behaviour of sunlight in water, and the role which light plays in controlling the productivity, and influencing the biological composition, of aquatic ecosystems have been important areas of scientific study for more than a century, and it was to meet the perceived need for a text bringing together the physical and biological aspects of the subject, that *Light and photosynthesis in aquatic ecosystems* was written. The book was well received, and is in use not only by research workers but also in university courses. In the eleven years since the first edition, interest in the topic has become, if anything, even greater than it was was before. This may be partly attributed to concern about global warming, and the realization that to understand the important role the ocean plays in the global carbon cycle, we need to improve both our understanding and our quantitative assessment of marine primary production. An additional, but related, reason is the great interest that has been aroused in the feasibility of remote sensing of oceanic primary productivity from space. The potentialities were just becoming apparent with the early CZCS pictures when the first edition was written. The continuing stream of further CZCS studies in the ensuing years, enormously enlarging our understanding of oceanic phytoplankton distribution, and the announced intention by space agencies around the world to put new and improved ocean scanners into space, have made this a particularly active and exciting field within oceanography.

But the light flux that is received from the ocean by the satellite-borne radiometers, and which carries with it information about the composition of the water, originates in fact as a part of the upwelling light flux within the ocean which has escaped through the surface into the atmosphere. To interpret the data we therefore need to understand the underwater light field, and how its characteristics are controlled by what is present within the aquatic medium.

In consequence of this sustained, even intensified, interest in underwater light, there is a continued need for a suitable text, not only for researchers, but also for use in university teaching. It is for this reason, the first edition being out of print, that I have prepared a completely revised version. Since marine bio-optics has been such an active field, a vast amount of literature had to be digested, but as in the first edition, I have tended to select specific papers mainly on the basis of their usefulness as illustrative examples, and many other equally valuable papers have had to be omitted from consideration. Even so, there are nearly twice as many literature references in this edition as there were in the first.

I would like to thank Mr K. Lyon of Orbital Sciences Corporation for providing illustrations of the SeaWiFS scanner and spacecraft for reproduction in this work. I would also like to thank the Ocean Optics Program of the Office of Naval Research (USA) for support during the writing of this book, under grant N00014-91-J-1366.

John Kirk

Canberra
July 1993

Part I

The underwater light field

1

Concepts of hydrologic optics

1.1 Introduction

The purpose of the first part of this book is to describe and explain the behaviour of light in natural waters. The word 'light' in common parlance refers to radiation in that segment of the electromagnetic spectrum – about 400 to 700 nm – to which the human eye is sensitive. Our prime concern is not with vision but with photosynthesis. Nevertheless, by a convenient coincidence, the waveband within which plants can photosynthesize corresponds approximately to that of human vision and so we may legitimately refer to the particular kind of solar radiation with which we are concerned simply as 'light'.

Optics is that part of Physics which deals with light. Since the behaviour of light is greatly affected by the nature of the medium through which it is passing, there are different branches of optics dealing with different kinds of physical systems. The relations between the different branches of the subject and of optics to fundamental physical theory, are outlined diagrammatically in Fig. 1.1. Hydrologic optics is concerned with the behaviour of light in aquatic media. It can be subdivided into limnological and oceanographic optics according to whether fresh, inland or salty, marine waters are under consideration. Hydrologic optics has, however, up to now been mainly oceanographic in its orientation.

1.2 The nature of light

Electromagnetic energy occurs in indivisible units referred to as *quanta* or *photons*. Thus a beam of sunlight in air consists of a continual stream of photons travelling at 3×10^8 m s^{-1}. The actual numbers of quanta involved are very large. In full summer sunlight, for example, 1 m^2 of horizontal surface receives about 10^{21} quanta s^{-1} of visible light.

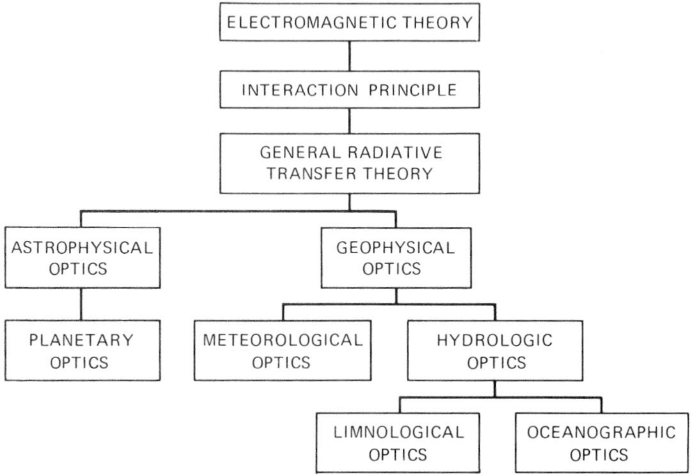

Fig. 1.1. The relation between hydrologic optics and other branches of optics (after Preisendorfer, 1976).

Electromagnetic radiation, despite its particulate nature, behaves in some circumstances as though it has a wave nature. Every photon has a wavelength, λ, and a frequency, ν. These are related in accordance with

$$\lambda = c/\nu \tag{1.1}$$

where c is the speed of light. Since c is constant in a given medium, the greater the wavelength, the lower the frequency. If c is expressed in m s^{-1}, and ν in cycles s^{-1}, then the wavelength, λ, is expressed in m. For convenience, however, wavelength is more commonly expressed in nanometres, a nanometre (nm) being equal to 10^{-9} m.

The energy, ϵ, in a photon varies with the frequency, and therefore inversely with the wavelength, the relation being

$$\epsilon = h\nu = hc/\lambda \tag{1.2}$$

where h is Planck's constant and has the value of 6.63×10^{-34} J s. Thus, a photon of wavelength 700 nm, from the red end of the photosynthetic spectrum, contains only 57% as much energy as a photon of wavelength 400 nm, from the blue end of the spectrum. The actual energy in a photon of wavelength λ nm is given by the relation

$$\epsilon = (1988/\lambda) \times 10^{-19} \text{ J} \tag{1.3}$$

A monochromatic radiation flux expressed in quanta s^{-1} can thus readily be converted to J s^{-1}, i.e. to watts (W). Conversely, a radiation flux, Φ, expressed in W, can be converted to quanta s^{-1} using the relation

$$\text{quanta s}^{-1} = 5.03\ \Phi\lambda \times 10^{15} \tag{1.4}$$

In the case of radiation covering a broad spectral band, such as for example the photosynthetic waveband, a simple conversion from quanta s^{-1} to W, or *vice versa*, cannot be carried out accurately since the value of λ varies across the spectral band. If the distribution of quanta or energy across the spectrum is known, then conversion can be carried out for a series of relatively narrow wavebands covering the spectral region of interest and the results summed for the whole waveband. Alternatively, an approximate conversion factor which takes into account the spectral distribution of energy that is likely to occur may be used. For solar radiation in the 400–700 nm band above the water surface, Morel & Smith (1974) found that the factor (Q: W) required to convert W to quanta s^{-1} was 2.77×10^{18} quanta s^{-1} W^{-1} to an accuracy of plus or minus a few per cent, regardless of the meteorological conditions.

As we shall discuss at length in a later section (§6.2) the spectral distribution of solar radiation under water changes markedly with depth. Nevertheless Morel & Smith found that for a wide range of marine waters the value of Q: W varied by no more than $\pm 10\%$ from a mean of 2.5×10^{18} quanta s^{-1} W^{-1}. As expected from eqn 1.4, the greater the proportion of long-wavelength (red) light present, the greater the value of Q: W. For yellow inland waters with more of the underwater light in the 550–700 nm region (see §6.2), by extrapolating the data of Morel & Smith we arrive at a value of approximately 2.9×10^{18} quanta s^{-1} W^{-1} for the value of Q: W.

In any medium, light travels more slowly than it does in a vacuum. The velocity of light in a medium is equal to the velocity of light in a vacuum, divided by the refractive index of the medium. The refractive index of air is 1.000 28, which for our purposes is not significantly different from that of a vacuum (exactly 1.0, by definition), and so we may take the velocity of light in air to be equal to that in a vacuum. The refractive index of water, although it varies somewhat with temperature, salt concentration and wavelength of light, may with sufficient accuracy be regarded as equal to 1.33 for all natural waters. Assuming that the velocity of light in a vacuum is 3×10^8 m s^{-1}, the velocity in water is therefore about 2.25×10^8 m s^{-1}. The frequency of the radiation remains the same in water but the wavelength diminishes in proportion to the decrease in velocity. When referring to monochromatic radiation, the wavelength we shall attribute to it is that which it has in a vacuum. Because c and λ change in parallel, eqns 1.2, 1.3

and 1.4 are as true in water as they are in a vacuum: furthermore, when using eqns 1.3 and 1.4, it is the value of the wavelength in a vacuum which is applicable, even when the calculation is carried out for underwater light.

1.3 The properties defining the radiation field

If we are to understand the ways in which the prevailing light field changes with depth in a water body, then we must first consider what are the essential attributes of a light field, in which changes might be anticipated. The definitions of these attributes, in part, follow the report of the Working Groups set up by the International Association for the Physical Sciences of the Ocean (1979), but are also influenced by the more fundamental analyses given by Preisendorfer (1976). A recent account of the definitions and concepts used in hydrologic optics is that by Mobley (1992).

We shall generally express direction within the light field in terms of the *zenith angle*, θ (the angle between a given light pencil, i.e. a thin parallel beam, and the upward vertical), and the *azimuth angle*, ϕ (the angle between the vertical plane incorporating the light pencil and some other specified vertical plane such as the vertical plane of the sun). In the case of the upwelling light stream it will sometimes be convenient to express a direction in terms of the *nadir angle*, θ_n (the angle between a given light pencil and the downward vertical). These angular relations are illustrated in Fig. 1.2.

Radiant flux. Φ, is the time rate of flow of radiant energy. It may be expressed in W (J s^{-1}) or quanta s^{-1}.

Radiant intensity, I, is a measure of the radiant flux per unit solid angle in a specified direction. The radiant intensity of a source in a given direction is the radiant flux emitted by a point source, or by an element of an extended source, in an infinitesimal cone containing the given direction, divided by that element of solid angle. We can also speak of radiant intensity at a point in space. This, the *field* radiant intensity, is the radiant flux at that point in a specified direction in an infinitesimal cone containing the given direction, divided by that element of solid angle. I has the units W (or quanta s^{-1}) steradian^{-1}.

$$I = \mathrm{d}\Phi/\mathrm{d}\omega$$

If we consider the radiant flux not only per unit solid angle but also per unit area of a plane at right angles to the direction of flow, then we arrive at the even more useful concept of radiance, L. Radiance at a point in space is the radiant flux at that point in a given direction per unit solid angle per unit area at right angles to the direction of propagation. The meaning of this

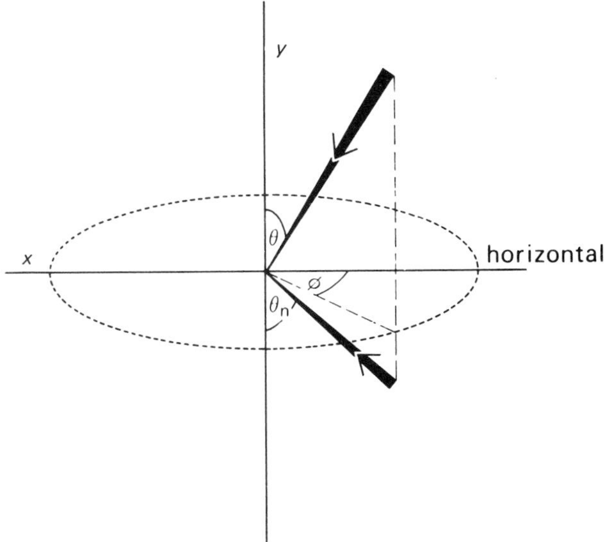

Fig. 1.2. The angles defining direction within a light field. The figure shows a downward and an upward pencil of light, both, for simplicity, in the same vertical plane. The downward pencil has zenith angle θ; the upward pencil has nadir angle θ_n, which is equivalent to a zenith angle of $(180° - \theta_n)$. Assuming the *xy* plane is the vertical plane of the Sun, or other reference vertical plane, then ϕ is the azimuth angle for both light pencils.

field radiance is illustrated in Figs. 1.3*a* and *b*. There is also *surface* radiance which is the radiant flux emitted in a given direction per unit solid angle per unit projected area (apparent unit area, seen from the viewing direction) of a surface: this is illustrated in Fig. 1.3*c*.

To indicate that it is a function of direction, i.e. of both zenith and azimuth angle, radiance is commonly written as $L(\theta, \phi)$. The angular structure of a light field is expressed in terms of the variation of radiance with θ and ϕ. Radiance has the units W (or quanta s^{-1}) m^{-2} steradian^{-1}.

$$L(\theta, \phi) = d^2\Phi/dS\cos\theta\, d\omega$$

Irradiance (at a point of a surface), E, is the radiant flux incident on an infinitesimal element of a surface, containing the point under consideration, divided by the area of that element. Less rigorously, it may be defined as the radiant flux per unit area of a surface.* It has the units W m^{-2}, or quanta (or photons) s^{-1} m^{-2}, or mol quanta (or photons) s^{-1} m^{-2}, where

* Terms such as 'fluence rate' or 'photon fluence rate', sometimes to be found in the plant physiological literature, are superfluous, and should not be used.

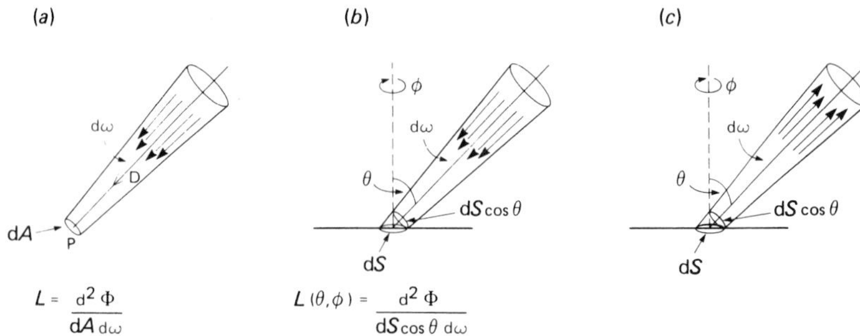

Fig. 1.3. Definition of radiance. (*a*) Field radiance at a point in space. The field radiance at P in the direction D is the radiant flux in the small solid angle surrounding D, passing through the infinitesimal element of area dA at right angles to D, divided by the element of solid angle and the element of area. (*b*) Field radiance at a point in a surface. It is often necessary to consider radiance at a point on a surface, from a specified direction relative to that surface. dS is the area of a small element of surface. $L(\theta, \phi)$ is the radiance incident on dS at zenith angle θ (relative to the normal to the surface) and azimuth angle ϕ: its value is determined by the radiant flux directed at dS within the small solid angle, dω, centred on the line defined by θ and ϕ. The flux passes perpendicularly across the area dS cos θ, which is the projected area of the element of surface, dS, seen from the direction θ, ϕ. Thus the radiance on a point in a surface, from a given direction, is the radiant flux in the specified direction per unit solid angle per unit projected area of the surface. (*c*) Surface radiance. In the case of a surface which emits radiation the intensity of the flux leaving the surface in a specified direction is expressed in terms of the surface radiance which is defined in the same way as the field radiance at a point in a surface except that the radiation is considered to flow away from, rather than on to, the surface.

1.0 mol photons is 6.02×10^{23} (Avogadro's number) photons. One mole of photons is frequently referred to as an *einstein*.

$$E = \mathrm{d}\Phi/\mathrm{d}S$$

Downward irradiance, E_d, and *upward irradiance*, E_u, are the values of the irradiance on the upper and the lower faces, respectively, of a horizontal surface. Thus, E_d is the irradiance due to the downwelling light stream and E_d is that due to the upwelling light stream.

The relation between irradiance and radiance can be understood with the help of Fig. 1.3*b*. The radiance in the direction defined by θ and ϕ is $L(\theta, \phi)$ W (or quanta s^{-1}) per unit projected area per steradian (sr). The projected area of the element of surface is dS cos θ and the corresponding element of solid angle is dω. Therefore the radiant flux on the element of surface within the solid angle dω is $L(\theta, \phi)$dS cos θ dω. The area of the element of surface is dS and so the irradiance at that point in the surface where the element is located, due to radiant flux within dω, is $L(\theta, \phi)$ cos θ dω. The total downward irradiance at that point in the surface is obtained by integrating with respect to solid angle over the whole upper hemisphere

$$E_{\mathrm{d}} = \int_{2\pi} L(\theta, \phi) \cos\theta \, \mathrm{d}\omega \tag{1.5}$$

The total upward irradiance is related to radiance in a similar manner except that allowance must be made for the fact that cos θ is negative for values of θ between 90 and 180°

$$E_{\mathrm{u}} = -\int_{-2\pi} L(\theta, \phi) \cos\theta \, \mathrm{d}\omega \tag{1.6}$$

Alternatively the cosine of the nadir angle, θ_{n} (see Fig. 1.2), rather than of the zenith angle, may be used

$$E_{\mathrm{u}} = \int_{-2\pi} L(\theta_{\mathrm{n}}, \phi) \cos\theta_{\mathrm{n}} \, \mathrm{d}\omega \tag{1.7}$$

The -2π subscript is simply to indicate that the integration is carried out over the 2π sr solid angle in the *lower* hemisphere.

The *net downward irradiance*, $\vec{E}$, is the difference between the downward and the upward irradiance

$$\vec{E} = E_{\mathrm{d}} - E_{\mathrm{u}} \tag{1.8}$$

It is related to radiance by the eqn

$$\vec{E} = \int_{4\pi} L(\theta, \phi) \cos\theta \, \mathrm{d}\omega \tag{1.9}$$

which integrates the product of radiance and cos θ over all directions: the fact that cos θ is negative between 90 and 180° ensures that the contribution of upward irradiance is negative in accordance with eqn 1.8. The net downward irradiance is a measure of the net rate of transfer of energy downwards at that point in the medium, and as we shall see later is a concept which can be used to arrive at some valuable conclusions.

The *scalar irradiance*, E_0, is the integral of the radiance distribution at a point over all directions about the point

$$E_0 = \int_{4\pi} L(\theta, \phi) \, \mathrm{d}\omega \tag{1.10}$$

Scalar irradiance is thus a measure of the radiant intensity at a point, which treats radiation from all directions equally. In the case of irradiance, on the other hand, the contribution of the radiation flux at different angles varies in proportion to the cosine of the zenith angle of incidence of the radiation: a phenomenon based on purely geometrical relations (Fig. 1.3, eqn 1.5), and sometimes referred to as the Cosine Law. It is useful to divide the scalar irradiance into a downward and an upward component. The *downward scalar irradiance*, E_{0d}, is the integral of the radiance distribution over the upper hemisphere

$$E_{0d} = \int_{2\pi} L(\theta, \phi)\, \mathrm{d}\omega \tag{1.11}$$

The *upward scalar irradiance*, E_{0u}, is defined in a similar manner for the lower hemisphere

$$E_{0u} = \int_{-2\pi} L(\theta, \phi)\, \mathrm{d}\omega \tag{1.12}$$

Scalar irradiance (total, upward, downward) has the same units as irradiance.

It is always the case in real-life radiation fields that irradiance and scalar irradiance vary markedly with wavelength across the photosynthetic range. This variation has a considerable bearing on the extent to which the radiation field can be used for photosynthesis. It is expressed in terms of the variation of irradiance or scalar irradiance per unit spectral distance (in units of wavelength or frequency, as appropriate) across the spectrum. Typical units would be W (or quanta s^{-1}) m^{-2} nm^{-1}.

If we know the radiance distribution over all angles at a particular point in a medium then we have a complete description of the angular structure of the light field. A complete radiance distribution, however, covering all zenith and azimuth angles, at reasonably narrow intervals, represents a large amount of data: with 5° angular intervals, for example, the distribution will consist of 1369 separate radiance values. A simpler, but still very useful, way of specifying the angular structure of a light field is in the form of the three average cosines – for downwelling, upwelling and total light – and the irradiance reflectance.

The average cosine for downwelling light, $\bar{\mu}_d$, at a particular point in the radiation field, may be regarded as the average value, in an infinitesimally small volume element at that point in the field, of the cosine of the zenith angle of all the downwelling photons in the volume element. It can be calculated by summing (i.e. integrating) for all elements of solid angle ($\mathrm{d}\omega$) comprising the upper hemisphere, the product of the radiance in that element of solid angle and the value of cos θ (i.e. $L(\theta, \phi)$ cos θ), and then dividing by the total radiance originating in that hemisphere. By inspection of eqns 1.5 and 1.11 it can be seen that

$$\bar{\mu}_d = E_d / E_{0d} \tag{1.13}$$

i.e. the average cosine for downwelling light is equal to the downward irradiance divided by the downward scalar irradiance.

The average cosine for upwelling light, $\bar{\mu}_u$, may be regarded as the average value of the cosine of the nadir angle of all the upwelling photons at a particular point in the field. By a similar chain of reasoning to the above, we conclude that $\bar{\mu}_u$ is equal to the upward irradiance divided by the upward scalar irradiance

$$\bar{\mu}_u = E_u/E_{0u} \tag{1.14}$$

In the case of the downwelling light stream it is often useful to deal in terms of the reciprocal of the average downward cosine, referred to by Preisendorfer (1961) as the *distribution function* for downwelling light, D_d, which can be shown[494] to be equal to the mean pathlength per vertical metre traversed, of the downward flux of photons per unit horizontal area per second. Thus $D_d = 1/\bar{\mu}_d$. There is, of course, an analogous distribution function for the upwelling light stream, defined by $D_u = 1/\bar{\mu}_u$.

The average cosine, $\bar{\mu}$, for the total light at a particular point in the field may be regarded as the average value, in an infinitesimally small volume element at that point in the field, of the cosine of the zenith angle of all the photons in the volume element. It may be evaluated by integrating the product of radiance and cos θ over all directions and dividing by the total radiance from all directions. By inspection of eqns 1.8, 1.9 and 1.10, it can be seen that the average cosine for the total light is equal to the *net* downward irradiance divided by the scalar irradiance

$$\bar{\mu} = \frac{\vec{E}}{E_0} = \frac{E_d - E_u}{E_0} \tag{1.15}$$

That $E_d - E_u$ should be involved (rather than, say, $E_d + E_u$) follows from the fact that the cosine of the zenith angle is negative for all the upwelling photons ($90° < \theta < 180°$). Thus a radiation field consisting of equal numbers of downwelling photons at $\theta = 45°$ and upwelling photons at $\theta = 135°$ would have $\bar{\mu} = 0$.

The remaining parameter which provides information about the angular structure of the light field is the *irradiance reflectance* (sometimes called the *irradiance ratio*), R. It is the ratio of the upward to the downward irradiance at a given point in the field

$$R = E_u/E_d \tag{1.16}$$

In any absorbing and scattering medium, such as sea or inland water, all these properties of the light field change in value with depth (for which we use the symbol z): the change might typically be a decrease, as in the case of irradiance, or an increase, as in the case of reflectance. It is sometimes useful to have a measure of the rate of change of any given property with depth.

All the properties with which we have dealt that have the dimensions of radiant flux per unit area, diminish in value, as we shall see later, in an approximately exponential manner with depth. It is convenient with these properties to specify the rate of change of the logarithm of the value with depth since this will be approximately the same at all depths. In this way we may define the *vertical attenuation coefficient* for

downward irradiance

$$K_d = -\frac{d \ln E_d}{dz} = -\frac{1}{E_d}\frac{dE_d}{dz} \quad (1.17)$$

upward irradiance

$$K_u = -\frac{d \ln E_u}{dz} = -\frac{1}{E_u}\frac{dE_u}{dz} \quad (1.18)$$

net downward irradiance

$$K_E = -\frac{d \ln (E_d - E_u)}{dz} = -\frac{1}{(E_d - E_u)}\frac{d(E_d - E_u)}{dz} \quad (1.19)$$

scalar irradiance

$$K_0 = -\frac{d \ln E_0}{dz} = -\frac{1}{E_0}\frac{dE_0}{dz} \quad (1.20)$$

radiance

$$K(\theta,\phi) = -\frac{d \ln L(\theta,\phi)}{dz} = -\frac{1}{L(\theta,\phi)}\frac{dL(\theta,\phi)}{dz} \quad (1.21)$$

In recognition of the fact that the values of these vertical attenuation coefficients are to some extent a function of depth they may sometimes be written in the form $K(z)$.

1.4 The inherent optical properties

There are only two things that can happen to photons within water: they can be absorbed or they can be scattered. Thus if we are to understand what happens to solar radiation as it passes into any given water body, we need some measure of the extent to which that water absorbs and scatters light. The absorption and scattering properties of the aquatic medium for light of any given wavelength are specified in terms of the absorption coefficient, the scattering coefficient and the volume scattering function. These have been referred to by Preisendorfer (1961) as *inherent* optical properties, because their magnitudes depend only on the substances comprising the aquatic medium and not on the geometric structure of the light fields that may pervade it.

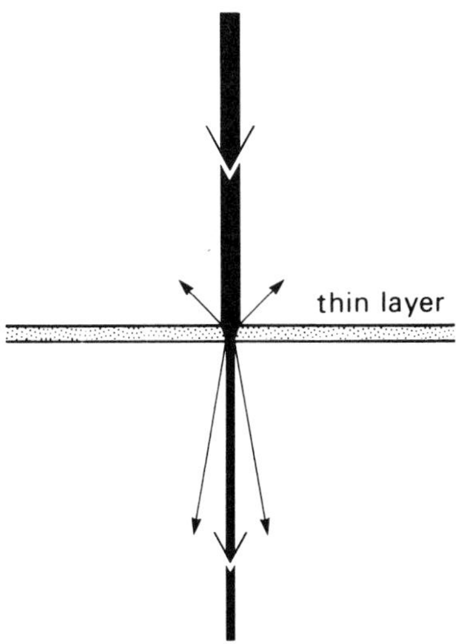

Fig. 1.4. Interaction of beam of light with a thin layer of aquatic medium. Of the light that is not absorbed, most is transmitted without deviation from its original path: some light is scattered, mainly in a forward direction.

They are defined with the help of an imaginary, infinitesimally thin, plane parallel layer of medium, illuminated at right angles by a parallel beam of monochromatic light (Fig. 1.4). Some of the incident light is absorbed by the thin layer. Some is scattered – that is, caused to diverge from its original path. The fraction of the incident flux which is absorbed, divided by the thickness of the layer, is the *absorption coefficient*, *a*. The fraction of the incident flux which is scattered, divided by the thickness of the layer, is the *scattering coefficient*, *b*.

To express the definitions quantitatively we make use of the quantities *absorptance*, *A*, and *scatterance*, *B*. If Φ_0 is the radiant flux (energy or quanta per unit time) incident in the form of a parallel beam on some physical system, Φ_a is the radiant flux absorbed by the system, and Φ_b is the radiant flux scattered by the system, then

$$A = \Phi_a/\Phi_0 \tag{1.22}$$

and

$$B = \Phi_b/\Phi_0 \tag{1.23}$$

i.e. absorptance and scatterance are the fractions of the radiant flux lost from the incident beam, by absorption and scattering, respectively. The sum of absorptance and scatterance is referred to as *attenuance*, *C*: it is the fraction of the radiant flux lost from the incident beam by absorption and scattering combined. In the case of the infinitesimally thin layer, thickness Δr, we represent the very small fractions of the incident flux which are lost by absorption and scattering as ΔA and ΔB, respectively. Then

$$a = \Delta A/\Delta r \tag{1.24}$$

and

$$b = \Delta B/\Delta r \tag{1.25}$$

An additional inherent optical property which we may now define is the *beam attenuation coefficient*, c. It is given by

$$c = a + b \tag{1.26}$$

and is the fraction of the incident flux which is absorbed and scattered, divided by the thickness of the layer. If the very small fraction of the incident flux which is lost by absorption and scattering combined is given the symbol ΔC (where $\Delta C = \Delta A + \Delta B$) then

$$c = \Delta C/\Delta r \tag{1.27}$$

The absorption, scattering and beam attenuation coefficients all have units of 1/length, and are normally expressed in m^{-1}.

In the real world we cannot carry out measurements on infinitesimally thin layers, and so if we are to determine the values of a, b and c we need expressions which relate these coefficients to the absorptance, scatterance and beam attenuance of layers of finite thickness. Consider a medium illuminated perpendicularly with a thin parallel beam of radiant flux, Φ_0. As the beam passes through it loses intensity by absorption and scattering. Consider now an infinitesimally thin layer, thickness Δr, within the medium at a depth, r, where the radiant flux in the beam has diminished to Φ. The change in radiant flux in passing through Δr is $\Delta \Phi$. The attenuance of the thin layer is

$$\Delta C = -\Delta\Phi/\Phi$$

(the negative sign is necessary since $\Delta\Phi$ must be negative)

$$\Delta\Phi/\Phi = -c\Delta r$$

Integrating between 0 and r we obtain

$$\ln\frac{\Phi}{\Phi_0} = -cr \tag{1.28}$$

or

$$\Phi = \Phi_0 e^{-cr} \tag{1.29}$$

indicating that the radiant flux diminishes exponentially with distance along the path of the beam. Eqn 1.28 may be rewritten

$$c = \frac{1}{r} \ln \frac{\Phi_0}{\Phi} \tag{1.30}$$

or

$$c = -\frac{1}{r} \ln (1 - C) \tag{1.31}$$

The value of the beam attenuation coefficient, c, can therefore, using eqn 1.30 or 1.31, be obtained from measurements of the diminution in intensity of a parallel beam passing through a known pathlength of medium, r.

The theoretical basis for the measurement of the absorption and scattering coefficients is less simple. In a medium with absorption but negligible scattering, the relation

$$a = -\frac{1}{r} \ln (1 - A) \tag{1.32}$$

holds, and in a medium with scattering but negligible absorption, the relation

$$b = -\frac{1}{r} \ln (1 - B) \tag{1.33}$$

holds, but in any medium which both absorbs and scatters light to a significant extent, neither relation is true. This can readily be seen by considering the application of these equations to such a medium.

In the case of eqn 1.33 some of the measuring beam will be removed by absorption within the pathlength r before it has had the opportunity to be scattered, and so the amount of light scattered, B, will be lower than that required to satisfy the equation. Similarly, A will have a value lower than that required to satisfy eqn 1.32 since some of the light will be removed from the measuring beam by scattering before it has had the chance to be absorbed.

In order to actually measure a or b these problems must be circumvented. In the case of the absorption coefficient, it is possible to arrange that most of the light scattered from the measuring beam still passes through approximately the same pathlength of medium and is collected by the detection system. Thus the contribution of scattering to total attenuation is made very small and eqn 1.32 may be used. In the case of the scattering coefficient there is no instrumental way of avoiding the losses due to absorption and so the absorption must be determined separately and appropriate corrections made to the scattering data. We shall consider ways of measuring a and b in more detail later (§§3.2 and 4.2).

The way in which scattering affects the penetration of light into the medium depends not only on the value of the scattering coefficient but also on the angular distribution of the scattered flux resulting from the primary scattering process. This angular distribution has a characteristic shape for any given medium and is specified in terms of the *volume scattering function*, $\beta(\theta)$. This is defined as the radiant intensity in a given direction from a volume element, $\mathrm{d}V$, illuminated by a parallel beam of light, per unit of irradiance on the cross-section of the volume, and per unit volume (Fig. 1.5*a*). The definition is usually expressed mathematically in the form

$$\beta(\theta)=\mathrm{d}I(\theta)/E\mathrm{d}V \tag{1.34}$$

Since, from the definitions in §1.3

$$\mathrm{d}I(\theta)=\mathrm{d}\Phi(\theta)/\mathrm{d}\omega$$

and

$$E=\Phi_0/\mathrm{d}S$$

where $\mathrm{d}\Phi(\theta)$ is the radiant flux in the element of solid angle $\mathrm{d}\omega$ oriented at angle θ to the beam, and Φ_0 is the flux incident on the cross-sectional area, $\mathrm{d}S$, and since

$$\mathrm{d}V=\mathrm{d}S.dr$$

where $\mathrm{d}r$ is the thickness of the volume element, then we may write

$$\beta(\theta)=\frac{\mathrm{d}\Phi(\theta)}{\Phi_0}\cdot\frac{1}{\mathrm{d}\omega\mathrm{d}r} \tag{1.35}$$

The volume scattering function has the units $\mathrm{m}^{-1}\,\mathrm{sr}^{-1}$.

Light scattering from a parallel light beam passing through a thin layer of medium is radially symmetrical around the direction of the beam. Thus, the light scattered at angle θ should be thought of as a cone with half-angle θ, rather than as a pencil of light (Fig. 1.5*b*).

From eqn 1.35 we see that $\beta(\theta)$ is the radiant flux per unit solid angle scattered in the direction θ, per unit pathlength in the medium, expressed as a proportion of the incident flux. The angular interval θ to $\theta+\Delta\theta$ corres-

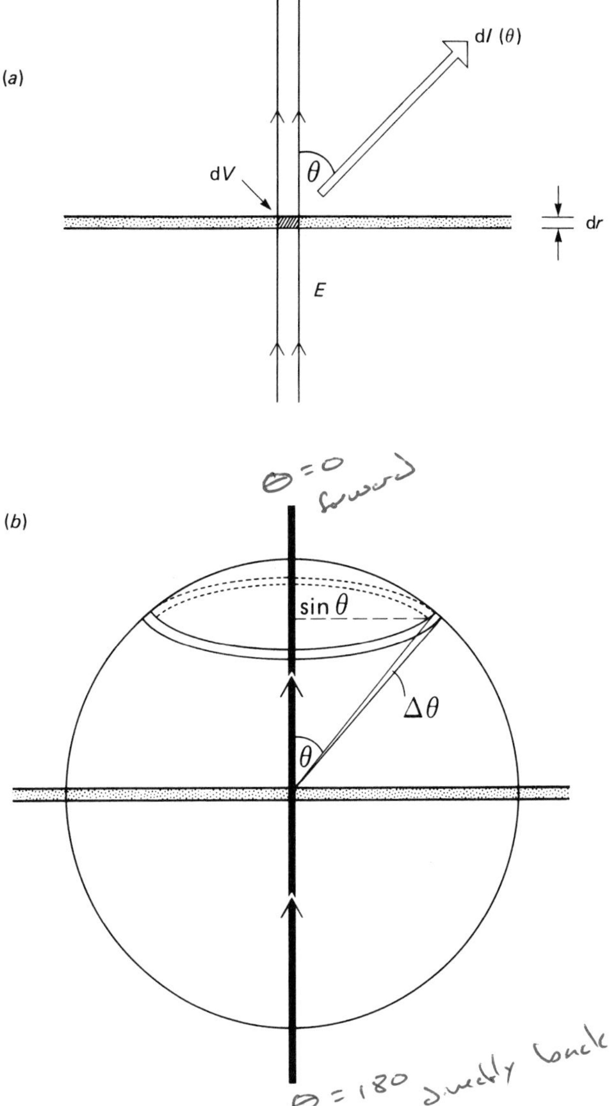

Fig. 1.5. The geometrical relations underlying the volume scattering function. (*a*) A parallel light beam of irradiance E and cross-sectional area dA passes through a thin layer of medium, thickness dr. The illuminated element of volume is dV. d$I(\theta)$ is the radiant intensity due to light scattered at angle θ. (*b*) The point at which the light beam passes through the thin layer of medium can be imagined as being at the centre of a sphere of unit radius. The light scattered between θ and $\theta + \Delta\theta$ illuminates a circular strip, radius $\sin\theta$ and width $\Delta\theta$, around the surface of the sphere. The area of the strip is $2\pi \sin\theta\Delta\theta$ which is equivalent to the solid angle (in steradians) corresponding to the angular interval $\Delta\theta$.

ponds to an element of solid angle, $d\omega$, equal to $2\pi \sin\theta\Delta\theta$ (Fig. 1.5*b*) and so the proportion of the incident radiant flux scattered (per unit pathlength) in this angular interval is $\beta(\theta)\, 2\pi \sin\theta\Delta\theta$. To obtain the proportion of the incident flux which is scattered in all directions per unit pathlength – by definition, equal to the scattering coefficient – we must integrate over the angular range $\theta = 0°$ to $\theta = 180°$

$$b = 2\pi \int_0^{\pi} \beta(\theta) \sin\theta \, d\theta = \int_{4\pi} \beta(\theta) d\omega \tag{1.36}$$

Thus, an alternative definition of the scattering coefficient is the integral of the volume scattering function over all directions.

It is frequently useful to distinguish between scattering in a forward direction and that in a backward direction. We therefore partition the total scattering coefficient, b, into a *forward scattering coefficient*, b_f, relating to light scattered from the beam in a forward direction, and a *backward scattering coefficient* (or simply, *backscattering coefficient*) b_b, relating to light scattered from the beam in a backward direction

$$b = b_b + b_f \tag{1.37}$$

We may also write

$$b_f = 2\pi \int_0^{\pi/2} \beta(\theta) \sin\theta \, d\theta \tag{1.38}$$

$$b_b = 2\pi \int_{\pi/2}^{\pi} \beta(\theta) \sin\theta \, d\theta \tag{1.39}$$

The variation of $\beta(\theta)$ with θ tells us the absolute amount of scattering at different angles, per unit pathlength in a given medium. If we wish to compare the *shape* of the angular distribution of scattering in different media separately from the absolute amount of scattering that occurs, then it is convenient to use the *normalized volume scattering function*, $\tilde{\beta}(\theta)$, sometimes called the scattering phase function, which is that function (units, sr^{-1}) obtained by dividing the volume scattering function by the total scattering coefficient

$$\tilde{\beta}(\theta) = \beta(\theta)/b \tag{1.40}$$

The integral of $\tilde{\beta}(\theta)$ over all solid angles is equal to 1. The integral of $\tilde{\beta}(\theta)$ up to any given value of θ is the proportion of the total scattering which occurs in the angular interval between 0° and that value of θ. We can also define normalized forward scattering and backward scattering coefficients, $\tilde{b}_f$ and $\tilde{b}_b$, as the proportions of the total scattering in forwards and backwards directions, respectively

$$\tilde{b}_f = b_f/b \tag{1.41}$$

$$\tilde{b}_b = b_b/b \tag{1.42}$$

Just as it is useful sometimes to express the angular structure of a light field in terms of a single parameter – its average cosine ($\bar{\mu}$) – so it can also be useful in the case of the scattering phase function to have a single parameter which provides some indication of its shape. Such a parameter is the *average cosine of scattering*, $\bar{\mu}_s$, which can be thought of as the average cosine of the singly scattered light field. Its value, for any given volume scattering function, may be calculated[494] from

$$\bar{\mu}_s = \frac{\int_{4\pi} \beta(\theta) \cos\theta \, d\omega}{\int_{4\pi} \beta(\theta) \, d\omega} \tag{1.43}$$

or (using eqn 1.40 and the fact that the integral of $\tilde{\beta}(\theta)$ over 4π is 1) from

$$\bar{\mu}_s = \int_{4\pi} \tilde{\beta}(\theta) \cos\theta \, d\omega \tag{1.44}$$

1.5 Apparent and quasi-inherent optical properties

The vertical attenuation coefficients for radiance, irradiance and scalar irradiance are, strictly speaking, properties of the radiation field since, by definition, each of them is the logarithmic derivative with respect to depth of the radiometric quantity in question. Nevertheless experience has shown that their values are largely determined by the inherent optical properties of the aquatic medium and are not very much altered by changes in the incident radiation field such as a change in solar elevation.[36] For example, if a particular water body is found to have a high value of K_d then we expect it to have approximately the same high K_d tomorrow, or next week, or at any time of the day, so long as *the composition of the water remains about the same*.

Vertical attenuation coefficients, such as K_d, are thus commonly used, and thought of, by oceanographers and limnologists as though they are optical properties belonging to the water, properties which are a direct measure of the ability of that water to bring about a diminution in the appropriate radiometric quantity with depth. Furthermore they have the same units (m^{-1}) as the inherent optical properties *a*, *b* and *c*. In recognition of these useful aspects of the various *K* functions, Preisendorfer (1961) has suggested that they be classified as *apparent optical properties* and we shall so treat them in this book. The reflectance, *R*, is also often treated as an apparent optical property of water bodies.

The two fundamental inherent optical properties – the coefficients for absorption and scattering – are, as we saw earlier, defined in terms of the behaviour of a parallel beam of light incident upon a thin layer of medium. Analogous coefficients can be defined for incident light streams having any specified angular distribution. In particular, such coefficients can be defined for incident light streams corresponding to the upwelling and downwelling streams that exist at particular depths in real water bodies. We shall refer to these as the *diffuse* absorption and scattering coefficients for the upwelling or downwelling light streams at a given depth. Although related to the normal coefficients, the values of the diffuse coefficients are a function of the local radiance distribution, and therefore of depth.

The *diffuse absorption coefficient* for the downwelling light stream at depth z, $a_d(z)$, is the proportion of the incident radiant flux which would be absorbed from the downwelling stream by an infinitesimally thin horizontal plane parallel layer at that depth, divided by the thickness of the layer. The diffuse absorption coefficient for the upwelling stream, $a_u(z)$, is defined in a similar way. Absorption of a diffuse light stream within the thin layer will be greater than absorption of a normally incident parallel beam because the pathlengths of the photons will be in proportion to $1/\bar{\mu}_d$ and $1/\bar{\mu}_u$, respectively. The diffuse absorption coefficients are therefore related to the normal absorption coefficients by

$$a_d(z) = \frac{a}{\bar{\mu}_d(z)} \tag{1.45}$$

$$a_u(z) = \frac{a}{\bar{\mu}_u(z)} \tag{1.46}$$

where $\bar{\mu}_d(z)$ and $\bar{\mu}_u(z)$ are the values of $\bar{\mu}_d$ and $\bar{\mu}_u$ which exist at depth z.

So far as scattering of the upwelling and downwelling light streams is concerned, it is mainly the backward scattering component which is of importance. The diffuse backscattering coefficient for the downwelling stream at depth z, $b_{bd}(z)$, is the proportion of the incident radiant flux from the downwelling stream which would be scattered backwards (i.e. upwards) by an infinitesimally thin, horizontal plane parallel layer at that depth, divided by the thickness of the layer: $b_{bu}(z)$, the corresponding coefficient for the upwelling stream is defined in the same way in terms of the light scattered downwards again from that stream. Diffuse total ($b_d(z)$, $b_u(z)$) and forward ($b_{fd}(z)$, $b_{fu}(z)$) scattering coefficients for the downwelling and upwelling streams can be defined in a similar manner. The following relations hold

$$b_d(z) = b/\bar{\mu}_d(z), \qquad b_u(z) = b/\bar{\mu}_u(z)$$

$$b_d(z) = b_{fd}(z) + b_{bd}(z), \qquad b_u(z) = b_{fu}(z) + b_{bu}(z)$$

Attenuation coefficients for the downwelling and upwelling light streams, $c_d(z)$ and $c_u(z)$, can also be defined

$$c_d(z) = a_d(z) + b_d(z), \qquad c_u(z) = a_u(z) + b_u(z)$$

The relation between a diffuse backscattering coefficient and the normal backscattering coefficient, b_b, is not simple but may be calculated from the volume scattering function and the radiance distribution existing at depth z. The calculation procedure is discussed later (§4.2).

Preisendorfer (1961) has classified the diffuse absorption, scattering and attenuation coefficients as *hybrid optical properties* on the grounds that they are derived both from the inherent optical properties and certain properties of the radiation field. I prefer the term *quasi-inherent optical properties*, on the grounds that it more clearly indicates the close relation between these properties and the inherent optical properties. Both sets of properties have precisely the same kind of definition: they differ only in the characteristics of the light flux that is imagined to be incident upon the thin layer of medium.

The important quasi-inherent optical property, $b_{bd}(z)$, can be linked with the two apparent optical properties, K_d and R, with the help of one more optical property, $\kappa(z)$, which is the average vertical attenuation coefficient in upward travel from their first point of upward scattering, of all the upwelling photons received at depth z.[492] $\kappa(z)$ must not be confused with, and is in fact much greater than, $K_u(z)$, the vertical attenuation coefficient (with respect to depth increasing downward) of the upwelling light stream. Using $\kappa(z)$ we link the apparent and the quasi-inherent optical properties in the relation

$$R(z) \simeq \frac{b_{bd}(z)}{K_d(z) + \kappa(z)} \tag{1.47}$$

At depths where the asymptotic radiance distribution is established (see §6.6) this relationship holds exactly. Monte Carlo modelling of the underwater light field for a range of optical water types[492] has shown that κ is approximately linearly related to K_d, the relationship at z_m (a depth at which irradiance is 10% of the subsurface value) being

$$\kappa(z_m) \simeq 2.5\, K_d(z_m) \tag{1.48}$$

1.6 Optical depth

As we have already noted, but will discuss more fully later, the downward irradiance diminishes in an approximately exponential manner with depth. This may be expressed by the equation

$$E_d(z) = E_d(0)\,e^{-K_d z} \tag{1.49}$$

where $E_d(z)$ and $E_d(0)$ are the values of downward irradiance at z m depth, and just below the surface, respectively, and K_d is the average value of the vertical attenuation coefficient over the depth interval 0 to z m. We shall now define the *optical depth*, ζ, by the eqn

$$\zeta = K_d z \tag{1.50}$$

It can be seen that a specified optical depth will correspond to different physical depths but to the same overall diminution of irradiance, in waters of differing optical properties. Thus in a coloured turbid water with a high K_d, a given optical depth will correspond to a much smaller actual depth than in a clear colourless water with a low K_d. Optical depth, ζ, as defined here is distinct from *attenuation length*, τ (sometimes also called optical depth or optical distance), which is the geometrical length of a path multiplied by the beam attenuation coefficient (c) associated with the path.

Optical depths of particular interest in the context of primary production are those corresponding to attenuation of downward irradiance to 10% and 1% of the subsurface values: these are $\zeta = 2.3$ and $\zeta = 4.6$, respectively. These optical depths correspond to the mid-point and the lower limit of the euphotic zone, within which significant photosynthesis occurs.

1.7 Radiative transfer theory

Having defined the properties of the light field and the optical properties of the medium we are now in a position to ask whether it is possible to arrive, on purely theoretical grounds, at any relations between them. Although, given a certain incident light field, the characteristics of the underwater light field are uniquely determined by the properties of the medium, it is nevertheless true that explicit, all-embracing analytical relations, expressing the characteristics of the field in terms of the inherent optical properties of the medium, have not yet been derived. Given the complexity of the shape of the volume scattering function in natural waters (see Chapter 4), it may be that this will never be achieved.

It is, however, possible to arrive at a useful expression relating the absorption coefficient to the average cosine and the vertical attenuation coefficient for net downward irradiance. In addition, relations have been derived between certain properties of the field and the diffuse optical properties. These various relations are all arrived at by making use of the equation of transfer for radiance. This describes the manner in which radiance varies with distance along any specified path at a specified point in the medium.

Assuming a horizontally stratified water body (i.e. with properties everywhere constant at a given depth), with a constant input of monochromatic unpolarized radiation at the surface, and ignoring fluorescent emission within the water, the equation may be written

$$\frac{\mathrm{d}L(z,\theta,\phi)}{\mathrm{d}r} = -c(z)L(z,\theta,\phi) + L^*(z,\theta,\phi) \qquad (1.51)$$

The term on the left is the rate of change of radiance with distance, r, along the path specified by zenith and azimuthal angles θ and ϕ, at depth z. The net rate of change is the resultant of two opposing processes: loss by attenuation along the direction of travel ($c(z)$ being the value of the beam attenuation coefficient at depth z), and gain by scattering (along the path dr) of light initially travelling in other directions (θ', ϕ') into the direction θ, ϕ (Fig. 1.6). This latter term is determined by the volume scattering function of the medium at depth z (which we write $\beta(z,\theta,\phi;\theta',\phi')$ to indicate that the scattering angle is the angle between the two directions θ,ϕ and θ',ϕ') and by the distribution of radiance, $L(z,\theta',\phi')$. Each element of irradiance, $L(z,\theta',\phi')\,\mathrm{d}\omega(\theta',\phi')$, (where $\mathrm{d}\omega(\theta',\phi')$ is an element of solid angle forming an infinitesimal cone containing the direction θ',ϕ') incident on the volume element along dr gives rise to some scattered radiance in the direction θ,ϕ. The total radiance derived in this way is given by

$$L^*(z,\theta,\phi) = \int_{2\pi} \beta(z,\theta,\phi;\theta',\phi')\, L(z,\theta',\phi')\, \mathrm{d}\omega(\theta',\phi') \qquad (1.52)$$

If we are interested in the variation of radiance in the direction θ,ϕ as a function of depth, then since $\mathrm{d}r = \mathrm{d}z/\cos\theta$, we may rewrite eqn 1.51 as

$$\cos\theta \frac{\mathrm{d}L(z,\theta,\phi)}{\mathrm{d}z} = -c(z)L(z,\theta,\phi) + L^*(z,\theta,\phi) \qquad (1.53)$$

By integrating each term of this equation over all angles

$$\int_{4\pi} \cos\theta \frac{\mathrm{d}L(z,\theta,\phi)}{\mathrm{d}z}\mathrm{d}\omega = -\int_{4\pi} c(z)\, L(z,\theta,\phi)\,\mathrm{d}\omega + \int_{4\pi} L^*(z,\theta,\phi)\,\mathrm{d}\omega$$

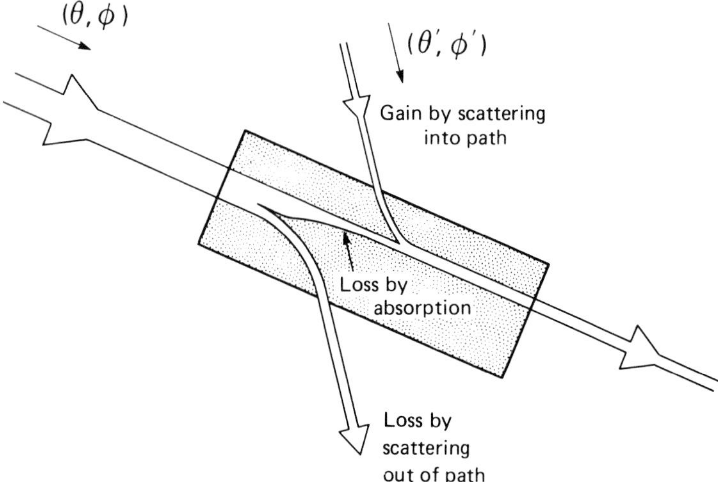

Fig. 1.6. The processes underlying the equation of transfer of radiance. A light beam passing through a distance, dr, of medium, in the direction θ, ϕ, loses some photons by scattering out of the path and some by absorption by the medium along the path, but also acquires new photons by scattering of light initially travelling in other directions (θ', ϕ') into the direction θ, ϕ.

we arrive at the relation

$$\frac{\mathrm{d}\vec{E}}{\mathrm{d}z} = -cE_0 + bE_0 = -aE_0 \tag{1.54}$$

originally derived by Gershun (1936).

It follows that

$$a = K_E \frac{\vec{E}}{E_0} \tag{1.55}$$

and

$$a = K_E \bar{\mu} \tag{1.56}$$

Thus we have arrived at a relation between an inherent optical property and two of the properties of the field. Eqn 1.56, as we shall see later (§3.2), can be used as the basis for determining the absorption coefficient of a natural water from *in situ* irradiance and scalar irradiance measurements.

Preisendorfer (1961) has used the equation of transfer to arrive at a set of relations between certain properties of the field and the *diffuse* absorption and scattering coefficients. One of these, an expression for the vertical attenuation coefficient for downward irradiance,

$$K_d(z) = a_d(z) + b_{bd}(z) - b_{bu}(z)R(z) \tag{1.57}$$

we will later (§6.7) find of assistance in understanding the relative importance of the different processes underlying the diminution of irradiance with depth.

2

Incident solar radiation

2.1 Solar radiation outside the atmosphere

The intensity and the spectral distribution of the radiation received by the Earth are a function of the emission characteristics and the distance of the Sun. Energy is generated within the Sun by nuclear fusion. At the temperature of about 20×10^6 K existing within the Sun, hydrogen nuclei (protons) fuse to give helium nuclei, positrons and energy. A number of steps is involved but the overall process may be represented by

$$4\,^1_1\mathrm{H} \rightarrow {}^4_2\mathrm{He} + 2\,^0_{+1}\mathrm{e} + 25.7\ \mathrm{MeV}$$

The energy liberated corresponds to the slight reduction in mass which takes place in the fusion reactions. Towards its periphery the temperature of the Sun greatly diminishes and at its surface it is only about 5800 K.

For any physical system with the properties of a black body (or 'full radiator') the amount of radiant energy emitted per unit area of surface and the spectral distribution of that radiation are determined by the temperature of the system. The radiant flux*, M, emitted per unit area, is proportional to the fourth power of the absolute temperature, in accordance with the Stefan–Boltzmann Law

$$M = \sigma T^4$$

where σ is 5.67×10^{-8} W m^{-2} K^{-4}. The Sun appears to behave approximately as a full radiator, or black body, and the radiant flux emitted per m^2 of its surface is about 63.4×10^6 W, corresponding (assuming a diameter of 1.39×10^6 km) to a total solar flux of about 385×10^{24} W. At the distance of the Earth's orbit (150×10^6 km) the solar flux per unit area facing the Sun is

* Throughout this chapter radiant flux and irradiance will be in energy units rather than quanta.

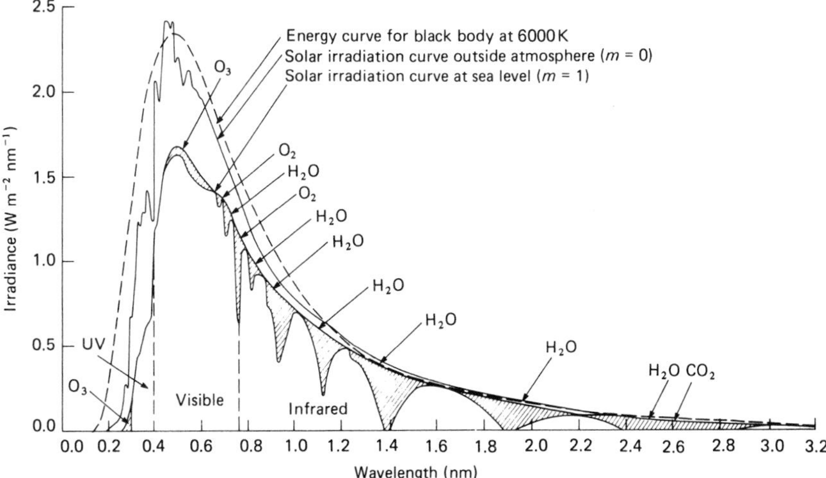

Fig. 2.1. The spectral energy distribution of solar radiation outside the atmosphere compared with that of a black body at 6000 K, and with that at sea level (zenith Sun). (By permission, from *Handbook of geophysics*, revised edition, U.S. Air Force, Macmillan, New York, 1960.)

about 1373W m^{-2}.[366] This quantity, the total solar irradiance outside the Earth's atmosphere, is sometimes called the *solar constant*.

The Earth, having a diameter of 12 756 km, presents a cross-sectional area of 1.278×10^8 km^2 to the Sun. The solar radiation flux onto the whole of the Earth is therefore about 1755×10^{14} W, and the total radiant energy received by the Earth from the Sun each year is about 5.53×10^{24} J.

For a full radiator, the spectral distribution of the emitted energy has a certain characteristic shape, rising steeply from the shorter wavelengths to a peak, and diminishing more slowly towards longer wavelengths. As the temperature, T, of the radiator increases, the position of the emission peak (λ_{max}) moves to shorter wavelengths in accordance with Wien's Law

$$\lambda_{max} = \frac{w}{T}$$

where w (Wien's displacement constant) is equal to 2.8978×10^{-3}m K. The Sun, with a surface temperature of about 5800 K has, in agreement with Wien's Law, maximum energy per unit wavelength at about 500 nm, as may be seen in the curve of solar spectral irradiance above the Earth's atmosphere in Fig. 2.1. It can be seen that the spectral distribution of the emitted solar flux is, compared to the theoretical curve for a full radiator at 6000 K, somewhat irregular in shape at the short-wavelength end. The dips in the curve are due to the absorption bands of hydrogen in the Sun's outer atmosphere. Photosynthetically available radiation (commonly abbreviated to PAR), 400–700 nm, constitutes 38% of the extraterrestrial solar irradiance.

2.2 Transmission of solar radiation through the Earth's atmosphere

Even when the sky is clear, the intensity of the solar beam is significantly reduced during its passage through the atmosphere. This reduction in intensity is due partly to scattering by air molecules and dust particles and partly to absorption by water vapour, oxygen, ozone and carbon dioxide in the atmosphere. With the Sun vertically overhead, the total solar irradiance on a horizontal surface at sea level is reduced by about 14% with a dry, clean atmosphere and by about 40% with a moist, dusty atmosphere, compared to the value above the atmosphere.[627] The proportion of the incident solar flux removed by the atmosphere increases as the solar elevation (the angle of the Sun's disc to the horizontal) decreases, in accordance with the increase in pathlength of the solar beam through the atmosphere. The atmospheric pathlength is approximately proportional to the cosecant of the solar elevation: it is, for example, twice as long with a solar elevation of 30° as with the Sun at the zenith.

Effect of scattering

Since the air molecules are much smaller than the wavelengths of solar radiation, the efficiency with which they scatter light is proportional to $1/\lambda^4$, in accordance with Rayleigh's Law. Scattering of solar radiation is therefore much more intense at the short-wavelength end of the spectrum, and most of the radiation scattered by the atmosphere is in the visible and ultraviolet ranges.

Some of the radiation scattered from the solar beam is lost to space, and some finds its way to the Earth's surface. In the case of 'pure' Rayleigh scattering (scattering entirely due to air molecules, with a negligible contribution from dust), there is as much scattering in a forward, as in a backward, direction (Fig. 2.2; Chapter 4). Thus, ignoring for simplicity the effects of multiple scattering, half the light scattered from a sunbeam in a dust-free atmosphere will be returned to space and half will continue (at various angles) towards the Earth's surface. The total solar energy reflected

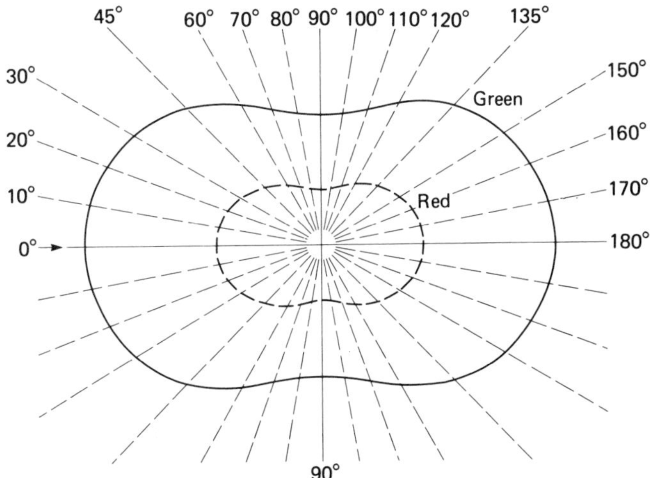

Fig. 2.2. Polar plot of intensity as a function of scattering angle for small particles ($r \simeq 0.025$ μm) for green ($\lambda \simeq 0.5$ μm) and red ($\lambda \simeq 0.7$ μm) light. (By permission, from *Solar radiation*, N. Robinson, Elsevier, Amsterdam, 1966.)

(by scattering) back to space by a cloudless dust-free atmosphere alone after allowing for absorption by ozone, is about 7%.[765]

The atmosphere over any part of the Earth's surface always contains a certain amount of dust, the quantity varying from place to place and with time at any given place. Dust particles scatter light, but are generally not sufficiently small relative to the wavelength of most of the solar radiation for scattering to obey the Rayleigh Law. They exhibit instead a type of scattering known as Mie scattering (see Chapter 4) which is characterized by an angular distribution predominantly in the forward direction, and a much weaker dependence on wavelength, although scattering is still more intense at shorter wavelengths. Although dust particles scatter light mainly in a forward direction, they do also scatter significantly in a backward direction; furthermore, a proportion of the light scattered forward, but at large values of scattering angle, will be directed upwards if the solar beam is not vertical. Since the particle scattering is additive to the air molecule scattering, a hazy (dusty) atmosphere reflects more solar energy to space than does a clean atmosphere.

That fraction of the solar flux which is scattered by the atmosphere in the direction of the Earth's surface constitutes skylight. Skylight appears blue because it contains a high proportion of the more intensely scattered, blue, light from the short-wavelength end of the visible spectrum. In a hazy atmosphere the increased scattering increases skylight at the expense of the direct solar flux. The proportion of the total radiation received at the Earth's surface which is skylight varies also with the solar elevation. As the atmospheric pathlength of the solar beam increases with decreasing solar elevation, so more of the radiation is scattered: as a consequence the direct solar flux diminishes more rapidly than does skylight.

At very low solar elevations (0°–20°) in the absence of cloud, the direct solar beam and the diffuse flux (skylight) contribute approximately equally to irradiance at the Earth's surface. As solar elevation increases, the irradiance due to the direct beam rises steeply but the irradiance due to skylight levels off above about 30°. At high solar elevations under cloudless conditions, skylight commonly accounts for 15–25% of the irradiance and the direct solar beam for 75–85%:[624] in very clear dry air the contribution of skylight can be as low as 10%.

Spectral distribution of irradiance at the Earth's surface

The scattering and absorption processes that take place within the atmosphere not only reduce the intensity, but also change the spectral distribution of the direct solar beam. The lowest curve in Fig. 2.1 shows the spectral distribution of solar irradiance at sea level for a zenith Sun and a clear sky. The shaded areas represent absorption, and so the curve forming the upper boundary of these shaded areas corresponds to the spectral distribution as it would be if there was scattering but no absorption. It is clear that the diminution of solar flux in the ultraviolet band (0.2–0.4 μm) is largely due to scattering, with a contribution from absorption by ozone. In the visible/photosynthetic band (0.4–0.7 μm), attenuation is mainly due to scattering, but with absorption contributions from ozone, oxygen and, at the red end of the spectrum, water vapour. In the long, infrared tail of the distribution, scattering becomes of minor importance and the various absorption bands of water vapour are mainly responsible for the diminution in radiant flux.

The proportion of infrared radiation removed from the solar beam by absorption during its passage through the atmosphere is variable since the amount of water vapour in the atmosphere is variable. Nevertheless it is generally true that a higher proportion of the infrared is removed than of the photosynthetic waveband. As a consequence, photosynthetically available radiation (0.4–0.7 μm) is a higher proportion of the solar radiation that reaches the Earth's surface than of the radiation above the atmosphere. PAR constitutes about 45% of the energy in the direct solar beam at the

Earth's surface when the solar elevation is more than 30°.[626] Skylight, consisting as it does of scattered and therefore mainly short-wave radiation, is predominantly in the visible/photosynthetic range.

Using the best available data for the spectral distribution of the extraterrestrial solar flux, Baker & Frouin (1987) have carried out calculations of atmospheric radiation transfer to estimate the PAR (in this case taken to be 350–700 nm) as a proportion of total insolation at the ocean surface, under clear skies, but with various types of atmosphere, and varying sun angle. They found E_d (350–700 nm)/E_d(total) for all atmospheres to lie between 45 and 50% at solar altitudes greater than 40°. The ratio increased with increasing atmospheric water vapour, because of a decrease in E_d(total). It was essentially unaffected by variation in ozone content, or in the aerosol content provided the aerosol was of a maritime, not continental, type. It was little affected by solar altitude between 40° and 90°, but decreased 1–3% as solar elevation was lowered from 40° to 10°.

Since the efficiency with which the photosynthetic apparatus captures light energy varies with wavelength, the usefulness of a given sample of solar radiation for primary production depends on the proportion of different wavelengths of light present, i.e. on its spectral distribution. Fig. 2.3 shows data for the spectral distribution of solar irradiance at the Earth's surface under clear sunny skies, within a few hours of solar noon at three different locations. As the solar elevation diminishes, the ratio of short- (blue) to long- (red) wavelength light in the direct solar beam decreases because of the intensified removal of the more readily scattered, short-wavelength light in the longer atmospheric path. On the other hand, as solar elevation diminishes, the relative contribution of skylight to total irradiance increases, and skylight is particularly rich in the shorter wavelengths: there is therefore no simple relation between solar elevation and the spectral distribution of total irradiance.

Effect of cloud

In addition to the effects of the gaseous and particulate components of the atmosphere, the extent and type of cloud cover are of great importance in determining the amount of solar flux which penetrates to the Earth's surface. We follow here the account given by Monteith (1973).

A few isolated clouds in an otherwise clear sky increase the amount of diffuse flux received at the Earth's surface but, provided they do not obscure the Sun, they have no effect on the direct solar beam. Thus, a small amount of cloud can increase total irradiance by 5–10%. A continuous sheet of

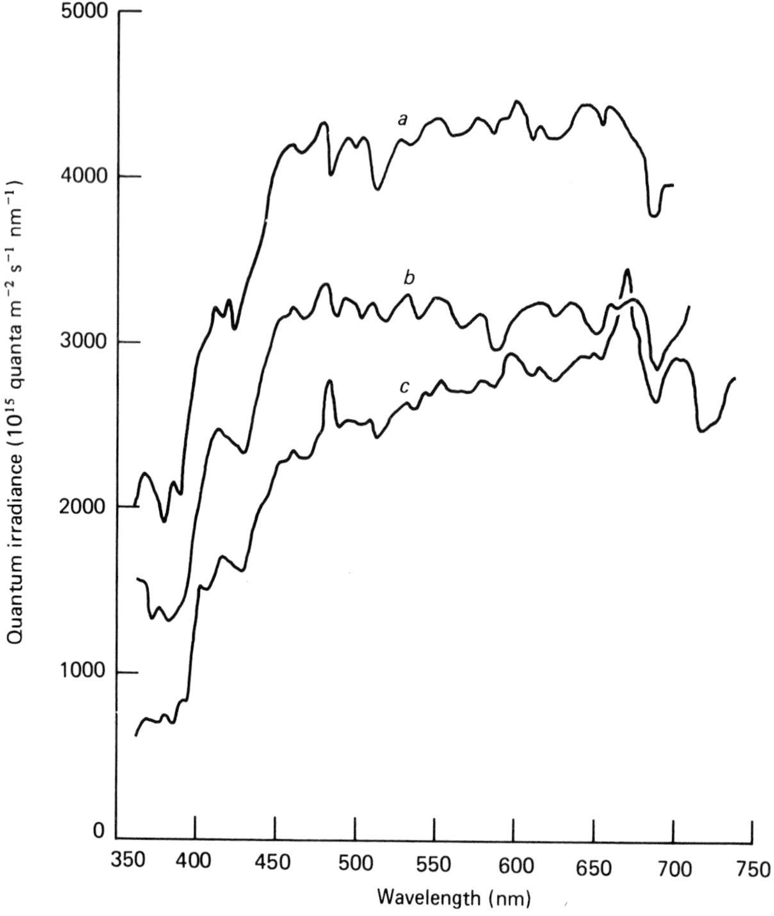

Fig. 2.3. Spectral distribution of solar quantum irradiance at the Earth's surface at three geographical locations (plotted from the data of Tyler & Smith, 1970). (*a*) Crater Lake, Oregon, USA (42°56′ N, 122°07′ W). Elevation 1882 m. 1100–1125 h, 5th August 1966. (*b*) Gulf Stream, Bahamas, Atlantic Ocean (25°45′ N, 79°30′ W). 1207–1223 h, 3rd July 1967. (*c*) San Vicente reservoir, San Diego, California, USA (32°58′ N, 116°55′ W). 0937–0958 h, 20th January 1967. All the measurements were made under clear skies.

cloud, however, will always reduce irradiance. Under a thin sheet of cirrus, total irradiance may be about 70% of that under a clear sky. A deep layer of stratus cloud, on the other hand, may transmit only 10% of the solar radiation, about 70% being reflected back to space by its upper surface and 20% being absorbed within it. On a day with broken cloud, the irradiance at a given point on the Earth's surface is intermittently varying from the full Sun's value to perhaps 20–50% of this as clouds pass over the Sun.

In desert regions there is rarely enough cloud to affect surface irradiance, but in the humid parts of the globe, cloud cover significantly reduces the average solar radiation received during the year. In much of Europe, for example, the average insolation (total radiant energy received m^{-2} day^{-1}) in the summer is 50–80% of the insolation that would be obtained on cloudless days.[624]

In recent years a large amount of new information on the distribution and optical characteristics of clouds around the Earth has become available from satellite remote sensing. The International Satellite Cloud Climatology Project (ISCCP) has been combining such data from geostationary and polar orbiting meteorological satellites from mid-1983 onwards. Bishop & Rossow (1991) have used the ISCCP data, together with modelling of radiation transfer through the atmosphere, to assess the effects of clouds on the spatial and temporal variability of the solar irradiance around the globe. The results show that the oceans are much cloudier than the continents, and receive a correspondingly lower solar irradiance. For example, in July 1983, approximately 9% of the ocean was perpetually cloud covered, contrasting with only 0.3% over land.

There are marked differences between regions of the ocean: the Northern and Southern ocean waters are almost perpetually cloud covered, but at an equatorial mid-Pacific location (1° N, 140° W) clear skies consistently prevail. In the northern hemisphere the Atlantic Ocean receives substantially more solar radiation than the Pacific Ocean in the summer, but there is little difference between the two oceans in the southern hemisphere. The irradiance at the ocean surface, as a percent of the clear sky value is about halved between 30° S and 60° S due to the persistent band of circumpolar cloudiness centred at 60° S. The Argentine Basin and Weddell Sea sectors of the western South Atlantic are, however – presumably due to proximity to land – consistently less cloudy than the rest of this circumpolar region, and are known to have greater levels of primary productivity. Bishop and Rossow suggest that solar irradiance is a major determinant of the rate of carbon fixation by phytoplankton in these nutrient-rich southern ocean waters.

Angular distribution of solar radiation under different atmospheric conditions

That part of the solar radiation at the Earth's surface which is the direct solar beam, of necessity consists of a parallel flux of radiation at an angle to the vertical equal to the solar zenith angle. Radiance in the direction of the solar disc is very high and falls off sharply just beyond the edge of the disc.

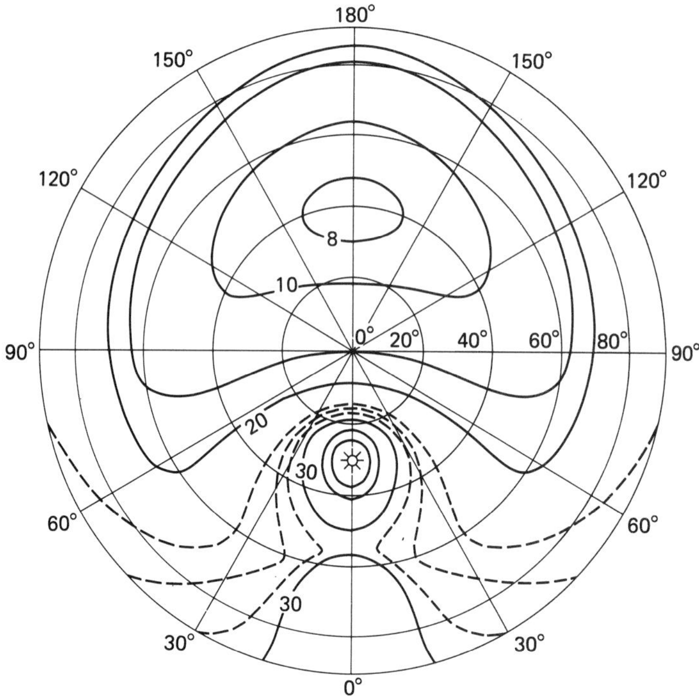

Fig. 2.4. Angular distribution of luminance (approximately proportional to radiance) in a clear sky. (By permission, from *Solar radiation*, N. Robinson, Elsevier, Amsterdam, 1966.)

The angular distribution of skylight from a clear sky is complex; it is shown diagrammatically in Fig. 2.4. Because scattering by dust particles, and to a lesser extent by air molecules, is strong in a forward direction, skylight is particularly intense at angles near to the angle of the Sun.[624] Radiance from the sky decreases markedly with increasing angular distance from the Sun; it reaches its minimum in that region of the celestial hemisphere approximately 90° from the Sun and then rises again towards the horizon.

Contrary to what casual observation might suggest, the radiance distribution beneath a heavily overcast sky is not uniform. In fact, the radiance at the zenith is 2–2½ times that at the horizon. To provide an approximation to the radiance distribution under these conditions, a Standard Overcast Sky has been defined[625] under which the radiance as a function of zenith angle, θ, is given by

$$L(\theta) = L(0)(1 + B\cos\theta)/(1 + B) \tag{2.1}$$

where B is in the range 1.1–1.4. The radiance under a sky with broken cloud must necessarily vary with angle in a highly discontinuous manner, and no useful general description can be given.

Average transmission characteristics of the atmosphere

A valuable summary of the effects of the atmosphere as a whole on the transmission of solar radiation from space to the Earth's surface has been provided by Gates (1962) in terms of the average fate of the radiant flux incident on the northern hemisphere. Over a year, 34% of the incoming solar radiation is reflected to space by the atmosphere; this is made up of 25% reflected by clouds and 9% scattered out to space by other constituents of the atmosphere. Another 19% of the incoming radiation is absorbed by the atmosphere: 10% within clouds, 9% by other components. This leaves 47% of the solar flux which, on average, reaches the Earth's surface. Of this 47%, 24% consists of the direct solar beam and 23% of diffuse light scattered from clouds (17%) and from the air (6%).

In the cloudy, dirty, moist atmosphere of London, about a third of the incident solar radiation per year is scattered back to space, about a third is absorbed within the atmosphere, and the remainder penetrates to the Earth's surface.[624] Of the 30% or so which is absorbed, 13% is by water vapour, 9% within clouds, and 8% by dust and smoke.

2.3 Diurnal variation of solar irradiance

For a given set of atmospheric conditions, the irradiance at any point on the Earth's surface is determined by the solar elevation, β. This rises during the day from zero (or its minimum value in the Arctic or Antarctic during the summer) at dawn to its maximum value at noon, and then diminishes in a precisely symmetrical manner to zero (or the minimum value) at dusk. The exact manner of the variation of β with time of day depends on the latitude, and on the solar declination, δ, at the time. The declination is the angle through which a given hemisphere (north or south) is tilted towards the Sun (Fig. 2.5). In summer it has a positive value, in winter a negative one: at the spring and autumn equinoxes its value is zero. Its maximum value, positive or negative, is 23°27′. Published tables exist in which the solar declination for the northern hemisphere may be found for any day of the year. Alternatively, a relation derived by Spencer (1971) may be used

$$\begin{aligned}\delta = 0.39637 - 22.9133 \cos \psi + 4.02543 \sin \psi \\ - 0.3872 \cos 2\psi + 0.052 \sin 2\psi\end{aligned} \tag{2.2}$$

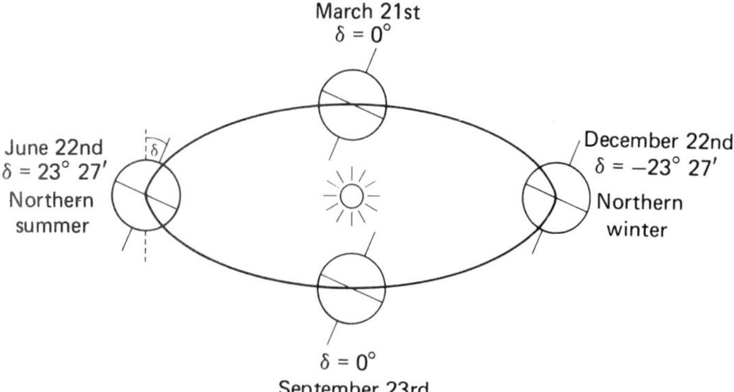

Fig. 2.5. Variation of the solar declination throughout the year.

where ψ is the date expressed as an angle ($\psi = 360°.d/365$; d = day number, ranging from 0 on January 1st to 364 on December 31st): both δ and ψ are in degrees. The declination for the southern hemisphere on a given date has the same numerical value as that for the northern hemisphere, but the opposite sign.

The solar elevation, β, at any given latitude, γ, varies with the time of day, τ, (expressed as an angle) in accordance with the relation

$$\sin\beta = \sin\gamma\,\sin\delta - \cos\gamma\,\cos\delta\,\cos\tau \qquad (2.3)$$

where τ is $360°.t/24$ (t being the hours elapsed since 00.00 h). If we write this in the simpler form

$$\sin\beta = c_1 - c_2\cos\tau \qquad (2.4)$$

where c_1 and c_2 are constants for a particular latitude and date ($c_1 = \sin\gamma\sin\delta$, $c_2 = \cos\gamma\cos\delta$), then it becomes clear that the variation of $\sin\beta$ with the time of day is sinusoidal. Fig. 2.6 shows the variations in both β and $\sin\beta$ throughout a 24-h period corresponding to the longest summer day (December 21st) and shortest winter day (June 21st) at the latitude of Canberra, Australia (35°S). For completeness, the values of β and $\sin\beta$ during the hours of darkness are shown: these are negative and correspond to the angle of the Sun below the horizontal plane. The variation of $\sin\beta$ is sinusoidal with respect to time measured within a 24-h cycle, but not with respect to hours of daylight.

The sine of the solar elevation β is equal to the cosine of the solar zenith angle θ. Irradiance due to the direct solar beam on a horizontal surface is, in accordance with the Cosine Law, proportional to $\cos\theta$. We might therefore

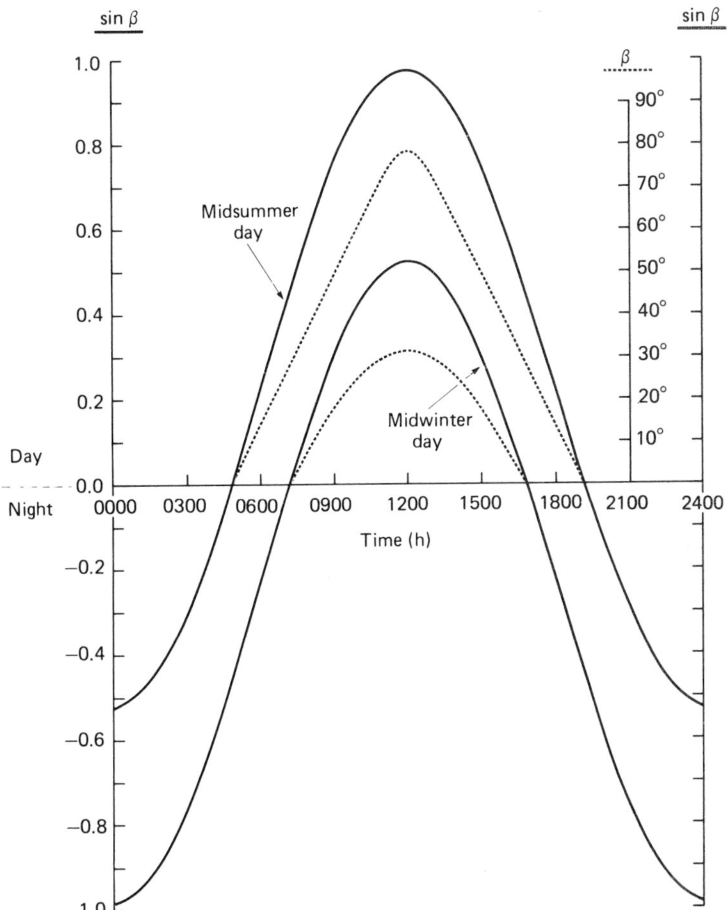

Fig. 2.6. Variation in solar elevation (β) and sin β during a 24-h period on the longest summer day and the shortest winter day at 35° latitude.

expect that in the absence of cloud, solar irradiance at the Earth's surface will vary over a 24-h period in much the same way as sin β (Fig. 2.6), except that its value will be zero during the hours of darkness. Exact conformity between the behaviour of irradiance and sin β is not to be expected since, for example, the effect of the atmosphere varies with solar elevation. The greater attenuation of the direct beam at low solar elevation due to increased atmospheric pathlength, although in part balanced by a (proportionately) greater contribution from skylight, does reduce irradiance early in the morning and towards sunset below the values which might be anticipated on the basis of eqn 2.4. This has the useful effect of making the curve of daily variation of irradiance approximately sinusoidal with respect to daylight hours, even though, as we have seen, the variation of sin β is strictly speaking only sinusoidal with respect to hours since 00.00 h. Fig. 2.7 gives some examples of the diurnal variation of total irradiance at different times of year and under different atmospheric conditions.

A smooth curve is of course only obtained when cloud cover is absent, or is constant throughout the day. Broken cloud imposes short-term irregularities on the underlying sinusoidal variation in irradiance (Fig. 2.7). If $E(t)$ is the total irradiance at time t h after sunrise, then by integrating $E(t)$ with respect to time we obtain the total solar radiant energy received per unit horizontal area during the day. This is referred to as the *daily insolation* and we shall give it the symbol, Q_s

$$Q_s = \int_0^N E(t)\mathrm{d}t \tag{2.5}$$

where N is the daylength.

If $E(t)$ has the units W m^{-2} and N is in s, then Q_s is in J m^{-2}. On a day of no, or constant, cloud cover the diurnal variation in $E(t)$ is expressed approximately by

$$E(t) = E_m \sin(\pi t/N) \tag{2.6}$$

where E_m is the irradiance at solar noon.[624] By integrating this expression in accordance with eqn 2.5, it can be shown that the daily insolation on such a day is related to the maximum (noon) irradiance by

$$Q_s = 2NE_m/\pi \tag{2.7}$$

Thus if, as for example in Fig. 2.7*c* (16th March), the maximum irradiance is 940 W m^{-2} in a 12-h day, then the total energy received is about 26 MJ.

On days with broken cloud, the degree of cloud cover is varying continuously throughout any given day. Nevertheless, in many parts of the world, the cloud cover averaged over a period as long as a month is approximately constant throughout the day.[947] Therefore, the diurnal variation in total irradiance averaged over a month will be approximately sinusoidal and will conform with eqn 2.6. From the average maximum irradiance, the average daily insolation can be calculated using eqn 2.7.

2.4 Variation of solar irradiance and insolation with latitude and time of year

At any given point on the Earth's surface the daylength and the solar elevation reach their maximum values in the summer and their minimum

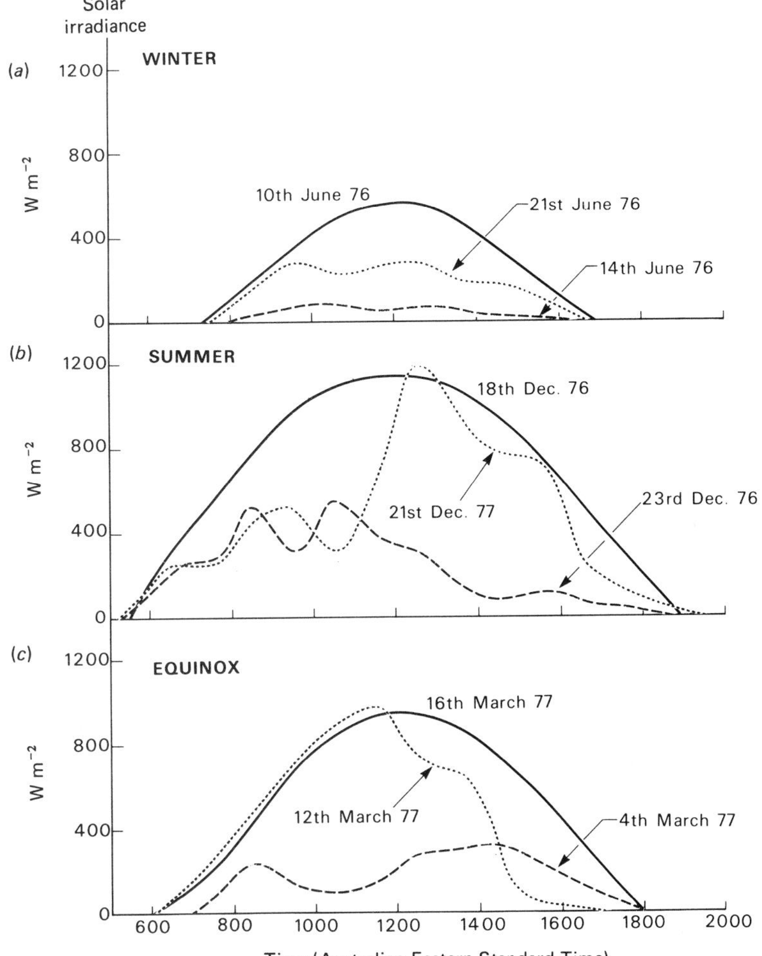

Fig. 2.7. Diurnal variation of total solar irradiance at different times of year and under different atmospheric conditions. The measurements were made at Krawaree, NSW, Australia (149°27′ E, 35°49′ S; 770 m above sea level) on dates close to (*a*) the shortest winter day, (*b*) the longest summer day, (*c*) the autumn equinox. (—— clear sky; ...... intermittent cloud; ------ generally overcast.) Data provided by Mr F. X. Dunin, CSIRO, Canberra.

values in the winter. Substituting into eqn 2.3 we can obtain expressions for the noon solar elevation on the longest summer day ($\delta = 23°27'$)

$$\sin \beta = 0.39795 \sin \gamma + 0.91741 \cos \gamma \qquad (2.8)$$

and the shortest winter day ($\delta = -23°27'$)

$$\sin \beta = -0.39795 \sin \gamma + 0.91741 \cos \gamma \qquad (2.9)$$

as a function of latitude. At the latitude of Canberra (35° S) for example, the corresponding solar elevations are 78.5 and 31.5°: at the latitude of London (51.5° N) they are 62 and 15°.

Maximum and minimum daylengths can also be obtained using eqn 2.3. If the time (expressed as an angle) at sunrise is τ_s, then (since $\sin \beta = 0$ at sunrise) at any time of year

$$\cos \tau_s = \tan \gamma \tan \delta \qquad (2.10)$$

Daylength expressed as an angle is $(360° - 2\tau_s)$, which is equal to $2 \cos^{-1}(-\tan \gamma \tan \delta)$. Expressed in hours, daylength is given by

$$N = 0.133 \cos^{-1}(-\tan \gamma \tan \delta) \qquad (2.11)$$

The longest day is therefore $0.133 \cos^{-1}(-0.43377 \tan \gamma)$ h and the shortest day is $0.133 \cos^{-1}(0.43377 \tan \gamma)$ h.

With increasing latitude, the solar elevation at noon and therefore the maximum value of solar irradiance (E_m) decrease, which in accordance with eqn 2.7, decreases the daily insolation. In the summer, however, this effect is counteracted by the increase in daylength (N in eqn 2.7) with increasing latitude, and the net result is that high-latitude regions can have a slightly greater daily insolation in midsummer than tropical regions. In the winter, of course, the high-latitude regions have shorter days (zero daylengths in the polar regions) as well as lower solar elevations, so their daily insolation is much less than at low latitudes. Throughout most of the year, in fact, the rule-of-thumb holds that the higher the latitude, the lower the daily insolation. Fig. 2.8 shows the change in daily insolation throughout the year, calculated ignoring atmospheric losses, for a range of latitudes. Fig. 2.9 shows the true (measured) daily insolation as a function of time of year at a southern hemisphere site, near Canberra, Australia, averaged over a 3-yr period. These data, unlike those in Fig. 2.8, include the effects of cloud and atmospheric haze.

2.5 Transmission across the air–water interface

To become available in the aquatic ecosystem, the solar radiation which has penetrated through the atmosphere must now find its way across the air–water interface. Some of it will be reflected back into the atmosphere. The proportion of the incident light which is reflected by a flat water surface

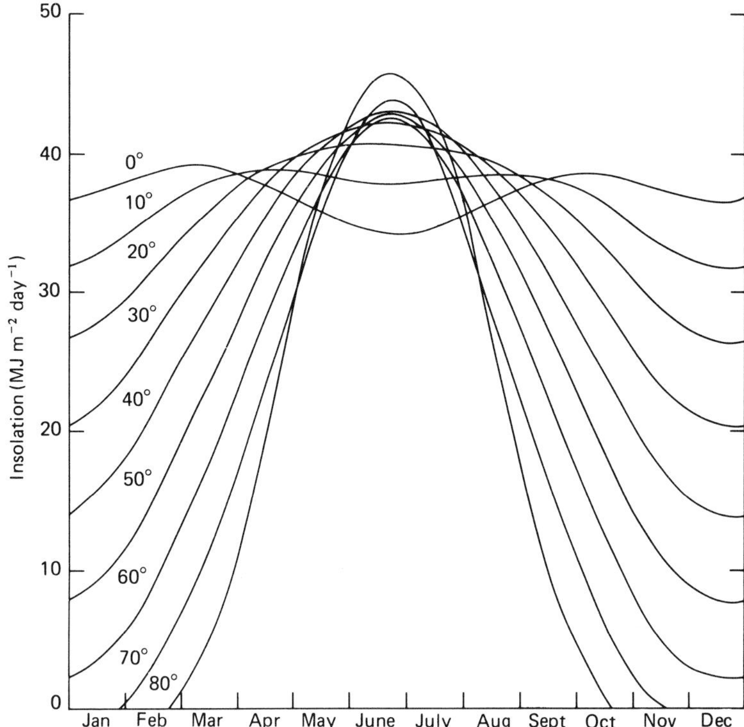

Fig. 2.8. Change in calculated daily insolation (ignoring the influence of the atmosphere) throughout the year at different latitudes in the northern hemisphere. The latitude is indicated above each curve. Plotted from data of Kondratyev (1954).

increases from 2% for vertically incident light towards 100% as the beam approaches grazing incidence. The dependence of reflectance, r, of unpolarized light on the zenith angle of the incident light in air (θ_a), and on the angle to the downward vertical (θ_w) of the transmitted beam in water is given by Fresnel's Equation

$$r = \frac{1}{2}\frac{\sin^2(\theta_a - \theta_w)}{\sin^2(\theta_a + \theta_w)} + \frac{1}{2}\frac{\tan^2(\theta_a - \theta_w)}{\tan^2(\theta_a + \theta_w)} \tag{2.12}$$

The angle, θ_w, in water is itself determined by θ_a and the refractive index, as we shall shortly see. The percentage reflectance from a flat water surface as a function of zenith angle is shown in Fig. 2.10, and in tabular form in Table 2.1. It will be noted that reflectance remains low, increasing only slowly, up to zenith angles of about 50°, but rises very rapidly thereafter.

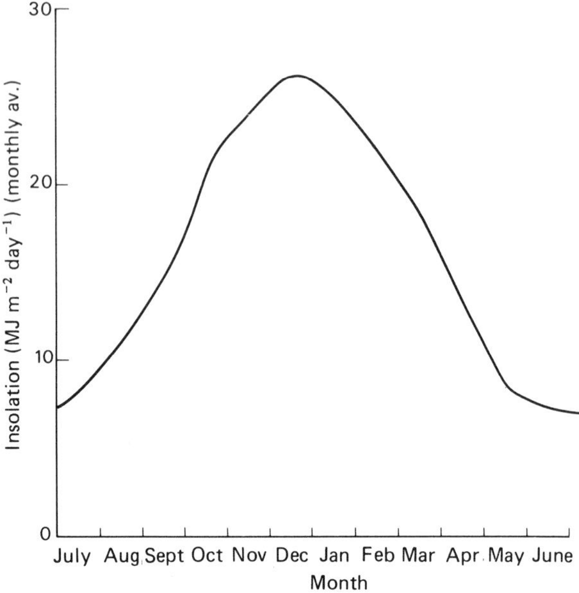

Fig. 2.9. Change in measured daily insolation throughout the year at the CSIRO Ginninderra Experiment Station, (35° S, 149° E). The curve corresponds to the daily average for each calendar month, averaged over the period 1978–80.

Roughening of the water surface by wind has little effect on the reflectance of sunlight from high solar elevations. At low solar elevations, on the other hand, reflectance is significantly lowered by wind since the roughening of the surface on average increases the angle between the light direction and the surface at the point of entry. The three lower curves in Fig. 2.10 show the effect of wind at different velocities on the reflectance.[327,26]

As wind speed increases the waves begin to break, and whitecaps are formed. The fraction of the ocean surface covered by whitecaps, W, is a function of windspeed, U (m s^{-1} at 10 m above surface), and can be represented as a power law

$$W(U) = AU^B \tag{2.13}$$

The coefficient A and the exponent B are functions of water temperature, but combining data from a range of water temperatures, Spillane & Doyle (1983) found the relationship

$$W(U) = 2.692 \times 10^{-5} U^{2.625} \tag{2.14}$$

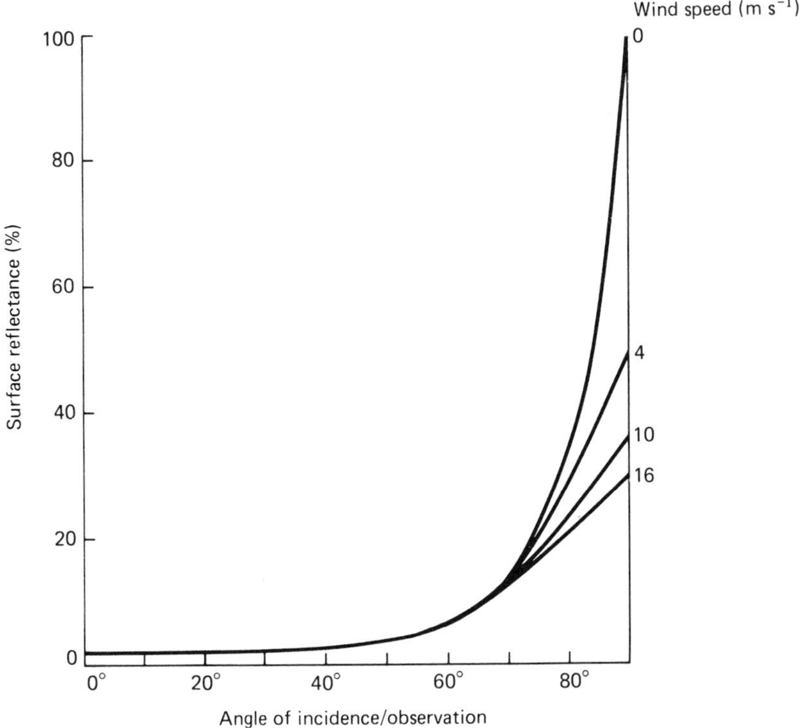

Fig. 2.10. Reflectance of water surface as a function of zenith angle of light (incident from above), at different wind speeds (data of Gordon, 1969; Austin, 1974*a*).

to give the best fit. Eqn 2.14 predicts, for example that whitecaps cover about 1% of the sea surface at a wind speed of 10 m s^{-1}, and 13% at 25 m s^{-1}. Freshly formed whitecaps consist of many layers of bubbles, and have a reflectance of about 55%.[994,865] They have, however, a lifetime of only 10–20 s in the ocean, and Koepke (1984) found that as they decay their reflectance decreases markedly due to thinning of the foam, and he estimated the effective reflectance to be on average only about 22%, and their contribution to total oceanic reflectance to be minor. Using eqn 2.14 we can calculate the ocean surface reflectance due to whitecaps to be only about 0.25% at a wind speed of 10 m s^{-1}, and ~3% at 25 m s^{-1}.

Because of the complex angular distribution of skylight, the extent to which it is reflected by a water surface is difficult to determine. If the very approximate assumption is made that the radiance is the same from all

Table 2.1. *Reflectance of unpolarized light from a flat water surface. The values of reflectance have been calculated using eqns 2.12 and 2.15, assuming that the water has a refractive index of 1.33*

Zenith angle of incidence, θ_a (degrees)	Reflectance (%)	Zenith angle of incidence, θ_a (degrees)	Reflectance (%)
0.0	2.0	50.0	3.3
5.0	2.0	55.0	4.3
10.0	2.0	60.0	5.9
15.0	2.0	65.0	8.6
20.0	2.0	70.0	13.3
25.0	2.1	75.0	21.1
30.0	2.1	80.0	34.7
35.0	2.2	85.0	58.3
40.0	2.4	87.5	76.1
45.0	2.8	89.0	89.6

directions, then a reflectance of 6.6% for a flat water surface is obtained.[422] For an incident radiance distribution such as might be obtained under an overcast sky, the reflectance is calculated to be about 5.2%.[715] Roughening of the surface by wind will lower reflectance for the diffuse light from a clear or an overcast sky.

As the unreflected part of a light beam passes across the air–water interface, it changes its direction to the vertical (while remaining, if the surface is flat, in the same vertical plane), due to refraction. The phenomenon of refraction is the result of the differing velocities of light in the two media, air and water. The change in angle (Fig. 2.11) is governed by Snell's Law

$$\frac{\sin \theta_a}{\sin \theta_w} = \frac{n_w}{n_a} \tag{2.15}$$

where n_w and n_a are the refractive indices of water and air, respectively. The ratio of the refractive index of water to that of air is a function of temperature, salt concentration and the wavelength of the light in question. For our purposes a value of 1.33 for n_w/n_a is close enough for both sea and fresh water at normal ambient temperatures and for light of any wavelength within the photosynthetic range. As Fig. 2.11 shows, the effect of refraction is to move the light closer to the vertical in the aquatic medium than it was in air. It will be noted that even light reaching the surface at grazing incidence (θ_a approaching 90°) is refracted downwards so that θ_w should not be greater

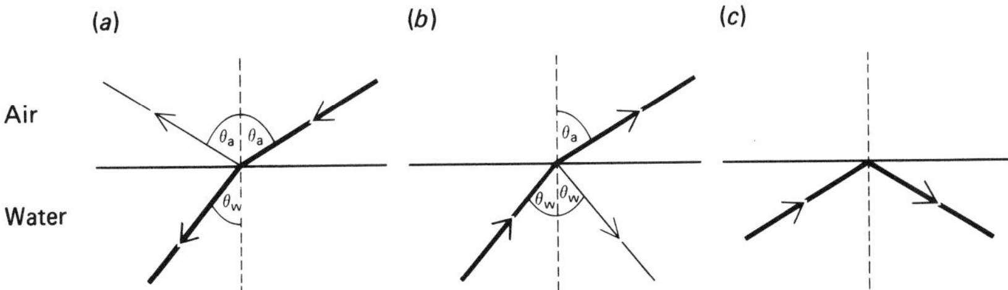

Fig. 2.11. Refraction and reflection of light at air–water boundary. (*a*) A light beam incident from above is refracted downwards within the water: a small part of the beam is reflected upwards at the surface. (*b*) A light beam incident from below at a nadir angle of 40° is refracted away from the vertical as it passes through into the air: a small part of the beam is reflected downwards again at the water–air boundary. (*c*) A light beam incident from below at a nadir angle greater than 49° undergoes complete internal reflection at the water–air boundary.

than about 49° for a calm surface. When the water is disturbed, some of the light will be found at angles greater than 49° after passing through the surface: nevertheless most of the downwelling light will be at values of θ_w between 0° and 49°. Statistical relations between the distribution of surface slopes and wind speed were derived by Cox & Munk (1954) from aerial photographs of the Sun's glitter pattern. Calculations using these relations show that, as might be expected, as wind speed increases, the underwater light becomes more diffuse.[334]

Snell's Law works in reverse also. A light beam passing upward within the water at angle θ_w to the downward vertical will emerge through the calm surface into the air at angle θ_a to the zenith, in accordance with eqn 2.15. There is, however, a very important difference between transmission from water to air, and from air to water, namely that in the former case complete internal reflection can occur. If, in the case of the upward-directed light beam within the water, θ_w is greater than 49°, then all the light is reflected down again by the water–air interface. Disturbance of the surface by wind decreases water-to-air transmission for upwelling light at angles within the range $\theta_w = 0°$ to $\theta_w = 49°$, but, by ensuring that not all the light is internally reflected, it increases transmission within the range $\theta_w = 49°$ to $\theta_w = 90°$.

3

Absorption of light within the aquatic medium

Having discussed how the solar radiation gets through the water surface, we shall now consider what happens to it within the water. Sooner or later most of the photons are absorbed: how this happens, and which components of the medium are responsible, form the subjects of this chapter.

3.1 The absorption process

The energy of a molecule can be considered to be part rotational, part vibrational, part electronic. A molecule can only have one of a discrete series of energy values. Energy increments corresponding to changes in a molecule's electronic energy are large, those corresponding to changes in vibrational energy are intermediate in size, and those corresponding to changes in rotational energy are small. This is indicated diagrammatically in Fig. 3.1. When molecules collide with each other in the liquid or gaseous state, or are in contact with each other in the solid state, there can be transfer of rotational or vibrational energy between molecules and this is accompanied by transitions from one rotational or vibrational energy level to another within each molecule.

Molecules can obtain energy from radiation as well as from other molecules. When a photon passes within the vicinity of a molecule, there is a finite probability that it will be captured by that molecule, i.e. be absorbed. If the photon is captured, then the energy of the molecule must increase by an amount corresponding to the energy of the photon. If the photon is of long wavelength (>20 μm), in the far infrared/microwave region of the spectrum, then its energy is low and its absorption can only cause a transition in the energy of the molecule from one rotational energy level to another. If the photon is in the infrared region (<20 μm), then its absorption will cause a transition from one vibrational level to another.

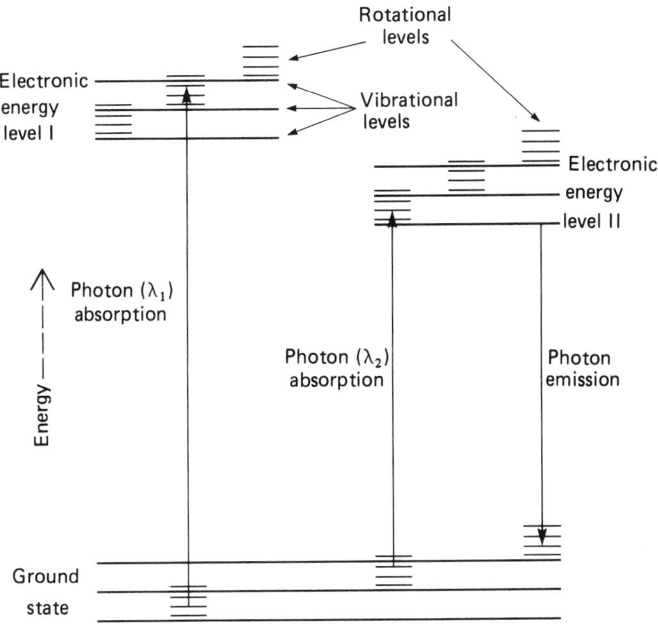

Fig. 3.1. Absorption of a photon raises an electron from the ground state to one of two possible excited singlet states, depending on the wavelength (and therefore energy) of the photon. Wavelengths λ_1 and λ_2 are each within a separate absorption band in the absorption spectrum of the molecule ($\lambda_2 > \lambda_1$).

Photons in the visible/photosynthetic part of the spectrum have sufficient energy to bring about, when they are absorbed, transitions from one electronic energy level (usually the ground state) to another. The first event in the absorption of light by the aquatic medium is, therefore, the capture of a photon by some molecule in the medium and the simultaneous transition of an electron in that molecule from the ground state to an excited state.

Within a complex molecule such as chlorophyll or any of the other photosynthetic pigments, there is usually more than one possible electronic energy transition which can occur. For example, in Fig. 3.1 excitation can occur up to electronic energy level I or to level II: to be more precise, excitation will occur to one of the many vibrational/rotational levels belonging to the given electronic energy level. Any given electronic transition is preferentially excited by light which has an amount of energy per photon corresponding to the energy required for the transition. Chlorophyll, as we shall see later, has two main absorption bands, in the red and in the blue regions of the spectrum. Absorption of a blue photon leads to transition to a substantially higher electronic energy level than absorption of a red photon. Because of the multiplicity of vibrational/rotational states, these two electronic energy states nevertheless overlap: i.e. the highest vibrational/rotational levels of the lower electronic state can have energies as high as the lowest level of the upper electronic state (Fig. 3.1). As a consequence, immediately after absorption of a blue photon there is a very rapid series of transitions downwards through the various rotational/vibrational levels (associated with transfer of small increments of rotational/vibrational energy to adjoining molecules) until the lower electronic energy state (referred to as the lowest excited singlet state) is reached.

It is the energy in this lowest excited singlet state which is used in photosynthesis and it is because an excited chlorophyll molecule usually ends up in this state anyway, that all *absorbed* visible quanta are equivalent, i.e. a red quantum, or even a green quantum if it succeeds in being absorbed, brings about as much photosynthesis as a blue quantum. This is also the reason why, in the context of primary production, it is more meaningful to express irradiance in quanta m^{-2} s^{-1} than in W m^{-2}.

In the case of light-absorbing molecules in the aquatic medium which are not part of the photosynthetic system, and which cannot therefore transfer their energy to a reaction centre for use in photosynthesis, there is likely to be a transition from the lowest excited singlet state to the corresponding triplet state by interaction with a paramagnetic molecule such as oxygen. In a singlet state, the two members of every pair of electrons in the molecule have opposite spin ($+\frac{1}{2}$ and $-\frac{1}{2}$) so that the resultant spin is zero: in a triplet state, the members of a pair of electrons in the molecule have the same spin so that the resultant spin is one. The excited triplet state is at a lower energy level and is also much longer lived than the excited singlet state, the average lifetime of the excited molecules being 10^{-5} to 10^{-4} s in the former case, compared to 10^{-9} to 10^{-8} s in the latter. Eventually, by interaction with a paramagnetic molecule such as oxygen, the excited molecule is returned to a singlet state by spin reversal and, by vibrational/rotational interactions with the surrounding molecules, it undergoes a downwards transition to one of the upper vibrational levels of the ground state. The electronic energy is thus dissipated as heat energy.

Excitation energy, whether in the photosynthetic system or in a molecule outside it, can be lost by re-emission of radiation. An electronically excited molecule, after undergoing the radiationless transition to the lowest level of the lowest excited singlet state, can then undergo a transition to one of the vibrational/rotational levels of the ground state by re-emitting a photon of light. This phenomenon is referred to as *fluorescence*. In living, photosynthesizing algal cells, only a very small proportion, ~1%, of the absorbed light is lost in this way. Most of the absorbed energy, whether captured initially by chlorophyll, carotenoid or biliprotein, is transferred by inductive resonance to the reaction centres where it is used to bring about biochemical changes (see Chapter 8). If photosynthesis is inhibited, say with DCMU (dichlorophenylmethylurea), then fluorescence increases to about 3% of the absorbed light. The non-photosynthetic, light-absorbing material of the aquatic medium can also re-emit some of its absorbed energy as fluorescence, but the fluorescence yield (quanta emitted/quanta absorbed) is again very low: most of the excited molecules are converted to the triplet state (see above), and from there to the ground state before they can emit a photon.

Thus, most of the light energy absorbed by the aquatic ecosystem, after existing for a very brief period as electronic excitation energy, ends up either as heat (vibrational/rotational energy distributed amongst all the molecules of the system) or as chemical energy in the form of photosynthetically produced biomass. Only a tiny proportion is turned back into light again by fluorescence, and even this is for the most part re-absorbed before it can escape from the system.

3.2 The measurement of light absorption

We saw in Chapter 1 how the light absorption properties of the aquatic medium are characterized in terms of the absorption coefficient, a. We saw also that although a is defined in terms of the behaviour of light passing through an infinitesimally thin layer of medium, its value can nevertheless be obtained from the observed absorptance, A, of a layer of medium of finite thickness, provided that the effects of scattering can be overcome.

Commercially available equipment (spectrophotometers) used for the measurement of light absorption, do so in terms of an optical parameter known as the *absorbance* or *optical density*: although there is no standard symbol we shall indicate it by D. Although absorbance is one of the most frequently measured parameters in biochemistry and chemistry it does not, unfortunately, appear to have a rigorous definition. It is commonly defined as the logarithm to the base 10 of the ratio of the light intensity, I_0, incident on a physical system to the light intensity, I, transmitted by the system

$$D = \log_{10} \frac{I_0}{I} \tag{3.1}$$

The meaning of the word 'intensity' is not usually given: we shall equate it to radiant flux, Φ. The incident beam is generally assumed to be parallel. The exact meaning of 'transmitted' is also not stated. If we make the simplest assumption that any light which re-emerges from the system, in any direction, is 'transmitted', then I includes all scattered light as well as that neither absorbed nor scattered. Thus, I is equal to the incident flux minus the absorbed flux ($\Phi_0 - \Phi_a$), from which, given that Φ_a/Φ_0 is absorptance, A, it follows that

$$D = \log_{10} \frac{1}{(1-A)} = -\log_{10}(1-A) \tag{3.2}$$

If we can arrange that all the scattered light is included in the measured value of I, then scattering may be ignored (except in so far as it increases the pathlength of the light within the system), and so eqn 1.32 may be considered to apply. Therefore,

$$D = 0.4343\, ar \tag{3.3}$$

i.e. the absorbance is equal to the product of the absorption coefficient of the medium and the pathlength (r) through the system, multiplied by the factor for converting base e to base 10 logarithms. The absorption coefficient may be obtained from the absorbance value using

$$a = 2.303\, D/r \tag{3.4}$$

In a typical spectrophotometer, monochromatic light beams are passed through two transparent glass or quartz cells of known pathlengths, one (the sample cell) containing a solution or suspension of the pigmented material, the other (the blank cell) containing pure solvent or suspending medium. The intensity of the light beam after passing through each cell is measured with an appropriate photoelectric device such as a photomultiplier. The intensities of the beams which have passed through the blank cell and the sample cell are taken to be proportional to I_0 and I, respectively, and so the logarithm of the ratio of these intensities is displayed as the absorbance of the sample.

When used in their normal mode (Fig. 3.2*a*), however, spectrophotometers have light sensors with a collection angle of only a few degrees. Thus any light scattered by the sample outside this collection angle will not be detected. Such spectrophotometers do not, therefore, measure true absorbance: indeed if their collection angle is very small, in effect they measure attenuance (absorption plus scattering, see §1.4). If, as is often the case, the sample measured consists of a clear coloured solution with negligible scattering relative to absorption, then scattering losses are trivial and the absorbance displayed by the instrument will not differ significantly

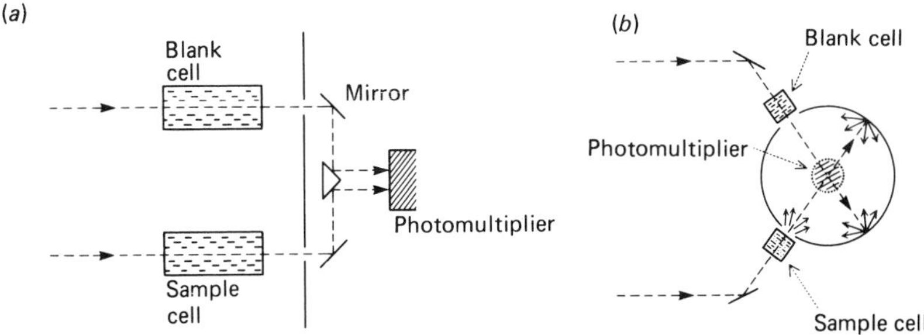

Fig. 3.2. Principles of measurement of absorption coefficient of non-scattering and scattering samples. The instrument is assumed to be of the double-beam type, in which a rotating mirror is used to direct the monochromatic beam alternately through the sample and the blank cells. (*a*) Operation of spectrophotometer in normal mode for samples with negligible scattering. (*b*) Use of integrating sphere for scattering samples.

from the true absorbance. If, however, the pigmented material is particulate, then a substantial proportion of the light transmitted by the sample may be scattered outside the collection angle of the photomultiplier, the value of I registered by the instrument will be too low, and the absorbance displayed will be higher than the true absorbance.

This problem may be solved by so arranging the geometry of the instrument that nearly all the scattered light is collected and measured by the photomultiplier. The simplest arrangement is to have the cells placed close to a wide photomultiplier so that the photons scattered up to quite wide angles are detected. A better solution is to place each cell at an entrance window in an integrating sphere (Fig. 3.2*b*): this is a spherical cavity, coated with white, diffusely reflecting material inside. By multiple reflection, a diffuse light field is set up within the sphere from all the light which enters it, whether scattered or not. A photomultiplier at another window receives a portion of the flux from this field, which may be taken to be proportional to I or I_0 in accordance with whether the beam is passing through the sample, or the blank cell. With this system virtually all the light scattered in a forward direction, and this will generally mean nearly all the scattered light, is collected.

Another procedure, which can be used with some normal spectrophotometers, is to place a layer of scattering material, such as opal glass, behind the blank and the sample cell.[812] This ensures that the blank beam and the sample beam are both highly scattered after passing through the cells. As a result, the photomultiplier, despite its small collection angle, harvests about the same proportion of the light transmitted through the sample cell, as of the light transmitted through the blank cell, and so the registered values of intensity are indeed proportional to the true values of I and I_0, and so can give rise to an approximately correct value for absorbance.

With all these procedures for measuring the absorbance of particulate materials, it is important that the suspension should not be so concentrated that, as a result of multiple scattering, there is a significant increase in the pathlength that the photons traverse in passing through the sample cell. This would have the effect of increasing the absorbance of the suspension and lead to an estimate of absorption coefficient which would be too high. Even with an integrating sphere, some of the light is scattered by the sample at angles too great for collection: a procedure for correction for this error has been described.[483]

An alternative approach, pioneered by Yentsch (1960), is to pass sea water, or fresh water, through a filter until enough particulate matter has accumulated, and then measure the spectrum of the material on the filter itself, without resuspending it. The problem with this method is that, as a consequence of multiple internal reflection within the particulate layer, there is very marked amplification of absorption (up to six-fold). To calculate the true absorption coefficients that this particulate material has, when freely suspended in the ocean or lake, the absorption amplification factor must be determined reasonably accurately, and this is not easy, especially since it is far from constant, varying from about 2.5 to 6.

A different experimental approach has been developed by Iturriaga & Siegel (1989), who use a monochromator in conjunction with a microscope to measure the absorption spectra of individual phytoplankton cells or other pigmented particles. If sufficient particles are measured, and their numbers in the water are known, then the true *in situ* absorption coefficients due to the particles can be calculated. While laborious, this does avoid the uncertainty associated with estimation of the amplification factor, in the previous method.

As we shall see later, the absorption coefficients of the different constituents of the aquatic medium can be determined separately. The value of the absorption coefficient of the medium as a whole, at a given wavelength, is equal to the sum of the individual absorption coefficients of all the components present. Furthermore, providing that no changes in molecular state, or physical aggregation take place with changes in concentration, then the absorption coefficient due to any one component is proportional to the concentration of that component (Beer's Law).

The total absorption coefficient of a natural water at a given wavelength can be determined from a measurement made within the water body itself:

an instrument designed for this purpose is known as an absorption meter. To make accurate measurements at those wavelengths where marine, and some inland, waters absorb only feebly, quite long pathlengths ($\geqslant$0.25 m) are needed. The simplest arrangement is to have, immersed in water, a light source giving a collimated beam and, at a suitable distance, a detector with a wide angle of acceptance. Most of the light scattered, but not absorbed, from the beam is detected: diminution of radiant flux between source and detector is thus mainly due to absorption and can be used to give a value for the absorption coefficient.

An alternative approach is to have, not a collimated beam, but a point source of light radiating equally in all directions.[47] The detector does not have to have a particularly wide angle of acceptance. For every photon which is initially travelling towards the detector and is scattered away from the detector, there will, on average, be another photon, initially not travelling towards the detector, which is scattered towards the detector. Thus diminution of radiant flux between source and detector is entirely due to absorption. An absorption meter based on this principle has been used to measure the absorption coefficient of water in Lake Baikal (Russia).[63] A problem with this and the previous type of absorption meter is that although scattering may not directly prevent photons reaching the detector, it increases the pathlength of the photons and thus increases the probability of their being absorbed, i.e. spuriously high values of absorption coefficient may be obtained. This error can be significant in waters with high ratios of scattering to absorption.

To avoid the necessity of comparing readings in the water body with readings in pure water or other standard medium, a variant on the second type of absorption meter has been developed which still uses a point light source but has two detectors at different distances from the source: from the difference in radiant flux on the two detectors, the absorption coefficient may be obtained.[269] By means of interference filters, this particular instrument carries out measurements of absorption coefficient at 50 nm intervals from 400 to 800 nm.

With all these absorption meters, if the light source is modulated and the detector is designed to measure only the modulated signal, then the contribution of ambient daylight to radiant flux on the detector can be eliminated.

In the case of an absorption meter using a collimated light beam, one possible solution to the problem of distinguishing between attenuation of the signal which is genuinely due to absorption from that which is due to photons in the measuring beam simply being scattered away from the detector, is to shine the beam down an internally reflecting tube[149,1019] containing the water under investigation. The principle is that photons which are scattered to one side will be reflected back again, and so can still be detected at the other end. In such an instrument, however, there is still some residual loss of photons by non-reflection at the silvered surface, and by backscattering, and so a correction term proportional to the scattering coefficient needs to be applied.[495]

An instrument based on completely different principles to any of the above is the integrating cavity absorption meter.[222,272,273] The water sample is contained within a cavity made of a translucent, and diffusely reflecting, material. Photons are introduced into the cavity uniformly from all round, and a completely diffuse light field is set up within it, the intensity of which can be measured. The anticipated advantages are, first, that since the light field is already highly diffuse, additional diffuseness caused by scattering will have little effect; second, because the photons undergo many multiple reflections from one part of the inner wall to another, the effective pathlength within the instrument is very long, thus solving the pathlength problem. A prototype instrument has given promising results.[273]

Absorption coefficients of natural waters can also be determined *in situ* from measurements of irradiance and scalar irradiance. The instrumentation for such measurements we shall discuss in a later chapter (see §5.1). Determination of absorption coefficient makes use of the relation (see §1.7)

$$a = K_E \vec{E}/E_0$$

where E_0 is the scalar irradiance, $\vec{E}$ is the *net* downward irradiance ($E_d - E_u$), and K_E is the vertical attenuation coefficient for $\vec{E}$. Thus a can be obtained from measurements of net downward and scalar irradiance at two, or a series of, depths (measurements at more than one depth are required to give K_E). Absorption meters based on these principles have been built.[381,864]

3.3 The major light-absorbing components of the aquatic system

Essentially all the light absorption which takes place in natural waters is attributable to four components of the aquatic ecosystem: the water itself, dissolved yellow pigments, the photosynthetic biota (phytoplankton, and macrophytes where present) and inanimate particulate matter (*tripton*). We shall now consider the spectral absorption properties of each of these, and their relative contributions to the absorption of photosynthetically available radiation.

Water

Pure water, although it appears colourless in the quantities we handle in everyday life, is in fact a blue liquid: the blue colour is clearly apparent under sunny conditions in oceanic waters or in coastal waters which are infertile and have little input from rivers. The colour of pure water arises from the fact that it absorbs only very weakly in the blue and the green regions of the spectrum, but its absorption begins to rise as wavelength increases above 550 nm and is quite significant in the red region: a 1-m thick layer of pure water will absorb about 35% of incident light of wavelength 680 nm.

Because of the very weak absorption it is very difficult to measure the absorption coefficient of pure water in the blue/green spectral region, and the values reported in the literature, determined by normal spectrophotometric procedures, vary widely. Smith & Baker (1981) have arrived at a set of values based partly on their own measurements of vertical attenuation coefficient for irradiance in the clearest ocean waters (K_d in such waters is slightly greater than a) and partly on what they consider to be the best laboratory measurements, especially those of Morel & Prieur (1977) in the photosynthetic region (380–700 nm).

The extent to which pure water absorbs light in the ultraviolet, including the ecologically relevant UV-B (280–320 nm) region, is a matter of some controversy. Liquid water does have an intense absorption band at 147 nm due to an electronic transition, but this, on theoretical grounds, would be expected to tail away to very low values in the 200–300 nm range. An anticipated exponential diminution of absorptivity with decreasing photon energy (the Urbach rule) implies, for example, an absorption coefficient of only $\sim 0.02\ m^{-1}$ at 207 nm, and still lower values at higher wavelengths.[734] While some workers over the period 1928–76 found absorption coefficients (200–300 nm) which appeared to be orders of magnitude higher than this, more recent studies by Quickenden & Irvin (1980), and Boivin *et al.* (1986), using very carefully purified water, indicate that absorption is indeed very low in the near UV. The previously reported high absorption coefficients are attributable to dissolved oxygen which does absorb UV quite strongly (and which is, of course, present in sea and lake water anyway), and to trace organic materials.

Literature values for the absorption coefficient of pure water over the range 280–800 nm are listed in Table 3.1, and Fig. 3.3 shows the absorption spectrum of pure water from the near ultraviolet, through the photosynthetic range, to the near infrared.

Table 3.1. *Absorption coefficients for pure water: 280–320 nm, Quickenden & Irvin (1980); 366 nm, Boivin et al. (1986); 380–700 nm, Morel & Prieur (1977); 700–800 nm, Smith & Baker (1981)*

λ (nm)	a (m^{-1})	λ (nm)	a (m^{-1})
280	0.0239[ab]	560	0.071
290	0.0140[ab]	570	0.080
300	0.0085[ab]	580	0.108
310	0.0082[ab]	590	0.157
320	0.0077[ab]	600	0.245
366	0.0055[a]	610	0.290
380	0.023	620	0.310
390	0.020	630	0.320
400	0.018	640	0.330
410	0.017	650	0.350
420	0.016	660	0.410
430	0.015	670	0.430
440	0.015	680	0.450
450	0.015	690	0.500
460	0.016	700	0.650
470	0.016	710	0.839
480	0.018	720	1.169
490	0.020	730	1.799
500	0.026	740	2.38
510	0.036	750	2.47
520	0.048	760	2.55
530	0.051	770	2.51
540	0.056	780	2.36
550	0.064	790	2.16
		800	2.07

Notes:
[a] Absorption coefficients derived from the published attenuation coefficients by subtracting estimated values[632,832] of scattering coefficients.
[b] These values were measured on deoxygenated water.

It can be seen that the absorption by water at the red end of the visible spectrum is really the tail end of a series of much stronger absorption bands in the infrared: there are even more intense absorption bands beyond 1000 nm. The two shoulders in the visible spectrum – the distinct one at ~604 nm and the weak one at ~514 nm – have been identified as corresponding to the fifth and sixth harmonics, respectively, of the O–H stretch vibration of liquid water;[905] the peaks at ~960 and ~745 nm in the infrared correspond to the third and fourth harmonics (the fundamental is at about 3 μm). Light absorption by pure water increases slightly with

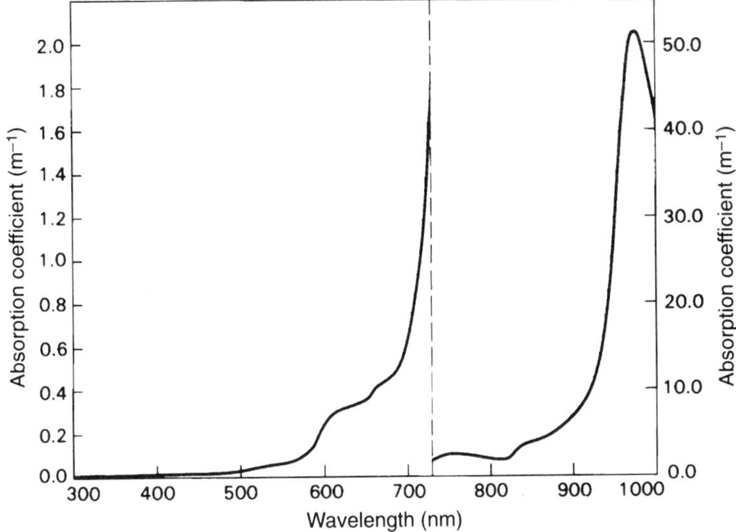

Fig. 3.3. Absorption spectrum of pure water. Absorption coefficient values have been taken from Table 3.1 for the range 310–790 nm, and from the data of Palmer & Williams (1974) for the range 790–1000 nm.

temperature: da/dT is $\sim +0.003\ m^{-1}\ {}^\circ C^{-1}$ between 10 °C and 30 °C over the wavelength range 400–600 nm,[386a] and is $\sim +0.009\ m^{-1}\ {}^\circ C^{-1}$ between 5 °C and 40 °C at 750 nm.[689a]

The contribution of water itself to the attenuation of PAR by absorption of quanta is of importance only above about 500 nm. The salts present in sea water appear to have no significant effect on absorption in the visible/photosynthetic range:[747,832] the nitrates and bromides, however, do cause a marked increase in absorption below 250 nm.[97,671]

Gilvin (yellow substance)

When plant tissue decomposes in the soil or in a water body, most of the organic matter is broken down by microbial actions within days or weeks to, ultimately, carbon dioxide and inorganic forms of nitrogen, sulphur and phosphorus. In the course of the decomposition process there is formed, however, a complex group of compounds loosely referred to as 'humic substances'. The chemistry of humic substances has been reviewed by Schnitzer (1978).

Degradation and other studies indicate that these substances are polymers consisting of aromatic rings which are joined by long-chain alkyl

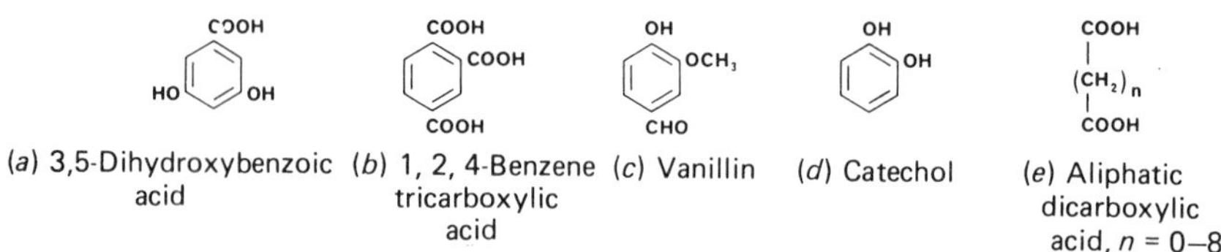

Fig. 3.4. Structures of some of the products liberated on chemical degradation of humic substances.

structures to form a flexible network.[802] Fig. 3.4 shows some of the many different compounds produced on oxidative chemical degradation of humic substances. A humic substance sample from the water of the Okefenokee Swamp, Georgia, USA, was found to have[919] an average atomic composition of $C_{74}H_{72}O_{46}N_{0.7}$.

Humic substances vary in size from freely soluble compounds with molecular weights (relative molecular masses) of a few hundred, to insoluble macromolecular aggregates with molecular weights in the hundreds of thousands and perhaps ranging up to the millions. Soil chemists classify humic substances into three main fractions on the basis of solubility behaviour. The soil is first extracted with dilute alkali. The humic material that does not dissolve is referred to as *humin*. Of the alkali-soluble fraction, some is precipitated on acidification: this is called *humic acid*. The humic material remaining in solution is called *fulvic acid*. In fact all three fractions are chemically very similar and differ mainly in molecular weight, humic acid molecules being larger than those of fulvic acid. They are yellow to brown in colour (the soluble ones giving rise to yellow-brown solutions), hydrophilic and acidic (due to the presence of carboxyl and phenolic groups). Fulvic acid has a higher content of oxygen-containing groups such as carboxyl and hydroxyl. The insolubility of the humin fraction may be due to its being firmly adsorbed or bonded to mineral particles and/or to its having a very high molecular weight.

While some humic material may be formed by oxidation and polymerization directly from the existing phenolic compounds (particularly lignin) in the decomposing plant tissue, it is also true that some saprophytic fungi excrete large amounts of phenolic substances when grown on carbohydrate, and these phenolic substances can undergo oxidation and polymerization to give humic-like material.[183] Thus it seems likely that some of the aromatic subunits of the humic materials originate in the plant and some are generated *de novo* during microbial breakdown. The relative distribution of the various lignin-derived aromatic subunits of dissolved humic substances in river and lake waters reflects the botanical composition of the dominant vegetation in the catchments. Ertel, Hedges & Perdue (1984) found that two Oregon water bodies had quite different phenolic composition in their dissolved humic acid, and these in each case correspond closely to the phenolic composition of the lignin in the, respectively, non-woody angiosperm- and woody gymnosperm-dominated catchments from which the waters were derived. In both water bodies the humic fraction yielded (on oxidation) four to six times more lignin phenols, relative to total organic carbon, than did the fulvic acid fraction.

The proportion of aromatic carbon in marine humic and fulvic acids is lower than in the corresponding fresh-water substances.[585] Marine fulvic acid appears in fact to contain only a very small proportion of aromatic residues and is predominantly aliphatic in nature.[356,585] Harvey *et al.* (1983, 1984) propose that fulvic and humic acids in sea water consist mainly of polymeric compounds formed by the oxidative crosslinking of polyunsaturated lipids derived from the biota. This is unlikely, however, to be the whole story, since total dissolved humic material from the eastern equatorial Pacific ocean has been shown to contain measurable amounts of lignin-derived phenolic residues.[611]

From the point of view of aquatic ecology, the significance of the soil humic material is that as water, originating as rainfall, drains through soil and into rivers and lakes, and ultimately into estuaries and the sea, it extracts from the soil some of the water-soluble humic substances and these impart a yellow colour to the water, with major consequences for the absorption of light, particularly at the blue end of the spectrum. James & Birge (1938) in the USA and Sauberer (1945) in Austria carried out extensive quantitative studies on the absorption spectra of lake waters with varying degrees of yellowness. The colour of humic substances is due to the presence of multiple double bonds, many of them conjugated, some in aromatic nuclei. In any sample of humic material there are numerous different chromophores, and consequently a multitude of electronic excitation levels which, because they overlap, give rise to a rather undifferentiated UV/visible absorption spectrum. Visser (1984) found that in surface waters derived from coniferous forest catchments in Quebec (Canada), the colour intensity per unit mass of the humic acid was nearly four times that of the fulvic acid: its concentration, however, was only about one seventh that of the fulvic acid, implying that in these waters, humic acid contributed about one third, and fulvic acid two thirds, of the total colour. As well as absorbing light, dissolved humic material in natural waters has a broad fluorescence emission band in the blue region.

The light absorption properties of these dissolved humic materials in

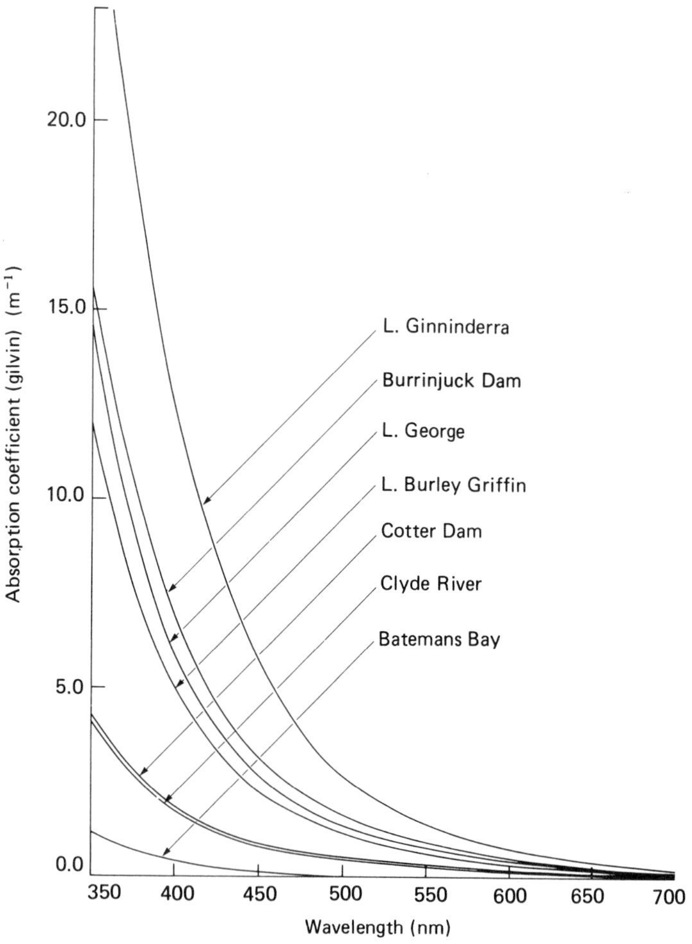

Fig. 3.5. Absorption spectra of soluble yellow material (gilvin) in various Australian natural waters (from Kirk, 1976*b*). The lowest curve (Batemans Bay, NSW) is for coastal sea water near the mouth of a river; the next curve (Clyde River, NSW) is for an estuary; the remainder are for inland water bodies in the southern tablelands of New South Wales/Australian Capital Territory. The ordinate scale corresponds to the true *in situ* absorption coefficient due to gilvin.

natural inland waters can be determined relatively easily by measuring the absorption spectrum of a water sample which has been filtered (0.2–0.4 μm pore size) in 5 cm or 10 cm pathlength cells.[478] Fig. 3.5 shows the absorption spectrum of this material from some Australian inland waters. These are typical humic substance absorption spectra, with absorption being very low or absent at the red end of the visible spectrum, and rising steadily with decreasing wavelength towards the blue: absorption in the ultraviolet region is higher still.

The presence of dissolved yellow material in inland waters is often easily apparent to the eye. Its presence in marine waters (where its concentration is much lower) is not so readily apparent, and the fact that it is important in these ecosystems too was pointed out by Kalle (1937, 1966). Fig. 3.5 also shows the absorption spectra of soluble yellow material in estuarine and coastal waters.

It seems likely that most of the dissolved yellow colour in inland waters is due to soluble humic substances leached from the soils in the catchment areas, and thus indirectly from the vegetation. Yellow material of the humic type can also be generated by decomposition of plant matter within the water: this could be of significance in productive water bodies. Much of the soluble humic material in river water is precipitated when it comes into contact with sea water.[422,817] Nevertheless, a fraction of the material remains in solution and most of the dissolved yellow colour in coastal sea water is due to humic substances derived from the land in river discharge. For example, Monahan & Pybus (1978) found that in regions with major river discharge off the west coast of Ireland, the concentration of soluble humic material diminished linearly with increasing salinity: the results indicated that essentially all the humic material in these coastal North Atlantic waters originated in the rivers. Since it is mainly the humic acid fraction which is lost in the estuaries,[232,262] the contribution of river inflow to yellow colour in the sea is mainly in the form of fulvic acid. The specific absorption (per unit mass at 440 nm) of marine fulvic acid is much lower than that of marine humic acid,[132] but this is made up for by the usually much higher concentration of fulvic acid, so that the two forms of dissolved humic material make comparable contributions to the absorption of light in the ocean.

As in fresh waters, some formation of dissolved yellow materials may take place within marine waters. Brown seaweeds, for example, actively excrete phenolic compounds which, probably as a consequence of oxidation and polymerization, give rise to yellow-brown materials of the humic type within the water.[818] It seems possible that this phenomenon might make a significant contribution to the amount of dissolved yellow material in the water near luxuriant brown algal beds.

It is not certain to what extent the yellow substance (present at low concentrations but still optically significant) in oceanic waters away from the coastal zone is land-derived humic material or is generated within the sea by decomposition of the phytoplankton. Højerslev (1979), on the basis of his measurements in the Baltic and North Atlantic, considers it unlikely that significant amounts of yellow substance are formed in the sea and attributes it, even in oceanic areas, to river discharge. Jerlov (1976), however, argues that the presence of yellow substance in the upwelling region west of South America proves its immediate marine origin, as this area is practically devoid of fresh-water supply from land drainage. Bricaud, Morel & Prieur (1981), on the basis of their measurements in a variety of waters, suggest that in the ocean away from regions of river discharge, the concentration of yellow substance is determined by biological activity averaged over a long period. Kopelevich & Burenkov (1977) observed a strong correlation between the concentration of yellow substance and the level of phytoplankton chlorophyll in productive oceanic waters. They proposed that oceanic yellow substance is of two kinds: a component resulting from the recent decomposition of phytoplankton, and a more stable component of much greater age. This latter 'conservative' component would predominate in oligotrophic waters and might, in agreement with Bricaud *et al.*, reflect average biological activity over a long time.

New evidence on the contribution of the terrestrial biosphere to the humic material in the ocean has come from measurement, in the oxidation products of humic substance from ocean water, of specific phenols derived from lignin, a structural polymer present only in land plants. Meyers-Schulte & Hedges (1986), using humic material from the Amazon River as a standard (entirely terrestrially derived), conclude on the basis of measurement of lignin-specific phenols, that about 10% of the humic material in ocean water from the eastern equatorial Pacific is terrestrially derived. Since this is the part of the world ocean least affected by direct river input, it seems likely that the percentage would be higher in other oceanic regions.

The humic material that ends up in the dissolved state in natural waters occurs in a wide range of molecular weights. Vapour-phase osmometry and small-angle X-ray scattering indicate molecular weights of 600–1000 for river fulvic acids and 1500–5000 for river humic acids.[5,585]

The dissolved yellow material in natural waters has been variously referred to as 'yellow substance', 'gelbstoff' (the same name, in German), 'yellow organic acids', 'humolimnic acid', 'fulvic acid', 'humic acid', etc. Dissolved yellow materials from different waters vary not only in molecular size but also in chemical composition.[452,883,884] In the context of light attenuation it would be useful to have a general name, applicable to all or any of these compounds regardless of their chemical nature, which simply indicates that they preferentially absorb light at the blue end of the spectrum. 'Yellow substance' or 'gelbstoff' are too non-specific and could apply to anything from butter to ferric chloride. I have suggested the word 'gilvin', a noun derived from the Latin adjective *gilvus* meaning pale yellow.[478] 'Gilvin' would thus be defined as a general term to be applied to those soluble yellow substances, whatever their chemical structure, which occur in natural waters, fresh or marine, at concentrations sufficient to contribute significantly to the attenuation of photosynthetically available radiation. I shall use the word 'gilvin' when referring to the dissolved yellow substances in natural waters, in the remainder of this book.

As the absorption spectra in Fig. 3.5 show, the concentration of gilvin varies markedly, not only between marine and fresh waters, but also amongst different inland waters. A convenient parameter by means of which the concentration of gilvin may be indicated is the absorption coefficient at 440 nm due to this material within the water: this we shall indicate by g_{440}. This wavelength is chosen because it corresponds approximately to the mid-point of the blue waveband peak that most classes of algae have in their photosynthetic action spectrum.

In many waters, certainly most inland waters, the gilvin absorption is sufficiently high for g_{440} to be measurable with reasonable accuracy, using 50 or 100 mm pathlength cells. When gilvin absorption is low, as in most marine waters, the absorption coefficient can be measured in the near-ultraviolet (350–400 nm) where absorption is higher, and g_{440} determined by proportion from a typical gilvin absorption spectrum, or using the approximate relationship.[97]

$$a(\lambda) = a(\lambda_0)e^{-S(\lambda - \lambda_0)} \tag{3.5}$$

where $a(\lambda)$ and $a(\lambda_0)$ are the absorption coefficients at wavelengths λ and λ_0 nm, respectively, and S is a coefficient describing the exponential slope of the absorption curve. For a wide range of seawaters, S in the surface layer has been found to vary from about 0.010 to 0.020 nm^{-1}, with a mean value in the region of 0.012–0.015 nm^{-1}.[97,132,515] For 12 New Zealand lakes, S varied from 0.015 to 0.020 nm^{-1}, with a mean of 0.0187 nm^{-1};[177] and for 22 Australian inland waters, S varied from 0.012 to 0.018 nm^{-1}, with a mean of 0.016 nm^{-1}.[495a]

Table 3.2 contains a selection of published data in the form of values of g_{440} for marine as well as fresh waters in various geographical regions. This compilation gives some idea of typical gilvin concentrations that may be expected in natural waters, but the lack of measurements in what are otherwise limnologically well-characterized parts of the world is apparent.

Table 3.2. *Values of absorption coefficient at 440 nm due to dissolved colour* (g_{440}) *and particulate colour* (p_{440}) *in various natural waters. Where several measurements have been taken, the mean value, the standard deviation, the range and the time period covered are in some cases indicated. For some waters, only the absorption coefficient for gilvin plus particles* (g + p) *was available. L. = lake, R. = river*

Water body	g_{440} (m^{-1})	p_{440} (m^{-1})	Reference
I. Oceanic waters			
Atlantic Ocean			
Sargasso Sea	~0	0.01	534
Off Bermuda	0.01*	—	405
Caribbean Sea	0.03($g+p$)		422
Gulf of Mexico – Gulf loop intrusion	0.005		132
Romanche Deep	0.05($g+p$)		422
Mauritanian upwelling	0.034–0.075*	—	97
Gulf of Guinea	0.024–0.113*	—	97
Pacific Ocean			
Galapagos Islands	0.02*	—	121
Galapagos Islands	0.16($g+p$)		422
Central Pacific	0.04($g+p$)		422
Off Peru	0.05*	—	121
Indian Ocean			
Oligotrophic water	0.02	—	518
Mesotrophic water	0.03	—	518
Eutrophic water	0.09	—	518
Mediterranean			
Western	0.0–0.03*	—	386

II. Shelf, coastal and estuarine waters			
Gulf of Mexico			
Yucatan shelf	0.006	—	132
Bay of Campeche	0.022	—	132
Cape San Blas	0.054	—	132
Mississippi plume	0.028	—	132
North America, Atlantic coast			
Rhode R. estuary	0.72	8.4	282
Chesapeake Bay (Rhode R. mouth)	0.27	0.80	282
Georgia salt marsh	1.52	—	990
Arabian Sea			
Cochin, India. 3 km off-shore	0.24†$(g+p)$		747
30 km off shore	0.10†$(g+p)$		747
Bay of Bengal			
Near Ganges mouths	0.37*	—	515
Yellow Sea	0.20–0.23*	—	386
Eastern Pacific			
Peruvian coast	0.29*	—	515
North Atlantic/North Sea/Baltic system			
W. Greenland	0.004*	—	386
North Atlantic	0.02*	—	451
Iceland	0.016*	—	386
Orkney–Shetland	0.016*	—	386
North Sea (Fladen Ground)	0.03–0.06*	—	386
Skagerrak	0.05–0.12*	—	386
Kattegat	0.12–0.27*	—	386
Baltic Sea	0.36–0.42*	—	386
South Baltic Sea	0.26	—	422
Bothnian Gulf	0.41*	—	422

Table 3.2 (*cont.*)

Water body	g_{440} (m^{-1})	p_{440} (m^{-1})	Reference
Mediterranean			
Villefranche Bay	0.060–0.161*	—	97
Marseilles drainage outfall	0.073–0.646*	—	97
R. Var mouth	0.136*	—	97
R. Rhone mouth	0.086–0.572*	—	97
Tyrrhenian Sea			
Gulf of Naples	0.02–0.20*	—	251
Northern Adriatic Sea			
Sacca di Goro (R. Po mouth)	0.32–3.43*	—	251
Venice Lagoon	0.44–0.73*	—	251
Southeast Australia			
(a) Jervis Bay			
3 stations	0.09–0.14*	0.03–0.04	495*a*
(b) Tasman Sea/Clyde R. system			
Tasman Sea	0.02*	—	479
Batemans Bay	0.18	—	478
Clyde R. estuary	0.64	—	479
(c) Gippsland (estuarine) lakes system			
L. King	0.58	0.25	495*a*
L. Victoria	0.65	0.22	495*a*
L. Wellington	1.14	2.27	495*a*
Latrobe R.	1.89	2.78	495*a*

New Zealand			
9 estuaries, North Island mouth sites, low water data	0.1–0.6	—	949
11 shelf stations, South Island	0.04–0.10	—	175
	0.07 av.		
Japan, Pacific coast			
17 km off Shimoda	0.024	0.133	500
5 km off Shimoda	0.011	0.095	500
Nabeta Bay	0.054	0.140	500
III. Inland waters			
Europe			
Rhine R.	0.48–0.73*	—	386
Donau R., Austria	0.85–2.02	—	386
Ybbs R., Austria	0.16*	—	386
Neusiedlersee, Austria	$\sim 2.0 \pm 0.4$*	—	193
(8 months)	1.4 − 2.8		
Blaxter L. (bog lake), England	9.65	—	604
Carmean Quarry, Ireland	0.23	—	429
Killea Reservoir, Ireland	0.5	—	429
Lough Neagh, N. Ireland	1.9	—	429
Lough Fea, Ireland	4.7	—	429
Lough Erne, Ireland	5.3	—	429
Loughnagay, Ireland	6.4	—	429
Lough Bradan, Ireland	17.4	—	429
Lough Napeast, Ireland	19.1	—	429
Loch Leven, Scotland	1.2	—	900
Africa			
L. George, Uganda	3.7	—	900
North America			
Crystal L., Wisc., USA	0.16	—	408

Table 3.2 (*cont.*)

Water body	g_{440} (m^{-1})	p_{440} (m^{-1})	Reference
Adelaide L., Wisc., USA	1.85	—	408
Otisco L., N.Y., USA	0.27	0.27	981
Irondequoit Bay, L. Ontario, USA	0.90	0.65	980
Bluff L., N.S., Canada	0.94	—	328
Punch Bowl, N.S., Canada	6.22	—	328
South America			
Guri Reservoir, Venezuela	4.84	—	558
Carrao R., Venezuela	12.44	—	558
Australia			
(a) Southern tablelands			
Cotter Dam	1.28–1.46	0.77	483, 495*a*
Corin Dam	1.19–1.61	0.11	483, 495*a*
L. Ginninderra	1.54 ± 0.78	0.16–0.58	478, 479, 483, 495*a*
(3-year range)	0.67–2.81		
L. George	1.80 ± 1.06	3.73–4.21	478, 479, 483, 495*a*
(5-year range)	0.69–3.04		
Burrinjuck Dam	2.21 ± 1.13	0.63–1.44	478, 479, 483, 495*a*
(5-year range)	0.81–3.87		
L. Burley Griffin	2.95 ± 1.70	2.91–2.96	478, 479, 483, 495*a*
(5-year range)	0.99–7.00		
Googong Dam	3.42	0.83	483
Queanbeyan R.	2.42	—	495*a*
Molonglo R.	0.44	—	495*a*
Molonglo R. below confluence with Queanbeyan R.	1.84	—	495*a*

Creek draining boggy ground	11.61	—	495*a*
(b) Murray–Darling system			
Murrumbidgee R., Gogeldrie Weir (10 months)	0.4–3.2	—	677
L. Wyangan	1.13	0.38	495*a*
Griffith Reservoir	1.34	3.73	495*a*
Barren Box Swamp	1.59	2.55	495*a*
Main canal, M.I.A.	1.11	5.35	495*a*
Main drain, M.I.A.	2.12	10.34	495*a*
Murray R., upstream of Darling confluence	0.81–0.85	—	677
Darling R., above confluence with Murray	0.7–2.5	—	677
(c) Northern Territory (Magela Creek billabongs)			
Mudginberri	1.11	1.13	498
Gulungul	2.28	1.68	498
Georgetown	1.99	18.00	498
(d) Tasmania (lakes)			
Perry	0.06	—	90
Ladies Tarn	0.40	—	90
Risdon Brook	0.98	—	90
Barrington	3.05	—	90
Gordon	8.29	—	90
(e) Southeast Queensland, coastal dune lakes			
L. Wabby	0.06	—	89
Basin L.	0.46	—	89
L. Boomanjin	2.59	—	89
L. Cooloomera	14.22	—	89

Table 3.2 (*cont.*)

Water body	g_{440} (m^{-1})	p_{440} (m^{-1})	Reference
(f) South Australia			
Mount Bold Reservoir	5.40	2.25	286
New Zealand			
Waikato R. (330 km, L. Taupo to the sea)			
L. Taupo (0 km)	0.070	0.033	174
Ohakuri (77 km)	0.22	0.32	174
Karapiro (178 km)	0.82	0.71	174
Hamilton (213 km)	0.97	0.98	174
Tuakau (295 km)	1.37	1.67	174
Lakes			
(means of monthly values over 11 months)			
Rotokakahi	0.09	—	177
Rotorua	0.23	—	177
Opouri	0.86	—	177
Hakanoa	1.84	—	177
D	4.87	—	177
Japan			
L. Kizaki	0.30	0.71	500
L. Fukami-ike	0.85	3.11	500

Note:
* Values measured at a wavelength less than 440 nm and converted to g_{440} on the basis of an appropriate gilvin absorption spectrum.
† Published values for c at 440 nm corrected for scattering.

In view of the great importance of gilvin in determining the penetration of light in natural waters, and in view of the ease (in most inland waters) with which its optical concentration can be measured,[478] it is to be hoped that the spectrophotometric assessment of dissolved colour will become a routine part of limnological methodology.

The data in Table 3.2 show that marine waters generally have much less dissolved colour than inland waters, and the greater the distance from land, the lower the concentration. The high concentration (for a marine water) within the Baltic Sea is noteworthy: it decreases from the Bothnian Gulf southwards, as the proportion of river water in the sea diminishes. The increase in concentration with distance from the sea upstream in estuarine systems can be seen in the data for Clyde River–Batemans Bay and Latrobe River–Gippsland Lakes (Australia).

Although gilvin is chemically rather stable – its concentration in a stored sample usually shows only small changes over a few weeks – its concentration within any inland water body changes, in either direction, with time, in accordance with rainfall events in different parts of the catchment and consequent changes in the concentration of gilvin in the inflowing waters. Some of the data in Table 3.2 give an indication of the extent of this variability. For example, in Lake Burley Griffin (Canberra, Australia) the value of g_{440} varied seven-fold over a 5-year period. Nevertheless, for any given water body, variation does tend to be around a certain mean value and the water body can usefully be regarded as typically high, low or intermediate in gilvin concentration. The category in which a particular lake, impoundment or river falls is determined by the drainage pattern, vegetation, soils and climate in the catchment.

The factors governing the concentration of gilvin in surface waters are not well understood and in view of the great influence of this material on aquatic primary production they deserve intensive investigation. One generalization that can be made is that gilvin concentration is high in drainage water from bogs or swamps: this can readily be seen, for example, in the peat bogs of northern Europe, and is shown quantitatively in the g_{440} values for an English bog lake and a creek draining boggy ground in the Australian southern tablelands (Table 3.2). Gilvin concentration is also high in the waters draining from humid tropical forests,[558] as the g_{440} values for two Venezuelan waters in Table 3.2 show. The lack of oxygen in the permanently, or frequently, waterlogged soil of such areas leads to a build-up of partially decomposed organic matter, and gilvin is derived from the soluble component of this. The other side of the coin is that water draining limestone-rich catchments tends to be low in gilvin. An inverse relationship between water colour and lake depth has been observed for North America.[329] The effects of vegetation type, soil mineralogy, agricultural practices, climate and other environmental parameters on gilvin concentration are not well understood.

Although, as we have noted, gilvin is chemically quite stable, it does undergo photochemical degradation by intense sunlight in the surface layer.[466,521] Kieber, Xianling & Mopper (1990) found the breakdown products to include a range of low-molecular-weight carbonyl compounds, such as pyruvate, formaldehyde and acetaldehyde, which would be readily utilized by aquatic microbes. Action spectra indicated that it is the UV-B region of the solar spectrum (280–320 nm) which is responsible. On the basis of their rate measurements, Kieber *et al.* estimate that the half-life of dissolved humic substances of riverine origin, in the oceanic mixed layer is 5–15 years. Mopper *et al.* (1991) present evidence suggesting that this photochemical pathway is in fact the main route for the degradation of the long-lived biologically refractory, dissolved organic carbon of the ocean.

Tripton

The inanimate particulate matter, or tripton, of natural waters is the fraction whose light absorption properties have received the least attention because they are so difficult to measure. At typical concentrations the material does not absorb light strongly but scatters quite intensely and so its absorption properties cannot be characterized by normal spectrophotometry with long-pathlength cells. To overcome these problems a procedure has been developed[483] in which the particulate matter from a natural water sample is collected on a filter and resuspended in a much smaller volume: the absorption spectrum of the concentrated material is then measured in a short-pathlength cell, using an integrating sphere, and from the absorbance values the absorption coefficients due to the particulate matter, in the original water body, may be calculated.

The total particulate fraction (*seston*, in limnological parlance) will of course include phytoplankton as well as tripton. If, however, the phytoplankton concentration is low then the particulate fraction spectrum can be attributed to tripton. Fig. 3.6 shows the absorption spectra of the particulate and soluble fractions of several Australian inland waters. In all cases except (*c*) and (*f*) the absorption by the particulate fraction is almost entirely due to tripton, although in two of these ((*d*) and (*g*)) a small 'shoulder' at about 670 nm due to phytoplankton chlorophyll may be seen.

The tripton absorption spectra all have much the same shape: absorption is low or absent at the red end of the spectrum and rises steadily as wavelength decreases into the blue and ultraviolet. These are typical humic substance absorption spectra and indeed have much the same shape as those of the dissolved yellow materials. Furthermore, typical tripton samples collected on a filter are brown in colour. The most plausible supposition is that the yellow-brown tripton colour is largely due to particulate humic material existing either bound to mineral particles or as free particles of humus. It seems likely that in inland waters it arises, together with the soluble humic material (gilvin), from the soils in the catchment. In productive waters or in oceanic waters well away from land drainage, some of the light-absorbing inanimate matter arises by decomposition of the phytoplankton. The detrital (non-living, particulate) fraction in sea water also has an absorption spectrum of the humic type (Fig. 3.7) but sometimes with shoulders due to the breakdown products of photosynthetic pigments.[99,647,769,404]

The *in situ* absorption coefficient due to particulate matter at 440 nm is a convenient general measure of particulate colour in any water. A suitable symbol would be p_{440}, which is analogous to g_{440} previously defined as a measure of soluble colour. Table 3.2 lists some observed values of p_{440} for various natural waters. In turbid waters containing large amounts of suspended material derived from soil erosion in the catchment or (in shallow waters) wind resuspension of sediments, non-living particulate absorption can exceed absorption due to dissolved colour. The waters of the Murrumbidgee Irrigation Area (Table 3.2) are examples of the first situation (see also Fig. 3.6*d*) and Lake George, NSW, is an example of the second (Fig. 3.6*g*, Table 3.2). Rivers frequently become more turbid and coloured with increasing distance downstream, from the headwaters to the estuary. A case in point is New Zealand's longest river, the Waikato. Davies-Colley (1987) found a progressive increase in both particulate (p_{440}) and soluble (g_{440}) colour in this river down its 330 km path from Lake Taupo to the sea (Table 3.2): the water is clear and blue-green as it leaves the oligotrophic lake, but is yellow and turbid when it finally enters the Tasman Sea.

It should be realized that p_{440} is not as good a general guide to particulate colour as g_{440} is to soluble colour. Whereas all gilvin spectra have approximately the same shape, the shape of the particulate fraction spectrum can change significantly in accordance with the proportion of humic material or phytoplankton (see below) in it. Nevertheless, in those waters (of common occurrence) with a substantial particulate humic component, it is a useful parameter.

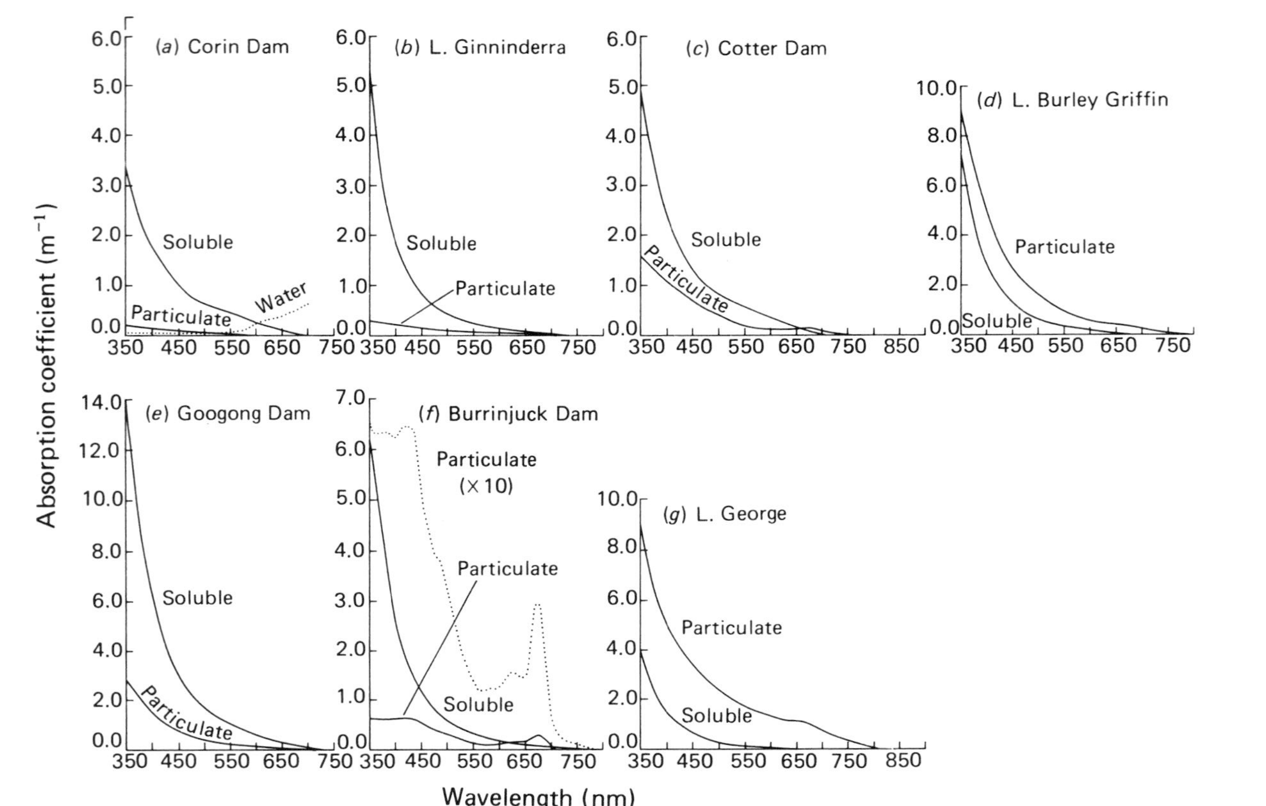

(a) Corin Dam
Soluble
Particulate
Water
(b) L. Ginninderra
Soluble
Particulate
(c) Cotter Dam
Soluble
Particulate
(d) L. Burley Griffin
Particulate
Soluble
(e) Googong Dam
Soluble
Particulate
(f) Burrinjuck Dam
Particulate
(×10)
Particulate
Soluble
(g) L. George
Particulate
Soluble
Absorption coefficient (m⁻¹)
Wavelength (nm)

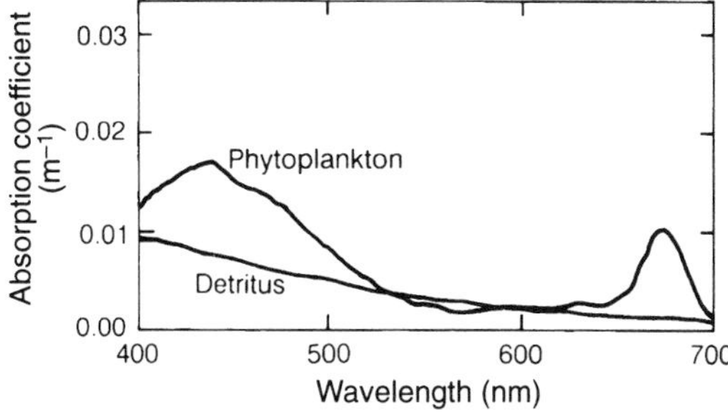

Fig. 3.7. Absorption spectra of detrital and phytoplankton particles from a mesotrophic station in the Sargasso Sea (after Iturriaga & Siegel, 1989).

Phytoplankton

The absorption of light by the photosynthetic pigments – chlorophylls, carotenoids, biliproteins – of the phytoplankton contributes to the attenuation of PAR with depth. Indeed in productive waters, the algae may be present in concentrations such that by self-shading they limit their own growth.

Light absorption by algal cells grown in laboratory culture has received a great deal of attention because of the use of such cultures as experimental material in fundamental photosynthesis research. What we need, however, is information on the light absorption properties of phytoplankton populations as they occur in nature since we cannot assume that the natural populations have absorption spectra identical to those of the same species grown in culture. The total concentration of pigments and the relative

Fig. 3.6. Comparison of the spectral absorption properties of the particulate and soluble fractions of various Australian inland waters in the southern tablelands of New South Wales/Australian Capital Territory (from Kirk, 1980*b*). The absorption spectrum of pure water is included in (*a*) for comparative purposes. The phytoplankton chlorophyll *a* contents (C_a) and nephelometric turbidities (T_n) of the different water bodies at the time of sampling were as follows: (*a*) Corin Dam, 8th June 1979, $C_a = 2.0$ mg m^{-3}, $T_n = 0.51$ NTU; (*b*) Lake Ginninderra, 6th June 1979, $C_a = 1.5$ mg m^{-3}, $T_n = 1.1$ NTU; (*c*) Cotter Dam, 8th June 1979, $C_a = 9.0$ mg m^{-3}, $T_n = 1.6$ NTU; (*d*) Lake Burley Griffin, 6th June 1979, $C_a = 6.3$ mg m^{-3}, $T_n = 17.0$ NTU; (*e*) Googong Dam, 21st June 1979, $C_a = 1.7$ mg m^{-3}, $T_n = 5.8$ NTU; (*f*) Burrinjuck Dam, 7th June 1979, $C_a = 16.1$ mg m^{-3}, $T_n = 1.8$ NTU; (*g*) Lake George, 28th November 1979, $C_a = 10.9$ mg m^{-3}, $T_n = 49$ NTU.

The ordinate scale corresponds to the true *in situ* absorption coefficient due to soluble or particulate colour: note that the scale in (*d*), (*e*) and (*g*) is different to that in (*a*), (*b*), (*c*) and (*f*).

amounts of the different pigments can all be affected by environmental variables such as nitrogen concentration in the medium, light intensity and spectral distribution, etc. Much research therefore needs to be done on the absorption spectra of natural phytoplankton populations: measurement techniques which overcome the problem of light scattering by the suspensions will of course have to be used. Ultimately what is required is not only the shape of the absorption spectrum, but the actual values of the absorption coefficients due to phytoplankton in the original water body. Data on phytoplankton populations consisting mainly of single known species would be particularly valuable.

Several studies of the absorption spectra of natural populations of marine phytoplankton have been carried out in recent years, in nearly every case – because of the low concentrations of cells present – using the method of collection, and measurement, on a filter,[99,469,503,557,617,647,1009] which as we noted earlier, does require the estimation of a large absorption amplification factor. Iturriaga & Siegel (1989) have used their technique of microspectrophotometry of individual cells to determine both the shape of the spectrum, and the *in situ* absorption coefficients, for natural phytoplankton populations in the Sargasso Sea (Fig. 3.7).

In the case of productive marine or fresh waters, it is feasible to measure absorption spectra of phytoplankton populations in suspension – if necessary, after a preliminary concentration step – using an integrating sphere, or opal glass,[335,483,597] and calculate the *in situ* absorption coefficients directly. Fig. 3.6*f* shows the spectrum (integrating sphere method) of the particulate fraction from Burrinjuck Dam, a eutrophic impoundment in New South Wales, Australia, at a time when it had a mixed bloom of *Melosira* sp. (a diatom) and *Anacystis cyanea* (a blue-green alga). The fraction consisted mainly of algal biomass and so the spectrum provides approximate values (somewhat too high in the blue due to the presence of some tripton) for the *in situ* absorption coefficients due to phytoplankton. Fig. 3.8 shows the particulate fraction spectrum – again largely due to phytoplankton – of the estuarine water of Lake King (Gippsland Lakes, Australia).

A quite different route to the determination of *in situ* phytoplankton absorption has been taken by Bidigare *et al.* (1987) and Smith *et al.* (1989). Making the plausible assumption that essentially all the undegraded photosynthetic pigments in the water column originate in living cells, they carried out a complete pigment analysis of the total particulate fraction, using HPLC; then, using literature data on the spectral properties of pigment–protein complexes, they calculated the absorption coefficients due to phytoplankton.

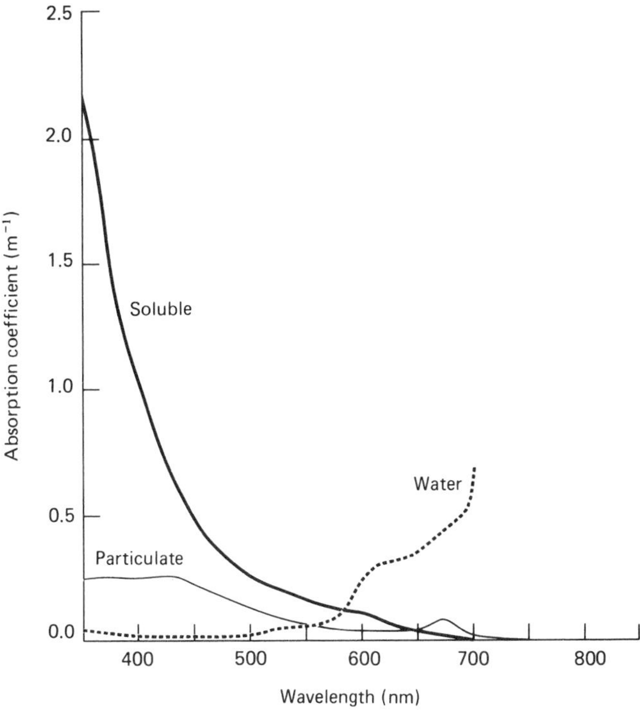

Fig. 3.8. Comparison of the spectral absorption properties of the different fractions in an estuarine water from southeast Australia – Lake King, Victoria (Kirk, unpublished data). Phytoplankton were present at a level corresponding to 3.6 mg chlorophyll *a* m^{-3} and the turbidity of the water was ~1.0 NTU.

It is possible to estimate the absorption coefficient of the medium at a given wavelength from the vertical attenuation coefficient for irradiance at that wavelength. By carrying out such calculations for pairs of stations (in the Atlantic off Northwest Africa) where the scattering coefficients were about the same but the phytoplankton population varied, Morel & Prieur (1977) were able to arrive at an absorption spectrum for the natural phytoplankton population present. The absorption coefficients corresponding to 1 mg chlorophyll *a* m^{-3} are plotted against wavelength in Fig. 3.9. The amount of light harvested by the phytoplankton component of the aquatic medium depends not only on the total amounts of the photosynthetic pigments present, but also on the size and shape of the algal cells or colonies within which the pigments are located. This subject is dealt with later (§9.3).

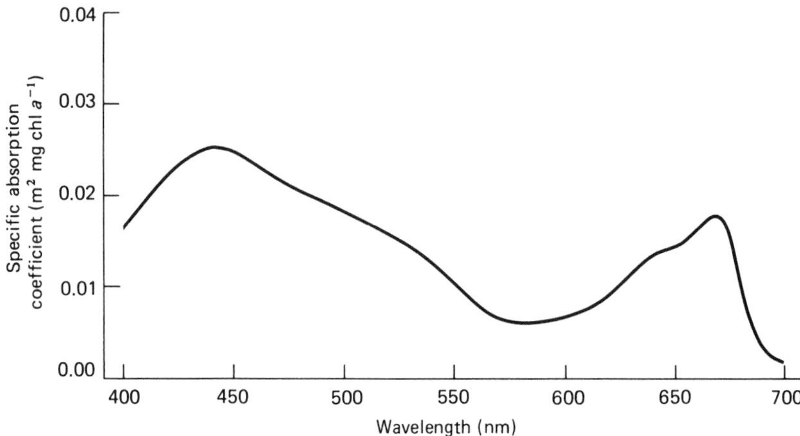

Fig. 3.9. Specific absorption coefficient (*in situ*), corresponding to 1 mg chlorophyll *a* m^{-3}, for oceanic phytoplankton (after Morel & Prieur, 1977).

Total absorption spectra

At any wavelength, the aquatic medium has a total absorption coefficient, which is the sum of the absorption coefficients of all the light-absorbing components, at that wavelength. The variation of this total absorption coefficient with wavelength is the absorption spectrum of the medium as a whole. For any given water body, the total absorption coefficient at each wavelength is obtained by adding together the known absorption coefficient of pure water at that wavelength and the absorption coefficients due to dissolved and particulate colour, determined as described above. Fig. 3.10 shows the total absorption spectra of five Australian waters, three inland, one estuarine and one marine. To give an approximate indication of how the total absorption spectrum might be made up, Fig. 3.11 shows, for an idealized, rather productive, ocean water, the individual absorption spectra of dissolved colour (gilvin), phytoplankton and detritus at plausible levels, and of water itself, together with the total absorption spectrum of the water due to all four components together.

3.4 Optical classification of natural waters

Natural waters vary greatly in the extent to which they transmit solar radiation and it is useful to have some broad indication of the optical character of a water without having to fully specify all the inherent optical properties. Jerlov (1951, 1976) has classified marine waters into a number of

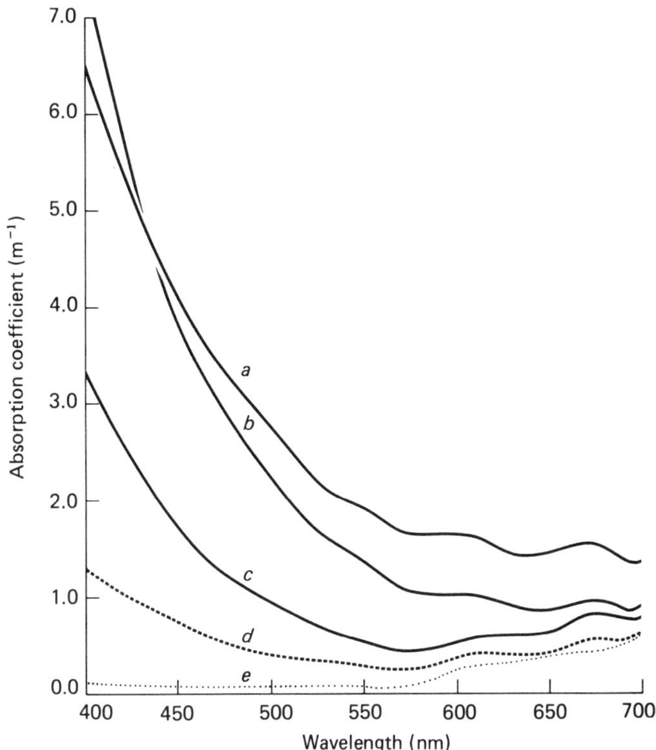

Fig. 3.10. Total absorption spectra of various natural waters in south-eastern Australia (Kirk, 1981*a* and unpublished data). (*a*) Lake George, NSW. (*b*) Lake Burley Griffin, ACT. (*c*) Burrinjuck Dam, NSW. (*d*) Lake King, Victoria. (*e*) Jervis Bay, NSW. Waters *a*, *b* and *c* are inland, *d* is estuarine, *e* is marine (Tasman Sea). Chlorophyll and turbidity data for waters *a–d* are given in the legends to Figs 3.6 and 3.8. The Jervis Bay water (*e*) contained 0.2 mg phytoplankton chlorophyll *a* m^{-3} and was optically intermediate between Jerlov oceanic water types I and III.

different categories on the basis of the curve of percent transmittance of downward irradiance against wavelength. He recognized three basic types of oceanic water – I, II and III – and nine types of coastal water (1 to 9), in order of decreasing transmittance: the spectral variation of percentage transmittance for some of these water types is shown in Fig. 3.12.

Jerlov's pioneering measurements were, however, made with broad-band colour filters, and the curves obtained with modern submersible spectroradiometers are in some cases in poor agreement with his.[629] Pelevin & Rutkovskaya (1977) have proposed that, instead, ocean waters be classified in terms of the vertical attenuation coefficient (base 10 logarithm)

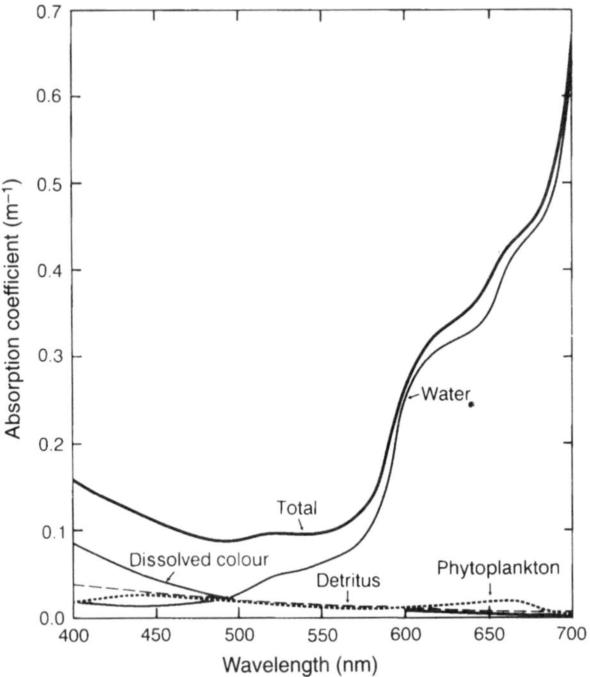

Fig. 3.11. Total absorption spectrum of an idealized, productive (1 mg chlorophyll *a* m^{-3}) oceanic water, together with spectra of the individual absorbing components.

for irradiance at 500 nm, multiplied by 100. Since the K_d versus λ curves vary in a fairly systematic manner from one oceanic water to another, the value of K_d (500 nm) would indeed convey quite a lot of information about any such water. However, given that all definitions of optical properties for natural waters are now standardized to use logarithms to the base e rather than base 10,[402] it would be preferable to modify Pelevin & Rutkovskaya's proposal so that the water type was specified by 100 K_d (500 nm), where K_d is based on $\log_e$, in accordance with the definition in eqn 1.17. This has the advantage that the number obtained is approximately equal (if attenuation is not too intense) to the percentage of downwelling light which is removed. For example, Jerlov's oceanic waters I and III, with irradiance transmittance values at 500 nm of 97.3% and 89%, corresponding to irradiance diminution of 2.7% and 11%, have 100 K_d (500 nm) values of 2.7 and 11.6.

Smith & Baker (1978*b*), on the basis of their measurements of the spectral variation of the vertical attenuation coefficient (K_d) for irradiance in various ocean waters, have concluded that in regions away from terrigenous influences, the attenuation (apart from that due to water) is mainly

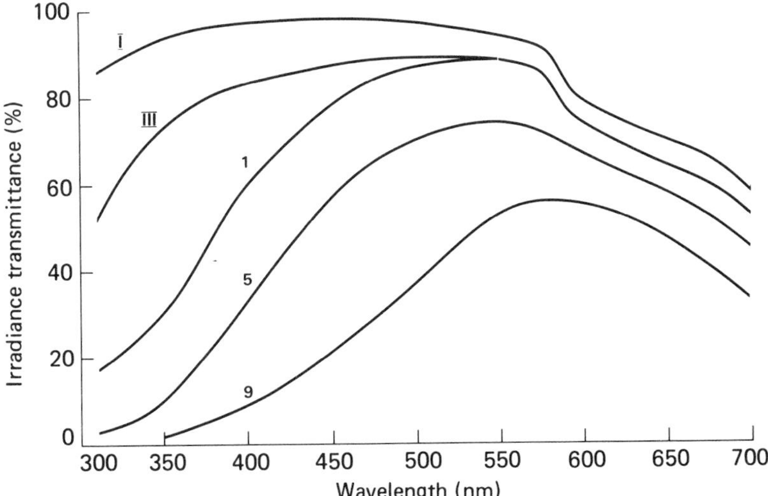

Fig. 3.12. Transmittance per metre of various optical types of marine water for downward irradiance. The waters are Jerlov's oceanic types I and III, and coastal types 1, 5 and 9 (data from Table XXVI of Jerlov, 1976).

due to the phytoplankton and the various pigmented detrital products that covary with it. They suggest that for such ocean waters, the total content of chlorophyll-like pigments provides a sufficient basis for optical classification, since on the basis of the pigment content, the curve of K_d against wavelength can be calculated.

A classification which has been found useful in the context of remote sensing of the ocean is that into 'Case 1' and 'Case 2' waters, put forward by Morel & Prieur (1977), and further refined by Gordon & Morel (1983). Case 1 waters are those for which phytoplankton and their derivative products (organic detritus and dissolved yellow colour, arising by zooplankton grazing, or natural decay of the algal cells) play a dominant role in determining the optical properties of the ocean. Case 2 waters are those for which an important or dominant contribution to the optical properties comes from resuspended sediments from the continental shelf, or from particles and/or dissolved colour in river runoff or urban/industrial discharge. Case 1 waters can range from oligotrophic to eutrophic, provided only that the particulate and coloured materials characteristic of Case 2 waters do not play a significant role. In Case 2 waters, phytoplankton and their derivative products may or may not also be present in significant amount.

A crude optical classification applicable mainly to inland waters has been proposed by Kirk (1980*b*) on the basis of measurements of the absorption spectra of the soluble and particulate fractions from Australian water bodies. In type G waters, the dissolved colour (gilvin), at all wavelengths in the photosynthetic range, absorbs light more strongly than the particulate fraction: examples are shown in Figs 3.6*a*, *b* and *e*. In type GA waters, gilvin absorbs more strongly than the particulate fraction throughout the shorter wavelength part of the spectrum, but the absorption coefficients of the particulate fraction exceed those of the soluble fraction at the red end of the spectrum due to the presence of substantial levels of algal chlorophyll: Figs 3.6*c* and *f* are examples. In type T waters, the particulate fraction, consisting mainly of inanimate material (tripton) absorbs light more strongly than the soluble fraction at all wavelengths: turbid waters with large amounts of suspended silt particles (Figs 3.6*d* and *g*) are in this category. In type GT waters, absorption by the soluble and particulate fractions is roughly comparable throughout the photosynthetic range: the waters in certain billabongs in tropical Australia have been observed to be in this category. In highly productive waters, the absorption coefficients due to algal biomass may exceed those due to dissolved colour (and water itself). Measurements by Talling (1970) of the spectral variation of the vertical attenuation coefficient of water from Loch Leven, Scotland, suggest that this eutrophic lake, dominated by *Synechococcus* sp., was in this category (type A) at the time: the waters of the upwelling area off West Africa, studied by Morel & Prieur (1975), also appear to be of type A. Most oceanic, and some coastal, waters (Fig. 3.10*e*) are so non-productive, and so free of silt and dissolved colour, that water itself is the dominant light absorber: they may be categorized as type W. Such waters are rare inland: one example is Crater Lake, Oregon, USA.[840] In certain estuarine and the more coloured coastal waters, the gilvin absorption at the blue end of the spectrum can be roughly similar in magnitude to the water absorption at the red end. Such waters, for example that in Fig. 3.8, may be classified as WG.

A water body can change from one optical type to another. Heavy rain in the catchment with consequent soil erosion could, for example, quickly change a type G water (gilvin-dominated) to a type T water (tripton-dominated); development of an algal bloom could change it to a type GA water. Nevertheless, some water bodies have water of a particular type most of the time. Bog lakes are typically type G all the time. Shallow, wind-exposed lakes with unconsolidated sediments are likely to be of type T all the time. Marine waters are, apart from the effects of the yearly phytoplankton cycle (in non-tropical areas), generally constant in their optical properties.

Prieur & Sathyendranath (1981) have proposed an optical classification scheme for sea water similar in some respects to that described above for fresh water. It is based on the relative importance of absorption by the three major components other than water: algal pigments, dissolved yellow materials and particulate matter other than phytoplankton. As a numerical indication of importance the absorption coefficient due to each component – C', Y' and P', respectively – is used. Different waters are classified as C' type, Y' type, P' type, or as hybrid types such as $C'Y'$, $C'P'$ or $P'Y'$. Most of the marine waters studied by these authors were of the C' type.

3.5 Contribution of the different components of the aquatic medium to absorption of PAR

Apart from the small amount of light scattered back out of the water, attenuation of PAR in water bodies is due to absorption, although the extent of this absorption within a given depth may be greatly amplified by scattering which increases the average pathlength of the photons within that depth. The relative contribution of different components of the system to this absorption at a given wavelength is in proportion to their absorption coefficients at that wavelength. Absorption coefficients for any particular component and the intensity of the incident solar radiation vary independently with wavelength throughout the photosynthetic range, and so the contribution of the different components to absorption of total PAR can only be assessed by carrying out the appropriate calculations at a series of narrow wavebands and summing the results. In this way it is possible to calculate what proportion of those photosynthetic quanta which are absorbed are captured by each of the different absorbing components of the system.[478,483] Table 3.3 presents the results of such calculations for 12 Australian water bodies.

It is clear that the three major components into which the light-absorbing material of the system has been divided – the water itself, the dissolved yellow substances (gilvin), and the particulate fraction (tripton/phytoplankton) – can all be substantial light absorbers. In type W waters, most of the photosynthetic quanta are captured by water itself. In oceanic waters not affected by terrigenous material (generally type W), those quanta not absorbed by water are captured mainly by the pigments of the living and dead phytoplankton.[830] In coastal waters, with low but significant amounts of gilvin, the contributions of the phytoplankton and the dissolved yellow material to absorption are likely to be small and comparable. In estuarine waters, such as Lake King (type WG), with more substantial levels of soluble yellow substances, the water and the gilvin capture most of the photosynthetic quanta. In the type G-inland waters (plenty of dissolved

Table 3.3. *Calculated distribution of absorbed photosynthetic quanta between the particulate fraction, soluble fraction and water in Australian water bodies. The calculations have in each case been carried out for the euphotic zone of the water body in question. Data from Kirk (1980b), Kirk & Tyler (1986) and unpublished*

		Quanta absorbed (% of total)		
Water body	Optical type	By water	By soluble fraction[a]	By particulate fraction[b]
Coastal – Oceanic				
Jervis Bay[c]	W	68.1	23.9	8.0
Estuarine				
Lake King[d]	WG	41.9	40.4	17.7
Inland impoundments				
Corin Dam[e]	G	34.8	60.0	5.2
Lake Ginninderra[e]	G	39.1	50.4	10.5
Googong Dam[e]	G	22.0	60.4	17.6
Cotter Dam[e]	GA	26.2	49.8	24.0
Burrinjuck Dam[c]	GA	28.2	45.5	26.3
Lake Burley Griffin[e]	T	19.4	22.2	58.4
Natural inland waters				
Latrobe River[d]	T	17.5	28.1	54.5
Lake George[c]	T	12.4	8.3	79.3
Gulungul billabong[f]	GT	20.0	39.7	40.3
Georgetown billabong[f]	T	5.9	7.5	86.6

Notes:
[a]Gilvin. [b]Tripton/phytoplankton. [c]New South Wales. [d] Victoria. [e]Australian Capital Territory. [f]Northern Territory.

colour, but low turbidity) described in Table 3.3, gilvin captures the most quanta, followed by water. In the type GA waters (dissolved colour, rather low turbidity, plentiful phytoplankton), gilvin is still the most important component but the particulate fraction now captures about as many quanta as does the water. In the type T waters (high turbidity due to tripton), most of the quanta are absorbed by the particulate fraction. The particulate fraction and gilvin between them take most of the quanta in type GT waters (intense soluble colour, high tripton turbidity). In highly productive, type A, waters, dominated by phytoplankton, we may assume that the algal biomass captures more photosynthetic quanta than the other components of the medium.

4

Scattering of light within the aquatic medium

We have seen that most of the solar photons which enter the water are absorbed. Many of these photons – most, in some waters – undergo scattering one or more times before they are absorbed. Scattering does not by itself remove light – a scattered photon is still available for photosynthesis. The effect of scattering is to impede the vertical penetration of light. It makes the photons follow a zig-zag path as they ricochet from one scattering particle to the next. This increases the total pathlength which the photons must follow in traversing a certain depth, and so increases the probability of their being captured by one of the absorbing components of the medium. In addition, some of the photons are actually scattered back in an upwards direction. Thus the effect of scattering is to intensify the vertical attenuation of the light.

In this chapter we shall consider the nature of the scattering process and the scattering properties of natural waters.

4.1 The scattering process

What do we mean by scattering? We say that a photon is scattered when it interacts with some component of the medium in such a way that it is caused to diverge from its original path. There are two kinds of scattering to be considered – density fluctuation scattering and particle scattering.

Density fluctuation scattering

In understanding the basis of density fluctuation scattering in liquids, it is helpful to begin with a consideration of molecular, or Rayleigh, scattering by gases such as air. According to the Rayleigh theory, within any particle, such as an air molecule, in a light field, a dipole is induced by the electrical vector of the field. As the dipole oscillates at the frequency of the exciting radiation, it emits radiation of the same frequency in all directions. It is this radiation which is the scattered light.

The Rayleigh molecular theory of scattering does not apply to liquids: the strong interactions between the molecules make it impermissible to consider the interaction of the radiation with molecules on an individual basis. In any liquid, however, the continual random motion of the molecules leads to localized microscopic fluctuations of density and therefore of dielectric constant and, in the Einstein–Smoluchowski theory, the interaction of the radiation field with these inhomogeneities – each of which can be regarded as a dipole – rather than with the individual molecules, is considered. The predicted angular distribution of scattering is similar to that given by the Rayleigh theory for gases, i.e. it is identical in the forward and backward directions (Figs 2.2 and 4.8). Again, as in Rayleigh scattering by gases, scattering by a pure liquid is predicted to vary inversely with the fourth power of the wavelength.

Particle scattering

The Rayleigh and Einstein–Smoluchowski theories of scattering apply only when the scattering centres are small relative to the wavelength of light: this is true in the case of gas molecules and of the tiny density fluctuations in pure liquids. Even the most pristine natural waters, however, are not, optically speaking, pure and they invariably contain high concentrations of particles – mineral particles derived from the land or from bottom sediments, phytoplankton, bacteria, dead cells and fragments of cells, etc. – all of which scatter light. The particles which occur in natural waters have a continuous size distribution which is roughly hyperbolic,[32] i.e. the number of particles with diameter greater than D is proportional to $1/D^{\gamma}$, where γ is a constant for a particular water body, but varies widely from 0.7 to 6 in different water bodies.[422] Although a hyperbolic distribution implies that smaller particles are more numerous than big ones, nevertheless most of the particle cross-sectional area that would be encountered by light in natural waters is due to particles of diameter greater than 2 μm[422] which is not small relative to the wavelengths of visible light, and so scattering behaviour different to the density fluctuation type must be anticipated. The smaller particles, although numerous, have a lower scattering efficiency.

A theoretical basis for predicting the light-scattering behaviour of spherical particles of any size was developed by Mie (1908). The physical basis of the theory is similar to that of Rayleigh in that it considers the

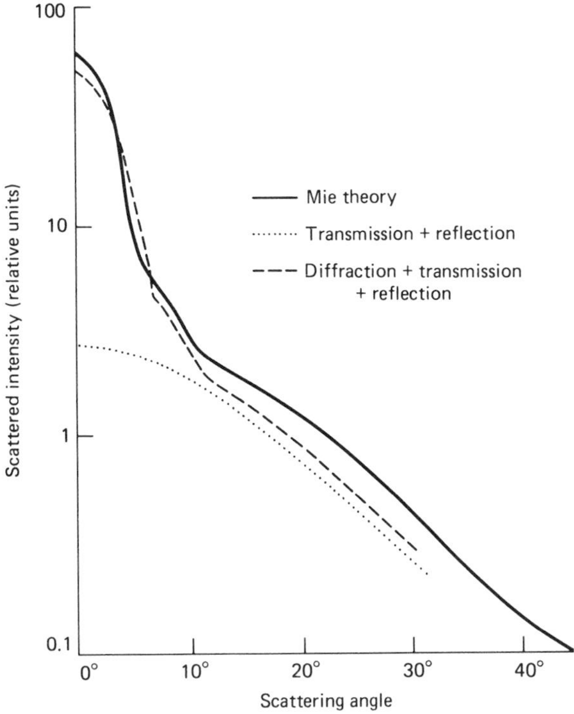

Fig. 4.1. Angular distribution of scattered intensity from transparent spheres calculated from Mie theory (Ashley & Cobb, 1958) or on the basis of transmission and reflection, or diffraction, transmission and reflection (Hodkinson & Greenleaves, 1963). The particles have a refractive index (relative to the surrounding medium) of 1.20, and have diameters 5–12 times the wavelength of the light. After Hodkinson & Greenleaves (1963).

oscillations set up within a polarizable body by the incident light field and the light re-radiated (i.e. scattered) from the body as a result of these oscillations. Instead of (as in Rayleigh theory) equating the particle to a single dipole, the Mie theory considers the additive contributions of a series of electrical and magnetic multipoles located within the particle. The advantage of the Mie theory is that it is all-embracing – for very small particles, for example, it leads to the same predictions as the Rayleigh theory; the disadvantage is that the analytical expressions are complex and do not lend themselves to easy numerical calculations. For particles larger than the wavelength of the light, Mie theory predicts that most of the scattering is in the forward direction within small angles of the beam axis (Fig. 4.1). A series of maxima and minima is predicted at increasing

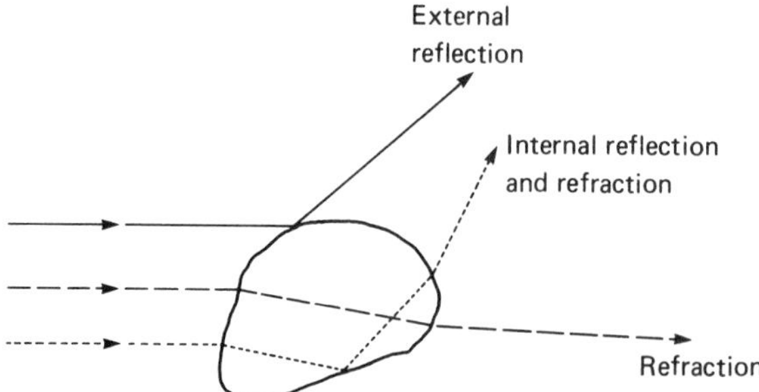

Fig. 4.2. Scattering of light by a particle: reflection and refraction processes.

scattering angle, but these are smoothed out when a mixture of particle sizes is present.

In the case of particles larger than a few wavelengths of light, a reasonable understanding of the mechanism of scattering can be obtained on the basis of diffraction and geometrical optics, without recourse to electromagnetic theory. When an object is illuminated by a plane light wave, the shadow of an object on a screen placed behind it is not quite precisely defined: just outside it a series of concentric faint dark bands will be present; and lighter concentric bands, where clearly some light is falling, will be present within the area of the geometric shadow. This phenomenon – *diffraction* – is due to interference (destructive in the dark rings, constructive in the light rings) between parts of the wave coming from different points around the edge of the illuminated object, and arriving simultaneously but out of phase (because of the different distances traversed) at particular points on the screen. In the case of a round object, superimposed on the circular shadow, there is a bright spot in the centre, surrounded by alternate dark and light rings. In fact most of the light diffracted by a particle is propagated in the forward direction within a small angle of the initial direction of the beam (giving rise to the bright spot). With increasing angular distance away from the axis, the diffracted intensity goes through a series of minima and maxima (dark and light rings) which diminish progressively in height.

Applying geometrical optics to these larger particles, it can readily be appreciated that some of the light will be reflected at the external surface and some will pass through the particle and undergo refraction, or internal reflection as well as refraction (Fig. 4.2). In all cases the photons involved will be made to deviate from their initial direction, i.e. will be scattered. Light scattered by these mechanisms is still predominantly in the forward direction: scattered intensity diminishes continuously with increasing angle but does not show the same degree of concentration at small angles as does scattering due to diffraction. Calculations by Hodkinson & Greenleaves (1963) for suspensions of spherical particles of mixed sizes show that most of the scattering at small angles (up to about 10–15°) can be attributed to diffraction, whereas most of the scattering at larger angles is due to external reflection and transmission with refraction (Fig. 4.1). The angular variation of scattering calculated on the basis of diffraction and geometrical optics is reasonably close to that derived by the Mie electromagnetic theory. Such discrepancies as exist may be largely due to the fact that as a result of the phase change induced in the transmitted ray by passage through a medium of higher refractive index, there are additional interference effects between the diffracted and the transmitted light:[945] this phenomenon is known as anomalous diffraction.

Any particle in a beam of light will scatter a certain fraction of the beam and the radiant flux scattered will be equivalent to that in a certain cross-sectional area of the incident beam. This area is the *scattering cross-section* of the particle. The efficiency factor for scattering, Q_{scat}, is the scattering cross-section divided by the geometrical cross-sectional area of the particle (πr^2 for a spherical particle of radius r). Similarly, in the case of an absorbing particle, the radiant flux absorbed is equivalent to that in a certain cross-sectional area of the incident beam: this area is the *absorption cross-section* of the particle. The efficiency factor for absorption, Q_{abs}, is the absorption cross-section divided by the geometrical cross-sectional area of the particle. The efficiency factor, Q_{att}, for attenuation (absorption and scattering combined) is thus given by

$$Q_{att} = Q_{scat} + Q_{abs} \tag{4.1}$$

The attenuation efficiency of a particle can be greater than unity: that is, a particle can affect the behaviour of more light in the incident beam than will be intercepted by its geometrical cross-section. This applies only to the scattering part of attenuation, i.e. it is possible for a particle to scatter, but not absorb, more light than its geometrical cross-section would intercept. In terms of electromagnetic theory we may say that the particle can perturb the electromagnetic field well beyond its own physical boundary. Mie theory, based as it is upon electromagnetism, can be used to calculate the absorption and scattering efficiencies of particles. The simpler anomalous diffraction theory of van de Hulst (1957) can also be used to calculate the scattering efficiency of particles with refractive index up to about twice that of the surrounding medium. The relation is

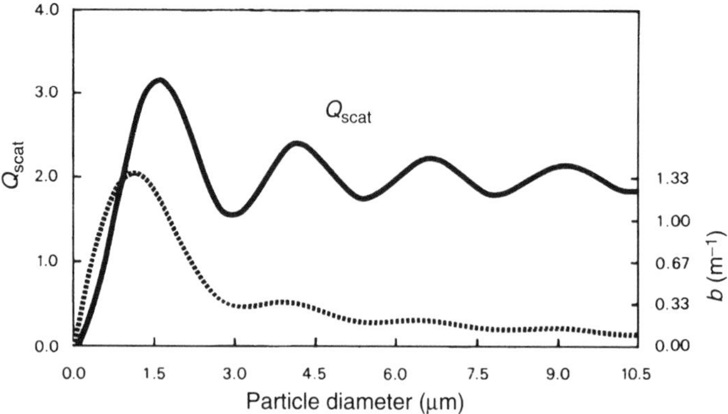

Fig. 4.3. Scattering efficiency of non-absorbing spherical particles as a function of size. The particles have a refractive index, relative to water, of 1.17. Wavelength = 550 nm. Continuous line, Q_{scat} for a single particle, calculated using the equation of van de Hulst (1957) – see text. Broken line, the scattering coefficient (b) for a suspension containing 1 g of particles m^{-3}.

$$Q_{att} = 2 - \frac{4}{\rho}\sin\rho + \frac{4}{\rho^2}(1 - \cos\rho) \qquad (4.2)$$

where $\rho = (4\pi a/\lambda)(m-1)$, m being the refractive index of the particle relative to that of the surrounding medium, and a being the radius of the particle. For a non-absorbing particle, $Q_{scat} = Q_{att}$. Fig. 4.3 shows the way in which the scattering efficiency for green light of a spherical non-absorbing particle of refractive index relative to water of 1.17 (a typical value for inorganic particles in natural waters), varies with particle size. It can be seen that scattering efficiency rises steeply from very low values for very small particles to about 3.2 at a diameter of 1.6 μm. With increasing diameter, it first decreases and then increases again and undergoes a series of oscillations of diminishing amplitude to level off at a Q_{scat} value of 2.0 for very large particles. A similar general pattern of variation of Q_{scat} with size would be exhibited by any scattering particle of the types found in natural waters at any wavelength in the photosynthetic range.

As diameter decreases below the optimum for scattering (e.g. from 1.6 μm downwards in Fig. 4.3) so efficiency for a single particle decreases. However, for a given mass of particles per unit volume, the number of particles per unit volume must increase as particle size decreases. It is therefore of interest to determine how the scattering coefficient of a particle suspension of fixed concentration by weight varies with particle size. The results of such a calculation for particles of density in the range typical of clay minerals at a concentration of 1 g m^{-3} is shown in fig. 4.3. As might be anticipated, because of the increase in particle number simultaneous with the decrease in diameter, total scattering by the suspension, expressed in terms of the value of b, does not decrease as precipitately with decreasing diameter below the optimum, as single particle scattering efficiency does, and the optimum particle diameter for suspension scattering (~ 1.1 μm) is lower than that for single particle scattering (~ 1.6 μm). As particle diameter increases beyond the optimum so scattering by the suspension shows a progressive decrease to very low values, with only minor, heavily damped, oscillations corresponding to the oscillations in Q_{scat} for the individual particles.[489]

4.2 Measurement of scattering

Beam transmissometers

In the absence of absorption, the scattering coefficient could in principle be determined by measuring the loss of intensity of a narrow parallel beam passing through a known pathlength of medium. If absorption as well as scattering occurs, then the parameter measured by the instrument would in fact be the beam attenuation coefficient, c, rather than the scattering coefficient. If it is possible also to measure the absorption coefficient, a, of the water at the appropriate wavelength, then the scattering coefficient, b, may be calculated ($b = c - a$).

Beam attenuation meters – or beam transmissometers as they are more commonly called – for the *in situ* measurement of c have long been important tools in hydrologic optics. In principle all they have to do is measure the proportion, C, of the incident beam which is lost by absorption and scattering in a pathlength, r, the beam attenuation coefficient being equal to $-[\ln(1-C)]/r$ (§1.4). In practice, the construction of instruments which accurately measure c is difficult. The problem is that most of the scattering by natural waters is at small angles. Therefore, unless the acceptance angle of the detector of the transmissometer is very small ($< 1°$), significant amounts of scattered light remain in the beam and so attenuation is underestimated. One attempted solution of this problem is to measure scattered light at some small angle greater than the acceptance angle of the transmissometer, and from this calculate the scattering contribution to the direct beam and subtract it from the direct beam intensity.[355]

The principles of design of beam transmissometers and the variation of the size of the error with the design parameters have been discussed by

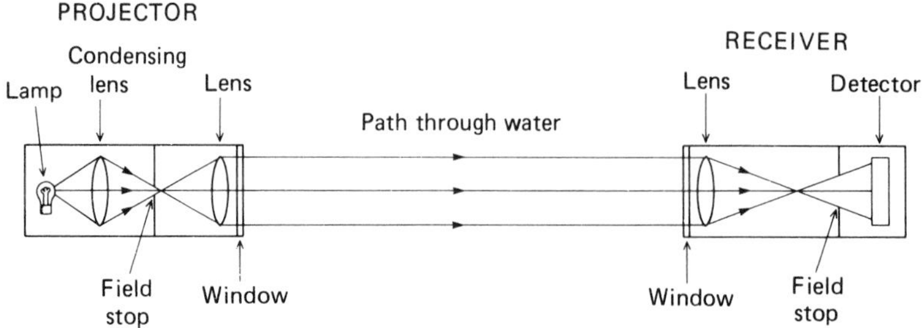

Fig. 4.4. Beam transmissometer optical system (adapted from Austin & Petzold, 1977). To confine measurements to a narrow spectral waveband, a filter can be included in the light path within the receiver.

Austin & Petzold (1977). One possible optical system is shown schematically in Fig. 4.4. Commercial instruments are available. Most beam transmissometers are monochromatic, but Borgerson *et al.* (1990) have developed a spectral version. Although *in situ* measurements are to be preferred, it is possible to measure the beam attenuation coefficient in the laboratory with a spectrophotometer, provided that a very small acceptance angle and long-pathlength cells are used. Bochkov, Kopelevich & Kriman (1980) have developed such an instrument for the shipboard determination of c of sea-water samples over the range 250–600 nm.

If, in any water body, the vertical attenuation coefficient for downward irradiance, K_d, and the beam attenuation coefficient are both measured for the same spectral waveband, then using certain empirical relations which have been found to exist between K_d, b and c,[929] or K_d, a and c,[700] it is possible to estimate b and thus determine a (from $c = a + b$), or estimate a and thus determine b. Gordon (1991) has described a calculation procedure by means of which, from near-surface measurements of c, K_d and irradiance reflectance (R), it is possible to estimate a, b and the backscattering coefficient, b_b. This method, like the others, makes use of empirically established relationships between these quantities, but has the advantage that it makes no assumption about the shape of the scattering phase function.

Variable-angle scattering meters

The scattering properties of natural waters are best determined by directly measuring the scattered light. The general principle is that a parallel beam of light is passed through the water, and the light scattered from a known

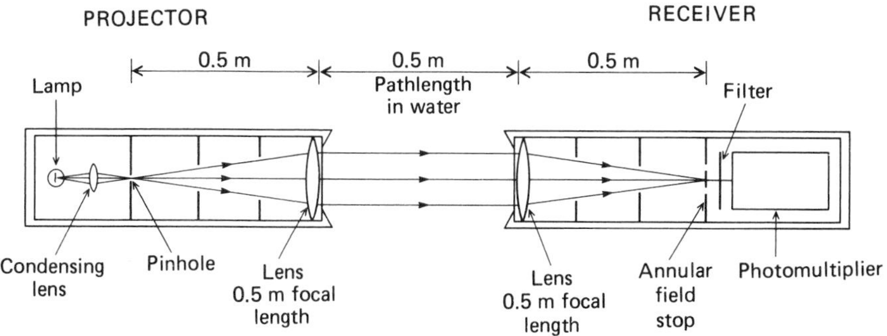

Fig. 4.5. Optical system of low-angle scattering meter (after Petzold, 1972). The vertical dimensions of the diagram are exaggerated.

volume at various angles is measured. In the ideal case, the volume scattering function, $\beta(\theta)$, is measured from 0 to 180°: this provides not only the angular distribution of scattering for that water but also, by integration, the total, forward and backward scattering coefficients (§1.4). Such measurements are in reality difficult to carry out and relatively few natural waters have been completely characterized in this way. The problem is that most of the scattering occurs at small angles (typically 50% between 0 and 2–6°) and it is hard to measure the relatively faint scattering signal so close to the intense illuminating beam. We shall consider *in situ* scattering meters first.

An instrument developed by Petzold (1972) for very low angles (Fig. 4.5) uses a highly collimated beam of light traversing a 0.5 m pathlength in water and then being brought to a focus by a long-focal-length lens. Light which has been neither scattered nor absorbed comes to a point in this plane. Light which has been caused to deviate by scattering arrives at this plane displaced a certain distance (proportional to the scattering angle) to one side. A field stop is placed in the focal plane, opaque except for a clear annular ring which allows only light corresponding to a certain narrow (scattering) angular range to pass through and be detected by a photomultiplier behind the stop. Three such field stops are used, each with the annular ring a different radial distance from the centre: these correspond to scattering angles of 0.085, 0.17 and 0.34°. To measure the intensity of the incident beam a fourth stop is used which has a calibrated neutral-density filter at the centre.

Kullenberg (1968) used a He–Ne laser to provide a collimated light beam traversing a pathlength of 1.3 m in water. At the receiving end of the instrument, the central part of the beam was occluded with a light trap and a

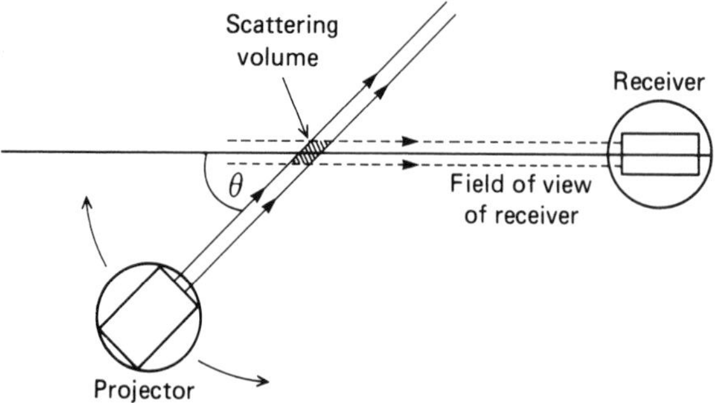

Fig. 4.6. Schematic diagram of optical system of general-angle scattering meter (after Petzold, 1972).

system of conical mirrors and annular diaphragms was used to isolate the light which had been scattered at 1, 2.5 or 3.5°.

Bauer & Morel (1967) used a central stop to screen off the collimated incident beam and all light scattered at angles up to 1.5°. Light scattered between 1.5 and 14° was collected by a lens and brought to a focus on a photographic plate: $\beta(\theta)$ over this angular range was determined by densitometry.

Instruments for measuring scattering at larger angles are constructed so that either the detector or the projector can rotate relative to the other. The general principle is illustrated in Fig. 4.6. It will be noted that the detector 'sees' only a short segment of the collimated beam of light in the water, and the length of this segment varies with the angle of view. Instruments of this type have been developed by Tyler & Richardson (1958) (20–180°), Jerlov (1961) (10–165°), Petzold (1972) (10–170°) and Kullenberg (1984) (8–160°).

The scattering properties of natural waters can be measured in the laboratory but, at least in cases in which scattering values are low, as in many marine waters, there is a real danger that the scattering properties may change in the time between sample collection and measurement. In the case of the more turbid waters commonly found in inland, estuarine and some coastal systems, such changes are less of a problem but it is still essential to keep the time between sampling and measurement to a minimum and to take steps to keep the particles in suspension. Commercial light scattering photometers were developed primarily for studying macromolecules and polymers in the laboratory, but have been adapted by a number of workers for the measurement of $\beta(\theta)$ in natural waters.[51,859] The water sample is placed in a glass cell illuminated with a collimated light beam. The photomultiplier is on a calibrated turntable and can be positioned to measure the light scattered at any angle within the range of the instrument (typically 20–135°). Laboratory scattering photometers for very small angles have been developed[25,862] and commercial instruments are available.

From the definition of volume scattering function (§1.4) it follows that to calculate the value of $\beta(\theta)$ at each angle it is necessary to know not only the radiant intensity at the measuring angle, but also the value of the scattering volume 'seen' by the detector (this varies with angle as we noted above), and the irradiance incident upon this scattering volume. In the case of laboratory scattering meters, scattering by a water sample can be related to that from a standard scattering medium such as pure benzene.[630]

Once the volume scattering function has been measured over all angles, the value of the scattering coefficient can be obtained by summation (integration) of $2\pi\beta(\theta)\sin\theta$ in accordance with eqn 1.36. The forward and backward scattering coefficients, b_f and b_b, are obtained by integration from 0 to 90°, and from 90 to 180°, respectively.

Fixed-angle scattering meters

As an alternative to measuring the whole volume scattering function in order to determine b, $\beta(\theta)$ can be measured at one convenient fixed angle, and by making reasonable assumptions about the likely shape of the volume scattering function in the type of water under study (see below), an approximate value of b can be estimated by proportion. In marine waters, for example, the ratio of the volume scattering function at 45° to the total scattering coefficient ($\beta(45°)/b$) is commonly in the range 0.021–0.035 sr^{-1}.[422] From an analysis of their own measurements of $\beta(\theta)$ and b in the Pacific and Indian Oceans and in the Black Sea, Kopelevich & Burenkov (1971) concluded that the error in estimating b from single-angle measurements of $\beta(\theta)$ is lower for angles less than 15°: 4° was considered suitable. A linear regression of the type

$$\log b = c_1 \log \beta(\theta) + c_2 \tag{4.3}$$

where c_1 and c_2 are constants, was found to give more accurate values for b than a simple proportionality relation such as $b = constant \times \beta(\theta)$. On the basis both of Mie scattering calculations, and analysis of literature data on volume scattering functions for ocean waters, Oishi (1990) concluded that there is an approximately constant ratio between the backscattering coefficient and the volume scattering function at 120°, so that b_b can be calculated from a scattering measurement at 120°, using the relationship

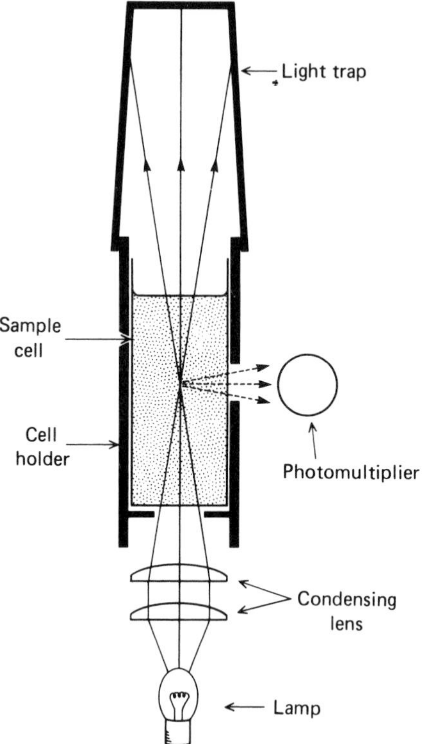

Fig. 4.7. Schematic diagram of optical system of nephelometric turbidimeter.

$$b_b \simeq 7\beta(120°) \tag{4.4}$$

As can readily be verified this works very well for the two volume scattering functions listed in Table 4.2 (below).

Turbidimeters

A specialized and simplified laboratory form of the fixed-angle scattering meter is the nephelometric turbidimeter. The word 'turbidity' has been used in a general sense to indicate the extent to which a liquid lacks clarity, i.e. scatters light as perceived by the human eye. In the most common type of instrument, a beam of light is directed along the axis of a cylindrical glass cell containing the liquid sample under study. Light scattered from the beam within a rather broad angle centred on 90° is measured by a photomultiplier located at one side of the cell (Fig. 4.7). The 'turbidity' (T_n)

of the sample in nephelometric turbidity units (NTU) is measured relative to that of an artificial standard with reproducible light-scattering properties. The standard can be a suspension of latex particles, or of the polymer formazin, made up in a prescribed manner. Turbidimeters, as at present constituted, do not attempt to provide a direct estimate of any fundamental scattering property of the water and the nephelometric turbidity units are essentially arbitrary in nature. Nevertheless, the turbidity measured in this way should be directly related to the average volume scattering function over an angular range centred on 90°, and so for waters of given optical type (e.g. waters with moderate to high turbidity due to inorganic particles) should bear an approximately linear relation to the scattering coefficient. Since turbidimetric measurements are so easily made, comparative studies on a variety of natural waters to determine the empirical relation between nephelometric turbidity and scattering coefficient would be valuable. Some existing indirect measurements (see below) suggest that for turbid waters, by a convenient coincidence, b/T_n is about 1 $m^{-1}NTU^{-1}$. Turbidimeters are however, not well suited for characterizing waters with very low scattering values, such as the clear oceanic types.

Indirect estimation of scattering properties

While few aquatic laboratories carry out measurements of the fundamental scattering properties of natural waters, many routinely measure underwater irradiance (§5.1). Since irradiance at any depth is in part determined by the scattering properties of the water, there is the possibility that information on the scattering properties might be derived from the measured irradiance values. For water with a specified volume scattering function and with incident light at a given angle, then at any given optical depth (see §1.6 for definition), the irradiance reflectance, R, and the average cosine $\bar{\mu}$, of the light are functions only of the ratio, b/a, of scattering coefficient to absorption coefficient. Conversely, if the value of reflectance at a certain optical depth is given, then the values of $\bar{\mu}$ and b/a are fixed, and in principle determinable. Kirk (1981*a*, *b*) used Monte Carlo simulation (§§5.5 and 6.7) to determine the relations between b/a for the medium and R and $\bar{\mu}$ at a fixed optical depth, for water with a normalized volume scattering function identical to that determined by Petzold (1972) for the turbid water of San Diego harbour. It was considered that these relations would be approximately valid for most natural waters of moderate to high turbidity. Using the computer-derived curves it is possible, given a measured value of irradiance reflectance at a specified optical depth, such as

$\zeta = 2.3$ (irradiance reduced to 10% of the subsurface value) to read off the corresponding values of $\bar{\mu}$ and b/a. The irradiance values are used to estimate K_E, the vertical attenuation coefficient for *net* downward irradiance (§1.3), and the value of a is calculated from the relation $a = \bar{\mu} K_E$(§1.7). Knowing a and b/a, the value of b may then be obtained.

As a test of the validity of this procedure, Weidemann & Bannister (1986) compared, for Irondequoit Bay, L. Ontario, estimates of a derived from irradiance measurements as indicated above, with estimates obtained by summing the measured absorption coefficients due to gilvin, particulate matter and water: agreement was good. Furthermore, estimates of $\bar{\mu}$ read off from the $\bar{\mu} = f(R)$ curve agreed with those obtained from the measured ratio of net downward, to scalar, irradiance (eqn 1.15): a similar finding was made by Oliver (1990) for a range of river and lake waters in the Murray–Darling basin, Australia.

Values of scattering coefficient obtained in this way by Kirk (1981*b*) for various bodies in Southeastern Australia were found to correlate very closely with the values of nephelometric turbidity (T_n), a parameter which (see above) we might reasonably expect to be linearly related to the scattering coefficient: the average ratio of b to T_n was 0.92 m^{-1} NTU^{-1}. Using literature data for Lake Pend Oreille in Idaho, USA,[717] the method was found to give a value of b differing by only 5% from that derived from the beam attenuation and absorption coefficients.

Di Toro (1978) using radiation transfer theory with a number of simplifying assumptions has arrived at an approximate relation between the scattering coefficient and a certain function of the reflectance and vertical attenuation coefficient for irradiance. For the turbid waters of San Francisco Bay, he observed a good linear relation between the values of b calculated in this way and the nephelometric turbidity: the actual ratio of scattering coefficient to turbidity was 1.1 m^{-1} FTU^{-1}, which is in reasonable agreement with the value of 0.92 m^{-1} NTU^{-1} (NTU and FTU being equivalent) obtained by Kirk (1981*b*). Combining these results with similar findings by numerous other workers[174,286,677,949,950,980,981] we arrive at the useful and convenient conclusion that, at least for waters of moderate to high turbidity, the scattering coefficient (m^{-1}) has approximately the same numerical value as the nephelometric turbidity (NTU).

Diffuse scattering coefficients

As we saw in our discussion of radiation transfer theory (§1.7), certain quasi-inherent optical properties of the medium – the diffuse scattering coefficients – play an essential role in our understanding of the underwater light field. These scattering coefficients have been defined previously (§1.5) in terms of the light scattered backwards or forwards by a thin layer of medium from an incident light field which is not a parallel beam at right angles (as in the case of the normal scattering coefficients) but has a radiance distribution identical to the downwelling or upwelling parts of a given underwater light field existing at a certain depth in a certain water body. There is no way at present of directly measuring the diffuse scattering coefficients. If, however, the radiance distribution at a given depth in the water body and the volume scattering function for the water are known, then it is possible to calculate the diffuse scattering coefficients.[482]

To calculate, for example, the backscattering coefficient for downwelling flux, the downwelling radiance distribution is distributed into a manageable number, e.g. 15, of angular intervals. Using the volume scattering function the proportion of the incident flux in each angular interval which is scattered in an upwards or a downwards direction by a thin layer of medium is calculated. The summed flux (for all incident angles) travelling upwards after scattering, expressed as a proportion of the total downwelling flux, divided by the thickness of the layer of medium, is $b_{bd}(z)$, the diffuse backscattering coefficient for downwelling flux at depth z m in that particular water body: $b_{bd}(z)$ is commonly two- to five-fold greater than b_b in natural water bodies.[482] The radiance distribution data necessary for such calculations may be obtained by direct measurement (§5.1), or derived from the inherent optical properties by computer modelling (§5.2).

An approximate estimate of the diffuse backscattering coefficient can be obtained much more simply from underwater irradiance measurements. Rearranging eqn 1.47 we obtain

$$b_{bd}(z) = R(z)[K_d(z) + \kappa(z)] \tag{4.5}$$

which, with help of eqn 1.48 can be rewritten

$$b_{bd}(z_m) \simeq 3.5R(z_m)K_d(z_m) \tag{4.6}$$

where z_m is the depth at which downward irradiance is 10% of the subsurface value. $R(z_m)$ and $K_d(z_m)$ are obtained from the irradiance data, and $b_{bd}(z_m)$ is calculated accordingly.

From the value of the diffuse backscattering coefficient derived in this way we can then go on to obtain an estimate of the normal backscattering and total scattering coefficients, b_b and b. Kirk (1989*a*) presents a Monte Carlo-derived curve, for optical water types covering the range $b/a = 0.5 - 30$, showing the ratio of diffuse to normal backscattering coefficient plotted against reflectance at depth z_m. From the observed value of reflectance, the ratio $b_{bd}(z_m)/b_b$ is read off, and is then used, in conjunction with eqn 4.6 to give b_b. Assuming $b \sim 53b_b$ (a reasonable estimate for most inland and coastal waters), the total scattering coefficient may then be obtained. Values of b calculated in this way are in good agreement with those obtained from irradiance data by the other indirect procedure, described in the previous section.

4.3 The scattering properties of natural waters

The scattering properties of pure water provide us with a suitable baseline from which to go on to the properties of natural waters. We shall make use here of a valuable review by Morel (1974) of the optical properties of pure water and pure sea water. For measurements of the scattering properties, water purified by distillation *in vacuo* or by repeated filtration through small-pore-size filters must be used: ordinary distilled water contains too many particles. Scattering by pure water is of the density fluctuation type, and so varies markedly with wavelength. Experimentally, scattering is found to vary in accordance with $\lambda^{-4.32}$ rather than λ^{-4} as predicted by density fluctuation theory alone: this is a result of the variation of the refractive index of water with wavelength. Pure sea water (35–38‰ salinity) scatters about 30% more intensely than pure water. Table 4.1 lists values of the scattering coefficient for pure water and pure sea water at a number of wavelengths. The volume scattering function of pure water or pure sea water has, as predicted by density fluctuation theory, its minimum at 90° and rises symmetrically towards greater or lesser angles (Fig. 4.8).

The scattering coefficients of natural waters are invariably much higher than those of pure water. Table 4.1 lists a selection of values from the literature. Even the lowest value – 0.016 m^{-1} at 546 nm for water from 1000 m depth in the Tyrrhenian Sea[631] – is 10 times as high as the value for pure water at that wavelength. Unproductive oceanic waters away from land have low values. Coastal and semi-enclosed marine waters have higher values due to the presence of resuspended sediments, river-borne terrigenous particulate material and phytoplankton. Resuspension of sediments is caused by wave action, tidal currents and storms. High levels of phytoplankton can give rise to quite high values of the scattering coefficient in oceanic upwelling areas, such as the Mauritanian upwelling off the west coast of Africa.[635] In arid regions of the continents large amounts of dust are carried up into the atmosphere by wind, and when subsequently redeposited in adjoining areas of the ocean, can substantially increase scattering

Table 4.1. *Scattering coefficient values for various waters*

Water	Wavelength (nm)	Scattering coefficient, b (m^{-1})	Reference
Pure water	400	0.058	632
	450	0.0035	632
	500	0.0022	632
	550	0.0015	632
	600	0.0011	632
Pure sea water	450	0.0045	632
	500	0.0019	632
Marine Waters			
Atlantic Ocean			
Sargasso Sea	633	0.023	522
	440	0.04	422
Caribbean Sea	655	0.06	422
3 oceanic stations	544	0.06–0.30	517
Bahama Islands	530	0.117	699
Mauritanian upwelling[a]	550	0.4–1.7	635
Mauritanian coastal[b]	550	0.9–3.7	635
Iceland coastal	655	0.1–0.5	379
Rhode River estuary (Chesapeake Bay)	720	1.7–55.3	282
Pacific Ocean			
Central (Equator)	440	0.05	422
Galapagos Islands	655	0.07	422
	440	0.08	422
126 oceanic stations (av.)	544	0.18	517
Offshore, Southern California	530	0.275	699
Kieta Bay (Solomon Is.)	544	0.54	517

Table 4.1 (*cont.*)

Water	Wavelength (nm)	Scattering coefficient, b (m^{-1})	Reference
Lagoon, Tarawa atoll (Gilbert Is.)	544	1.04	517
San Diego Harbour (California)	530	1.21–1.82	699
Indian Ocean			
164 oceanic stations (av.)	544	0.18	517
Tasman Sea			
Australia			
Jervis Bay, NSW			
Entrance, 30 m deep	450–650	0.25	700
Inshore, 15 m deep	450–650	0.4–0.6	700
New Zealand			
9 estuaries, North Island mouth sites, low water data	400–700	1.1–4.8	949
Mediterranean Sea			
Tyrrhenian Sea (1000 m depth)	546	0.016	631
Western Mediterranean	655	0.04	380
Bay of Villefranche	546	0.1	631
North Sea			
Fladen Ground	655	0.07–0.13	382
English Channel	546	0.65	631
Baltic Sea			
Kattegat	655	0.15	422
South Baltic	655	0.20	422
Bothnian Gulf	655	0.28	422

Black Sea			
33 stations (av.)	544	0.41	517
Inland Waters			
North America			
L. Pend Oreille	480	0.29	717
L. Ontario, coastal	530–550	1.5–2.5	115
Irondequoit Bay, L. Ontario	400–700	1.9–5.0	980
Otisco L., N.Y.	400–700	0.9–4.6	217, 981
Onondaga L., N.Y.	400–700	2.2–11.0	218
Owasco L., N.Y.	400–700	1.0–4.6	217
Seneca R., N.Y.	400–700	3.1–11.5	217
Woods L., N.Y.	300–770	0.13–0.20	117
Dart's L., N.Y.	300–770	0.19–0.25	117
Australia			
(a) Southern tablelands			
Corin Dam	400–700	1.5	485
Burrinjuck Dam	400–700	2.0–5.5	485
L. Ginninderra	400–700	4.4–21.6	485
L. Burley Griffin	400–700	2.8–52.6	485
L. George	400–700	55.3, 59.8	485
(b) Murray–Darling system			
Murrumbidgee R., Gogeldrie Weir	400–700	9–58	677
Murray R., upstream of Darling confluence	400–700	13.0	677
Darling R., above confluence with Murray R.	400–700	27.8–90.8	677
(c) Northern Territory (Magela Creek billabongs)			
Mudginberri	400–700	2.2	498
Gulungul	400–700	5.7	498
Georgetown	400–700	64.3	498
(d) Tasmania (lakes)			
Perry	400–700	0.27	90

Table 4.1 (*cont.*)

Water	Wavelength (nm)	Scattering coefficient, b (m^{-1})	Reference
Risdon Brook	400–700	1.8–2.7	90
Barrington	400–700	1.1–1.4	90
Pedder	400–700	0.6–1.3	90
Gordon	400–700	1.0	90
(e) Southeast Queensland, coastal dune lakes			
Basin	400–700	0.6	89
Boomanjin	400–700	1.1	89
Wabby	400–700	1.5	89
(f) South Australia			
Mount Bold Reservoir	400–700	5.7–6.8	286
New Zealand			
Waikato R. (330 km, L. Taupo to the sea)			
L. Taupo (0 km)	400–700	0.4	174
Ohakuri (77 km)	400–700	1.0	174
Karapiro (178 km)	400–700	1.2	174
Hamilton (213 km)	400–700	1.9	174
Tuakau (295 km)	400–700	6.3	174
Lakes			
Rotokakahi	400–700	1.5	950
Rotorua	400–700	2.1	950
D	400–700	3.1	950

Notes:
[a] Waters rich in phytoplankton.
[b] High concentration of resuspended sediments.

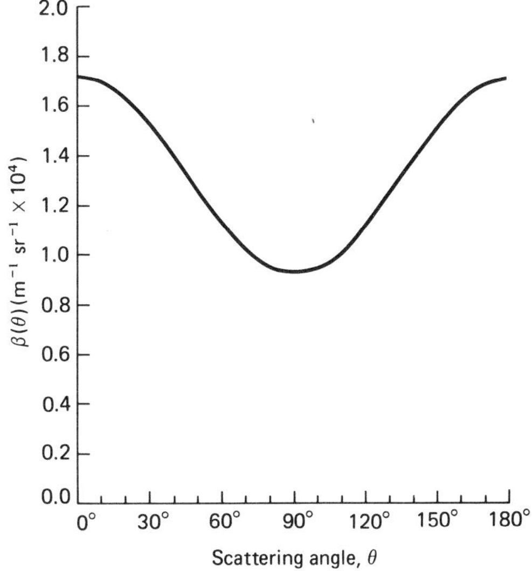

Fig. 4.8. Volume scattering function of pure water for light of wavelength 550 nm. The values are calculated on the basis of density fluctuation scattering, assuming that $\beta(90°) = 0.93 \times 10^{-4}$ m^{-1} sr^{-1} and that $\beta(\theta) = \beta(90°)(1 + 0.835 \cos^2 \theta)$ (following Morel, 1974).

in the water.[516] Scattering coefficient values are on average higher in inland and estuarine waters than in the open sea. Indeed the very high values that can occur in some turbid inland waters (Table 4.1) are unlikely to be equalled in the sea because high ionic strength promotes the aggregation and precipitation of colloidal clay minerals. Resuspension of sediments by wind-induced turbulence, especially in shallow waters, can substantially increase scattering in inland water bodies.[662] Filter-feeding zooplankton, by packaging suspended clay particles into more rapidly settling faecal pellets, can substantially increase the rate of clearing of turbid lake water.[306]

Volume scattering functions for natural waters differ markedly in shape from that of pure water. They are invariably characterized, even in the clearest waters, by an intense concentration of scattering at small forward angles. This, as we saw earlier, is typical of scattering by particles of diameter greater than the wavelength of light, and scattering in natural waters is primarily due to such particles. Fig. 4.9 shows the volume scattering functions for a clear oceanic water and a moderately turbid, harbour water. They are quite similar in shape but there is a noticeable difference at angles greater than 90°: scattering by the oceanic water shows a greater tendency to increase between 100° and 180°. This is because at these

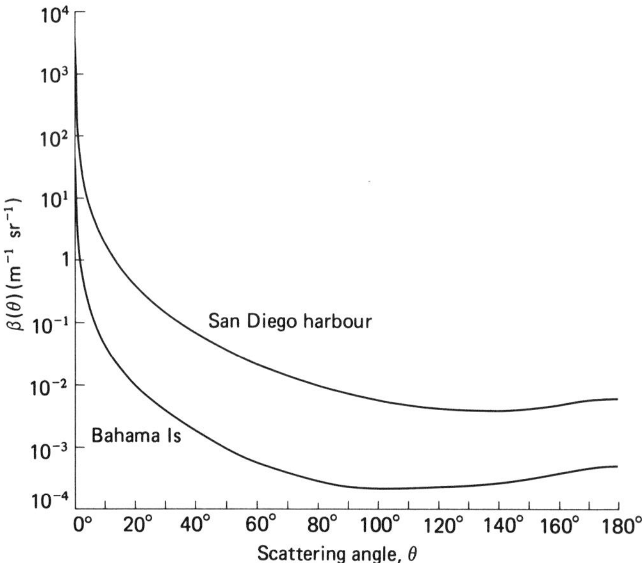

Fig. 4.9. Volume scattering functions for a clear, and a moderately turbid, natural water. The curves for the clear oceanic water ($b=0.037$ m^{-1}) of Tongue-of-the-Ocean, Bahama Islands, and the somewhat turbid water ($b=1.583$ m^{-1}) of San Diego harbour, have been plotted from the data of Petzold (1972).

larger angles density fluctuation scattering (which exhibits this kind of variation in this angular range – Fig. 4.8) becomes a significant proportion of the total in the case of the clear oceanic water (see below), but remains insignificant compared to particle scattering in the case of the harbour water.

A knowledge of the shape of the volume scattering function is essential for calculations on the nature of the underwater light field. Table 4.2 lists values for the normalized volume scattering function ($\tilde{\beta}(\theta)=\beta(\theta)/b$), and the cumulative scattering (between 0° and θ, as a proportion of the total) at a series of values of θ, for Tongue-of-the-Ocean (Bahamas) and San Diego harbour water.[699] It will be noted that backscattering ($\theta>90°$) constitutes 4.4% of total scattering in the case of the clear oceanic water but only 1.9% in the turbid harbour water. Waters of a given broad optical type appear, on the basis of published measurements to date, to have $\tilde{\beta}(\theta)$ curves of rather similar shape. We may therefore take these data sets as being reasonably typical for clear oceanic, and moderately/highly turbid waters, respectively. Above a certain minimum level of turbidity, such that particle scattering is dominant at all angles, we would not expect the normalized

Table 4.2. *Volume scattering data for green (530 nm) light for a clear and a turbid natural water.* $\tilde{\beta}(\theta)$ = *normalized volume scattering function* ($\beta(\theta)/b$). *Cumulative scattering* = $2\pi\int_0^\theta \tilde{\beta}(\theta)\sin\theta\, d\theta$. *Data from Petzold (1972)*
(a) Atlantic Ocean, Bahama Islands. b = *0.037 m*$^{-1}$
(b) San Diego harbour, California, USA, b = *1.583 m*$^{-1}$

	(*a*)		(*b*)	
Angle θ (°)	$\tilde{\beta}(\theta)$ (sr^{-1})	Cumulative scattering 0°→θ	$\tilde{\beta}(\theta)$ (sr^{-1})	Cumulative scattering 0°→θ
1.0	67.5	0.200	76.4	0.231
2.0	21.0	0.300	23.6	0.345
3.16	9.0	0.376	10.0	0.431
5.01	3.91	0.458	4.19	0.522
6.31	2.57	0.502	2.69	0.568
7.94	1.70	0.547	1.72	0.616
10.0	1.12	0.595	1.09	0.664
15.0	0.55	0.687	0.491	0.750
20.0	0.297	0.753	0.249	0.806
25.0	0.167	0.799	0.156	0.848
30.0	0.105	0.832	0.087	0.877
35.0	0.072	0.857	0.061	0.898
40.0	0.051	0.878	0.0434	0.916
50.0	0.0276	0.906	0.0234	0.940
60.0	0.0163	0.925	0.0136	0.956
75.0	0.0093	0.943	0.0073	0.971
90.0	0.0066	0.956	0.00457	0.981
105.0	0.0060	0.966	0.00323	0.987
120.0	0.0063	0.975	0.00276	0.992
135.0	0.0072	0.984	0.00250	0.995
150.0	0.0083	0.992	0.00259	0.997
165.0	0.0110	0.997	0.00323	0.999
180.0	0.0136	0.100	0.00392	1.000

volume scattering function to alter its shape with increasing turbidity, since the shape of the $\tilde{\beta}(\theta)$ curve is determined by the intrinsic scattering properties of the particles, not by their concentration. This is why the San Diego harbour data may reasonably be applied to much murkier water. Indeed the $\tilde{\beta}(\theta)$ data in Table 4.2(*b*) are applicable to the majority of natural waters other than the very clear oceanic ones: Timofeeva (1971) has presented evidence that $\tilde{\beta}(\theta)$ is virtually the same for most natural waters. However, in the specialized situation of scattering being dominated by organic particles with a low refractive index, as might be the case in oceanic upwelling areas or eutrophic inland waters, a somewhat different set of $\tilde{\beta}(\theta)$ data might be appropriate.

The contribution of density fluctuation scattering to total scattering varies not only with the type of water body but also with the angular range of scattering and wavelength of light under consideration. In the moderately or very turbid waters such as are typically found in inland, estuarine and some coastal water bodies, scattering at all angles and all visible wavelengths is predominantly due to the particles present. In contrast, in the very clear waters of the least fertile parts of the ocean, density fluctuation scattering can be a significant component of total scattering at short wavelengths and the major component at large scattering angles. In the Sargasso Sea, for example, Kullenberg (1968) found that density fluctuation scattering accounted for 3% of the total scattering coefficient at 655 nm and 11% at 460 nm. For scattering angles from 60–75 to 180°, however, the greater part of the scattering at both wavelengths was due to density fluctuation: in the case of blue (460 nm) light it accounted for only 0.3% of the scattering at 1°, but contributed 89% at 135°. Because of its importance as a component of backscattering in oceanic waters, density fluctuation scattering makes a disproportionately large contribution to the upwelling light stream. Morel & Prieur (1977) pointed out that even for a water with a total scattering coefficient of 0.29 m^{-1} (rather turbid by oceanic standards), while density fluctuation scattering accounts for only 1% of the total scattering coefficient, it contributes about 33% of the backscattering. Thus, even in the not-so-clear oceanic waters, density fluctuation scattering initiates a substantial proportion of the upwelling light flux.

Compared with density fluctuation scattering, particle scattering is rather insensitive to wavelength. Since, even in a very clear oceanic water, density fluctuation scattering makes up only a small proportion of the total scattering coefficient, the dependence of this scattering on $\lambda^{-4.3}$ does not bring about a marked variation of the value of b for natural waters with wavelength. There is, however, some evidence that in particle-dominated natural waters, the value of b varies approximately in accordance with λ^{-1}; i.e. shorter wavelengths are scattered more intensely.[631] When backscattering alone is considered, then in oceanic waters, in which as we saw above density fluctuation scattering is a major contributor for $\theta > 90°$, a more marked inverse dependence on wavelength is to be expected.

Since total light scattering in natural waters is dominated by the particulate contribution, it increases broadly in proportion to the concentration of suspended particulate matter. For example, Jones & Wills (1956), using suspensions of kaolin (which would scatter but not significantly absorb light), found a linear relation between the beam attenuation coefficient (approximately equal in this system to the scattering coefficient) for green (550 nm) light and the concentration of suspended matter. Approximately linear relations between scattering coefficient and concentration of suspended matter have been observed for natural aquatic particulates by other workers.[70,877] The constant of proportionality can, however, vary from one kind of suspended matter to another:[207] the refractive index and the size distribution of the particles both influence the relation between b and sediment concentration.

The concentration of particulate matter, and therefore the intensity of scattering, in inland and estuarine waters, and to some extent in coastal waters, is strongly influenced, not only by the nature of the physical environment (climate, topography, soil type, etc.) but also by the uses to which the land is put. The more thickly the ground is covered with vegetation, the less erosion and so the less transference of soil particles to surface waters occurs. Erosion is very low under undisturbed forest and is generally not high from permanent grassland. Logging activities and overgrazing of pastures can cause serious erosion even in these environments. Ground which is broken up for crop planting is susceptible to erosion until such time as the crop canopy is established. The extent to which erosion occurs in these different terrestrial systems depends not only on the extent to which the ground is protected by vegetation, but also on the nature of the soil (some types being more easily eroded than others), the average slope (the greater the slope, the greater the erosion), and the intensity of the rainfall (a short torrential downpour causes more erosion than the same amount of rain falling over a longer period). The average lifetime of the soil particles in suspension once they have been transferred into the aquatic sphere is quite variable and depends on their size and mineral chemistry, and on the ionic composition of the water: some fine clay particles for example, particularly in waters of low electrolyte content, can remain in suspension for long periods with drastic consequences for light attenuation in the water bodies which contain them.

It is clear that to understand the underwater light field, we must know not only about the water itself but also about the surrounding land forms which to a large degree confer upon the aquatic medium the particular optical properties it possesses.

4.4 The scattering properties of phytoplankton

Phytoplankton cells and colonies scatter, as well as absorb, light and can make a significant contribution to the total scattering behaviour of the aquatic medium, but to an extent that varies from one species to another:

Table 4.3. *Specific scattering coefficients for phytoplankton*

Organism	Wavelength (nm)	b_c (m^2 mg chl a^{-1})	Reference
Marine			
Tetraselmis maculata	590	0.178	640
Hymenomonas elongata	590	0.078	640
Emiliania huxleyi	590	0.587	640
Platymonas suecica	590	0.185	640
Skeletonema costatum	590	0.535	640
Pavlova lutheri	590	0.378	640
Prymnesium parvum	590	0.220	640
Chaetoceros curvisetum	590	0.262	640
Isochrysis galbana	590	0.066	640
Synechocystis sp.	590	0.230	636
Freshwater			
Scenedesmus bijuga	550	0.107	176
Chlamydomonas sp.	550	0.044	176
Nostoc sp.	550	0.113	176
Anabaena oscillarioides	550	0.139	176
Natural (freshwater) populations			
Irondequoit Bay (L. Ontario, USA)			
Av. mixed population	400–700	0.08	980
Cyanobacterial bloom	400–700	0.12	980
L. Hume (Murray R., Australia)			
Melosira granulata dominant	400–700	0.11	677
L. Mulwala (Murray R., Aust.)			
M. granulata dominant	400–700	0.22	677

this has been studied in detail, both experimentally and theoretically, by Morel, Bricaud and coworkers.[95,98,636,640] A convenient parameter in terms of which to compare the scattering propensities of different species is the specific scattering coefficient, b_c, which is the scattering coefficient that would be exhibited by cells of a given species suspended at a concentration corresponding to 1 mg chlorophyll a m^{-3}: it has the units m^2 mg chl a^{-1}.

Table 4.3 lists values for b_c measured on laboratory cultures of a range of marine and freshwater phytoplankton species, and on some natural populations. Algae such as diatoms (*S. costatum*) and coccolithophores (*E. huxleyi*) in which a substantial proportion of the total biomass consists of mineralized cell walls, or scales, scatter more light per unit chlorophyll than, for example, naked flagellates (*I. galbana*). Also, blue-green algae with gas vacuoles scatter light much more intensely than those without.[286]

Like the mineral and detrital particles which carry out the greater part of the scattering in most natural waters, algal cells have a scattering phase function which is strongly peaked at small forward angles,[879] but the backscattering ratio (b_b/b, the proportion of the total scattering which is in a backwards direction, $\theta > 90°$) is much lower (0.0001–0.004) for the living cells[98,879] than for the mineral and detrital particles (~0.019). This is a consequence[98] of the low refractive index (relative to water) of the living cells (1.015–1.075)[3,133,636] compared with that of the inorganic particles (1.15–1.20).[422] The backscattering ratio is greater in the small (picoplankton) cells, such as cyanobacteria, than in the larger eukaryotic cells.[879]

In the ocean, satellite remote sensing shows that blooms of coccolithophores have a high reflectance, indicating efficient upward, and therefore backward, scattering which might seem to contradict the generalization that phytoplankton are weak backward scatterers. It is thought, however, that the intense upward scattering from these blooms originates with the numerous detached coccoliths, rather than from the living cells themselves.[388,636]

Within the last decade the powerful new technique of flow cytometry, which makes use of light scattering from individual cells, has been adapted for the study of phytoplankton populations.[3,140,560,696,850,861,1005] The sample fluid, e.g. ocean water, is injected coaxially into a stream of particle-free sheath fluid. The liquid passes through a capillary flow chamber which is traversed by an intense argon-ion laser beam operating typically at 488 or 514 nm. The dimensions of, and rate of flow through, the flow chamber are such that the individual cells pass through the laser beam one at a time, and as they do so they scatter the incident light, and also exhibit fluorescent emission in response to absorption of light by their photosynthetic pigments. Light scattered forward (in the range 1.5–19°) and at ~90°, and fluorescent emission in the orange (530–590 nm) and/or red (>630 nm) wavebands, are measured. Different sizes and pigment classes of phytoplankton have different combinations of scattering and fluorescence signals, and so the technique can be used for enumerating, characterizing, and following the development of, natural phytoplankton populations.

5

Characterizing the underwater light field

Having described the nature of the incident radiation flux presented to the aquatic biosphere and the influences to which the light is subject within the water, we shall now consider the kind of light field that results. We shall begin, in this chapter, by examining the ways in which the properties of this light field are studied. The physical definitions of these properties are given in Chapter 1.

5.1 Irradiance

Irradiance meters

The most frequently and easily measured property of the underwater light field is irradiance. Knowledge of this parameter is valuable, not only because it provides information about how much light is available for photosynthesis, but also because irradiance plays a central role in the theory of radiation transfer in water. An irradiance meter, since it is meant to measure the radiant flux per unit area, must respond equally to all photons that impinge upon its collector, regardless of the angle at which they do so. With any given meter this can be tested by observing the way in which its response to a parallel light stream varies with its angle to that light stream.

As the angle of the radiant flux to the collector changes, the area of the collector projected in the direction of that radiant flux changes, and the proportion of the flux intercepted by the collector changes correspondingly (Fig. 5.1). Thus, the response of an irradiance meter to parallel radiant flux (wider than the collector) should be proportional to the cosine of the angle (θ) between the normal to the collector surface and the direction of the flux. An irradiance collector which meets this criterion is known as a cosine collector.

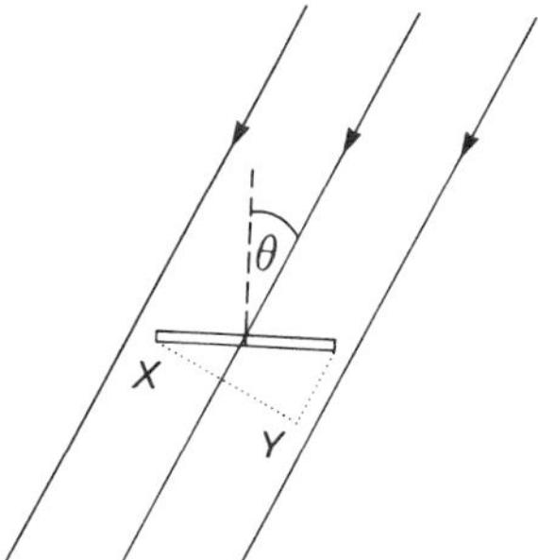

Fig. 5.1. The area which the collector presents to the beam is linearly related to XY, which is proportional to $\cos\theta$.

The collector of an irradiance meter usually consists of a flat disc of translucent diffusing plastic, although opal glass may be used when measurements below 400 nm are required.[423] It has been found that a more faithful cosine response to light incident at large values of θ is obtained if the disc projects slightly up above the surrounding housing, allowing some collection of light through the edge.[827] Some of the light that penetrates into the diffusing plastic is scattered backwards and a part of this succeeds in passing up through the diffuser surface again. Because the difference between the refractive indices of plastic and water is less than that between those of plastic and air, there is less internal reflection of this light within the collector and therefore a greater loss of light when the collector is immersed in water than when it is in air: this is known as the immersion effect. To take account of this, if an irradiance meter has been calibrated in air, the underwater readings should be multiplied by the appropriate correction factor for that instrument. This factor is usually in the region of 1.3–1.4 in the case of a plastic collector.

Within the irradiance sensor, beneath the collecting disc, there is a photoelectric detector. In the older instruments this was usually a selenium photovoltaic cell: nowadays a silicon photodiode, which has a broader spectral response, is more commonly used. If a very narrow waveband of light is measured, as in certain spectroradiometers (see below), then a photomultiplier may be used as the detector to provide the necessary sensitivity. The electrical signal from the photodetector is transmitted by cable to the boat where, after amplification if required, it is displayed on a suitable meter or digital readout.

In all natural waters most of the light at any depth is travelling in a downwards direction. Measurement of the irradiance of the upwelling flux is nevertheless worthwhile. This is partly because, in some waters, upwelling flux is a significant component (up to ~20%) of the total available radiant energy, but even more so because the ratio of upwelling to downwelling flux (irradiance reflectance, $R = E_u/E_d$) can provide information about the inherent optical properties of the water (see §§4.2, 6.4) and because the properties of the upwelling flux are of central importance to the remote sensing of the composition of natural waters. Thus, the more we know about this flux and its relation to other characteristics of the medium the better.

The holder in which the sensor is lowered into the water (Fig. 5.2) should thus be made so that the sensor can be facing upwards to measure downwards irradiance (E_d), or downwards to measure upward irradiance (E_u). To minimize shading effects, the sensor is always lowered on the sunny side of the boat. For measurements in rivers, the sensor must be attached to a rigid support.

Measurements made with an uncorrected wide-band detector are unreliable because of the spectral variation of sensitivity: part of the diminution of irradiance with depth could be due to a shift of the wavelength of the predominating light to a part of the spectrum where the detector is less sensitive. It is therefore desirable to confine the measured light to a specific waveband with an appropriate filter in the sensor: narrow-band interference filters are best for this purpose. The alternative, and in the context of photosynthesis the best, solution (short of determining the complete spectral distribution of irradiance), is to use a meter designed to respond equally to all quanta within the photosynthetic range regardless of wavelength, but to be insensitive to quanta outside this range. In this way a measure of the total photosynthetically available radiation (PAR) is obtained.

It is better that the meter should be designed to respond equally to quanta regardless of wavelength rather than to respond equally to a given amount of radiant energy regardless of wavelength, because the usefulness of the prevailing light for photosynthesis is more closely related to the flux of quanta than to the flux of energy. This is because once a quantum has been absorbed by a cell, it makes the same contribution to photosynthesis regardless of its wavelength (see §3.1), although the probability of its being absorbed in the first place of course varies with wavelength. An absorbed red quantum for example is just as useful as an absorbed blue quantum even though it may contain only two-thirds of the energy.

An irradiance meter designed in this way is commonly referred to as a quanta meter. Since no existing photodetector in its normal state responds

Fig. 5.2. Submersible irradiance sensor (Li–Cor 192S Underwater Quantum Sensor: Li–Cor Inc., Lincoln, Nebraska, USA) in a lowering frame. The sensor can be placed in the frame facing upwards or downwards.

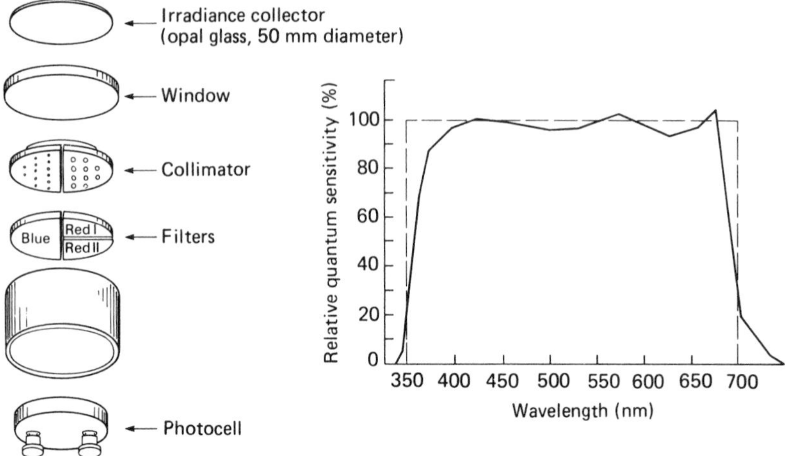

Fig. 5.3. Exploded view of a quanta meter together with its quantum sensitivity curve (after Jerlov & Nygard, 1969).

equally to all quanta throughout the photosynthetic range, ways of adjusting the relative response in different parts of the spectrum must be found. Jerlov & Nygard (1969) used a combination of three colour filters, each covering a different part of the photodetector (Fig. 5.3). The filters were covered with an opaque disc with a number of small holes drilled in it: the number, size and position of these holes determined the amount of light reaching each filter. By a suitable combination of this variable attenuation with the differing absorption properties of the filters it was possible to obtain an approximately constant quantum sensitivity throughout the photosynthetic range. Commercial instruments using filter combinations with a silicon photodiode, and measuring the total quantum flux in the range 400–700 nm, are now available.

The photosynthetically available radiation at any depth can now readily be measured with a submersible quanta meter. Smith (1968) suggested that the vertical attenuation coefficient for downwards irradiance of PAR is the best single parameter by means of which different water bodies may be characterized with respect to the availability of photosynthetically useful radiant energy within them. As we shall see in the next chapter, K_d for PAR is not exactly constant with depth even in a homogeneous water body. Nevertheless the variation is not great and knowledge of, say, the average value of K_d(PAR) over a certain depth range such as the euphotic depth, is very useful. It is determined by measuring downwelling PAR at a series of depths within the depth interval of interest and making use of the approximate relation (see §§6.1, 6.3)

$$\ln E_d(z) = -K_d z + \ln E_d(0) \tag{5.1}$$

where $E_d(z)$ and $E_d(0)$ are the values of downward irradiance at z m and just below the surface, respectively. The linear regression coefficient of $\ln E_d(z)$ with respect to depth gives the value of K_d. Or, less accurately, K_d may be determined from measurements at just two depths, z_1 and z_2

$$K_d = \frac{1}{z_2 - z_1} \ln \frac{E_d(z_1)}{E_d(z_2)} \tag{5.2}$$

Measurement problems in the field

There are three main sources of error in the measurement of underwater irradiance in the field: wave action, fluctuations in surface-incident flux due to drifting clouds, and perturbation of the light field by the ship. The convex part of a surface wave acts as a converging lens and will focus the incident light at some depth within the water. As the wave moves along, so this zone of intense light travels with it. This is essentially the same phenomenon as that which produces the well-known moving patterns of bright lines to be seen on the bottom of a swimming pool on a sunny day. A stationary upward-pointing irradiance meter will be subjected to a series of intense pulses of radiation of duration typically of the order of some milliseconds as the waves move past overhead. There are, in addition, rapid negative fluctuations in irradiance caused by the de-focussing effect of the concave part of the wave.

An irradiance meter traversing a series of depths will also be subject to these rapid temporal fluctuations in intensity. Dera & Gordon (1968) showed that the average fractional fluctuation in irradiance for a given surface wave field increased with depth to a maximum and then decreased, eventually to insignificant levels. For example, at a shallow coastal site, where K_d(525 nm) was 0.59 m^{-1}, the average fractional fluctuation in E_d(525 nm) rose to a maximum value of ~67% at about 0.5 m depth, and then decreased progressively with depth, falling to ~5% at about 3 m. It was found that the clearer the water, the greater the depth at which significant rapid temporal fluctuation could be detected. The upwelling light field, by contrast, is hardly at all affected by this wave focussing phenomenon. At the above site, for example, the average fractional fluctuation in E_u was only ~5% over the whole depth range.

To extract good values for the vertical attenuation coefficient, some kind of smoothing of the data must be carried out. At its simplest, in the case of a manual instrument with a meter readout, the operator can concentrate on taking observations at any given depth, only between the 'blips'. Alternatively, a simple electronic damping circuit can be used to smooth out the fluctuations.[179,481] In the case of sophisticated rapid-profiling instruments which continuously record irradiance, complete with all the wave-induced rapid fluctuations, as they descend,[836] a variety of mathematical smoothing procedures can subsequently be applied to the stored data.[834]

Fluctuations in the light field due to clouds differ from those due to waves firstly in that they are much slower, and secondly in that their effects are manifest throughout the whole illuminated water column and affect the upwelling and downwelling light streams to the same extent. The procedure most commonly used to overcome this problem is to monitor the incident solar flux continuously with a reference irradiance meter on deck, and use the data so obtained to adjust (for purposes of determining vertical attenuation coefficients, or reflectance) the concurrently obtained underwater irradiance values as appropriate. Davies-Colley, Vant & Latimer (1984) found that a more satisfactory (less variable) correction for changing ambient light was achieved if the reference meter was placed within the water at some fixed depth, preferably the one at which irradiance was reduced to 10–20% of the surface value.

Computer modelling of the light field[314] indicates that the third problem – perturbation of the field by the ship – is relatively unimportant for measurements of downward irradiance made on the sunny side of the vessel under clear skies, but that errors under overcast conditions can be significant, and measurements of upward irradiance are strongly influenced by the ship's presence under either type of illumination. To solve this problem, techniques have been developed for deploying the instrument at some distance from the ship.[977] Since the seriousness of the problem is a function of the size of the boat, it is of more significance for oceanographers than for limnologists. Calculations by Gordon & Ding (1992) show that in measurement of the upward irradiance (or radiance), self-shading by the instrument itself can cause significant error. The error is greatest in strongly absorbing waters, and under vertical sun, and increases with the diameter of the instrument.

A measurement problem of a different nature to those discussed above is that of obtaining a representative set of measurements within a realistic time period when, as is invariably the case in the sea, the area of interest is very large. To address this problem, Aiken (1981, 1985) and Aiken & Bellan (1990) have developed the Undulating Oceanographic Recorder, an instrument platform which is towed 200–500 m behind the ship. It is designed to follow an undulating trajectory within the water: for example, when towed at 4–6 m s^{-1} (8–12 knots) it moves between near-surface depths and ~70 m with an undulation pitch length of ~1.6 km. The platform is fitted with a suite of sensors measuring a range of oceanographic parameters, including downwelling and upwelling light in a number of wavebands across the photosynthetic range. The distance from the towing vessel is such that ship shadow and wake problems are eliminated. The data can be logged at, for example, 10-s intervals for durations of 11 h, so that essentially continuous information over long stretches of ocean can be accumulated.

Secchi depth

A crude visual method of estimating K_d, commonly used before the ready availability of photoelectric instruments and still sometimes used today, is based on the device known as the Secchi disc. A white disc, of diameter 20–30 cm, is lowered into the water and the depth at which it just disappears from view is noted. This is referred to as the *Secchi disc transparency*, or as the *Secchi depth*, Z_{SD}. On the basis of their measurements in marine waters, Poole & Atkins (1929) made the empirical observation that the Secchi depth is approximately inversely proportional to the vertical attenuation coefficient for downwelling irradiance and pointed out that Z_{SD} could therefore be used to estimate K_d: the value obtained will be that applicable to a rather broad waveband corresponding roughly to the spectral sensitivity curve of the human eye. To calculate K_d, the relation $K_d = 1.44/Z_{SD}$ may be used.[390]

Using the contrast transmittance theory of Duntley & Preisendorfer, Tyler (1968) concluded that the reciprocal of the Secchi depth is proportional to $(c + K_d)$, the sum of the beam attenuation and vertical attenuation coefficients, rather than to K_d alone. On theoretical grounds, Tyler arrived at the relation $(c + K_d) = 8.69/Z_{SD}$; from measurements in turbid coastal sea water, Holmes (1970) obtained $(c + K_d) = 9.42/Z_{SD}$. For a number of New Zealand lakes covering a wide range of optical properties, Vant & Davies-Colley (1984, 1988) have also found an approximately linear relationship between the reciprocal of Secchi depth and $(c + K_d)$.

Since, in natural waters, c is usually substantially greater than K_d, the Secchi depth is determined more by c than by K_d. Within some waters, for example in much of the ocean, variations in attenuation and transparency are predominantly due to changes in one component of the system, such as the phytoplankton. As a consequence, K_d and c will tend roughly to covary, and this will account for the approximate constancy of the inverse relationship between Z_{SD} and K_d which is sometimes observed in such cases. *A priori* considerations suggest, however, that use of the Secchi disc could sometimes give highly inaccurate values for K_d. If, for example, as a consequence of increased levels of particles, light scattering in a water body increased much more than absorption, then c would increase much more than K_d and use of the relation $K_d = 1.44/Z_{SD}$ would overestimate K_d. There is, in fact, now ample field evidence that the product, K_dZ_{SD}, is very far from constant, especially in inland waters: it has been found to vary as much as five-fold in New Zealand lakes,[178] and seven-fold in Alaskan lakes:[510] in agreement with theory, Secchi depth is particularly sensitive to turbidity.

A comprehensive account of the physical and physiological basis of the Secchi disc procedure was given by Preisendorfer (1986*a*, *b*), and the subject has also been reviewed by Højerslev (1986). Preisendorfer concluded that the primary function of the Secchi disc, and indeed its only legitimate *raison d'etre*, is to provide a simple visual index of water clarity in terms of Z_{SD}, or the inferred quantity $(c + K_d)$ (which is $\simeq 9/Z_{SD}$, and which could reasonably be referred to as the *contrast attenuation coefficient*[489,951]). In the same vein, Davies-Colley & Vant (1988), on the basis of their own extensive studies on the relationship between Secchi depths and other optical properties in 27 New Zealand lakes, propose that this device should be recognized as measuring image attenuation, as distinct from diffuse light attenuation.

5.2 Scalar irradiance

Although some multicellular aquatic plants have their photosynthetic tissue so arranged as to achieve maximum interception of light incident from above, to the randomly oriented cells of the phytoplankton all light, from whatever direction it comes, is equally useful. Scalar irradiance (E_0), as we saw earlier (§1.3) is defined as the integral of radiance distribution at a point over all directions about the point: it is equivalent to the total radiant flux per m^2 from all directions at a given point in the medium. In the context of photosynthesis, information on the scalar irradiance is therefore somewhat to be preferred to data on irradiance.

The technology of measurement of scalar irradiance is essentially the same as that of irradiance with the exception of the collector. Since this must respond equally to light from all directions, it has to be spherical. Fig. 5.4 illustrates the principles involved in collectors used for scalar irradiance and for its downwelling and upwelling components. Like the collector of a normal irradiance meter, the spherical collector is made of diffusing plastic or opal glass so that it collects all incident photons with equal efficiency, regardless of the angle at which they encounter its surface. The light which penetrates the spherical collector can be transmitted to the photodetector via another, flat collector in the bottom of the sphere[381] or, in the case of a

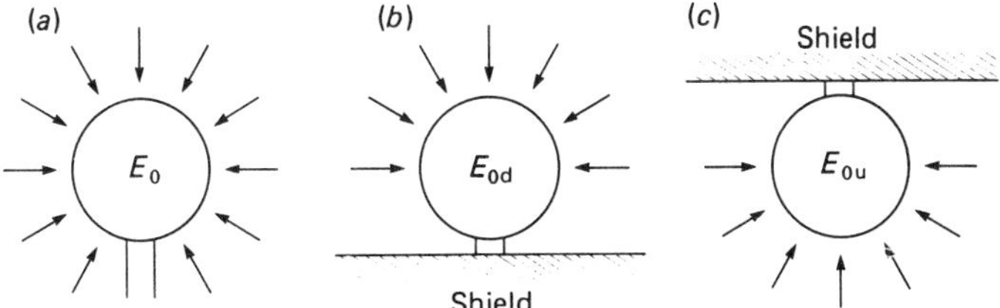

Fig. 5.4. Principles involved in collection of light for determination of (*a*) scalar irradiance, (*b*) downward scalar irradiance and (*c*) upward scalar irradiance. In the cases of downward and upward scalar irradiance, the collector must have a shield wide enough relative to the collector to ensure that very little upward and downward light, respectively, reaches the collector.

solid diffusing sphere, can be transmitted from the centre of the sphere by a fibre-optic bundle or quartz light-conducting rod.[81] Scalar irradiance meters can use a filter to confine the measured light to a narrow waveband,[381] or can be designed as quanta meters covering the whole (400–700 nm) photosynthetic range.[81,863] Fig. 5.5 shows two commercially available quantum (PAR) scalar sensors.

Højerslev (1975) describe a meter with two spherical collectors but with only half of each sphere exposed. One exposed hemisphere faces upwards, the other faces downwards. It is possible to show that the upper hemisphere collects radiant flux in proportion to $\frac{1}{2}(E_0+\vec{E})$ and the lower hemisphere collects flux in proportion to $\frac{1}{2}(E_0-\vec{E})$, where $\vec{E}$ is the net downward, or vector, irradiance (E_d-E_u). The signals from the two collectors are recorded separately, and by adding and subtracting these measurements, the scalar irradiance and net downward irradiance, respectively, are determined. The information provided by this meter can also be used to determine the absorption coefficient of the aquatic medium at the wavelength determined by the interference filter incorporated in the instrument. From measurements of $\vec{E}$ above and below a given depth, K_E, the vertical attenuation coefficient for net downward irradiance is obtained. The absorption coefficient is then calculated from $a=K_E\vec{E}/E_0$ (§1.7) using the values of $\vec{E}$ and E_0 at the depth in question.

Karelin & Pelevin (1970) developed a meter with three collectors: upward and downward irradiance collectors and a spherical collector for scalar irradiance. Any one of five colour filters could, by remote control, be placed in front of the photodetector. They used the data to determine the absorption coefficients of the medium, as well as characterizing the underwater light field.

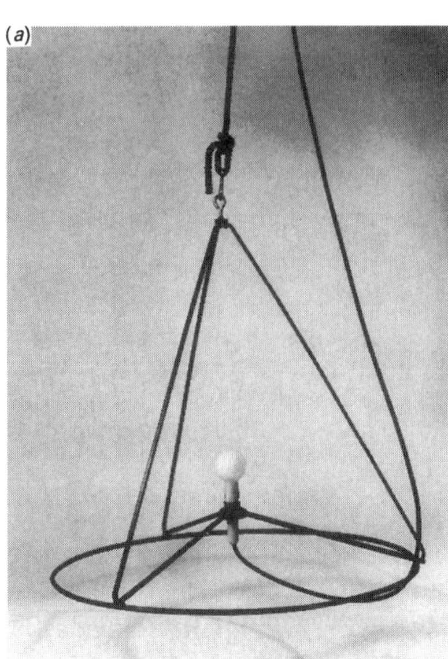

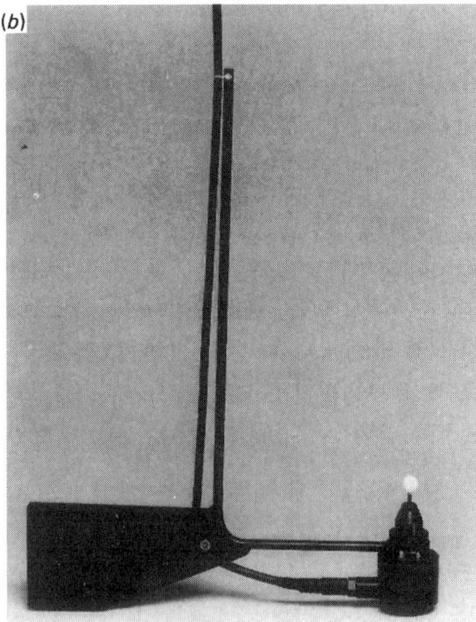

Fig. 5.5. Submersible scalar irradiance sensors. (*a*) *Li–Cor* Spherical Quantum Sensor LI–193SB, in lowering frame (Li–Cor Inc., Lincoln, Nebraska, USA). (*b*) *Biospherical* Quantum Scalar Irradiance Sensor QSP-200, in lowering frame (courtesy of Biospherical Instruments Inc., San Diego, California, USA).

5.3 Spectral distribution of irradiance

Any photon with wavelength from 400 to 700 nm is in principle *available* for photosynthesis: however, the probability of a given photon being captured by the photosynthetic biomass within a water body will vary markedly with wavelength in accordance with the absorption spectrum of the array of photosynthetic pigments present. It is quite possible for two different light fields to have the same irradiance for PAR but, because of their differing spectral distributions, to differ significantly in their usefulness to a particular type of alga. Thus, a better assessment of the value of a particular light field for photosynthesis can be made if the way in which irradiance or scalar irradiance varies across the spectrum is known. An instrument which measures this is known as a spectroradiometer.

A spectroradiometer is an irradiance or scalar irradiance meter in which a variable monochromator is interposed between the collector and the photodetector. In one commercially available spectroradiometer, monochromaticity is achieved by means of 12 separate silicon photodiodes, each with its own interference filter, and measuring irradiance at 12 wavebands each of 10 nm in the 400–700 nm range.[68] These are scanned electronically, so that a complete scan can be carried out in 6 ms. In another version of the instrument, simultaneous measurements of upwelling radiance in several wavebands are also made.[68,836]

To provide data of greater spectral resolution than is possible with interference filters (10–18 nm), a grating monochromator can be used[48,501,943] to give a continuous scan of spectral irradiance (resolution ~5 nm) across the whole photosynthetic range, and an instrument of this type is commercially available.[561] To scan across the complete spectrum using a grating monochromator and a single photodetector takes of the order of some seconds, which can be a disadvantage in a fluctuating light field. In an instrument developed at the Indian National Institute of Oceanography,[188] the spectrum dispersed from the grating is focussed across a linear photodiode array which is scanned electronically, and in this way high spectral resolution (~3 nm) is combined with a very short scan time (12 ms).

As a result of the world-wide depletion of stratospheric ozone, and the development of a hole in the ozone layer over Antarctica, there is now greatly increased interest in the penetration of ultraviolet radiation into the sea, but this is difficult to measure both because of the low levels of energy in the UV under water, and of the very rapid change of irradiance with wavelength in this spectral region. To specifically address this problem,

Smith *et al.* (1992) have developed a new underwater spectroradiometer which measures irradiance from 250 to 350 nm with 0.2 nm resolution, and from 350 to 700 nm with 0.8 nm resolution.

5.4 Radiance distribution

To understand fully the underwater light field we need a detailed knowledge of the angular distribution of radiant flux at all depths. Some information on the angular structure of the field can be derived from the measurements of irradiance discussed above. Irradiance reflectance ($R = E_u/E_d$) is a crude measure of angular structure. More information is contained in the three average cosines: $\bar{\mu}$ for the total light field, $\bar{\mu}_d$ for the downwelling stream, and $\bar{\mu}_u$ for the upwelling stream, all of which can be derived from irradiance and scalar irradiance values

$$(\bar{\mu} = \vec{E}/E_0, \bar{\mu}_d = E_d/E_{0d}, \bar{\mu}_u = E_u/E_{0u}).$$

For a complete description, however, we need the value of radiance at all vertical and azimuthal angles, i.e. the radiance distribution. Radiance is a measure of the radiant intensity (per unit projected area) in a specified direction at a particular point in the field, and so should be measured with a meter which can be oriented in that direction and which ideally should have an infinitesimally small acceptance angle. In reality, to ensure that the meter collects enough light to measure with reasonable accuracy, particularly in the dim light fields existing at greater depths, acceptance angles of 4–7° may be used. The simplest type of radiance meter is known as a Gershun tube photometer, after the Russian physicist Gershun, who made notable contributions to our understanding of the structure of light fields in the 1930s. It uses a cylindrical tube to limit the acceptance angle; the photodetector, with a filter in front of it, is at the bottom (Fig. 5.6). To achieve the necessary sensitivity, a photomultiplier would normally be used as the detector.

The most difficult problem encountered in measurement of underwater radiance is that of accurately controlling the zenith and azimuth angle of the photometer tube: quite complex systems involving built-in compasses, servo mechanisms and propellers to rotate the instrument are required.[422,787,937] Another problem is that to determine a complete radiance distribution with a radiance meter is likely to take quite a long time, during which the position of the Sun will change. For example, if radiance is measured at intervals of 10° for the zenith angle (0–180°) and 20° for the azimuth angle (0–180° is sufficient, the distribution being symmetrical

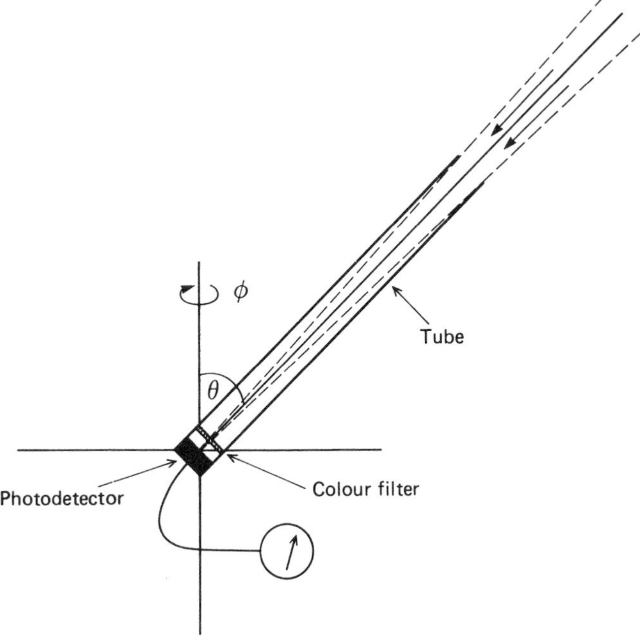

Fig. 5.6. Schematic diagram of a radiance meter.

about the plane of the Sun) then 190 separate values must be recorded. Smith, Austin & Tyler (1970) have developed a photographic technique by means of which radiance distribution data can be recorded quickly. Their instrument consists of two cameras placed back to back, each equipped with a 180°-field-of-view lens. One camera photographs the upper and the other photographs the lower hemisphere. By densitometry of the film, the relative radiance values over all zenith and azimuth angles can be determined at whatever degree of resolution is required, at a later time. In a further refinement of the same principle, Voss (1989) has developed an electro-optic radiance distribution camera system, in which the radiance distribution is recorded, not on film, but with a 260 × 253 pixel electro-optic charge injection device.

5.5 Modelling the underwater light field

The nature of the underwater light field resulting from a given incident light field is determined by the inherent optical properties of the aquatic medium. In principle, therefore, it should be possible, if we know the inherent optical properties, to calculate the properties of the underwater light field. Such a calculation procedure could be used, for example, to explore in more detail than would be practicable by measurement the exact way in which the nature of the light field depends on the inherent optical properties. It could be used as a substitute for *in situ* measurement in environments in which such measurement might be difficult. It could also be used for predicting the effects on the underwater light field of anticipated changes in the optical properties of the water, resulting perhaps from the activities of man, such as, for example, the discharge of wastewater into a water body.

In reality, the complex behaviour of the photon population in the water caused by the combined effects of scattering and absorption precludes the establishment of an explicit analytical relation between the properties of the field and those of the medium. There are, however, computer modelling procedures by means of which it is possible, by making physically realistic assumptions about the ways in which the behaviour of light at any point in a water body is determined by the scattering and absorption properties of the medium, to calculate the nature of the light field that will be set up throughout the whole water column.

The simplest calculation procedure, and the one which has been used most, is the Monte Carlo method. Its application to the behaviour of solar photons in the ocean was pioneered by Plass & Kattawar (1972), and it has also been put to valuable use by Gordon and coworkers.[315–319] The principles are described in some detail in Kirk (1981*a*, *c*). A brief account will be given here.

The behaviour of individual photons within a scattering/absorbing medium is stochastic in nature. The lifetime of, and the geometrical path followed by, any given photon are governed by its random encounters with absorbing molecules or algal cells and with scattering particles. The inherent optical properties of the medium – the absorption and scattering coefficients, the volume scattering function – are not only (in accordance with their definitions) measures of the proportion of an incident flux absorbed, scattered and so on per unit depth by a thin layer of medium, but are also related in simple ways to the probability that any given photon will, within a certain distance, be absorbed, scattered, or scattered at a certain angle.

The Monte Carlo method takes advantage of the statistical nature of photon behaviour and uses a computer to follow the fate of a large number of photons, one at a time, passing into an imaginary body of water (corresponding, if desired, to some real body of water) with specified optical properties. Random numbers are used in conjunction with appropriate cumulative frequency distributions (based on the optical properties) to choose pathlengths between each interaction with the medium, to decide whether the interaction is one of scattering or absorption, to select the scattering angle, and so on. Photons are introduced from above the surface, at an angle or a selection of angles appropriate for the lighting conditions (direct Sun, overcast sky, etc.) under consideration. Refraction at the surface is allowed for. If it is wished to take account of waves, then the surface slope/wind frequency distribution data of Cox & Munk (1954) can be used. Each photon is followed, its trajectory being recalculated after each scattering event until it is absorbed or scattered up through the surface again. Within the imaginary water body there is a series of depth markers. For every single trajectory of each photon the computer records which depth markers the photon passes, in what direction (upwards or downwards) and at what angle. When a large number (say 10^6) of photons has been processed, the computer can calculate from the accumulated data the values at each depth of irradiance and scalar irradiance (upwards and downwards), reflectance, average cosines (downwards, upward, total) and radiance distribution (at a set of azimuth angles or averaged over all azimuth angles). In this way a complete description of the underwater light field is obtained.

It will very often be the case that something much less than a complete description of the light field is required: perhaps just the vertical attenuation coefficient for irradiance in the photosynthetic waveband (K_d[PAR]), or the subsurface reflectance ($R[0, \lambda]$) in certain wavebands, or the visual clarity of the water body expressed in terms of the Secchi depth (Z_{SD}). In such cases, rather than carrying out the full computer simulation, much simpler calculations can be carried out using certain empirical relationships – expressing K_d, or R, or Z_{SD} as approximate functions of the inherent optical properties – that have arisen out of computer simulation of the light field in waters of various optical types (see §6.7).

As we noted above, one of the practical applications of modelling of the underwater light field is the prediction of the effects of discharge of wastewater on the optical character of the receiving water body.[491,493] Laboratory measurements of absorption and scattering on the wastewater and the receiving water, together with data on anticipated dilution rates, will make it possible to calculate the values of a and b in the water body before and after discharge of the wastewater. With this information, together with an appropriate scattering phase function, the light field that would exist (under some standard meteorological conditions) in the water body with and without the added wastewater can be calculated by Monte Carlo modelling. Alternatively, if it is only certain aspects of the field, such as K_d(PAR), $R(0, \lambda)$ or Z_{SD}, which are required, the empirical relationships referred to above may be used. In this way an industrial water user, for example, can assess what effect its effluent will have on certain key indicators of optical water quality in a surface water body, before that effluent is ever discharged.

6
The nature of the underwater light field

Having in earlier chapters considered the nature of the solar radiation flux incident on the surface of the aquatic ecosystem, and the influences to which the light is subject once it enters the water, we shall now discuss the kind of underwater light field that results.

6.1 Downward irradiance – monochromatic

As a result of absorption and scattering of the solar flux, the downward irradiance, E_d, of the light field diminishes with depth. In Fig. 6.1, E_d for greenish-yellow light, expressed as a percentage of the value just below the surface, is plotted against depth in a fresh-water impoundment. Irradiance diminishes in an approximately exponential manner in accordance with

$$E_d(z) = E_d(0)\ e^{-K_d z} \tag{6.1}$$

or

$$\ln E_d(z) = -K_d z + \ln E_d(0) \tag{6.2}$$

where $E_d(z)$ and $E_d(0)$ are the values of downward irradiance at z m and just below the surface, respectively, and K_d is the vertical attenuation coefficient for downward irradiance.

In Fig. 6.2*a–d*, the logarithm of E_d is plotted against depth for red, green and blue light in one oceanic and three inland waters. The graphs are approximately linear, in accordance with eqn 6.2, but significant divergence from linearity is in some cases apparent. In the oceanic water for example, it can be seen that the green and the blue light are attenuated more rapidly below about 10 m than above. This may be attributed to the downward flux becoming more diffuse, less vertical in its angular distribution with depth, with a consequent increase in attenuation (see §§6.6, 6.7).

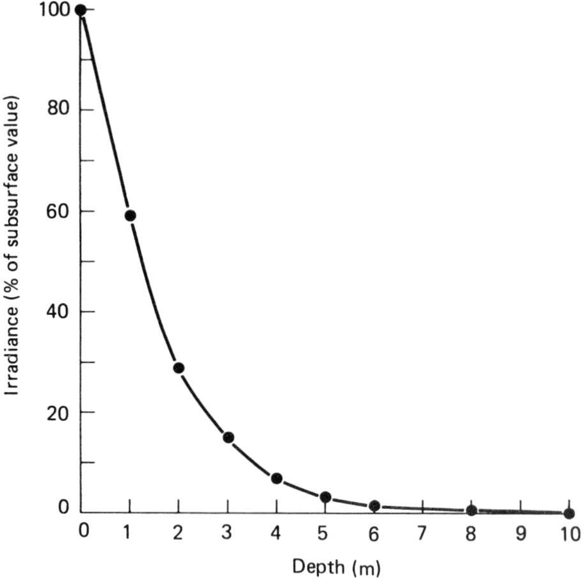

Fig. 6.1. Exponential diminution of downward irradiance of greenish-yellow (580 nm) light with depth in a fresh-water body (Burrinjuck Dam, NSW, Australia, 8th December 1977). (Kirk, unpublished data.)

The relative rates of attenuation in the different wavebands are determined largely by the absorption spectrum of the aquatic medium. In the tropical Pacific Ocean (Fig. 6.2*a*), and in non-productive oceanic waters generally, where water itself is the main absorber, blue and green light both penetrate deeply and to about the same extent, while red light, which water absorbs quite strongly, is attenuated much more rapidly. In the productive waters of oceanic upwelling areas, blue light is attenuated more strongly than green light, due to absorption by phytoplankton pigments,[635] but still not as strongly as red light. In coastal waters, which contain more yellow substances and phytoplankton than normal oceanic waters, green is again the most penetrating waveband. Only in the most coloured coastal waters, influenced by major river discharge (Jerlov's types 7–9), however, is blue light attenuated as strongly as red light.

In contrast to the sea, in fresh water the blue waveband is usually the most strongly attenuated (Fig. 6.2*b*–*d*), because of the higher levels of yellow substances that typically occur in inland waters. Green is usually the most penetrating waveband in inland waters, followed by red (Fig. 6.2*b*). When the concentration of yellow materials is high, however, red light may

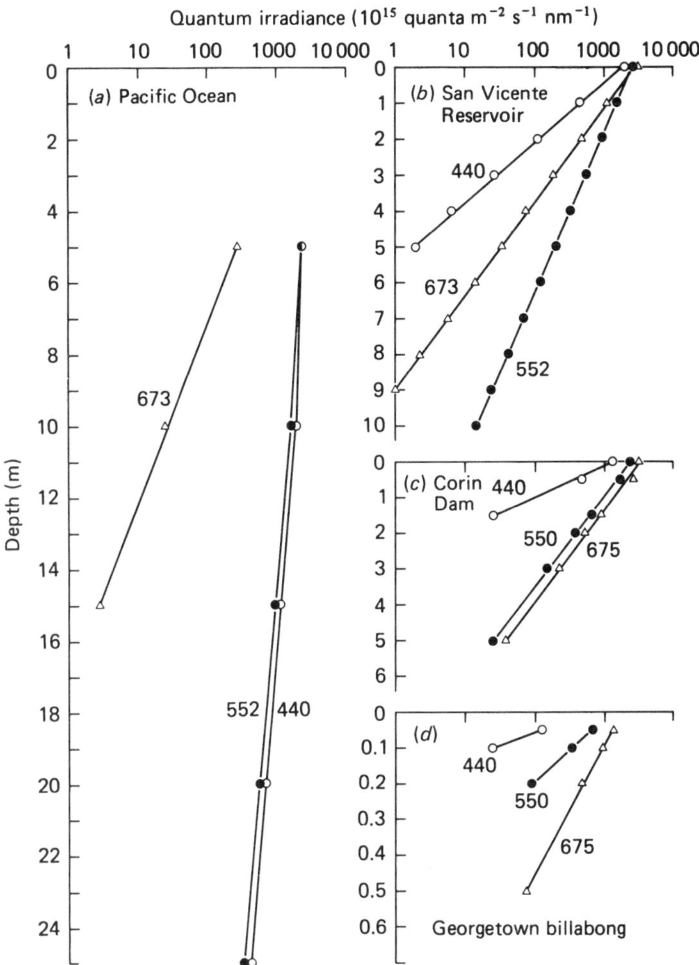

Fig. 6.2. Depth profiles for downward irradiance of blue (○), green (●) and red (△) light in a tropical oceanic and three inland fresh-water systems. (*a*) Pacific Ocean, 100 km off Mexican coast (plotted from data of Tyler & Smith, 1970). (*b*) San Vicente Reservoir, California, USA (plotted from data of Tyler & Smith, 1970). (*c*) Corin Dam, ACT, Australia (Kirk, unpublished data). (*d*) Georgetown billabong, Northern Territory, Australia (plotted from unpublished data of P. A. Tyler).

All measurements were taken under sunny conditions. Note the expanded depth scale in (*d*). The wavelength (nm) at which measurements were made is indicated alongside each graph.

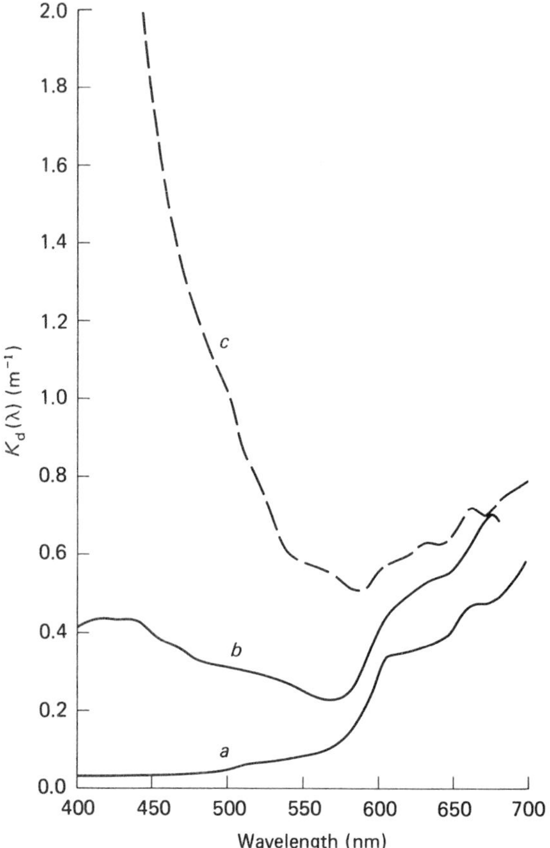

Fig. 6.3. Spectral variation of vertical attenuation coefficient, $K_d(\lambda)$ for downward irradiance in (*a*) the Gulf Stream (Atlantic Ocean) off the Bahama Islands (plotted from data of Tyler & Smith, 1970). (*b*) The Mauritanian upwelling area (8.0–9.9 mg chl *a* m^{-3}) off the West African coast (plotted from data of Morel, 1982). (*c*) Burrinjuck Dam, southern tablelands of NSW, Australia (plotted from data of Kirk, 1979).

penetrate as far as green (Fig. 6.2*c*), and in very yellow waters red light penetrates best of all (Fig. 6.2*d*).

Fig. 6.3 compares the spectral variations of the vertical attenuation coefficient for irradiance across the photosynthetic range in an unproductive oceanic water, a productive (upwelling) oceanic water and an inland impoundment.

As the solar altitude decreases, so the pathlength of the unscattered solar beam within the water, per metre depth, increases in proportion to cosec α, where α is the angle of the solar beam within the water. We may therefore expect K_d to increase with diminishing solar altitude. Such an effect is observed but is small:[422] it is only of significance in the upper layer of very clear waters. Baker & Smith (1979) found that in an inland impoundment, K_d at several wavelengths varied by not more than 5% between solar altitudes of 80 and 50°, and by not more than 18% between 80 and 10°.

6.2 Spectral distribution of downward irradiance

The data in Fig. 6.2 show that attenuation of solar radiation in the whole photosynthetic waveband takes place at widely different rates in different parts of the spectrum. As a consequence, the spectral composition of the downwelling flux changes progressively with increasing depth. Fig. 6.4*a* shows the changes down to 25 m depth in the clear oceanic water of the Gulf Stream. Most of the light below about 15 m is confined to the blue–green, 400–550 nm, waveband with the peak occurring in the blue region at about 440–490 nm. In the coastal/estuarine water of Batemans Bay, Australia, attenuation due to yellow substances in the blue region is comparable to that due to water at the red end of the spectrum, so that at 4 m depth, while there is still substantial radiant flux throughout the photosynthetic range, the distribution peaks markedly at about 570 nm (Fig. 6.4*b*). In European coastal waters, Halldal (1974) found spectral distributions rather similar to that in Fig. 6.4*b*, with a marked peak at about 570 nm.

In inland waters with their, usually, higher concentration of yellow substances, the rapid attenuation at the short-wavelength end of the spectrum means a virtually complete removal of blue light within quite shallow depths. In Lake Burley Griffin, Australia, for example, at a time of low water turbidity (Fig. 6.4*c*) there was essentially no blue light still present below about 2 m depth. In the lower half of the euphotic zone (defined in the following section) in such waters, the photosynthetically available flux typically consists of a broad band extending from the green to the red, often with a peak in the yellow at about 580 nm. Near the surface of the water there are significant levels of blue light available for photosynthesis, but taking the euphotic zone as a whole, the total amount of blue light available for photosynthesis is greatly reduced in such waters (see §6.5).

In those coloured, turbid waters in which there are high levels of humic material in the particulate fraction as well as in the soluble state, green light is also attenuated rapidly (Fig. 6.2*d*), with the result that the downward flux in the lower part of the, very shallow, euphotic zone consists largely of

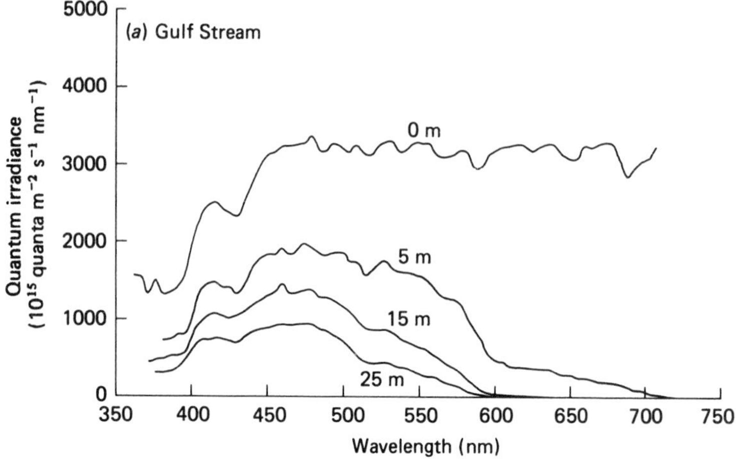

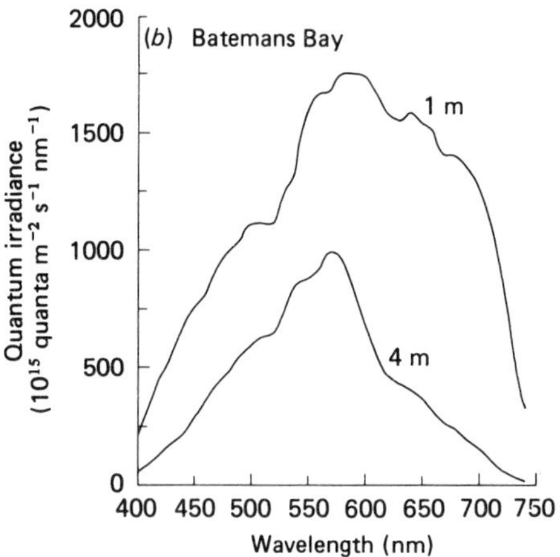

orange–red (600–700 nm) light. In the spectral distribution for Lake Burley Griffin under turbid conditions, shown in Fig. 6.4*d*, the peak is right over at the far-red end of the spectrum.

6.3 Downward irradiance – PAR

As a broad indication of the availability of light for photosynthesis in an aquatic ecosystem, information on the penetration of the whole photosynthetic waveband is of great value. As solar radiation penetrates a water body, it becomes progressively impoverished in those wavelengths which the aquatic medium absorbs strongly and relatively enriched in those

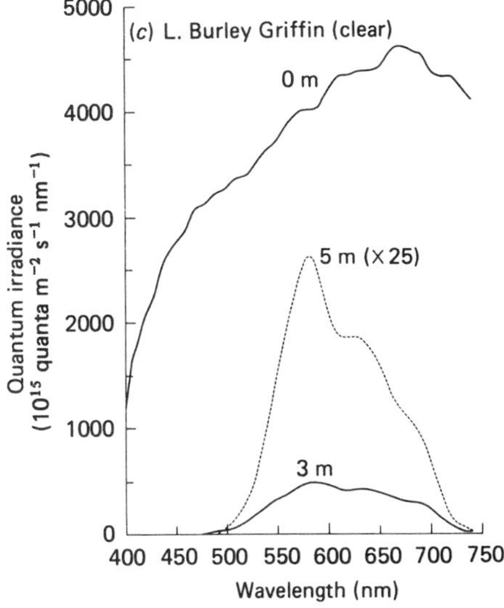

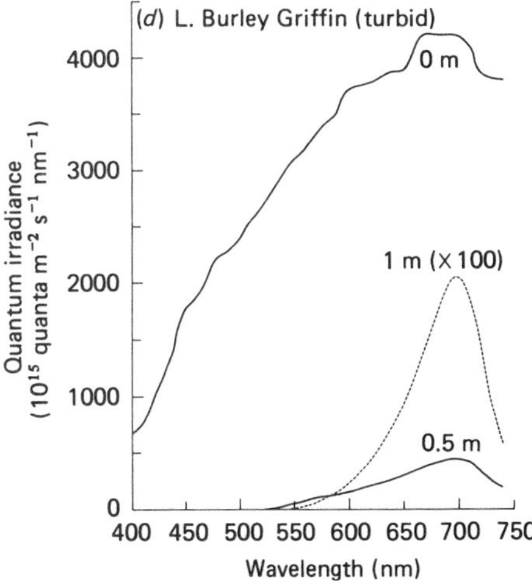

Fig. 6.4. Spectral distribution of downward irradiance in marine and inland water. (*a*) The Gulf Stream (Atlantic Ocean) off the Bahama Islands (plotted from data of Tyler & Smith, 1970). (*b*) Batemans Bay, NSW, Australia (after Kirk, 1979). (*c*) Lake Burley Griffin, ACT, Australia, 29th September, 1977; water comparatively clear (turbidity = 3.7 NTU) (after Kirk, 1979). (*d*) Lake Burley Griffin, 6th April 1978; water turbid (turbidity = 69 NTU) (after Kirk, 1979).

wavelengths which are absorbed weakly. We would therefore expect the attenuation coefficient for total photosynthetically available radiation to be higher in the upper few metres and to fall to a lower value with increasing depth. This change in the rate of attenuation of PAR with depth can readily be observed in most marine waters and the clearer inland systems: two of the curves in Fig. 6.5 – for the Tasman Sea, and for a relatively clear lake – show the increase in slope of the log E_d curve with increasing depth. The curve eventually becomes approximately linear, indicating that the downward flux is now confined to wavebands all with about the same, relatively low, attenuation coefficient. In oceanic waters the light in this region is predominantly blue–green (Fig. 6.2*a*), whereas in inland waters the penetrating waveband is likely either to be green (Fig. 6.2*b*), to extend from the green to the red (Fig. 6.2*c*) or to be predominantly red (Fig. 6.2*d*).

A countervailing tendency, which exists at all wavelengths, is for attenuation to increase with depth as a result of the downward flux becoming more diffuse, due to scattering. By counteracting the effect of changes in spectral composition, it may partly explain why graphs of log E_d against depth for turbid waters are so surprisingly linear (Fig. 6.5, L. Burley Griffin), and lack the biphasic character seen in the clearer waters. However, since high turbidity is commonly associated with increased absorption at the blue end of the spectrum (see §3.3) it is also true that in such waters the blue waveband is removed at even shallower depths than usual and so the change in slope of the curve occurs quite near the surface and is not readily detectable.

Even when, as in the clearer waters, the graph of log E_d against depth is noticeably biphasic, the change of slope is usually not very great. Thus, the attenuation of total PAR with depth is nearly always approximately, and often accurately, exponential in agreement with eqns 6.1 and 6.2. Attenuation of PAR in a given water body can therefore generally be characterized by a single value of K_d, or, at worst, by two values, one above and one below the change in slope. The vertical attenuation coefficient for downward irradiance of PAR provides a convenient and informative parameter in terms of which to compare the light-attenuating properties of different water bodies. Table 6.1 presents a selection of values, including some obtained by summation of spectral distribution data across the photosynthetic range. Oceanic waters have the lowest values of K_d(PAR) as might be expected from their low absorption and scattering. Inland waters, with rare exceptions such as Crater Lake in Oregon, USA, have much higher values, with coastal and estuarine waters coming in between. The highest values are found in very turbid waters (e.g. L. George and Georgetown billabong in

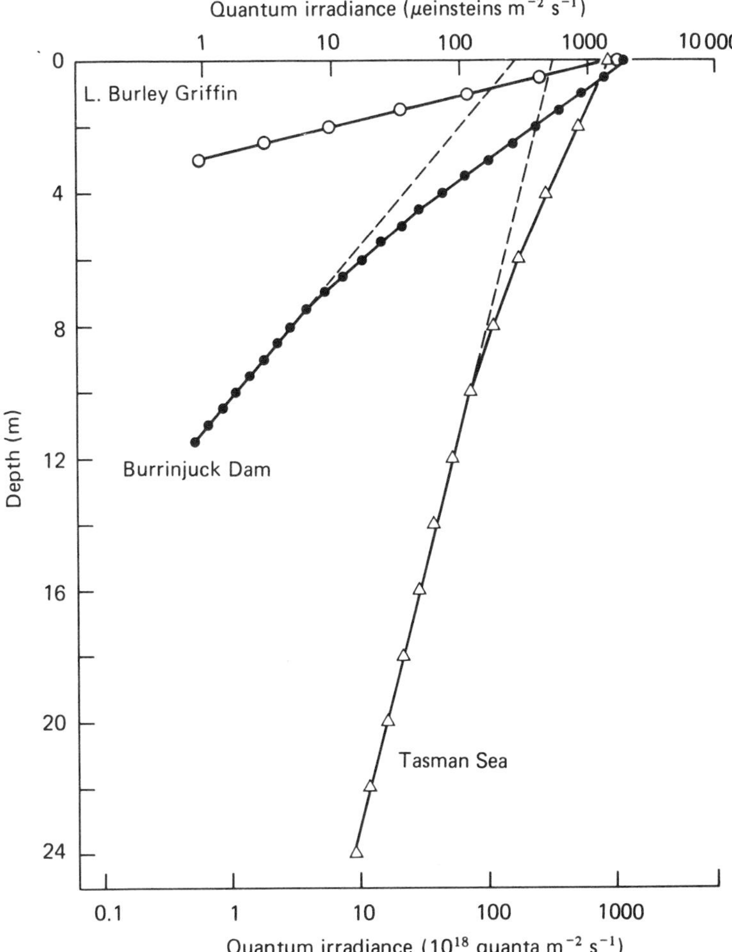

Fig. 6.5. Attenuation of downward quantum irradiance of PAR with depth in a coastal water (Tasman Sea, off Batemans Bay, NSW) and two inland waters (Lake Burley Griffin, ACT; Burrinjuck Dam, NSW) in Australia (Kirk, 1977*a*, and unpublished). The marked decrease in rate of attenuation in Burrinjuck Dam below about 7 m is particularly noteworthy: spectroradiometric measurements showed that most of the light below this depth was confined to the 540–620 nm (green–yellow) waveband.

Australia) in which the suspended tripton strongly absorbs, as well as scatters the light. High values ($> 2.0 \text{ m}^{-1}$) may also be associated with dense algal blooms (Sea of Galilee – *Peridinium*; L. Simbi, Kenya – *Spirulina*), with intense soluble yellow colour but low scattering (L. Pedder, Tasmania), or with a combination of high soluble colour and scattering (L. Burley

Table 6.1. *Vertical attenuation coefficients for downward quantum irradiance of PAR in some marine and fresh waters. Where several measurements have been taken, the mean value, the standard deviation, the range and the time period covered are in some cases indicated*

Water body	K_d(PAR) (m^{-1})	Reference
I Oceanic waters		
Atlantic Ocean		
Sargasso Sea	0.03	939
Sargasso Sea		
5 m	0.098	837
35 m	0.073	837
75 m	0.046	837
Gulf Stream, off Bahamas	0.08	943
Tropical East Atlantic		
Guinea Dome	0.08–0.096	635
Mauritanian upwelling, offshore	0.16–0.38	635
Mauritanian upwelling, coastal	0.20–0.46	635
Pacific Ocean		
Off Oahu, Hawaii	0.032	66
100 km off Mexico	0.11	943
Eastern North Pacific (33° N, 142° W) (2-week period)	0.112–0.187	819
II Coastal and estuarine waters		
Europe		
North Sea		
Offshore Netherlands	0.41	755
Dogger Bank	0.06–0.15	755
Ems–Dollard estuary (Netherl./Germany)		
Inner region	~7	158

Outer region	~1	158
Bjornafjord, Norway	0.15	423
Shannon estuary, Ireland		
Upper	1.8–8.6	605
Middle	1.7–4.5	605
Lower	0.35–1.8	605
N. America		
Gulf of California	0.17	943
Chesapeake Bay, Rhode R. mouth	1.10–2.05	282
Delaware R. estuary	0.6–5.0	886
Hudson R. estuary, N.Y.		
av. 10 stations, July	2.02	882
San Francisco Bay		
Shallows, inner estuary	10–13	152
Outer estuary	~1	152
Georgia Embayment		
St Catherine's Sound	2.9	670
8 km offshore	1.8	670
30 km offshore	0.27	670
60 km offshore	0.09	670
Fraser R., Strait of Georgia (Canada)		
River mouth	0.8	354
Porlier Pass	0.27	354
Australia/New Zealand		
Tasman Sea, coastal New South Wales	0.18	479
Port Hacking estuary, NSW , Australia	0.37	803
Clyde R. estuary, NSW, Australia	0.71	479
Australian coastal sea lakes, NSW		
Lake Macquarie	0.55 ± 0.09	804
Tuggerah Lakes	1.25 ± 0.18	804

Table 6.1 (*cont.*)

Water body	K_d(PAR) (m^{-1})	Reference
New Zealand estuaries		
9 estuaries, North Island, mouth sites, low water data	0.3–1.1	949
III Inland waters		
N. America		
Great Lakes		
L. Superior	0.1–0.5	425
L. Huron	0.1–0.5	425
L. Erie	0.2–1.2	425, 866
L. Ontario	0.15–1.2	425
Irondequoit Bay (L. Ontario)	1.03 ± 0.11	980
Finger lakes, N.Y.		
Otisco	0.564 ± 111	216
Seneca	0.468 ± 0.075	216
Skaneateles	0.238 ± 0.029	216
Crater L., Oregon	0.06	943
San Vicente reservoir, Calif.	0.64	943
L. Minnetonka, Minn.	0.7–2.8	608
McConaughy reservoir, Nebr.	1.6 (av.)	768
Yankee Hill reservoir, Nebr.	2.5 (av.)	768
Pawnee reservoir, Nebr.	2.9 (av.)	768
Alaskan lakes		
44 clear lakes, little colour	0.31 ± 0.12	510
21 clear lakes, yellow	0.70 ± 0.07	510
23 turbid lakes, little colour	1.63 ± 1.51	510

Europe		
L. Zurich (10-month period)	0.25–0.65	795
Esthwaite Water, England	0.8–1.6	352
Loch Croispol, Scotland	0.59	855
Loch Uanagan, Scotland	2.35	855
Forest lakes, Finland		
Nimeton	3.45	444
Karkhujarvi	2.49	444
Tavilampi	1.75	444
Sea of Galilee	0.5	205
Sea of Galilee (*Peridinium* bloom)	3.3	205
Africa		
L. Simbi, Kenya	3.0–12.3	609
L. Tanganyika	0.16 ± 0.02	364
Volcanic lakes, Cameroon		
Barombi Mbo	0.148	507
Oku	0.178	507
Wum	0.305	507
Beme	0.353	507
South African impoundments		
Hartbeespoort	0.67	971
Rust de Winter	1.70	971
Bronkhorstspruit	4.23	971
Hendrik Verwoerd	13.1	971
Australia		
(a) Southern Tablelands		
Corin Dam	0.87	495*a*
L. Ginninderra	1.46 ± 0.68	479, 495*a*
(3 year range)	0.84–2.74	
Burrinjuck Dam	1.65 ± 0.81	479, 495*a*
(6 year range)	0.71–3.71	

Table 6.1 (*cont.*)

Water body	K_d(PAR) (m^{-1})	Reference
L. Burley Griffin	2.81 ± 1.45	479, 495*a*
(6 year range)	0.86–6.93	
L. George	15.1 ± 9.3	479, 495*a*
(5 year range)	5.7–24.9	
(b) Murray–Darling system		
Murrumbidgee R., Gogeldrie Weir (10 months)	1.4–8.0	677
Murray R., upstream of Darling confluence	1.85–2.16	677
Darling R., above confluence with Murray	2.78–8.6	677
(c) Snowy Mountains impoundments		
Blowering	0.48	804
Eucumbene	0.38	804
Jindabyne	0.49	804
Talbingo	0.46	804
(d) Southeast Queensland coastal dune lakes		
Wabby	0.48	89
Boomanjin	1.13	89
Cooloomera	3.15	89
(e) Northern Territory (Magela Creek billabongs)		
Mudginberri	1.24	498
Gulungul	2.21	498
Georgetown	8.50	498

(f) Tasmania (lakes)		
Perry	0.21	90
Ladies Tarn	0.41	90
Barrington	1.23	90
Gordon	1.86	90
Pedder	2.39	90
New Zealand (lakes)		
Taupo	0.14	179
Rotokakahi	0.32	178
Ohakuri	0.40	178
Rotorua	0.90	178
D	2.30	178
Hakanoa	12.1	178

Griffin, Australia). In shallow lakes, resuspension of bottom sediments by wind-induced wave action can increase the attenuation coefficient several-fold, and if the sediments contain a substantial proportion of clay particles then the increased attenuation can last for a week or so after the initial storm event.[365]

A useful, if approximate, rule-of-thumb in aquatic biology is that significant phytoplankton photosynthesis takes place only down to that depth, z_{eu}, at which the downwelling irradiance of PAR falls to 1% of that just below the surface. That layer within which E_d(PAR) falls to 1% of the subsurface value is known as the *euphotic zone*. Making the assumption that K_d(PAR) is approximately constant with depth, the value of z_{eu} is given by $4.6/K_d$. This, as we have seen, is a reasonable assumption for the more turbid waters and so will give useful estimates of the depth of the euphotic zone in many inland, and some coastal, systems. In the case of those clear marine waters in which there is a significant increase in slope of the log E_d(PAR) *versus* depth curve, a value of K_d(PAR) determined in the upper layer could give rise to a substantial underestimate of the euphotic depth.

Another useful reference depth is z_m, the mid-point of the euphotic zone. This, by definition, is equal to $\frac{1}{2}z_{eu}$: given the approximately exponential nature of the attenuation of PAR with depth, it follows that $z_m \simeq 2.3/K_d$, and corresponds to that depth at which downward irradiance of PAR is reduced to 10% of the value just below the surface.

6.4 Upward irradiance

As a result of scattering within the water, at any depth where there is a downward flux there is also an upward flux. This is always smaller, usually much smaller, than the downward flux but at high ratios of scattering to absorption can contribute significantly to the total light available for photosynthesis. Furthermore, in any water the upwelling light is of crucial importance for the remote sensing of the aquatic environment (Chapter 7), since it is that fraction of the upward flux which penetrates the surface that is detected by the remote sensors.

At any depth, the upward flux can be regarded as that fraction of the downward flux at the same depth which at any point below is scattered upwards and succeeds in penetrating up to that depth again before being absorbed or scattered downwards. Thus we might expect the irradiance of the upward flux to be linked closely to the irradiance of the downward flux; this is found to be the case. Fig. 6.6 shows the upward and downward irradiance of PAR diminishing with depth in parallel together in an

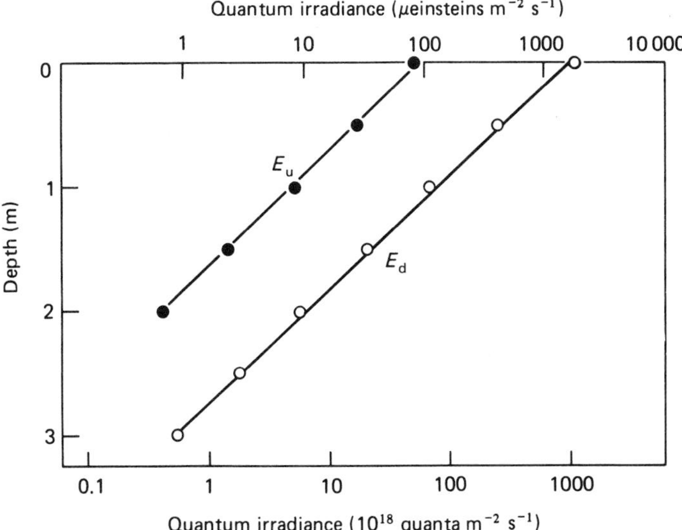

Fig. 6.6. Parallel diminution of upward (●) and downward (○) irradiance of PAR with depth in an Australian lake (Burley Griffin, ACT) (after Kirk, 1977*a*).

Australian lake. Changes in downward irradiance associated with variation in solar altitude, or cloud cover would also be accompanied by corresponding changes in upward irradiance. Given the close dependence of upward irradiance on downward irradiance, it is convenient to consider any effects that the optical properties of the water may have on the upwelling flux in terms of their influence on the ratio of upwelling to downwelling irradiance, E_u/E_d, i.e. irradiance reflectance, R.

A small proportion of the upward flux originates in forward scattering, at large angles, of downwelling light that is already travelling at some angular distance from the vertical. As solar altitude decreases and the solar beam within the water becomes less vertical, there is an increase in that part of the upward flux which originates in forward scattering. The shape of the volume scattering function is such that this more than counterbalances the fact that an increased proportion of backscattered flux is now directed downwards. The net result is that irradiance reflectance increases as solar altitude decreases, but the effect is not very large. In the Indian Ocean, R for 450 nm light at 10 m depth increased from 5.2% to 7.0% as the solar altitude decreased from 80 to 31°.[422] For Lake Burley Griffin in Australia, irradiance reflectance of the whole photosynthetic waveband (400–700 nm) just below the surface increased from 4.6% at a solar altitude of 75° to 7.9%

at a solar altitude of 27°. However, at 1 m, by which depth in this turbid water the light field is well on the way to reaching the asymptotic state (see §6.6), R increased only from 7.4 to 8.6%.[479]

Although as we have seen, there is a contribution from forward scattering, most of the upwelling flux originates in backscattering. Thus we might expect R to be approximately proportional to the backscattering coefficient, b_b, for the water in question. As the upwards-scattered photons travel up from the point of scattering to the point of measurement, their numbers are progressively diminished by absorption and – less frequently – by further backscatterings which redirect them downwards again. We may therefore expect that reflectance will vary inversely with the absorption coefficient, a, of the water. We might also expect that the dominant tendency of reflectance to increase with backscattering will be somewhat lessened by the contribution of backscattering to the diminution of the upward flux.

The actual manner in which irradiance reflectance varies with the inherent optical properties of the medium has been explored by mathematical modelling of the underwater light field. A simplified version of radiative transfer theory[211,644] leads to the conclusion that R is proportional to $b_b/(a+b_b)$, for media (such as most natural waters) in which $b_b \ll a$. This is the kind of relation that might be anticipated on the qualitative grounds outlined above. In fact, since b_b is generally so much smaller than a, we might expect that to a reasonable approximation, $R(0)$ should be proportional simply to b_b/a. Numerical modelling of the underwater light field for waters of various optical types, by Monte Carlo and other methods,[319,484,729] reveals that this is indeed the case and we can write

$$R(0)=C(\mu_0)b_b/a \tag{6.3}$$

The constant of proportionality, $C(\mu_0)$, is itself a function of solar altitude, which we can express in terms of μ_0, the cosine of the zenith angle of the refracted solar beam, below the surface. For any given water body it is the case that reflectance increases as solar altitude decreases,[316,319,488,641] i.e. $C(\mu_0)$ increases as μ_0 decreases, and indeed can be expressed as an approximate linear function of $(1-\mu_0)$,[488,494] $1/\mu_0$[426] or $[(1/\mu_0)-1]$,[316] in, for example, a relationship such as

$$C(\mu_0)=M(1-\mu_0)+C(1.0) \tag{6.4}$$

where $C(1.0)$ is the value of $C(\mu_0)$ for zenith sun ($\mu_0=1.0$) and M is a coefficient whose value is determined by the shape of the scattering phase function.[316,494] It turns out to be the case for zenith sun that the constant of proportionality in eqn 6.3, i.e. $C(1.0)$, is approximately equal to 0.33,[319,484,729] and this remains true for waters with a wide range of scattering phase functions.[494]

Morel & Prieur (1977) compared their measurements of R across the photosynthetic spectrum with the values calculated using eqn 6.3, for a variety of oceanic and coastal waters. For clear blue oceanic waters, agreement between the observed and the calculated curves of spectral distribution of R was good. For upwelling oceanic waters with high phytoplankton levels and for turbid coastal waters, agreement was satisfactory from 400 to 600 nm, but in some cases not good at longer wavelengths. Part of the problem in productive waters was a chlorophyll fluorescence emission peak at 685 nm (see §7.5) in the upwelling flux which increased the observed reflectance over the calculated values in this region.

That part of the upwelling flux just below the surface which is directed approximately vertically upwards is of particular significance for remote sensing. The subsurface radiance in a vertically upward direction we shall refer to as L_u. The angular distribution of the upwelling flux is such that upward radiance does not change much with nadir angle in the range 0–20°. Thus, a measured value of $L_u(\theta)$ within this range, or an average value over this range, can be taken as a reasonable estimate of L_u. L_u, like E_u, varies for a given water, in parallel with E_d, and we shall refer to the ratio L_u/E_d as the *radiance reflectance*.

The value of radiance reflectance, like that of irradiance, is a function of the inherent optical properties of the water. Given a relation between E_u/E_d and inherent optical properties, such as that embodied in eqn 6.3, if we know the ratio of E_u to L_u, we can relate L_u/E_d to b_b and a. The simplifying assumption is often made that the radiance distribution of the upwelling flux is identical to that above a Lambertian reflector (same radiance values at all angles). If this were so, then E_u would be equal to πL_u. In fact, the radiance distribution is not Lambertian (see §6.6 and Fig. 6.13) and measurements in Lake Pend Oreille at a solar altitude of 57°[937] showed that E_u was equal to 5.08 L_u near the surface in this water body.[27] Monte Carlo modelling calculations (Kirk, unpublished) have yielded values of E_u/L_u, just below the surface, of about 4.9 for waters with b/a values in the range 1.0 to 5.0, at a solar altitude of 45°. Thus, for intermediate solar altitudes we may reasonably assume that $E_u/L_u \simeq 5$, and the ratio is not likely to be very different at other solar altitudes. An equation relating L_u/E_d to b_b and a may be derived from eqn 6.3 by dividing the constant of proportionality (C) by 5. Monte Carlo calculations for waters with b/a values in the range 1.0 to 5.0 give rise to the relation

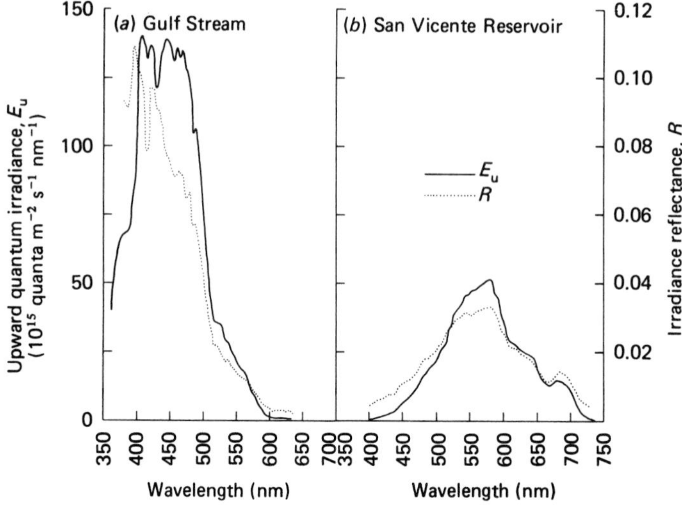

Fig. 6.7. Spectral distribution of upward irradiance and irradiance reflectance in an oceanic and an inland water (plotted from data of Tyler & Smith, 1970). (*a*) Gulf Stream (Atlantic Ocean) off Bahama Islands, 5 m depth. (*b*) San Vicente Reservoir, San Diego, California, USA, 1 m depth.

$$L_u/E_d \simeq 0.083\ b_b/a \tag{6.5}$$

for a solar altitude of 45°.

The spectral distribution of the upwelling flux must depend in part on that of the downwelling flux, but, as eqns 6.3 and 6.5 show, it is also markedly influenced by the variation in the ratio of b_b to a across the spectrum. In clear oceanic waters, for example, R can be as high as 10% at the blue (400 nm) end of the photosynthetic spectrum where pure water absorbs weakly but backscatters relatively strongly (see §4.3), and as low as 0.1% at the red (700 nm) end where water absorbs strongly.[644] Fig. 6.7 shows the spectral distributions of upward irradiance and irradiance reflectance in a clear oceanic water and in an inland impoundment. The upwelling flux in the oceanic water consists mainly of blue light in the 400–500 nm band. In productive oceanic waters with high levels of phytoplankton, the photosynthetic pigments absorb much of the upwelling blue light and so the peak of the upwelling flux is shifted to 565–570 nm in the green.[644] There is also a peak at 685 nm due to fluorescence emission by phytoplankton chlorophyll. In the inland water (Fig. 6.7*b*), yellow substances and phytoplankton absorb most of the blue light and a broad band, peaking at about 580 nm, with most of the quanta occurring between 480 and 650 nm is

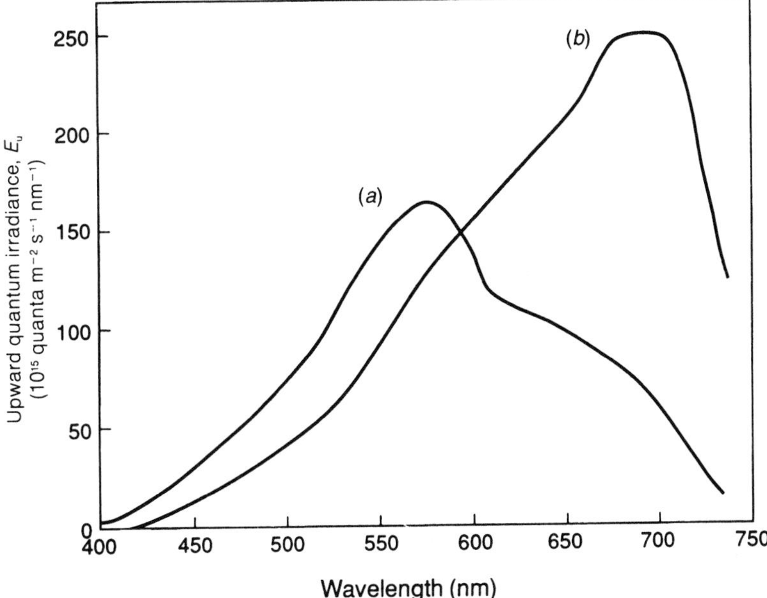

Fig. 6.8. Spectral distribution of subsurface upward irradiance in Lake Ginninderra, ACT, Australia. (*a*) 20th April 1983. Appearance – clear, green. $b = 3.2\ m^{-1}$. $a_{440} = 1.22\ m^{-1}$. (*b*) 15th August 1984. Appearance – turbid, brown. $b = 28.2\ m^{-1}$. $a_{440} = 23.1\ m^{-1}$.

observed: the chlorophyll fluorescence emission at about 680 nm can be seen in this curve.

Of the upwelling light flux which reaches the surface, about half is reflected downwards again, and the remainder passes through the water/air interface to give rise to the *emergent flux* (§7.2). It is this flux, combined in varying proportions with incident light reflected at the surface, that is seen by a human observer looking at a water body, and its intensity and spectral distribution largely determine the perceived visual/aesthetic quality of the water body.[181,487,491,493] Fig. 6.8 shows the spectral distributions of the subsurface upwelling flux in an Australian lake when it had on one occasion a clear, green appearance, and on another, a turbid, brown appearance. In the first case the spectral distribution peaked in the green–yellow region at about 575 nm, as a consequence of absorption at the blue end of the spectrum by moderate levels of humic substances, and at the red end by water itself. In the second case the upwelling flux had a greater total irradiance, due to intense scattering by suspended soil particles, and a peak in the red region at 675–700 nm, resulting from strong absorption in the blue and green regions by high concentrations of soluble and particulate humic substances. Rivers derived from glaciers are characteristically milky white or grey in appearance, due to the presence of high concentrations of mineral particles ('glacial flour'), but little organic material.

The apparent colour of a water body is determined by the chromaticity coordinates of the flux received by the observer, and these can be calculated from the spectral distribution using the C.I.E standard colorimetric system (see Jerlov, 1976, for further details). Davies-Colley *et al.* (1988) have carried out such calculations using upwelling spectral distribution data, for 14 New Zealand lakes, and suggest that this is a potentially useful tool for water resource managers with a concern for the aesthetic quality of the water bodies for which they are responsible. A comprehensive treatment of colour and clarity in natural water bodies in the context of human use of such waters may be found in the book by Davies-Colley, Vant & Smith (1994).

In the oceanic waters studied by Tyler & Smith (1970), irradiance reflectance for total PAR at 5 m depth varied from about 2 to 5%. For slightly to very turbid inland water bodies in southeast Australia, irradiance reflectance values for PAR just below the surface were usually between 4 and 10%, but values as low as 2% and as high as 19% were observed, the higher values being associated with higher turbidity.[479,482] The reflectance values increased somewhat with depth (see §6.6), the maximum value observed so far being about 34%. In inland waters with low scattering, but intense colour due to high concentrations of soluble yellow substances, reflectance of PAR can be very low. In a series of lakes of this type in Tasmania, irradiance reflectance for PAR just below the surface ranged from about 1.2% down to 0.14%.[90]

Where there is a bloom of coccolithophores – haptophycean algae whose cells are covered with highly scattering calcareous scales (coccoliths) – the reflectance of the ocean is greatly increased. In a coccolithophore bloom in the Gulf of Maine, Balch *et al.* (1991) measured subsurface reflectance values in the blue–green waveband of 5–7% at one station, and 22–39% at another. The high reflectance values appeared to be due primarily to large numbers of detached coccoliths suspended in the water, rather than to the whole cells.

In clear ocean waters with little colour, Raman scattering of the predominantly blue-green downwelling light stream gives rise, because of the associated shift to longer wavelengths (see p. 201), to a faint diffuse scattered light field in the 520–700 nm range.[594a,869a,887a] While this is of little importance for photosynthetic primary production, and makes only a

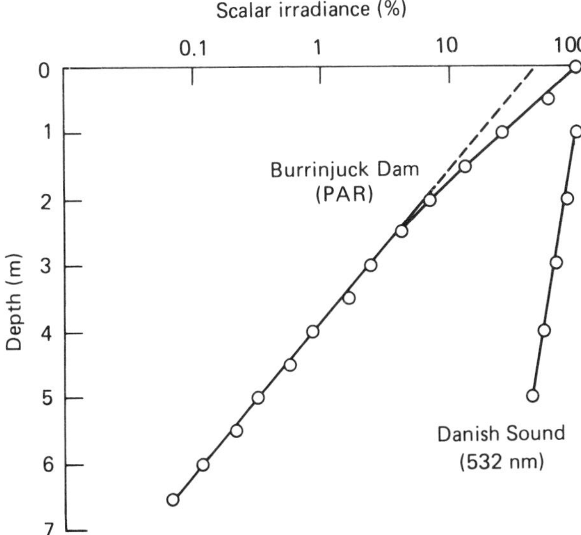

Fig. 6.9. Attenuation of scalar irradiance with depth. The data for Burrinjuck Dam (an Australian inland impoundment) were obtained with a scalar quantum irradiance sensor measuring the full 400–700 nm waveband (Kirk, unpublished measurements). The data for the Danish Sound (Baltic Sea) are measurements by Højerslev (1975) at 532 nm.

small contribution to the downwelling light field, it can contribute significantly to the upwelling light stream in the lower region of the euphotic zone, and is the probable cause of anomalous increases in reflectance at greater depths, which have sometimes been observed.

In water bodies which are sufficiently shallow for significant light to reach the bottom, then unless the bottom is very dark in colour there is an increase in E_u near the bottom due to reflection from it.

6.5 Scalar irradiance

For randomly oriented phytoplankton cells, photons are equally useful in photosynthesis regardless of the direction from which they come. Scalar irradiance, E_0, is therefore the best all-round measure of the availability of light for photosynthesis at a given depth. For monochromatic light scalar irradiance, like downward irradiance, diminishes with depth in an approximately exponential manner, as the linear graph (Fig. 6.9) of log E_0 against depth for green light in the Danish Sound[381] shows. When scalar irradiance is measured with a quanta meter which responds to the whole photosynthetic waveband, the variation of log E_0 with depth in fairly clear water is biphasic, the rate of attenuation being higher near the surface than lower down. This can be seen in the data for Burrinjuck Dam, Australia, in Fig. 6.9. The explanation is the same as that for the analogous phenomenon observed with downward irradiance (Fig. 6.5): the more strongly absorbed wavelengths are being removed in the upper layer, leaving the weakly absorbed wavelengths to penetrate lower down.

Although scalar irradiance is the best parameter in terms of which to express the availability of light for photosynthesis, the most commonly measured parameter is downward irradiance, E_d. It is of interest to examine the relation between them. As might be expected, since scalar irradiance includes both the upwelling and the downwelling light, and since it represents all angular directions equally (whereas downward irradiance, in accordance with the cosine law, is progressively less affected by light flux as its zenith angle increases), at any point in the water the value of scalar irradiance is always higher than that of downward irradiance.

From the radiance distribution data for Lake Pend Oreille, USA,[717,937] it is possible to calculate that in the 480 nm waveband the ratio E_0/E_d is 1.30–1.35 for various depths within the euphotic zone. The higher the ratio of scattering to absorption, the more diffuse the underwater light field becomes, and the greater the difference between E_0 and E_d. Fig. 6.10 shows the way E_0/E_d at the mid-point of the euphotic zone (z_m) increases as the ratio of the scattering coefficient to the absorption coefficient increases: E_0/E_d is equal to 1.5, 1.75 and 2.0 for b/a values of 6, 10 and 18, respectively. Since the absorption coefficient of natural waters can vary markedly with wavelength, the ratios of scalar to downward irradiance will vary across the spectrum. Taking the whole photosynthetic waveband, for clear oceanic waters the average b/a ratio is low enough to give rise to E_0/E_d values of up to about 1.2. For typical inland and some coastal waters, however, with b/a values in the region of 4 to 10, or even up to 20 or 30 in the most extreme cases, we might commonly expect E_0/E_d to be in the region of 1.4–1.8, rising to 2.0–2.5 in the very turbid waters.

Thus, if it is wished to determine the absolute amount of light available for photosynthesis at a given depth in a water body, a measurement of downward irradiance may seriously underestimate this, particularly in turbid waters. On the other hand, the vertical attenuation coefficient, K_d, for downward irradiance is close in value to the vertical attenuation coefficient, K_0, for scalar irradiance. Monte Carlo calculations (Kirk, unpublished) show that for media with b/a ranging from 0.3 to 30, K_d/K_0 varies only between about 1.01 and 1.06. The measured value of K_d can therefore be taken as a reasonable estimate of the value of K_0, and used to predict the attenuation of scalar irradiance with depth.

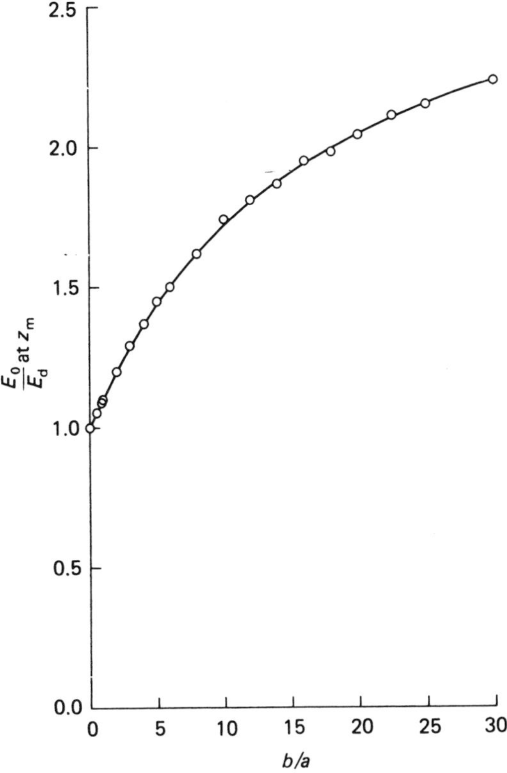

Fig. 6.10. Ratio of scalar to downward irradiance at the mid-point of the euphotic zone as a function of the ratio of scattering to absorption coefficients. Data obtained by Monte Carlo calculation for vertically incident monochromatic light using the method of Kirk (1981*a*, 1981*c*).

It would be useful to have, for any given water body, a measure of the total amount of light available for photosynthesis at a given instant, throughout the whole water column, 1 m square. We shall give this parameter the symbol, Q_t: it has the units joules or quanta. Making the simplifying approximate assumption that K_0 does not vary with depth, and also that no light reaches the bottom, it can be shown that

$$Q_t = \frac{E_0(0)}{cK_0} \tag{6.6}$$

where $E_0(0)$ is the scalar irradiance just below the surface and c is the speed of light in the medium. Strictly speaking, this equation applies only to monochromatic light (for which the variation of K_0 with depth is small) but may be regarded as very approximately applicable to PAR. Thus if one water body has a K_0 (or K_d, since $K_0 \simeq K_d$) value twice that in another water body, it has about half the total amount of light available for photosynthesis.

6.6 Angular distribution of the underwater light field

As sunlight penetrates a water body, its angular distribution begins to change – to become less directional, more diffuse – immediately the light penetrates the surface, as a result of scattering of the photons. The greater the depth, the greater the proportion of photons which have been scattered at least once. The angular distribution produced is not, however, a function of scattering alone: the less vertically a photon is travelling, the greater its pathlength in traversing a given depth, and the greater the probability of its being absorbed within that depth. Thus the more obliquely travelling photons are more intensely removed by absorption and this effect prevents the establishment of a completely isotropic field. The resultant angular distribution is determined by this interaction between the absorption and the scattering processes. Eventually the angular distribution of light intensity takes on a fixed form – referred to as the *asymptotic radiance distribution* – which is symmetrical about the vertical axis and whose shape is determined only by the values of the absorption coefficient, the scattering coefficient, and the volume scattering function.[210,422,717,730,928] That such an equilibrium radiance distribution would be established at great depth was predicted on theoretical grounds independently by Whitney (1941) in the USA and Poole (1945) in Ireland, and a rigorous proof of its existence was given by Preisendorfer (1959). Early measurements by Jerlov & Liljequist (1938) in the Baltic Sea showed some movement of the radiance distribution towards a symmetrical state with increasing depth. The particularly accurate measurements of Tyler (1960) down to 66 m in Lake Pend Oreille, USA, demonstrated the very close approach of the radiance distribution to the predicted symmetrical state at this near asymptotic depth.

The progression of the angular structure of the underwater light field towards the asymptotic state can be seen from the change in the radiance distribution with increasing depth. This is illustrated in Fig. 6.11 which is based on the measurements of Tyler (1960) in Lake Pend Oreille. Fig. 6.11*a* is for radiance at different vertical angles in the plane of the Sun. Near the surface the light field is highly directional with most of the flux coming from the approximate direction of the Sun. With increasing depth the peak in the radiance distribution becomes broader as its centre shifts towards $\theta = 0°$. The final, asymptotic, radiance distribution would be symmetrical about

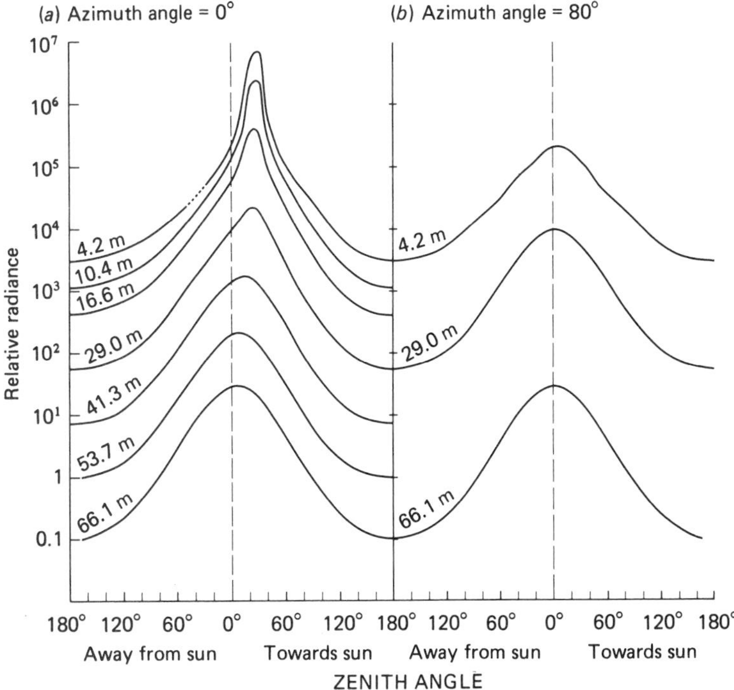

Fig. 6.11. Radiance distribution of underwater light field at different depths. Plotted from measurements at 480 nm made by Tyler (1960) in Lake Pend Oreille, USA, with solar altitude 56.6°, scattering coefficient 0.285 m^{-1} and absorption coefficient 0.117 m^{-1}. (*a*) Radiance distribution in the plane of the Sun. (*b*) Radiance distribution in a plane nearly at right angles to the Sun.

the zenith. It is noticeable that in the plane of the Sun this final state had not quite been reached even at 66 m. Fig. 6.11*b* shows that in a plane almost at right angles to that of the Sun, the radiance distribution is nearly symmetrical about the vertical even near the surface. At intermediate azimuth angles, radiance distributions intermediate between those in Fig. 6.11*a* and *b* exist.

Fig. 6.12 shows the near-asymptotic radiance distribution in Lake Pend Oreille in the form of a polar diagram. If in a natural water the ratio of scattering to absorption increases to high levels, then the shape of such a polar diagram of the asymptotic radiance distribution tends towards a circle (i.e. the light field approaches the completely diffuse state). If, however, scattering decreases to low values relative to absorption, then the polar diagram takes on the form of a narrow, downward pencil.[716]

In the context of photosynthesis, what is significant about a radiance

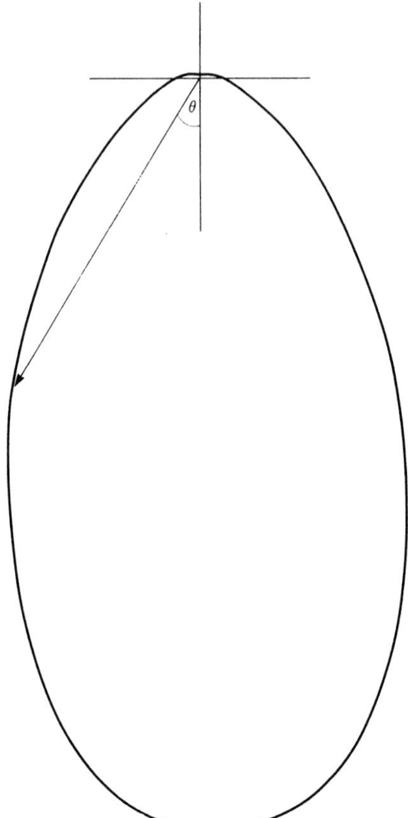

Fig. 6.12. Near-asymptotic radiance in Lake Pend Oreille, USA, plotted as a polar diagram. The data are the same as those in the lowest curve of Fig. 6.11*b*.

distribution is its relevance to the rapidity of attenuation of light intensity with depth. The azimuthal distribution of radiance at each vertical angle has no bearing on this. It is therefore legitimate and, in the interests of simplicity, advantageous to express the angular distribution of the light field at any depth in terms of the radiance averaged over all azimuth angles at each vertical angle. Fig. 6.13 shows a series of such radiance distributions at increasing depth, calculated by the Monte Carlo procedure, for PAR in an Australian lake. The much more rapid approach to a final symmetrical distribution in this case is partly the result of averaging over all azimuth angles, and partly because of the increased scattering relative to absorption in this lake relative to Lake Pend Oreille.

An even simpler way of expressing the angular distribution of the light

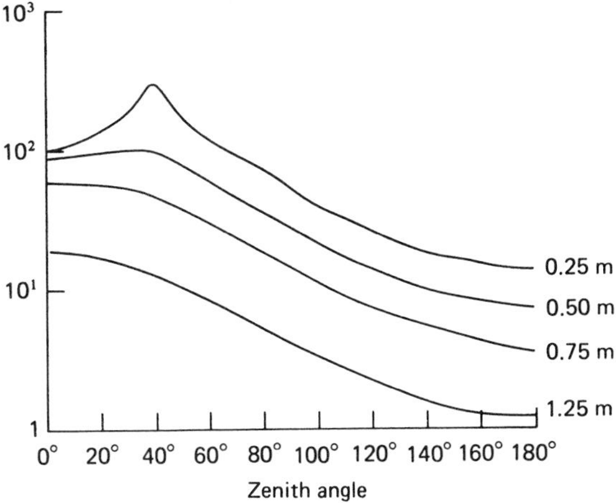

Fig. 6.13. Radiance distribution of underwater light field averaged over all azimuth angles in a turbid lake. Radiance values for total PAR in Lake Burley Griffin, Australia, were obtained by Monte Carlo calculation using the method of Kirk (1981*a*, 1981*c*) on the basis of the measured absorption and scattering properties (absorption spectrum – Fig. 3.9*b*; scattering coefficient – 15.0 m^{-1}; solar elevation – 32°).

field is in terms of the three average cosines (see §1.3) and the irradiance reflectance. In Fig. 6.14 the values of these parameters, obtained by Monte Carlo calculation[484,486] are plotted against the optical depth ($\zeta = K_d z$) in water in which $b/a = 5.0$. The average cosine and the average downward cosine initially both diminish sharply in value but then begin to level off as the angular distribution of the light approaches the asymptotic radiance distribution. There is no significant further change in the angular distribution beyond the depth ($z_{eu}, \zeta = 4.6$) at which irradiance has been reduced to 1% of the subsurface value, and indeed most of the change takes place before z_m ($\zeta = 2.3$) is reached.

It will be noted that for vertically incident light $\bar{\mu}_d$ starts (at $z = 0$) from a value of slightly less than 1.0. This is because the downwelling light just below the surface includes not only those photons which have just passed down through the surface (and these do have $\bar{\mu}_d = 1.0$), but also those photons which were part of the upwelling stream and which have just been reflected downwards from the surface. These latter photons have a $\bar{\mu}_d$ much less than 1.0 and so, although they constitute only a small proportion of the total, they bring the average value of $\bar{\mu}_d$ significantly below 1.0.

For light incident on the surface at angles other than the vertical, the behaviour is much the same as for perpendicularly incident light except that

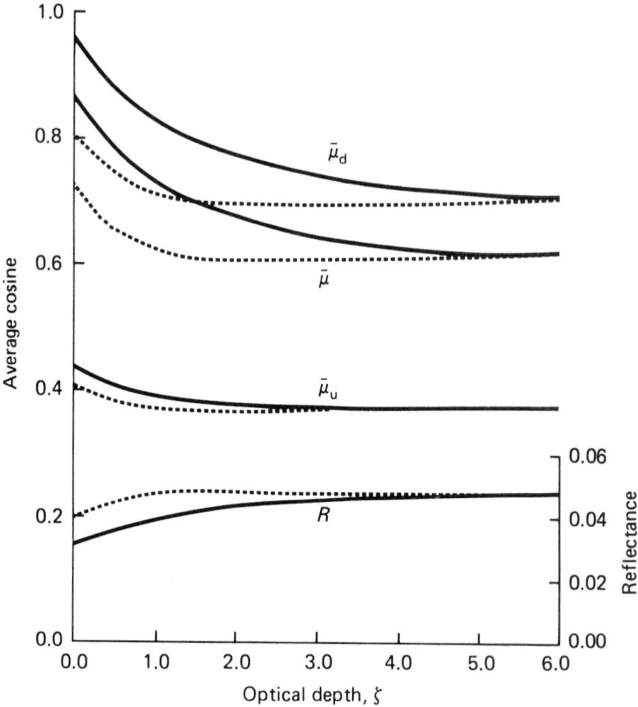

Fig. 6.14. Variation of average cosines for downwelling ($\bar{\mu}_d$), upwelling ($\bar{\mu}_u$) and total ($\bar{\mu}$) flux, and irradiance reflectance (R) with optical depth ($\zeta = K_d z$) in water with $b/a = 5$. Data obtained by Monte Carlo calculation.[484] (Vertically incident light ——. Light incident at 45° ------.)

$\bar{\mu}$ and $\bar{\mu}_d$ just below the surface have lower values, and the light field reaches its asymptotic state at shallower depths: compare the curves for $\theta = 0°$ and $\theta = 45°$. In the case of a non-perpendicular incident beam it should be realized that although the angular distribution averaged over all azimuths has settled down to almost its final form at z_{eu}, the downward radiance distribution in the vertical plane of the Sun is still significantly different from the asymptotic distribution: this can be seen in Fig. 6.11*a* in the curves for 29 m ($z_{eu} \simeq 28$ m).

The angular distribution of the upwelling stream acquires its final form a very short distance below the surface. In Lake Pend Oreille, USA, $\bar{\mu}_u$ varied only between 0.37 and 0.34 at all depths between 4.2 and 66.1 m.[940] Monte Carlo calculations[484] indicate that for waters with b/a values ranging from 0.1 to 20, $\bar{\mu}_u$ is typically in the range 0.35 to 0.42 throughout the euphotic zone. That the upwelling stream should have these characteristics is not

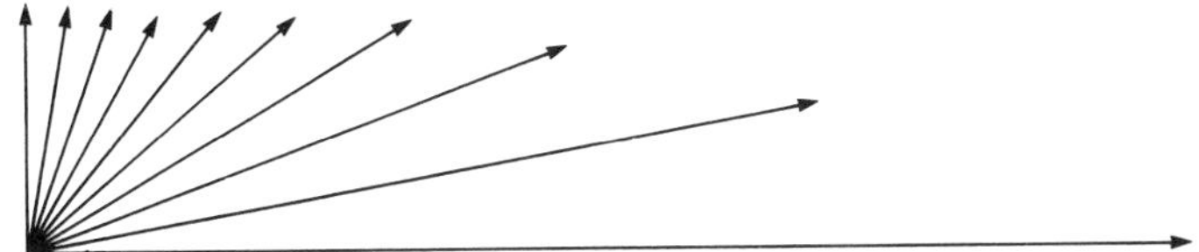

Fig. 6.15. Vector diagram of radiance distribution of upwelling light at 4.24 m depth in Lake Pend Oreille, USA. Radiance measurements of Tyler (1960) averaged over all azimuth angles. Radiance vectors are at nadir angle intervals of 10°.

surprising since it consists predominantly of backscattered light: backscattering does not vary strongly with angle and so rather similar upward radiance distributions are produced whatever the downward radiance distribution.

The radiance distribution of the upwelling light stream near the surface in Lake Pend Oreille is shown in the form of a polar diagram in Fig. 6.15. If, as is sometimes assumed (see §6.4), the upwelling flux has the same radiance distribution as that above a Lambertian reflector (same radiance at all angles), all the radiance vectors in Fig. 6.15 would be the same length. It is clear from the data that the upward radiance distribution is far from Lambertian.

As might be expected from the changes in angular distribution with depth, irradiance reflectance increases with depth but levels off at a final value concurrently with the settling down of $\bar{\mu}$ and $\bar{\mu}_d$ to their final values (Fig. 6.14). Thus, in natural water bodies the depth at which the asymptotic radiance distribution (averaged over all azimuth angles) is established, can be found by determining at what depth the value of irradiance reflectance ceases to increase.

6.7 Dependence of properties of the field on optical properties of the medium

We have noted earlier that the diminution of light intensity with depth is due to a combination of absorption and scattering processes, and also that the angular distribution of light intensity is determined by interaction between absorption and scattering. We shall now examine the quantitative relations that exist between the inherent absorption and scattering properties of the medium and the angular structure and vertical attenuation of irradiance in the light field that is established.

Angular structure

The shape of the volume scattering function, as defined by $\tilde{\beta}(\theta)$, the normalized volume scattering function (§1.4), can have a major influence on the character of the light field. In the majority of natural waters, however, scattering is particle-dominated to such an extent that $\tilde{\beta}(\theta)$ curves are rather similar in shape from one water to another. To begin with, therefore, we shall not consider the effects of variation in $\tilde{\beta}(\theta)$, and shall assume that the waters under consideration have a typical particle-dominated volume scattering function, similar to that measured by Petzold in San Diego harbour (Fig. 4.9).

To give the conclusions as much generality as possible, our analysis will be in terms of optical depth ($\zeta = K_d z$) rather than actual depth. This makes it permissible to confine our attention to the ratio of scattering, to absorption coefficients, b/a, rather than to their absolute values. A convenient optical depth at which to study the effects of changes in b/a is z_m, the mid-point of the euphotic zone ($\zeta = 2.3$). Fig. 6.16 shows the total and downward average cosines, and reflectance at z_m, as a function of b/a, determined by Monte Carlo calculation.[484] As scattering increases relative to absorption, so the underwater light field at this depth becomes less vertical, more diffuse, as shown by the decrease in $\bar{\mu}$ and $\bar{\mu}_d$. Reflectance increases almost linearly with b/a over much of the range, but the curve as a whole has a slightly sigmoid character. If the zenith angle of the incident light is changed from 0 to 45°, it makes very little difference to the reflectance, and not much to the average cosine except at low values of b/a.

In addition to delineating the quantitative dependence of the angular structure of the light field on b/a, Fig. 6.16 illustrates a useful general conclusion. Namely, that for water with a given $\tilde{\beta}(\theta)$, the relations between $\bar{\mu}$, $\bar{\mu}_d$, $\bar{\mu}_u$, R at a specified optical depth, and b/a, are fixed at any given angle of incident light, and indeed are largely independent of the angle of incident light. The fixed nature of the relation between $\bar{\mu}$, R and b/a provides the basis for a method of estimating the absolute values of both b and a for any real water body, using only measurements of underwater irradiance (see §4.2).

Vertical attenuation of irradiance

In the absence of scattering ($b = 0$), K_d, the vertical attenuation coefficient for downward irradiance, is determined only by the absorption coefficient and the zenith angle, θ, of the light beam within the water, in accordance

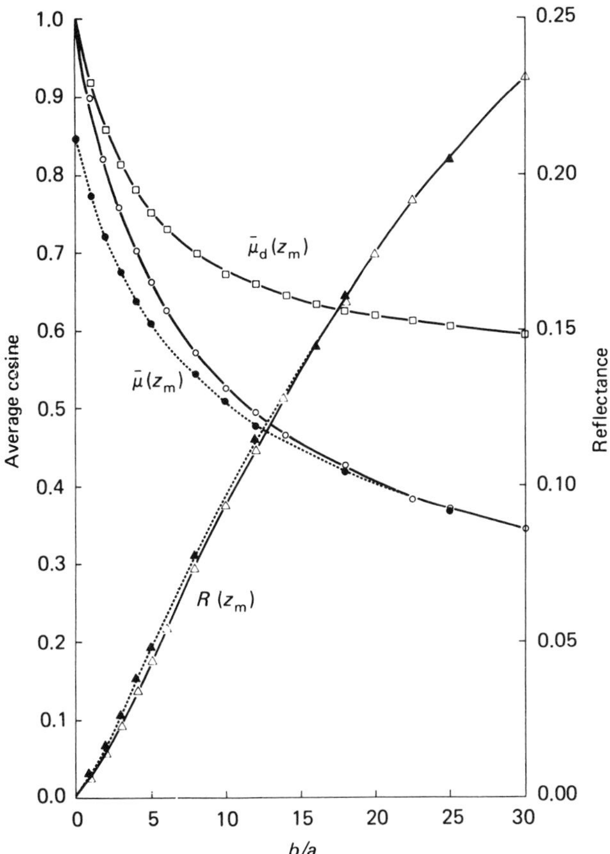

Fig. 6.16. Average cosines and irradiance reflectance at mid-point (z_m) of euphotic zone as a function of b/a. Data obtained by Monte Carlo calculation.[484] (Vertically incident light ——. Light incident at 45° ------.)

with $K_d = a/\cos\theta$. In a scattering medium K_d is increased, partly because of the altered angular distribution of the downwelling light, and partly because of upward scattering of the downwelling stream. Concurrently with the progressive change in angular distribution with depth, so the value of K_d increases with depth, levelling off at that optical depth at which the asymptotic radiance distribution becomes established. Fig. 6.17 shows K_d as a function of optical depth for vertically incident light in a medium for which $b/a = 5.0$.

While K_d always increases with depth when the incident light is parallel, it is possible that the converse might be true if the incident light was diffuse, e.g. from an overcast sky. If, as could be the case, especially in water with a

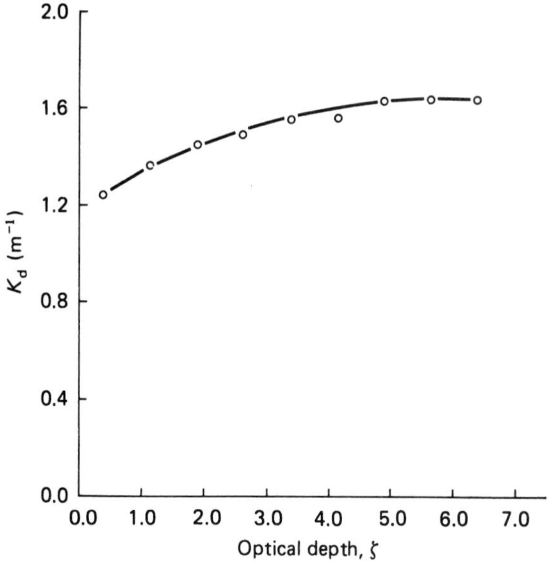

Fig. 6.17. Vertical attenuation coefficient for downward irradiance as a function of optical depth for $b/a=5$, and $a=1.0$ m^{-1}. Data obtained by Monte Carlo calculation for vertically incident light.[484]

low value of b/a, the final asymptotic radiance distribution was more vertical than the initial radiance distribution, then K_d would decrease with depth.

The relation between K_d and the absorption and scattering coefficients may, for a given incident light field and a given $\tilde{\beta}(\theta)$, be expressed with complete generality by expressing K_d/a as a function of b/a. The absolute values of a, b and K_d do not have to be specified: it is the ratios which are important. A certain ratio of b to a gives rise to a specific ratio of K_d to a, regardless of the actual value of any of the coefficients. The calculated value of K_d/a at z_m for vertically incident light is plotted against b/a in Fig. 6.18. K_d starts off equal to a when $b=0$, and rises progressively as b increases, in a linear manner to begin with, but curving over at high ratios of b to a. The values of $K_d(z_m)/a$ over the whole range up to $b/a=30$ were found[484] to fit very closely to a curve specified by the equation

$$\frac{K_d(z_m)}{a}=\left[1+\frac{Gb}{a}\right]^{\frac{1}{2}} \tag{6.7}$$

or

$$K_d(z_m)=(a^2+Gab)^{\frac{1}{2}} \tag{6.8}$$

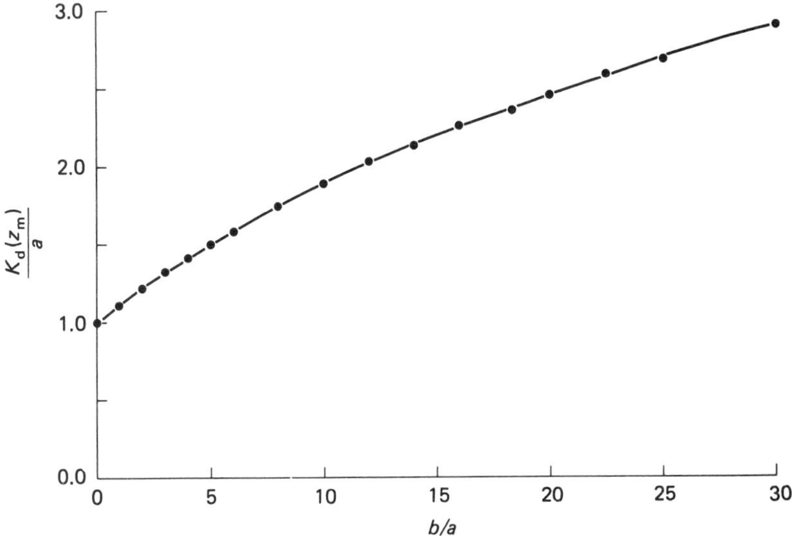

Fig. 6.18. Ratio of vertical attenuation coefficient for downward irradiance (at z_m) to absorption coefficient as a function of b/a. Data obtained by Monte Carlo calculation for vertically incident light.[484]

G is a coefficient which can be regarded as determining the relative contribution of scattering to vertical attenuation or irradiance, and its value is determined by the shape of the scattering phase function. For a water with the San Diego harbour $\tilde{\beta}(\theta)$, as in Fig. 6.18, G has the value 0.256. Similar equations to 6.7 and 6.8 can be written in which $K_d(z_m)$ is replaced by $K_d(\text{av})$, the average value of K_d through the zone within which downward irradiance is reduced to 1% of that penetrating the surface: in this case the coefficient, G, has a slightly different value (0.231). Although $K_d(z_m)$ is a more precisely defined, and theoretically satisfactory, form of K_d, we shall nevertheless from here on direct our attention mainly to $K_d(\text{av})$, since this is the K_d most commonly measured in the field. $K_d(z_m)$ and $K_d(\text{av})$ are in fact normally close to each other in value, and show the same kinds of dependence on the inherent optical properties of the medium.

The way in which K_d varies with solar altitude is conveniently expressed in terms of its dependence on μ_0, the cosine of the refracted solar beam just beneath the surface. This dependence is due not only to the anticipated change, with angle, in the pathlength of the photons per vertical metre traversed, which gives rise to a dependence of K_d on $(1/\mu_0)$, but in addition to the fact that the coefficient G also varies with solar angle in accordance with

$$G(\mu_0) = g_1\mu_0 - g_2 \tag{6.9}$$

where g_1 and g_2 are constants for a particular scattering phase function.[488,494] Thus we can write

$$\frac{K_d}{a} = \frac{1}{\mu_0}\left[1 + G(\mu_0)\frac{b}{a}\right]^{\frac{1}{2}} \tag{6.10}$$

and

$$K_d = \frac{1}{\mu_0}[a^2 + G(\mu_0)ab]^{\frac{1}{2}} \tag{6.11}$$

or the corresponding forms, such as

$$K_d = \frac{1}{\mu_0}[a^2 + (g_1\mu_0 - g_2)ab]^{\frac{1}{2}} \tag{6.12}$$

in which we substitute for $G(\mu_0)$. For water bodies with the San Diego phase function, in the version of eqn 6.12 applicable to K_d(av), the constants have the values $g_1 = 0.425$ and $g_2 = 0.19$. In the version for $K_d(z_m)$, $g_1 = 0.473$ and $g_2 = 0.218$. Using these values of g_1 and g_2, eqn 6.12 can be used to calculate K_d from μ_0, a and b for most of the waters that limnologists and coastal oceanographers deal with, and is of considerable predictive value in relation to the optical water quality of these aquatic ecosystems.[491,493]

In the various regions of the open ocean, the scattering phase function can differ significantly in shape from the phase function of coastal and inland waters. The relationships embodied in eqns 6.7 to 6.12 nevertheless still apply.[494] It is the value of the coefficient $G(\mu_0)$ which varies with the shape of the phase function. A useful measure of the shape of the scattering phase function is $\bar{\mu}_s$, the average cosine of scattering (see §1.4), and we can use this to illustrate the nature of the dependence of $G(\mu_0)$ on the characteristics of the phase function. It is helpful to bear in mind that $\bar{\mu}_s = 1.0$ corresponds to all the photons being scattered forward without any change in direction, and that a progressive decrease in $\bar{\mu}_s$ from 1.0 towards zero corresponds to photons being scattered through wider and wider angles. To simplify matters we can confine our attention initially to $G(1.0)$, the value of G for vertically incident light ($\mu_0 = 1.0$). Monte Carlo calculations of the light field in a wide range of optical water types[494] show (Fig. 6.19), as we might expect, that the dependence of vertical attenuation on scattering, expressed through $G(1.0)$, diminishes steeply as the scattering becomes increasingly concentrated within narrow forward angles ($\bar{\mu}_s$ increases towards 1.0), and disappears altogether when $\bar{\mu}_s = 1.0$. Linear regression gives

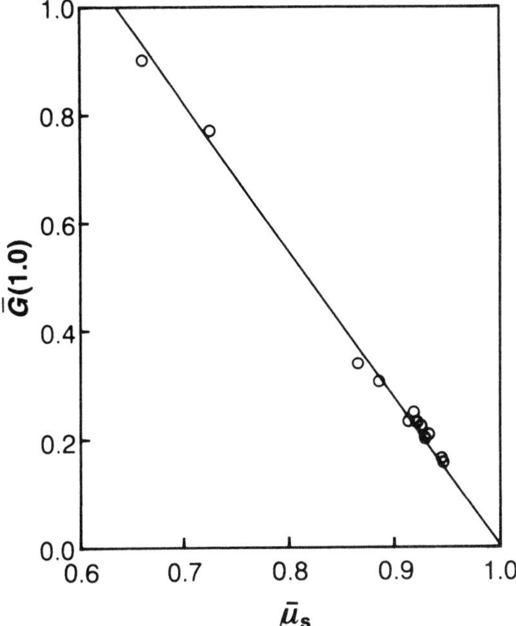

Fig. 6.19. Variation of $G(1.0)$ with the average cosine of scattering of the water.

$$G(1.0) = -2.401\bar{\mu}_s + 2.430 \tag{6.13}$$

with $r^2 = 0.997$. $G(1.0)$ can also be satisfactorily represented as a positive linear function of the reciprocal of $\bar{\mu}_s$

$$G(1.0) = 1.533(\bar{\mu}_s)^{-1} - 1.448 \tag{6.14}$$

with $r^2 = 0.979$.

As an indication of the extent to which scattering intensifies vertical attenuation of light it may be noted from Fig. 6.18 that b/a ratios of about 5 and 12 increase K_d by 50 and 100%, respectively. Comparable effects can be shown for real water bodies. In Table 6.2, data which show the effect of scattering on attenuation of the whole photosynthetic waveband in four Australian inland waters of varying turbidity are presented. The table compares the values of K_d for PAR calculated from the total absorption spectrum of the water, assuming no scattering, with those actually measured within the water with an irradiance (PAR) meter. The ratio of the observed to the calculated value of K_d is a measure of the extent to which scattering intensifies light attenuation. The increase in K_d(PAR) due to

Table 6.2. *Effects of light scattering on vertical attenuation coefficient* (K_d) *for irradiance of PAR (400–700 nm), in four water bodies of differing turbidity, in southeast Australia.* K_d *values calculated from the measured absorption spectra are compared with those obtained* in situ *with a quanta meter*[484]

Water body	Turbidity (NTU)	K_d (Calculated) (m^{-1})	K_d (observed) (m^{-1})	Effect of scattering K_d(obs.)/K_d(calc.)
Burrinjuck Dam	1.8	0.775	0.90	1.16
L. Ginninderra	4.6	0.547	0.90	1.65
L. Burley Griffin	17.4	1.333	2.43	1.82
L. George	49.0	1.742	5.67	3.25

scattering ranged from 16% in Burrinjuck Dam, which was rather clear at the time, to more than three-fold in the very turbid water of Lake George.

In our consideration of the ways in which scattering affects attenuation we can now make use of eqn 1.57, derived from radiative transfer theory by Preisendorfer (1961), which expresses the vertical attenuation coefficient for downward irradiance as a function of the diffuse absorption and backscattering coefficients.

$$K_d(z) = a_d(z) + b_{bd}(z) - b_{bu}(z)R(z)$$

Although this relation is not, in view of the difficulty of measuring the diffuse coefficients, of everyday practical use, it is conceptually valuable in helping us to understand the nature and relative importance of the radiation transfer mechanisms underlying the attenuation process. It tells us that attenuation of downward irradiance is the result of three different processes, represented by the three different terms on the right-hand side of the equation. Absorption from the downwelling stream is accounted for by the diffuse absorption coefficient $a_d(z)$ which is equal to $a/\bar{\mu}_d(z)$, and so must increase as scattering causes $\bar{\mu}_d$ to decrease. Light is also removed from the downwelling stream by upward scattering of the photons: this process is represented by the diffuse backscattering coefficient for the downwelling stream, $b_{bd}(z)$. Opposing these two processes is the downward scattering of photons from the upwelling stream, which acts to increase the downward flux: this is represented by the term, $-b_{bu}(z)R(z)$, where $b_{bu}(z)$ is the diffuse backscattering coefficient for the upwelling stream and $R(z)$ is irradiance reflectance.

For any given values of a and b, and having specified $\tilde{\beta}(\theta)$ and the incident

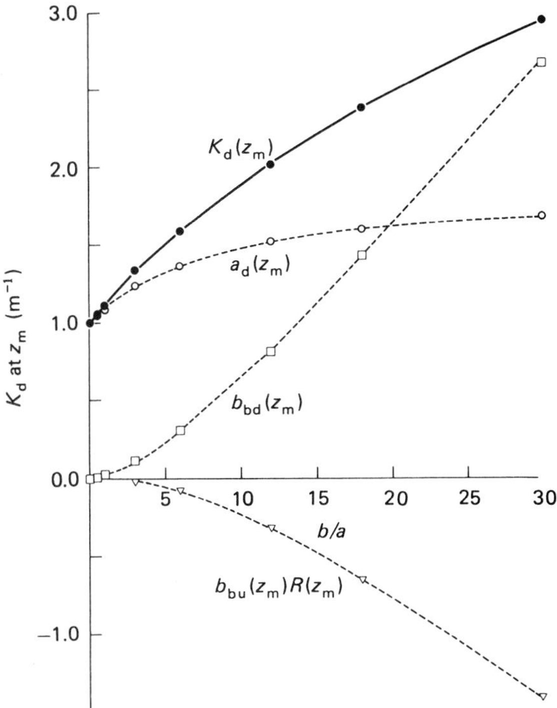

Fig. 6.20. Changes in the different components of the attenuation processes as a function of b/a. The calculated values of the three right-hand terms in eqn 1.57 are plotted alongside the corresponding values of K_d at z_m (data from Kirk, 1981*a*).

light field, then if the radiance distribution at depth z m is calculated by a Monte Carlo, or other, procedure, it is possible to calculate the diffuse optical properties at that depth.[482] Thus all the terms on the right-hand side of eqn 1.57 may be determined for a series of values of b and a, and so the relative importance of the three different processes can be evaluated for different types of water.

Fig. 6.20 shows the change in value of each of these three terms, compared to the value of K_d, at z_m, as the ratio of scattering to absorption coefficients increases from 0.0 to 30.0. At low values of scattering, up to b/a ratios of about 3, attenuation is almost entirely due to absorption, and furthermore the scattering-induced increase in attenuation is mainly a consequence of the increase in absorption resulting from the changed angular distribution of the downwelling flux. Upward scattering of the downwelling stream contributes only slightly to attenuation.

When b/a has risen to about 7, upward scattering of the downwelling stream accounts for about 25% of all the attenuation and contributes as much to the increase in attenuation as does the increase in absorption of the downwelling light. At b/a values of about 11 and 20, upward scattering of the downwelling light accounts for 0.5 and 1.0 times as much attenuation, respectively, as does absorption, and at higher b/a values it becomes the major mechanism for attenuation. The downward scattering of the upwelling stream, which acts to diminish attenuation, has little effect over the lower part of the range of b/a values studied, but becomes significant from about $b/a = 6$ onwards, and counteracts a large part of the attenuation due to the other two processes at b/a values in the range 12 to 30.

6.8 Partial vertical attenuation coefficients

In any medium containing a number of absorbing and scattering components, the contributions of the different components to any one of the absorption and scattering coefficients (normal, diffuse, forward, back) are additive, e.g. if there are n components, then

$$a_{\text{total}} = a_1 + a_2 + \ldots + a_n$$

and similar relations hold for b, $a_d(z)$, $b_{bd}(z)$ etc. Consequently, for an aquatic medium containing n absorbing/scattering components we can expand eqn 1.57 to

$$\begin{aligned} K_d(z) = {} & a_d(z)_1 + a_d(z)_2 + \ldots + a_d(z)_n \\ & + b_{bd}(z)_1 + b_{bd}(z)_2 + \ldots + b_{bd}(z)_n \\ & - b_{bu}(z)_1 R - b_{bu}(z)_2 R - \ldots - b_{bu}(z)_n R \end{aligned}$$

Rearranging, we obtain

$$\begin{aligned} K_d(z) = {} & [a_d(z)_1 + b_{bd}(z)_1 - b_{bu}(z)_1 R] \\ & + [a_d(z)_2 + b_{bd}(z)_2 - b_{bu}(z)_2 R] \\ & + \ldots + [a_d(z)_n + b_{bd}(z)_n - b_{bd}(z)_n R] \end{aligned}$$

which can be written in the form

$$K_d(z) = K_d(z)_1 + K_d(z)_2 + \ldots + K_d(z)_n \qquad (6.15)$$

Thus the vertical attenuation coefficient for downward irradiance in a natural water can be partitioned into a set of partial vertical attenuation coefficients, each corresponding to a different component of the medium. The partial vertical attenuation coefficient for the ith component is given by

$$K_d(z)_i = a_d(z)_i + b_{bd}(z)_i - b_{bu}(z)_i R \qquad (6.16)$$

It is important to remember that although the contributions of the different components of the medium to total attenuation of irradiance can be simply added together in accordance with eqn 6.15, the nature of their contributions to attenuation can vary markedly from one component to another. Consider, for example, a water in which component j is dissolved yellow colour, and component $j+1$ consists of suspended mineral particles, scattering intensely but with little intrinsic colour. For the jth component, $K_d(z)_j$ will consist mainly of the absorption term, $a_d(z)_j$, in eqn 6.16, whereas for the $(j+1)$th component, $K_d(z)_{j+1}$ will consist mainly of the backscattering term, $b_{bd}(z)_{j+1}$.

It is also important to remember that the value of $K_d(z)_i$, and consequently the contribution of the ith component to $K_d(z)$, is not a linear function of the concentration of that component. Any substantial change in the absorption and/or scattering characteristics of the medium resulting from a change in concentration of one of the components will inevitably affect the radiance distribution at depth z. The absorption and scattering coefficients on the right-hand side of eqn 6.16, being quasi-inherent rather than inherent optical properties (see §1.5) are functions of the radiance distribution as well as of the concentration of the component in question. Consequently, although their values will increase with the concentration of the component, the increase will not be linear with concentration. Furthermore, R, being a function of the radiance distribution, will also change as the concentration of any component changes. In short, $K_d(z)_i$ and $K_d(z)$ will certainly increase as the amount of any of the components of the medium increases, but except for small increments in concentration, the increases will not be linearly related to the increase in concentration.

7

Remote sensing of the aquatic environment

In any water body some of the light which penetrates the water is caused, by scattering within the water, to pass up through the surface again. Of this emergent flux, 90% originates within the depth (equal to $1/K_d$) in which downward irradiance falls to 37% ($1/e$) of the subsurface value.[325] It can be regarded as a sample derived from the underwater light field, and so by studying it with appropriate detection instruments above the surface, information about that field and therefore about the optically significant components of the medium, can be obtained. There is not much point in having detection instruments just above the surface: they would be more useful below. If, however, this emerging radiant flux can be studied by remote sensing instruments, located a considerable distance above the surface, in an aeroplane or space satellite, then the considerable advantage is gained that information about the underwater environment over a large area can be obtained in a short time. This makes it possible to acquire a synoptic view of a large aquatic ecosystem, within a fraction of the time and effort that would be involved in carrying out measurements over the same area from a surface vessel.

It will readily be appreciated, however, that a price must be paid. Measurements of the emergent flux, from a great distance, cannot be as accurate, or yield as much information, as measurements carried out within the water itself. We shall now consider the kinds of measurements that can be made, the correction procedures that must be carried out, and the nature of the information that may be obtained. We shall then go on to examine some of the studies that have been carried out so far.

7.1 The upward flux and its measurement

A photometer in an aeroplane or satellite, pointed down at the ocean or any other water body can receive light originating in four different ways: by

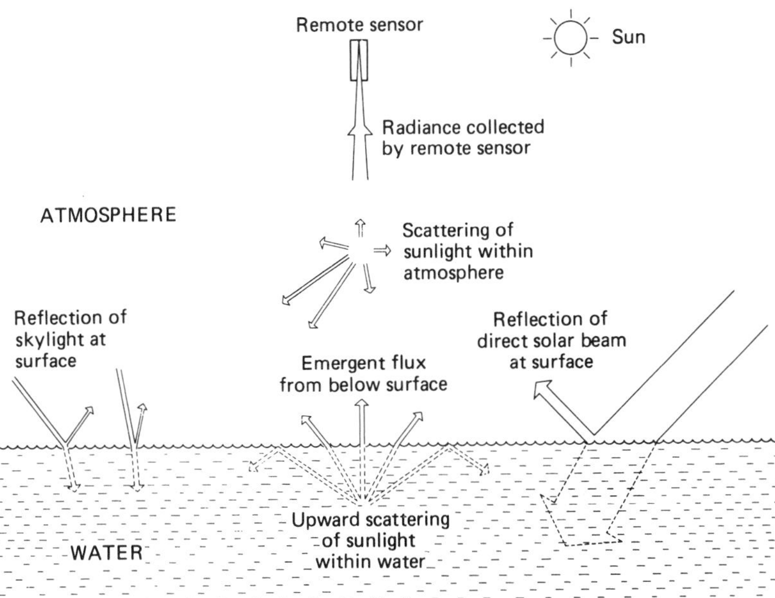

Fig. 7.1. The different origins of the light received by a remote sensor above a water body.

scattering from below the surface, by reflection of skylight at the surface, by reflection of the direct solar beam at the surface, and by scattering within the atmosphere (Fig. 7.1). Only the first of these four light fluxes, which we shall refer to as the emergent flux, contains information about the underwater light field and the composition of the aquatic medium. The essential problem, therefore, is to quantify the emergent flux in the presence of the other light fluxes.

The photometer used to measure the light flux above the water surface could in principle be either an irradiance meter receiving all the upwelling light through an angle of 180° with its cosine collector, or could be a radiance meter with a narrow angle of acceptance. In practice the sensor is usually a radiance meter since the irradiance meter has the serious disadvantages that it has to receive all the upwelling light including the reflected solar beam, and also that it cannot be directed at specific areas on the surface.

To avoid receiving the surface-reflected solar beam, the radiance meter can be directed at a part of the surface well outside the solar glitter pattern. A more difficult problem is that of accounting for that part of the measured radiance which originates within the atmosphere. Depending on wavelength, anything from 80 to 100% of the radiance received by a satellite-borne radiance meter, directed at the ocean surface outside the Sun glitter pattern, originates in atmospheric Rayleigh (air molecule) and aerosol (particle) scattering. Ways of correcting for this flux are considered in a later section. This problem is largely eliminated when measurements are taken from low-flying aircraft (see Fig. 7.8*a*).

Skylight also originates in atmospheric scattering. Correction for the contribution of reflected skylight to the measured upwelling flux can be lumped in with the total atmospheric scattering correction.[320] An alternative procedure is to carry out the measurements at an angle of 53° to the vertical.[148] Light reflected from a plane surface at this angle (Brewster's angle) is polarized and can be eliminated by placing a polarized filter oriented at right angles to the major axis of polarization over the aperture of the radiance meter. A consequence of viewing at this angle, however, is that the effective air mass through which the flux from within the water must travel is increased by a factor of 1.67:[26] this would be a serious disadvantage in the case of a satellite or high-flying aircraft, but not in the case of a low-flying aircraft.

Measurement systems – general considerations

Remote sensing photometers used in low-altitude aircraft normally have a fixed direction of view. They measure upward radiance at a series of points along a linear track determined by the flight path of the aircraft. For two-dimensional mapping, the aircraft must traverse the area of interest many times. The spectral distribution of the radiance at each point is determined using interference filters or a spectroradiometer.

With increasing altitude the area that it is practicable to 'view' increases. To take advantage of this, remote sensing from satellites and higher-altitude aircraft uses spatially scanning photometers, which collect information, not just from the thin line immediately below the trajectory of the satellite or aircraft, but from a broad swath of the Earth's surface which can be anything from a few kilometres to hundreds of kilometres wide. There are two ways of achieving this, sometimes referred to respectively as 'pushbroom' scanning and 'whiskbroom' scanning.[822] In a pushbroom scanner the instrument optics form an image of a thin transverse strip of the Earth's surface extending right across the swath at right angles to the line of flight, and present it across a line array of photodetector elements. Each of these detector elements thus receives radiation from one of the elements of area at the surface, which between them form the strip across the swath.

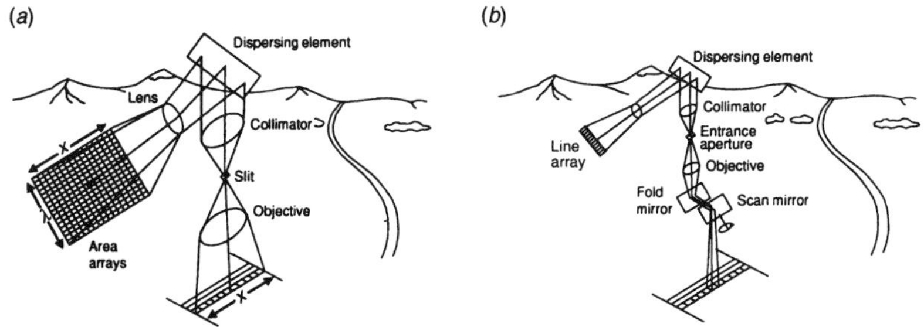

Fig. 7.2. Imaging spectrometry in remote sensing. (*a*) Pushbroom scanner with two-dimensional array of photodetectors. (*b*) Whiskbroom scanner with line array of photodetectors. (After HIRIS Instrument Panel Report. NASA, 1987).

In fact, what is required for each element of area at the surface is a sample not of the total radiation but of the radiant flux in each of some specified set of spectral wavebands. The radiation from the transverse strip at the surface can, by means of dichroic beam splitters (partially reflective mirrors which reflect radiation below a certain wavelength and let the longer wavelength energy through), be subdivided into a number of separate fluxes each of which can be directed through a different spectral filter onto its own line array of photodetector elements. A more radical solution is to spectrally disperse the radiant flux with a prism or grating and form an image on a two-dimensional array of photodetectors, one axis corresponding to position of the individual surface element across the swath, the other to wavelength (Fig. 7.2*a*). In principle, in this way, a complete spectrum of the flux from each element of area can be obtained.

Remote sensing radiometers of the whiskbroom type scan by means of a rotating or oscillating mirror which ensures that the direction of viewing is continually moving from one side of the swath to the other, at right angles to the satellite/aircraft trajectory (Fig. 7.2*b*). The scene reflected in the mirror is viewed through a telescope with a very small acceptance angle (0.002–0.05°) so that at any given instant only a small unit element of the Earth's surface is in view. The flux collected by the telescope can be partitioned into a set of spectral bands using beam splitters and filters, and a corresponding set of photodetectors, or be dispersed with a grating or prism onto a line array of detector elements to provide a more complete spectral distribution.

With either the pushbroom or whiskbroom type of radiometer, a series of contiguous but non-overlapping, square image elements, corresponding to a linear sequence of elements of area at the surface, extending across the swath, is recorded. By the time the next side-to-side scan begins the satellite or aircraft has moved on so that the next left-to-right sequence of image elements adjoins, but does not overlap, the previous sequence. The accumulated set of image elements constitutes a mosaic corresponding to that strip of the Earth's surface defined by the path of the satellite/aircraft and its maximum lateral angle of view.

For each image element, i.e. for each element of surface, the photometer has recorded a set of radiance values corresponding to the number of spectral bands with which it operates. The radiance values are encoded digitally, and a key parameter for any remote sensing instrument is the number of bits to which digitization is carried out: this can vary from 6 bits giving a measurement range of up to 64 different radiance values, to 12 bits giving 4096 different values. This information, after being relayed to a receiving station in the case of a satellite, is stored on tape. The data can then be used, after appropriate computer processing, to prepare a map of any part of the Earth's surface that lay beneath the satellite's orbit, or aircraft's path. The continuous swath covered by a satellite is broken down for mapping purposes into individual sections which might be some hundreds of kilometres long.

Each small element of the map, referred to as a *pixel*, corresponds to one of the square unit elements of surface viewed by the scanning photometer. The intensity, or the colour within a selected colour scale, of each pixel in the final map displayed on the computer screen can be directly related to the radiance value in any one of the spectral bands, or can be determined by the value of some parameter (e.g. chlorophyll concentration) derived by calculation from more than one of the radiance values.

Low altitude (0–600 m) line-of-flight systems

We consider here instruments which measure radiance only along a linear path, i.e. which do not simultaneously scan from side to side.

For completeness we shall deal first here with radiometric sensing of water bodies carried out from boats, even though this is, of course, strictly speaking not 'remote' sensing. As a quantitative measure of ocean colour, Jerlov (1974) defined the *Colour Index* to be the ratio of nadir radiance (i.e. vertically upwards flux, nadir angle $\theta_n = 0°$) within the water in the blue at 450 nm to that in the green at 520 nm, and developed a colour meter consisting of a pair of downward pointing Gershun tubes with a blue and a green interference filter, respectively, to measure it. The meter was designed to be lowered manually from shipboard to any depth. Neuymin *et al.* (1982)

chose a somewhat different pair of wavelengths in the blue and green for the colour index, namely 440 and 550 nm, which has the advantage that this ratio is particularly sensitive to algal pigments, and consequently this is the pair of wavelengths most favoured for remote sensing of oceanic phytoplankton (see later). As it happens, Neuymin *et al.* chose to define their colour index in terms of the green to blue ratio (in the upwelling flux), rather than vice versa, and so their index *increases* with phytoplankton content. Their instrument is located at 5–6 m depth in a shaft running through the vessel, and can provide a continuous record of the colour index of the sea along the ship's track.

Bukata, Jerome & Bruton (1988) have developed a shipboard radiometer mounted on a 4.5 m boom from the bow, at 4.5 m above the water surface. Nadir radiance from the lake water surface is measured in four broad wavebands in the blue, green, red and near-infrared regions, together with downward solar irradiance in the same four wavebands. Measurements are taken continually from the moving vessel at half-second intervals.

Gitelson & Kondratyev (1991) sought to combine both 'remote' and sea-level sensing by flying 10–15 m above the sea surface in a helicopter from which they simultaneously made radiometric measurements and took water samples. Upward and downward radiance and irradiance, in nine 10-nm spectral channels between 430 and 750 nm were measured with a hand-held spectrometer requiring less than 1 s to complete its readings.

Where a spectroradiometer is being used from an aircraft it is important that the spectral scan is completed quickly to minimize the effects of changes in the emergent flux from point to point along the flight path. Clarke *et al.* (1970) used a spectroradiometer with a 3° × 0.5° field of view which scanned from 400 to 700 nm in 12 s. The instrument was operated at the Brewster angle in conjunction with a polarizer, to eliminate reflected skylight. Neville & Gower (1977) used a spectroradiometer[966] in which the spectrum produced by a diffraction grating was distributed over an array of 256 silicon diodes, each of which continuously detected the radiant flux in its own narrow region of the spectrum. It was possible to read out the complete spectrum, 380–1065 nm, at 2 s intervals.[85] This instrument was also operated at the Brewster angle. McKim, Merry & Layman (1984) used a spectroradiometer in which a 500-channel spectrum, covering the range 400–1100 nm, is obtained by distributing the spectrum from the diffraction grating onto a silicon diode array on the inner face of a vidicon tube, the diode array being read with an electron beam. A complete spectrum is recorded every 320 ms. At an operating altitude of 600 m and a speed of 200 km h^{-1}, the 'footprint' of the instrument is ~18 m square.

At the other extreme, spectrally speaking, useful information can be obtained by airborne remote sensing in two or three wavebands, provided these are carefully chosen. Arvesen, Millard & Weaver (1973) developed a differential radiometer which carried out simultaneous measurements of upward radiance at 443 and 525 nm, and continuously compared one with the other (in effect, measuring Jerlov's Colour Index). This had the advantage that changes in incident light intensity or variations in surface water roughness had similar effects on the flux in both wavebands and so were automatically corrected. Changes in phytoplankton concentration, however, which would be expected specifically to affect the ratio of upward flux in these two wavebands (see §7.5), markedly affected the signal.

To address the need for an affordable aircraft ocean colour instrument, NASA and NOAA developed the Ocean Data Acquisition System (ODAS).[343] This measures nadir radiance in three 15-nm bands centred on 460, 490 and 520 nm, these bands being chosen specifically to make possible the use of a previously developed[129] spectral curvature algorithm for remote sensing of phytoplankton (see below). The three radiances are sampled 10 times per second and averaged over 1 s. With the aircraft flying at 50 m s^{-1} at an altitude of ~150 m, the footprint of the instrument is about 5 × 50 m. Position is measured accurately at 20 s intervals with Loran-C navigation.

Medium/high altitude (2–20 km) spatially scanning systems

A spatially scanning photometer – the Ocean Colour Scanner (OCS) – was developed by the NASA Goddard Space Flight Centre for remote sensing of ocean colour from a U-2 aircraft at altitudes of 18 to 20 km.[396,471] This has 10 spectral bands, ranging from 433 to 772 nm, with bandwidths of about 20 nm. It scans by means of a rotating mirror at 45° and uses a telescope with a field of view of ~0.2°, corresponding, at the U-2 operating altitude of 19.8 km, to a unit element of area at sea level of about 75 m square. The swath covered is 25 km wide.

The M2S Multispectral Scanner is designed for use at intermediate altitude. It has 10 spectral bands covering the range 380 to 1060 nm, bandwidths being about 40 nm. From an altitude of 3 km, the swath width at the surface is 8.5 km and the unit element of area viewed is 8 m square.[434] The NASA Multichannel Ocean Colour Sensor (MOCS)[129] measures radiance in 20 contiguous 15-nm wide bands between 400 and 700 nm. It has a spatial resolution of 4 × 2 mrad, corresponding to 20 × 10 m at the surface, and a field of view of 17.1°, corresponding to a swath width of ~1.5 km, when viewed from a plane at 5 km altitude. The Daedalus Airborne

Thematic Mapper has 11 channels, with five broad, approximately contiguous, bands in the visible region covering the range 420 to 690 nm, and another six bands in the infrared.[999] It has a particularly wide field of view (86°), and an angular resolution of 2.5 mrad, corresponding to a geometric resolution at sea level of 10 m square from a plane at 4 km altitude.

The NASA Airborne Visible and Infrared Imaging Spectrometer (AVIRIS),[654a] which became operational in 1987, is the most advanced airborne remote sensing radiometer in current use. Its optical system is of the type illustrated in Fig. 7.2*b*, using line arrays of silicon and indium antimonide detectors to provide 224 contiguous spectral bands covering the visible/infrared region from 0.41 to 2.45 μm, with ~10 nm resolution. It is designed to be flown at ~20 km in a high altitude ER-2 aircraft: with its angular resolution of 1 mrad, its ground-instantaneous-field-of-view is 20 m. The total field of view is 30°, giving a swath width of 11 km.

Satellite systems

Several Landsat satellites[815] have been placed in orbit. All contain the MSS Multispectral Scanner with spectral bands 4 (500–600 nm), 5 (600–700 nm), 6 (700–800 nm) and 7 (800–1100 nm). The Landsat optical system was designed primarily for remote sensing of terrestrial regions. Its sensitivity is inadequate to detect the subtle variations in reflectance that occur in the aqueous parts of the globe, and in particular it has no spectral band in the crucially important blue (400–500 nm) region of the spectrum. A mirror oscillating through an angle of 2.9° causes the direction of view to scan from side to side. From its altitude of 920 km it covers a swath 185 km wide, and the unit element of area viewed is about 80 m square.

Landsat D (launched 1982, altitude 705 km) carries, as well as an MSS scanner, a new scanner known as the Thematic Mapper, which has a band in the blue (450–520 nm) as well as bands in the green (520–600 nm), red (630–690 nm) and near-infrared (760–900 nm) wavelengths plus three infrared bands. The unit element of area viewed is 30 m square. Although primarily designed for prediction of crop production, the Thematic Mapper may prove also to be of value for remote sensing of the aquatic environment, but has not been much used for this purpose yet.

Two other spaceborne sensors, not designed for studies on water body colour and composition, but which have occasionally been used for this purpose, are the Advanced Very High Resolution Radiometer (AVHRR), and the high resolution radiometer (HRV) in the System Probatoire d'Observation de la Terre (SPOT). AVHRR on the NOAA TIROS-N

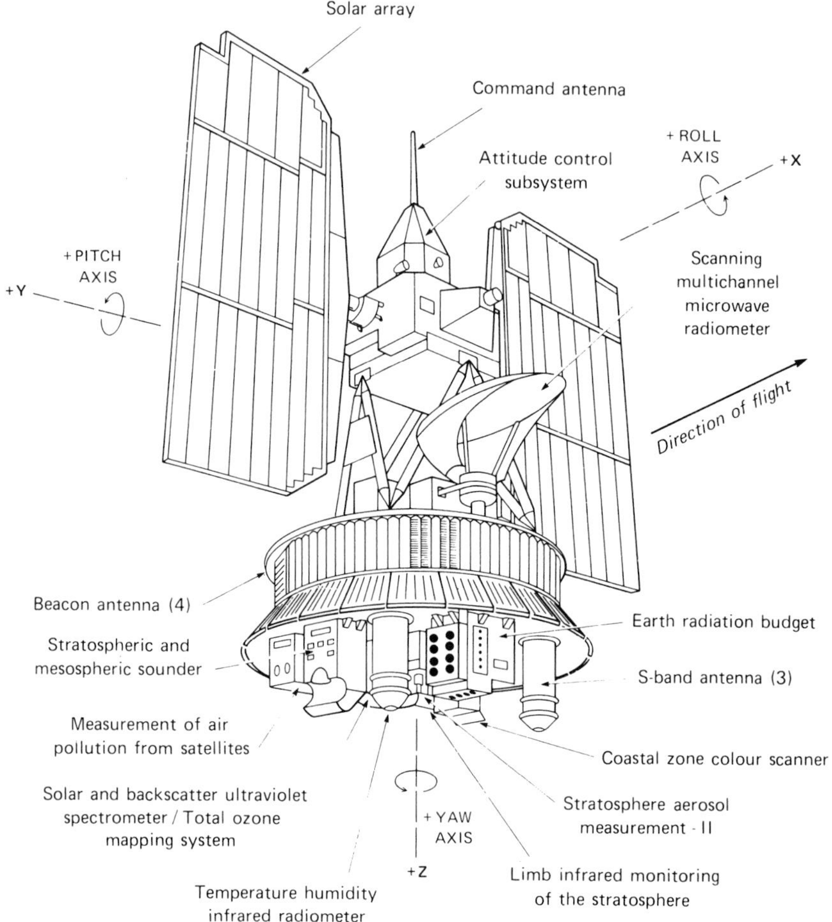

Fig. 7.3. The Nimbus-7 satellite with its various sensors (NASA-PHOTO).

satellite has a band in the red (580–680 nm) and in the near-infrared (720–1000 nm) and four other bands in the near-infrared (for sea surface temperature measurement). Its resolution at sea level is 1 km, with a 3000 km swath. HRV on the French SPOT platform has a green (500–590 nm), a red (610–690 nm) and a near-infrared (790–900 nm) band, and is distinguished by its high ground-level resolution of 20 m (with a 60 km swath), and by being pointable ($\pm 26°$) both across and along track.

In October 1978 the Nimbus-7 satellite (Fig.7.3) was launched, carrying the Coastal Zone Colour Scanner (CZCS) which was designed specifically for remote sensing of the marine environment: its optical system is shown

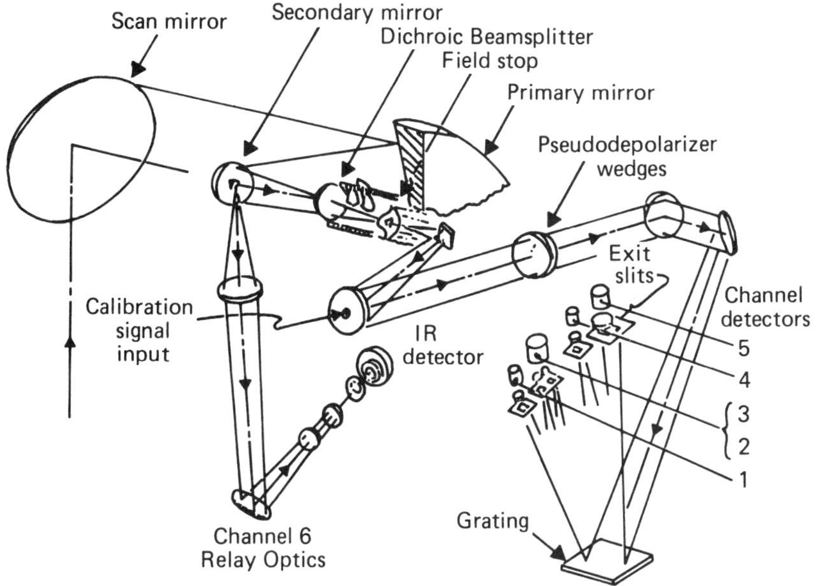

Fig. 7.4. Optical system of the Coastal Zone Colour Scanner in the Nimbus-7 satellite (by permission, from Hovis, 1978).

in Fig.7.4. The CZCS has four bands in the visible region, each 20 nm wide, centred on 443, 520, 550 and 670 nm, one band in the near infrared (700–800 nm) and another band in the infrared (10.5–12.5 μm) for temperature sensing. To facilitate avoidance of Sun glitter the sensor can be tilted so that it scans at up to 20° from the nadir, ahead of or behind the spacecraft. It has a mirror at 45° rotating at 8.1 revolutions per second, at which is directed a telescope which gives an instantaneous field of view of 865 μrad (0.05°) corresponding, at the satellite altitude of 955 km, to a unit element of area 825 m square at the surface.[395,396] It covers a continuous swath 1636 km wide, which for mapping purposes is broken down into individual sections which might typically be 700 km along the spacecraft's track. The radiometric sensitivity of the CZCS is 60-fold higher than that of the MSS on Landsat and has 8-bit digitization. A simple threshold, using the high reflectance of clouds and land in the 700–800 nm waveband, is used to distinguish these areas from open water: it is only to these latter picture elements that the atmospheric correction procedure described later is applied.

The CZCS eventually ceased to function in 1986. During its life it revolutionized our knowledge of the global distribution of ocean colour, and more importantly, of those oceanic constituents, particularly phytoplankton, which affect ocean colour. The archived data accumulated from this scanner during its years of operation are still being worked on. The implications of this new synoptic view of the oceanic biosphere are so far-reaching for our understanding of the marine ecosystem, and in particular for the global carbon cycle, in which oceanic primary production plays a crucial role, that the urgent necessity to replace the CZCS is widely acknowledged. While at the time of writing there is no comparable sensor in orbit, a number are in process of construction, and two of these are briefly described below.

The first to be launched (1994) is SeaWiFS (Sea-viewing-Wide-Field-of-view-Sensor), a joint EOSAT/NASA project.[229] The scanner is of the whiskbroom type with a rotating telescope, and measurement is carried out in eight wavebands isolated with dichroic beam splitters and interference filters. There are six 20-nm spectral bands in the visible region, centred on 412, 443, 490, 510, 555 and 670 nm, and two bands for atmospheric correction in the near infrared at 745–785 nm (blocked between 760 and 770 nm to minimize interference from the atmospheric oxygen absorption band) and 843–887 nm. As the name indicates, SeaWiFS has a very wide field of view – ±58.3° to either side of the track – which, from its orbit height of 705 km, gives a swath width of 2800 km. To avoid sun glitter, the instrument can be directed at the nadir, or at 20°, along track or behind. The sea level resolution is 1.13 km, and the radiance values are 10-bit digitized. Fig. 7.5 shows the instrument design, and in Fig. 7.6 it is shown deployed on the 'SeaStar' spacecraft on which it will in fact (unlike CZCS on Nimbus-7) be the only instrument. SeaStar will be placed in orbit by the relatively small Pegasus rocket which is itself launched from a plane.

The European Space Agency is developing the Medium Resolution Imaging Spectrometer (MERIS)[234] as one of the instruments for launching on the European Polar Platform (POEM-1) in 1998. The optical system, of which there are six identical units within the instrument (see below), covers the spectral range 400 to 1050 nm, but measurements will in fact be confined to 15 spectral bands – which can be selected on command from the ground – within this range. It is currently proposed to have eight bands in the visible, centred on 410, 445, 490, 520, 565, 620, 665 and 682.5 nm, each 10 nm wide except for the 682.5 nm (chlorophyll fluorescence peak) band which is 5 nm wide. Of the seven bands in the near-infrared (700–1050 nm) four will be used to calculate the atmospheric contribution to the radiance measured in the visible bands above the atmosphere (see §7.3), and the remainder for water vapour measurement. Scanning is of the pushbroom type, and the

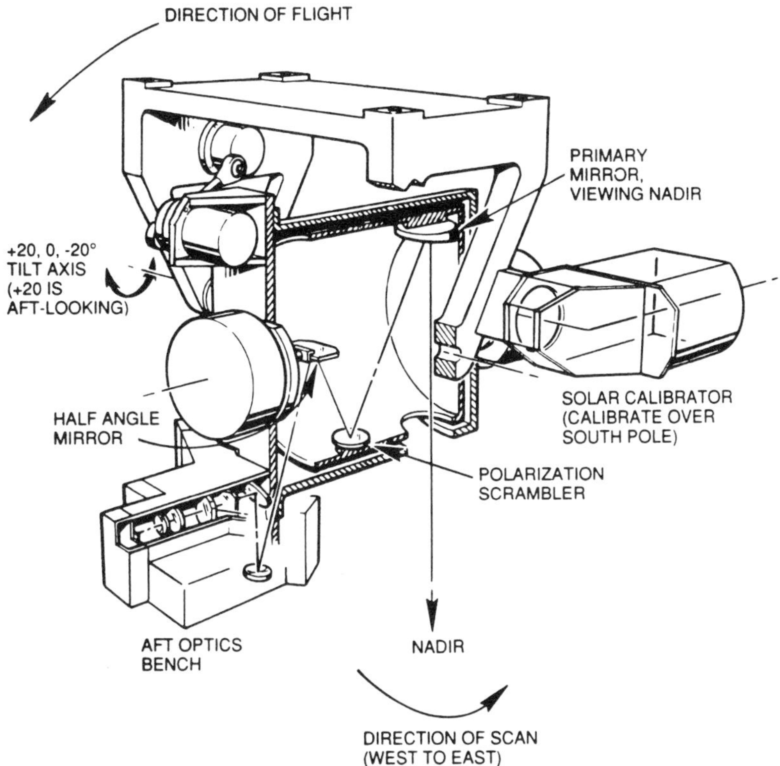

Fig. 7.5. Design of the SeaWiFS instrument for remote sensing of the ocean from space. By permission, Orbital Sciences Corporation, USA.

two-dimensional photodetector arrays are silicon matrix CCDs (charge-coupled device), and the radiance values are 12-bit digitized.

The instrument field of view is 82° centred about the nadir. This wide field of view is shared between the six identical optical modules, each of them receiving 14° of the total field. The spatial resolution at sea level is 250 m when the instrument is operated in full resolution mode (e.g. in coastal waters) and 1000 m in reduced resolution mode (e.g. over the ocean). From the orbit height of ~830 km the swath width is 1450 km, and complete coverage of the Earth's surface is carried out every three days.

7.2 The emergent flux

The particular light flux which is of the greatest interest in the present context is the upwelling light flux just below the surface. However, the flux

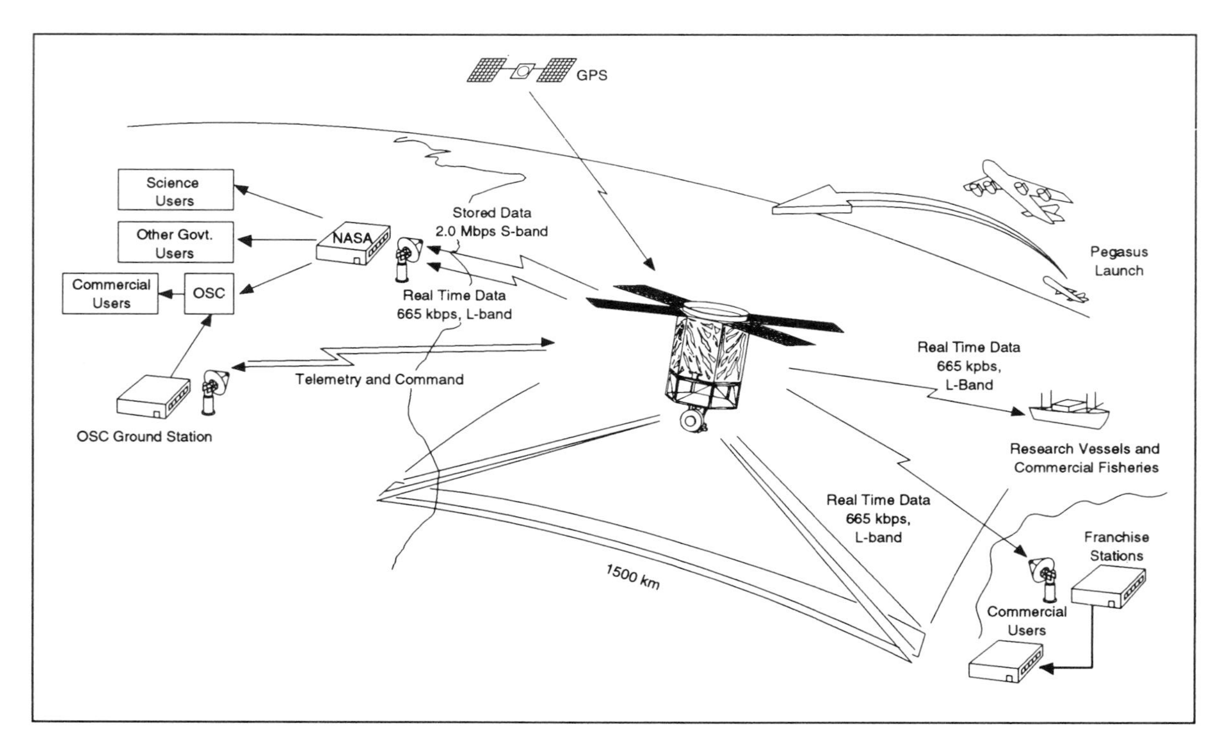

Fig. 7.6. SeaWiFS instrument shown deployed on the **Seastar**™ spacecraft. By permission, Orbital Sciences Corporation, USA.

which, after due correction, is remotely sensed is the emergent flux – that part of the upwelling flux which succeeds in passing up through the surface. How are the two related?

Although about half the total upwelling light flux is reflected downwards again at the water/air interface, this does not represent a serious loss, since it is mainly the flux at larger angles which undergoes reflection. Remote sensing normally involves measurement of radiance at a specific angle rather than total upward irradiance, this angle being not more than 58° from the nadir. Thus remote sensing is normally concerned with that upwelling flux which, below the surface, is at angles of 0–40° to the vertical, and under calm conditions only 2–6% of this is reflected at the water/air boundary. Reflection in this angular range is somewhat increased under rough conditions: at the smaller angles losses remain insignificant but can rise to 16–27% for angles at the upper end of this range.[26]

As the light passes through the water/air boundary it undergoes refraction which, in accordance with Snell's Law (§2.5) increases its angle to the vertical. A further consequence of refraction is that the flux contained within a small solid angle, $d\omega$, below the surface spreads out to a larger solid angle, $n^2 d\omega$ (where n is the refractive index), above the surface. Because of this effect, the value of emergent radiance at any given angle is about 55% of the corresponding subsurface radiance from which it is derived. Combining this effect with the much smaller effect of internal reflection, Austin (1980) proposes a factor of 0.544 for relating radiance just above the surface, $L_w(\theta', \phi)$, to the corresponding radiance just below the surface, $L_u(\theta', \phi)$

$$L_w(\theta', \phi) = 0.544 L_u(\theta, \phi) \tag{7.1}$$

θ being the nadir angle within the water and θ' the angle in air after refraction at the surface.

If $L_w(\theta', \phi)$ can be determined from remote sensing measurements, then it can be multiplied by 1.84 to give $L_u(\theta, \phi)$, and thus provide information about the underwater light field. Upward radiance within the water at nadir angles up to about 30° varies only slightly with azimuth angle (see §6.6). Thus, whatever the azimuth angle of observation, provided the nadir angle of observation is not more than about 42°, the estimated subsurface radiance value, $L_u(\theta, \phi)$ can be taken as the value at that subsurface nadir angle averaged over all azimuth angles, $L_u(\theta)$. Furthermore, $L_u(\theta)$ does not change much with nadir angle in the range 0–30°, and so the value of $L_u(\theta)$ can be taken as an estimate of the radiance, L_u, in the vertically upward direction.

7.3 Correction for atmospheric scattering and solar elevation

We have already noted that at least 80% of the upward radiance measured by a photometer in a satellite, or in a very high altitude (~20 km) aircraft, above the ocean originates by scattering of the solar beam by air molecules and by aerosol particles (dust, water droplets, salt, etc.) within the atmosphere – this is known as the *path radiance*. To arrive at a value of L_w from the measured radiance, the path radiance must be removed, and furthermore the attenuation of the emergent flux during its passage through the turbid atmosphere to the sensor must be allowed for.

As discussed in a later section, it is the ratio of the emergent flux to the incident downward flux, rather than the absolute value of emergent flux alone, which is directly related to the optical properties of the aquatic medium. It is therefore desirable to have an estimate of the downward irradiance at the surface of the water: this will be determined by solar elevation and by attenuation of the solar beam through the atmosphere, and so both of these will have to be taken into account.

Landsat

In the case of Landsat studies on aquatic systems, no generally applicable method of correcting for these atmospheric effects appears to have been devised. The use of Landsat for remote sensing of water composition has hitherto involved correlating increases in radiance in particular wavebands with increases in concentration of particular types of suspended particles, the increased radiance being due to increased backscattering. In some cases correlations between the uncorrected radiances and concentration of a particular component have been used. The problem in such a case is that on another occasion, when atmospheric conditions have changed, the previously derived relation will no longer apply. In studies on Lake Superior, Canada, for example, the curve of measured radiance against turbidity shifted up and down and changed somewhat in slope on different days, in accordance with changes in atmospheric conditions.[797]

A simple, but very approximate, procedure for reducing atmospheric effects is the *dark pixel correction method*.[760,761,766] This involves locating the darkest pixel in the scene, and subtracting the radiance value for this from the radiance values for all the other pixels. The rationale is that the darkest pixel will have the same atmospheric contribution as all the other pixels, but must have the smallest contribution from scattering within the water. Subtraction of the darkest pixel radiance from all the other pixel radiances should thus remove their atmospheric contribution, and what is left is a function of the *difference* in particle scattering between the various parts of the water body, and that part with the darkest pixel. To be useful, this method does, however, require that there should exist somewhere in the scene an area of water with a particles content that is at least relatively low: it also assumes constant atmospheric conditions throughout the scene, a reasonable assumption in the case of lakes, the water bodies to which this procedure has normally been applied.

Up to this point we are still dealing in terms of radiance, measured above the atmosphere, but as we have noted, for relating to the optical properties of the medium, the reflectance of the water body would be preferable, and this will depend on solar irradiance at the surface. A first-order correction for the variation in reflectance due to changing solar elevation may be made by dividing the measured radiance by the cosine of the solar zenith angle,[2,520,648] which in effect normalizes radiance values to a standard downward irradiance, and so links them more closely, but still far from exactly, to reflectance.

Verdin (1985) attempted to determine the actual reflectance values of a water body from Landsat data, with the help of atmospheric radiative transfer calculations. An essential component of the procedure is that there should be an area within the scene which can confidently be assumed to contain clear oligotrophic water. For each waveband a reflectance value for this area is assumed, on the basis of literature data. A plausible initial assumption concerning the prevailing atmospheric turbidity is made from which, by radiative transfer calculation, a value for path radiance is arrived at and tested to see whether, taken together with the assumed water body reflectance, it accounts for the satellite-measured radiance from the clear water. The calculation is repeated in an iterative manner, progressively altering atmospheric turbidity, until a satisfactory value for path radiance is arrived at, and this is assumed to apply throughout the scene. The radiative transfer calculations also yield values for the transmittance of solar radiation down through the atmosphere, and of radiance from the water surface back to the satellite. Thus, for every pixel in the scene, the value of the water-leaving radiance, L_w, and of the incident solar irradiance, can be calculated, and from these the reflectance can be determined.

MacFarlane & Robinson (1984) have also sought to determine actual reflectance values from Landsat data, for English coastal water. They used an atmospheric correction method similar to that developed for the CZCS (see below), and found it to give superior results to the simpler procedures, such as darkest pixel correction, division by cosine of solar zenith angle etc.

Nimbus-7

Remote sensing of water composition with Landsat involves relating *increased* radiance values to the concentration of different kinds of suspended particles, and useful results can be obtained by subtracting comparatively crude estimates of path radiance from the observed radiance values. The Coastal Zone Colour Scanner (CZCS) carried by the Nimbus-7 satellite, in contrast, is required to detect decreases in radiance in some wavebands as well as increases in others, and a much better estimate of the atmospheric scattering contribution to the measured radiance is needed. Ways of achieving this are outlined below.

The correction procedure of Gordon *et al.* (1983), based on earlier work by Gordon (1978) and Gordon & Clark (1981), takes as its starting point the proposition that the total radiance, L^{λ}, observed by the sensor at wavelength λ, is the sum of the radiance due to Rayleigh scattering (L_R^{λ}), that due to aerosol scattering (L_A^{λ}), and that backscattered out of the ocean (L_W^{λ}) attenuated by the atmospheric transmittance factor t^{λ} during its passage from the surface to the satellite

$$L^{\lambda} = L_R^{\lambda} + L_A^{\lambda} + t^{\lambda} L_W^{\lambda} \tag{7.2}$$

L_R^{λ} includes the Rayleigh component of the skylight reflected towards the sensor by the surface, as well as that backscattered from the direct solar beam. L_A^{λ} similarly, includes the aerosol component of the reflected skylight as well as that light scattered back by aerosol particles towards the sensor from the solar beam. The atmospheric transmission factor is for diffuse rather than direct transmittance because when the sensor is viewing a given element of area, some of the flux is coming from neighbouring elements.

The correction procedure goes on to take advantage of the fact that water absorbs light strongly in the red/near infrared region, and so essentially all the radiance measured at this end of the spectrum can be attributed to scattering within the atmosphere.[171] For oceanic waters, which are usually very clear, the 670 nm band may be used. In more turbid waters, however, such as those of inland water bodies or near the coast, the concentration of particulate matter can be high enough to backscatter significant amounts of 670 nm light out of the water[116] and in these cases a more strongly absorbed, near infrared waveband would be preferable.

For the 670 nm radiance, since $L_W^{670} \simeq 0$. eqn 7.2 becomes

$$L^{670} = L_R^{670} + L_A^{670} \tag{7.3}$$

Since the contribution to scattering by air molecules is essentially constant, the radiance at the sensor's viewing angle due to Rayleigh scattering is a known function of solar elevation and wavelength, and so L_R^{670} and L_R^λ, can be calculated and are the same for all pixels. The value of L_A^{670} is then obtained from the measured L^{670}, using eqn 7.3.

The approximate assumption is now made that the radiance due to aerosol scattering at wavelengths λ, while not identical to that at 670 nm, is at least proportional to it, i.e. as L_A^{670} varies with atmospheric turbidity from one part of the scene to another, so L_A^λ will vary in parallel with it. The ratio, L_A^λ/L_A^{670}, we shall refer to as $S(\lambda, 670)$. Substituting for L_A^λ in eqn 7.2, and rearranging, we obtain

$$L_W^\lambda = (1/t^\lambda)[L^\lambda - L_R^\lambda - S(\lambda, 670)L_A^{670}] \tag{7.4}$$

an equation for the emergent radiance at wavelength λ, with the remaining unknowns, t^λ and $S(\lambda, 670)$.

The diffuse transmittance of the atmosphere for wavelength λ, t^λ, is a function of absorption by ozone, Rayleigh scattering by air molecules, and scattering and absorption by aerosol particles. The first two of these can be calculated and are the same for all pixels. Gordon *et al.* (1983) propose, partly on the grounds that this is *diffuse* transmittance, and most aerosol scattering is in a forward direction anyway, that the effect of aerosol on transmission of the water-leaving radiance to the sensor may be disregarded for all atmospheres in which this kind of remote sensing is likely to be attempted. Thus, t^λ can be calculated for all wavebands.

To determine the coefficient $S(\lambda, 670)$ in eqn 7.4, Gordon *et al.* make use of the observation from field measurements[321] that in clear oceanic water with less than 0.25 mg m^{-3} of phytoplankton chlorophyll, the water-leaving radiance values at 520, 550 and 670 nm under sunny conditions are quite constant and predictable functions of solar elevation. Thus if an area of such water can be identified somewhere within the scene, then the values of L_W within these wavebands are already known, and substitution into eqn 7.4, together with the appropriate values of t^λ, L^λ, L_R^λ and L_A^{670}, will give the values of $S(520, 670)$ and $S(550, 670)$. Now the coefficient S, which is the ratio of aerosol path radiance at wavelength λ to that at wavelength 670 nm, combines two different elements: $F_0'(\lambda)/F_0'(670)$, the ratio of extraterrestrial solar irradiance in the two wavebands, corrected for absorption by ozone, and $\epsilon(\lambda, 670)$, which is the ratio for the two wavebands, of aerosol path radiance per unit incident irradiance

$$S(\lambda, 670) = \epsilon(\lambda, 670)\frac{F_0'(\lambda)}{F_0'(670)} \tag{7.5}$$

Therefore using the appropriate values of $F_0'(520)$, $F_0'(550)$ and $F_0'(670)$, the values of $\epsilon(520, 670)$ and $\epsilon(550, 670)$ may be determined from the already-determined values of $S(520, 670)$ and $S(550, 670)$ for the clear water pixels. The significance of this is that provided the *type* of aerosol, although not of course the concentration, is constant throughout the scene (often a reasonable assumption over the ocean) then these ratios of aerosol path radiance (per unit incident irradiance) in a given pair of wavebands, apply to every pixel in the scene. Thus, for every pixel, once L_A^{670} has been determined, L_A^{520} and L_A^{550} can be calculated.

There remains the problem of determining L_A^{443} for each pixel. Advantage here is taken of the fact that aerosol scattering varies with wavelength approximately in accordance with

$$\epsilon(\lambda_1, \lambda_2) = \left[\frac{\lambda_1}{\lambda_2}\right]^{-\alpha} \quad (7.6)$$

where α is known as the Ångstrøm exponent. A mean value of α is obtained from $\epsilon(520, 670)$ and $\epsilon(550, 670)$, and then used to determine $\epsilon(443, 670)$. $S(443, 670)$ is calculated with eqn 7.5, and then used to obtain L_A^{443}.

By the above procedure, all the terms on the right-hand side of eqn 7.4 can be determined, and the value of L_W^{λ}, the emergent radiance at wavelength λ, can be obtained for every pixel in the satellite image. If desired this can be converted to the subsurface upward radiance, L_u^{λ}, by dividing by 0.544 (§7.2). If, to eliminate the effects of variation in lighting and viewing geometry, it is wished to determine the reflectance at or below the surface, this can be calculated for the appropriate solar elevation, using the atmospheric transmission information derived from the satellite's own measurements, together with reasonable assumptions about Rayleigh scattering and absorption by ozone.[959,96]

The atmospheric correction procedure of Gordon *et al.* assumes, as we have noted, that in the 670 nm waveband the emergent flux from the ocean is so low that it can be disregarded. Even for oceanic waters, however, this is not always a satisfactory solution, and Smith & Wilson (1980) and Bricaud & Morel (1987) have developed atmospheric correction procedures which, by actually estimating L_W^{670}, make this assumption unnecessary. Their calculation procedures initially set L_W^{670} to zero and then proceed, essentially in the manner of Gordon *et al.*, to arrive at values of L_W^{550} and L_W^{443}. An empirical relationship, between these two radiances and L_W^{670}, based on accumulated field data, is then used to arrive at a new value of L_W^{670}. This new, non-zero, value is used as the starting point of a calculation of new values of L_W^{550} and L_W^{443}, which can then once again be used to up-date the value of L_W^{670}. The iteration continues until successive values of L_W^{670} agree to within some preset proportion.

Correction for atmospheric aerosol scattering could be improved with the help of radiance data at wavebands in the near infrared, where water absorbs so strongly that the emergent flux from the sea could indeed be ignored. To achieve this, measurements in such wavebands are being carried out in the new generation of spaceborne ocean colour sensors: SeaWiFS has two such near-infrared bands, and MERIS has four.

High-altitude aircraft

Some remote sensing studies on water composition carried out from high-altitude aircraft have not attempted to correct for the contribution of atmospheric scattering to the measured radiance. They have relied instead on multiregression analysis to detect the correlation between total radiances in various wavebands, and the changing water composition parameters.[432,433]

Kim, McClain & Hart (1979) used a correction procedure analogous to that described above for the Nimbus-7 measurements, to correct their data for Californian coastal waters obtained with the Ocean Colour Scanner from a U-2 aircraft at 19.8 km altitude. They assumed that all the radiance measured at 778 nm was due to atmospheric scattering. From the 778 nm radiance, using a mathematical model, they calculated, for all the other wavebands used by their scanner, the expected contribution of atmospheric scattering to observed radiance, and these values were subtracted from the observed values.

7.4 Relation between remotely sensed radiances and water composition

While values for upward radiance are of some interest in themselves, their main significance lies in what they can tell us about the optical properties of the medium, and thus about the concentrations of the optically significant constituents in the water. A possible route by which such information may be acquired makes use of the approximate relations, derived from theoretical modelling of the underwater light field (§6.4, eqns 6.3 and 6.5), that exist between radiance reflectance (L_u/E_d), or irradiance reflectance (E_u/E_d) and the ratio of backscattering coefficient to absorption coefficient.

Whichever relation is used, it is necessary to have a value for the downward irradiance, E_d, just below the surface. Knowing the solar elevation, making reasonable assumptions about Rayleigh scattering and ozone absorption, and assigning a plausible value to aerosol scattering (based, if possible on the remote sensor's own measurements as described previously) the atmospheric transmittance may be calculated[959] and so the downward irradiance on the surface determined. From this, making due allowance for reflection at the surface, a value for E_d could be obtained. In the case of remote sensing from low-flying aircraft, downward irradiance can be measured on the aircraft and assumed to be equal to that just above the water surface.

L_u is obtained by multiplying L_W by 1.84 (see §7.2). E_u is obtained by multiplying L_u by 5, or some slightly different factor depending on the solar elevation (see §6.4). If, as could be the case in low-altitude studies, the remote sensor measures irradiance, the value of E_u may be obtained by dividing the observed value of upward irradiance above the surface by 0.52 to allow for the approximately 48% reflection of diffuse upwelling irradiance at the water/air boundary.[26]

Using the appropriate equation relating L_u/E_d or E_u/E_d to b_b/a, the ratio of the backscattering coefficient to the absorption coefficient could be estimated. If, on the basis of other information, a plausible value could be selected for b_b, then an approximate value for a could be arrived at. The possibility of directly estimating b_b from remotely sensed reflectance in the far red/near infrared region is discussed later. If measurements are made at three wavelengths, and reasonable assumptions are made about the manner in which b_b and the values of a for the different absorbing components (water, phytoplankton, non-living colour), vary with wavelength, then the actual concentrations of phytoplankton and non-living pigmented material can be estimated: the theoretical basis for such calculations is given by Morel (1980). Ways of calculating phytoplankton concentration from above-surface reflectance measurements have also been described by Pelevin (1978) and Kopelevich & Mezhericher (1979).

Most remote sensing studies of water composition have made use, not of the *a priori* deductive method outlined above, but of an empirical approach in which relations have been sought between remotely measured radiances in particular wavebands, or functions of one or more radiances or reflectances, and the concentrations of specific components such as phytoplankton and suspended solids determined by *in situ* measurements within water bodies. Once such relations are established it should in principle be possible to use them to determine the optical properties and composition of water bodies from remote sensing data alone. This has been a very active field of research in recent years, and some representative examples are discussed below.

A useful concept in this context is that of a *retrieval variable*,[909] which is any function of the reflectance (or radiance) values in one or more wavebands, $X[R(\lambda_i)]$, that is empirically found to be related to the concentration of some component of the water by a simple algorithm. Thus, the problem in remote sensing is usually to identify a suitable X, and to develop a simple equation, by means of which the concentration of the constituent of interest may be calculated from X.

Suspended solids

One of the water quality parameters which has been most extensively studied in this way is the concentration of total suspended solids (TSS). Increases in the amount of suspended particles will, at wavelengths where the particles do not absorb strongly, increase the backscattering coefficient of the water more than the absorption coefficient, and so, in accordance with eqns 6.3 and 6.5, increase the emergent flux.

In the case of reservoirs in Mississippi, it was found that the irradiance reflectance measured above the water surface increased with the concentration of suspended (inorganic) sediment at all wavelengths from 450 to 900 nm (Fig. 7.7).[762] The best, linear, correlation between reflectance and suspended solids concentration was found between 700 and 800 nm. Correlations between suspended solids or nephelometric turbidity (proportional to suspended solids) and radiances in particular wavebands have been obtained with the Landsat satellites. Klemas, Borchardt & Treasure (1973) found that in Delaware Bay the distribution of MSS 600–700 nm radiance corresponded best with sediment load in the upper 1 m of the water; the 500–600 nm radiance was more subject to interference from atmosphere haze while the 700–800 nm waveband did not penetrate sufficiently into the water column. Similarly, in Landsat imagery of Kenyan coastal waters, the 700–800 nm band revealed only high sediment concentrations near the surface, whereas the 600–700 nm and 500–600 nm bands were sensitive to lesser concentrations lower in the water column.[93] Suspended solids values for the Bay of Fundy, Nova Scotia, Canada, correlated with particular functions of upward radiance in the 500–600, 600–700 and 700–800 nm bands, but the latter two bands gave the highest correlation coefficients.[648] In the case of Lake Ontario, Canada, a good linear relation was observed between radiance reflectance in the 600–700 nm band and measured turbidity.[113] For two turbid oxbow lakes of the Mississippi River (USA), Moon Lake (Mississippi) and L. Chicot (Arkansas), it was found that of the four Landsat MSS bands, the 700–800 nm band was the best for monitoring suspended sediment loads.[758,761,798]

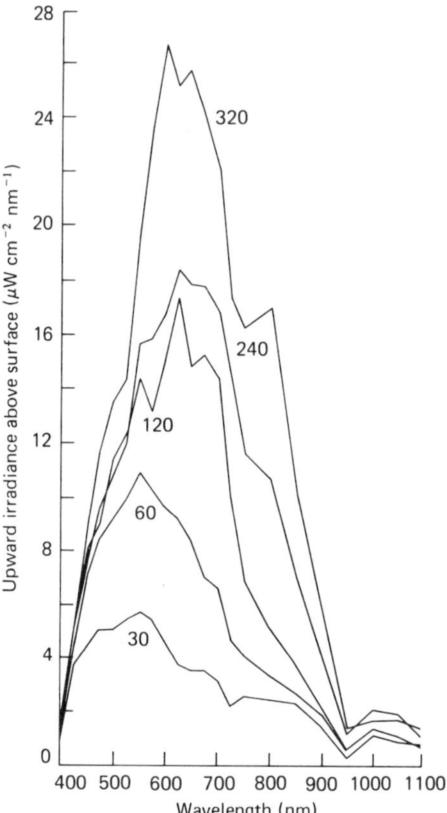

Fig. 7.7. Intensity and spectral distribution of upwelling radiation measured 20–50 cm above the surface of fresh-water bodies containing various levels of suspended solids (after Ritchie, Schiebe & McHenry, 1976). The concentration of suspended solids in mg l^{-1} is indicated next to each curve.

Munday & Alföldi (1979) found that Landsat radiance values correlated somewhat better with the logarithm of TSS than with TSS alone, in the Bay of Fundy. Some algorithms for suspended sediments have been developed using combinations of Landsat radiances rather than the radiance values in a single waveband.[464,758,759,931] The quest for a universal algorithm for suspended sediments[391,759] is, however, never likely to succeed, no matter what combination of bands is used. The main reason is that the scattering efficiency of suspended particles *per unit weight* is very much a function of size (Fig. 4.3), and average particle size of suspended sediments is quite variable, both with time and place. There are additional reasons related to the absorption properties of the surrounding medium and of the particles themselves. Since reflectance varies inversely with a, at the same time as it increases with b_b, a reproducible linear relation between emergent radiance and TSS is only to be expected if the total absorption coefficient of the aquatic medium remains constant while the concentration of suspended solids varies. This need not be the case, and variations in absorption coefficient from place to place or over a period of time, within the water body under study, would affect the reproducibility of any empirically derived relation between L_W and TSS, and so affect the accuracy with which the concentration of suspended solids can be determined from remotely sensed data. In the red and near infrared wavebands, however, absorption is mainly due to water itself, and so fluctuations in absorption should not be large.

The spectral distribution of light reflected by suspended inanimate particulate matter varies in accordance with its own absorption properties, which are determined by its chemical nature. For example, suspensions of white clay and red silt have quite different reflectance spectra.[666] Sydor (1980), on the basis of his measurements of the spectral distribution of light scattered upwards from suspensions of particular types (red clay, mine tailings) concluded that evidence on the identity as well as concentration of the suspended solids may be derived using the 500–600, 600–700 and 700–800 nm bands of Landsat.

The work that we have discussed so far, and indeed the majority of studies on TSS that have been carried out, have made use of Landsat MSS data, but there have been some studies with other sensors. Ritchie, Cooper & Schiebe (1990) found Thematic Mapper data to be as good as , but no better than, MSS data for estimating suspended sediments in Moon Lake (Miss.), despite the narrower spectral band width of the former. They consider that the major justification for using Thematic Mapper data for sediment studies would be that its higher spatial resolution (30 m) would make it possible to monitor smaller lakes. The SPOT multispectral scanner has a ground-level resolution of 20 m. For the turbid waters of Green Bay, L. Michigan, Lathrop & Lillesand (1989) found that for extra-atmospheric reflectance in all three bands (green, red, near-infrared) the best correlation was with the logarithm of suspended sediment concentration: some improvement was obtained by regressing the logarithm of TSS against a combination of the three reflectances.

Stumpf & Pennock (1989) have developed a method for determining suspended sediment concentrations in estuaries using the red and near-infrared channels of the AVHRR. They use a correction procedure similar to that developed for CZCS data (above) to remove atmospheric effects, and work in terms of the calculated values for reflectance.

The coastal Zone Colour Scanner itself has not been very much used for monitoring suspended sediments because it was designed for observation of relatively clear waters, and its channels are saturated by the high radiance values coming from waters above a certain level of turbidity.[13,278] Tassan & Sturm (1986) have sought to devise a retrieval variable for suspended sediments in coastal waters, for use with CZCS data, which will have low sensitivity both to the presence of phytoplankton pigments and to the atmospheric correction uncertainty. They propose the variable

$$X_s = [R(550) - R(670)]^a [R(520)/R(550)]^b$$

where the $R(\lambda)$ values are the subsurface irradiance reflectances at the indicated wavelengths, and the exponents a and b can have the values $0.5 \leqslant a \leqslant 1.5$ and $-2.5 \leqslant b \leqslant -0.7$. The retrieval variable so calculated is used in the algorithm

$$\log \text{TSS} = A + B \log X_s$$

the coefficients A and B being determined by best fit to *in situ* measurements of reflectance and sediment concentration. For the Adriatic Sea, offshore from Venice, with $a = 1$ and $b = -1$, field data gave $A = 2.166$ and $B = 0.991$.

As the concentration of suspended sediments in water bodies increases, so the shape of the spectral distribution of reflectance changes: in particular, on either side of the main peak in the $R(\lambda)$ spectrum, the gradient of the curve becomes steeper as the reflectance peak increases in height. It had already been found[129,331] that changes in spectral curvature of reflectance can be used to remotely sense phytoplankton concentration. Chen, Curran & Hanson (1992) have demonstrated the feasibility of the same approach for remote sensing of TSS: strong correlations between $\mathrm{d}R(\lambda)/\mathrm{d}\lambda$ and sediment concentration were found in a number of spectral ranges within the visible and near-infrared regions. The particular advantage of this approach – compared to deriving TSS simply from $R(\lambda)$ – was that the correlation was less affected by the variable specular reflectance from a roughened sea surface. Its use does, however, require data which are either spectrally continuous across the wavelength range, or which come from a small set of closely spaced narrow wavebands.

Phytoplankton – increased reflectance

One empirical approach to the remote sensing of phytoplankton is to treat it as a special case of suspended solids and to use the increased reflectance in the near infrared associated with the increased biomass. In the Landsat MSS red (600–700 nm) band, high concentrations of algae are associated with decreased radiance because of the chlorophyll absorption in this region. Strong (1974) observed in the case of Landsat pictures of an algal bloom on Utah Lake, USA, that there was a contrast reversal in going from the red to the infrared (700–800 nm or 800–1100 nm) bands: areas dark in the red image showed up as light against a darker background in the infrared images. Bukata *et al.* (1974) found that the reflected radiance in the Landsat 700–800 nm band correlated well with the logarithm of the chlorophyll concentration in Lake Ontario, Canada, and some nearby eutrophic water bodies. The sensitivity of this method is, however, comparatively low (>10 mg chl *a* m^{-3}) and it is perhaps most applicable to highly productive inland waters which develop algal blooms. Suspended inorganic particles can also increase reflectance in the 700–800 nm band. A comparison of the distribution of 600–700 nm waveband reflectance (which correlates better with inorganic turbidity than with biomass) with that of 700–800 nm reflectance is necessary to ensure that suspended sediments are not mistaken for phytoplankton.[112]

Dekker, Malthus & Seyhan (1991), using a multispectral scanner (the Programmable Multispectral Imager) from an aircraft at 1000 m over eutrophic lakes in the Netherlands, found a good linear negative correlation between phytoplankton chlorophyll and the ratio of red (673–687 nm) to near-infrared (708–715 nm) radiance values. Mittenzwey *et al.* (1992), using a shipboard spectroradiometer positioned just above the water surface, found for productive (3–350 mg chl *a* m^{-3}) lakes and rivers around Berlin, a very clearcut positive correlation ($r^2=0.98$) between phytoplankton chlorophyll concentration and the ratio of a near-infrared (705 nm) to a red (670 nm) reflectance. The basis of these observations is no doubt, once again, the combination of high biomass reflectance in the near-infrared, with absorption by the chlorophyll peak in the red.

The reflectance of water bodies in the green waveband also increases with phytoplankton concentration (see next section). Lathrop & Lillesand (1986), using Thematic Mapper data for Lake Michigan, found good linear relationships between the logarithm of chlorophyll concentration and the logarithm of radiance in the green band (520–600 nm), for the central lake (oligotrophic) and Green Bay (eutrophic). The relationships were, however, very different for the two parts of the lake, the central lake showing a much higher green radiance for a given level of phytoplankton than the bay, possibly due to a higher level of yellow humic substances in the latter water body, leading to greater absorption of green light. Clearly there is no prospect for a universal phytoplankton algorithm based on green reflectance, but locally applicable algorithms of this type may be of some use.

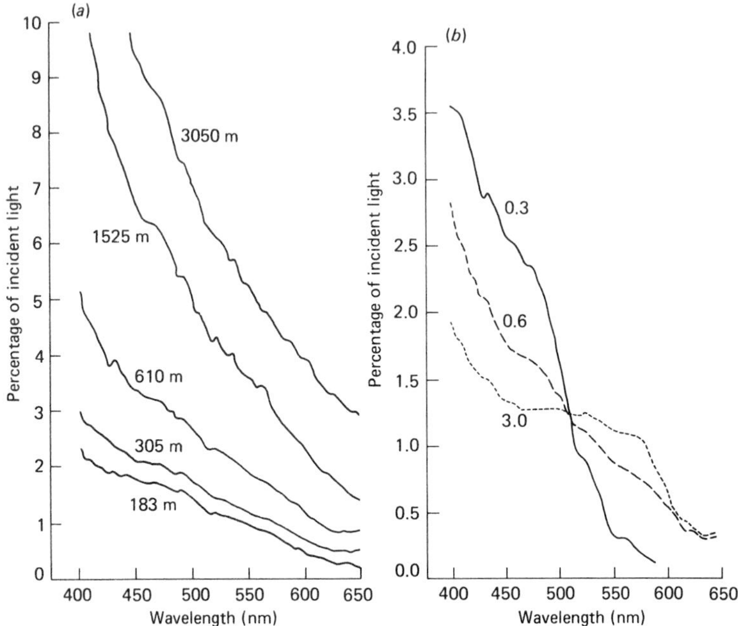

Fig. 7.8. Spectral distribution of reflectance above the Atlantic ocean off New England, USA (after Clarke *et al.*, 1970). Measurements were made at zenith angle of 53°, using a polarizer to eliminate reflected skylight. (*a*) Dependence of reflectance on altitude: measurements were made at five altitudes over the same location (1° latitude east of Cape Cod). The large, and increasing, contribution of atmospheric scattering to measured reflectance at altitudes above 305 m is readily apparent. (*b*) Dependence of reflectance on phytoplankton content. Measurements were made at low altitude (305 m). The phytoplankton chlorophyll level (mg chl *a* m^{-3}) is indicated above each curve. The three curves are for water on the northern edge of the Gulf Stream (0.3 mg chl *a* m^{-3}), on the edge of the continental shelf (0.6 mg chl *a* m^{-3}), and over the productive Georges Shoals (3.0 mg chl *a* m^{-3}).

Phytoplankton – increased absorption

The best approach to the remote sensing of phytoplankton is undoubtedly that which makes use of the increased absorption, and therefore reduced radiance reflectance, at the blue end of the spectrum caused by the photosynthetic pigments. Clarke *et al.* (1970) measured from a low-flying aircraft, the spectral distribution of the emergent flux in different regions of the Northwest Atlantic Ocean with chlorophyll *a* concentrations varying from <0.1 to 3.0 mg m^{-3}. The general tendency (Fig. 7.8*b*) was that as the phytoplankton concentration increased, reflectance decreased in the blue

(400–515 nm) and increased in the green (515–600 nm). The increased reflectance in the green may be attributed to the fact that phytoplankton, being refractive particles, increase scattering at all wavelengths, but in this spectral region absorb only weakly. Empirical methods of remotely estimating phytoplankton concentration generally take advantage of both the decreased reflectance in the blue and the increased reflectance in the green, by working in terms of the ratios or differences between reflectances in these two wavebands.

Morel & Prieur (1977) on the basis of their measurements of the spectral distribution of the upwelling and downwelling light flux in various oceanic waters found that the ratio of underwater irradiance reflectance (related in the manner discussed above to the reflectance estimated by remote sensing above the water) at 440 nm to that at 560 nm decreases as the phytoplankton pigment content increases. If data for waters with high scattering coefficients due to suspended sediments are omitted then an approximately linear relation is obtained between the logarithm of R_{440}/R_{560} and the logarithm of the chlorophyll plus phaeopigment (chlorophyll breakdown products) concentration over the range 0.02 to 20 mg total pigment m^{-3}.[634]

Gordon *et al.* (1983) have used the effect of phytoplankton on the spectral distribution of the upwelling flux to develop an algorithm by means of which L_W values obtained from the Coastal Zone Colour Scanner in the Nimbus-7 satellite can be used to calculate phytoplankton pigment concentration. They measured the upward nadir radiance within the water (which as we saw earlier is directly related to the emergent flux) at 440, 520 and 550 nm in marine waters with chlorophyll *a* (plus phaeopigment) concentrations ranging from 0.03 to 78 mg m^{-3}. The data conformed reasonably well to relations of the type

$$\log_{10} C = \log_{10} a + b \log_{10} R_i \quad (7.7)$$

where C is the chlorophyll *a* (plus phaeopigment) concentration in mg m^{-3}, and R_i is the 440/550, 440/520 or 520/550 ratio of radiances. In the case of the 440/550 nm pair, for example, in Case 1 waters $\log_{10} a$ is 0.053 and b is -1.71, i.e. as phytoplankton concentration increases so the ratio of upwelling radiance at 440 nm to that at 550 nm decreases. By substitution of a pair of values of upwelling radiance into the appropriate form of eqn 7.7 the chlorophyll (plus phaeopigment) concentration can be calculated. For chlorophyll concentrations below 0.6 mg m^{-3} the radiances at 443 and 550 nm are most useful. At higher concentrations, however, L_W^{443} becomes too small for accurate determination by the CZCS, and the radiances at 520 and 550 nm are used instead. Approximate agreement was found between the variation of chlorophyll concentration determined by surface sampling along a ship course of several hundred km in the Gulf of Mexico with that calculated for the same course from CZCS data:[323] similar results were obtained in the Atlantic Bight and Sargasso Sea.[322]

An algorithm such as eqn 7.7, with appropriate values of $\log_{10} a$ and b, can be applied to all Case 1 oceanic waters, i.e. waters whose optical properties are almost entirely determined by phytoplankton and its associated debris and breakdown products. In Case 2 waters, whose optical properties are substantially affected by river-borne turbidity and dissolved yellow colour, or by resuspended sediments, relationships such as eqn 7.7 still apply,[104,326,618] but the values of the two constants are different from those for Case 1 waters, and will also differ from one Case 2 water to another. Thus there is no universal Case 2 algorithm, and so locality-specific algorithms must be sought for these waters.

In Case 2 waters with significant amounts of land-derived gilvin, it is impossible with CZCS data to distinguish the lowered reflectance in the blue caused by this soluble yellow colour from that due to phytoplankton. The next generation of ocean-viewing spaceborne radiometers, such as SeaWiFS and MERIS, are all designed to carry out measurements at the extreme short-wave end of the visible spectrum, at ~410 nm, where phytoplankton absorption has somewhat diminished (relative to that at 440 nm), but gilvin absorption is even higher. It is anticipated that the availability of this short-wavelength band will increase the feasibility of distinguishing phytoplankton from dissolved yellow colour.[131,791]

The Landsat Thematic Mapper has a blue, as well as a green and a red, waveband. The very limited amount of data so far available suggests that while it is not likely to be suitable for measuring the low levels of phytoplankton in oligotrophic ocean waters, it might provide useful data on the higher levels occurring in coastal waters, by means of a blue–green algorithm[215,465,473] in accordance with eqn 7.7.

The major uncertainty in the use of the remotely sensed blue and green radiances for phytoplankton estimation is the determination of the very large atmospheric correction. This problem can be avoided by using low-altitude aircraft. In waters with significant levels of suspended sediments, a blue–red algorithm may work better.[220] Borstad *et al.* (1980) eliminated atmospheric effects by flying at a height of 100 m and minimized sky reflection by taking measurements of radiance at the Brewster angle, with a polarizing filter over the aperture of their scanning spectrometer. For the relatively eutrophic British Columbian coastal waters they obtained an approximately linear relation between the ratio of radiance reflectance at 560 nm to that at 440 nm and the chlorophyll *a* concentration measured at

3 m depth, over the range 2 to 13 mg m^{-3}. Viollier *et al.* (1980) carried out low-altitude airborne measurements of reflectance at 466 and 525 nm, together with measurements of chlorophyll, in the tropical East Atlantic Ocean and in the North Sea. They obtained an approximately linear relation between $(R_{525}-R_{466})$ and the logarithm of chlorophyll concentration over the range 0.2 and 3 mg m^{-3}.

Phytoplankton – changes in spectral curvature and slope

As the concentration of phytoplankton in a water body changes, as well as the major changes in the blue, green, red and near-infrared components of the upwelling light flux that we have discussed so far, there are localized changes in the shape of the spectral distribution curve that can be quantitated with a radiometer having the requisite spectral resolution. Spectral curvature or slope in different regions of the spectral distribution may thus provide us with another set of retrieval variables for phytoplankton, additional to those based on the broad blue/green, red/near-infrared changes.

This approach was first developed by Grew (1981) using the airborne Multichannel Ocean Colour Sensor over US East Coast waters, and applying to the data a spectral curvature parameter of the form

$$G(\lambda)=\frac{L(\lambda)^2}{L(\lambda-\Delta\lambda)L(\lambda+\Delta\lambda)} \tag{7.8}$$

where $L(\lambda)$ is the radiance at a central wavelength and the other two radiances are at wavelengths at an interval $\Delta\lambda$ on either side. With $\lambda=490$ nm and $\Delta\lambda=30$ nm, $G(\lambda)$ proved to be a good retrieval variable for phytoplankton. These findings were confirmed and extended by Campbell & Esaias (1983) using an algorithm for phytoplankton chlorophyll of the form

$$\log_{10} C=a-b\log_{10} G(490) \tag{7.9}$$

The coefficients a and b, although dependent on aircraft altitude, are insensitive to other extraneous effects such as solar elevation. Campbell & Esaias show that log $G(\lambda)$ is a measure of the curvature of the graph of log $L(\lambda)$ in the range $\lambda\pm\Delta\lambda$, and point out that the basis for the sensitivity of this algorithm to phytoplankton is that the spectral reflectance curve of pure water has a large negative slope at ~490 nm, and that as phytoplankton pigments increase this progressively decreases towards zero (see Fig. 7.8*b*).

Hoge & Swift (1986), using the $G(\lambda)$ retrieval variable, with $\Delta\lambda=30$ nm, and testing the results against chlorophyll levels independently determined by laser-induced chlorophyll fluorescence (see later), found in continental shelf waters that the spectral curvature algorithm for phytoplankton chlorophyll worked well with the central wavelength placed not only anywhere within the 460–510 nm interval (carotenoid, and chlorophyll Soret, absorption bands), but also within two spectral regions in the red, 645–660 nm (chlorophyll long-wavelength absorption peak) and 680–695 nm (chlorophyll fluorescence).

The change, with varying phytoplankton concentration, in the shape of the spectral distribution of water-leaving radiance, can also be characterized in terms of the change in slope ($dL/d\lambda$) of the curve in any suitable spectral region. This would require the measurement of differences in radiance values over short-wavelength intervals ($\Delta L[\lambda] = L[\lambda + \Delta\lambda] - L[\lambda]$). Less measurement precision is, however, required if a radiance ratio rather than a radiance difference is determined, and for this reason an alternative measure of the slope, namely the ratio of radiances at two closely spaced wavelengths ($L[\lambda]/L[\lambda + \Delta\lambda]$) may, for engineering reasons, be preferable. Hoge, Wright & Swift (1987) measured this ratio as a function of wavelength over the range 410–770 nm, with $\Delta\lambda = 11.25$ nm, using a low altitude (150 m) airborne spectroradiometer over coastal, shelf-slope and oceanic waters, while simultaneously measuring chlorophyll by laser-induced fluorescence. These closely-spaced radiance ratios were found to be highly correlated with phytoplankton chlorophyll, in a broad spectral band (430–560 nm) through the blue and green regions (carotenoid, chlorophyll Soret, absorption) and also in the red, notably at ~675 nm (chl absorption) and at ~690 nm (chl fluorescence). To recover chlorophyll values from these measured radiance ratios, appropriate algorithms in accordance with eqn 7.7 are used.

Phytoplankton – fluorescence

The methods for remote sensing of phytoplankton described above make use of the fact that algal cells absorb light in a certain region of the spectrum. However, algal cells also emit light: about 1% of the light a photosynthesizing cell absorbs is re-emitted as fluorescence, with a peak at about 685 nm. This fluorescence is too feeble to be detected in the downwelling light stream within the water but can show up as a distinct peak in the spectral distribution of the upwelling stream or in the curve of apparent reflectance against wavelength (see Fig. 6.7).[644,942]

This peak can also be detected in the spectral distribution of the emergent flux. Calculations indicate that the increased fluorescence associated with an increase in phytoplankton chlorophyll of 1 mg m^{-3} in the water would lead to an additional upward radiance of 0.03 W m^{-2} sr^{-1} μm^{-1} above the water.[253,254] Neville & Gower (1977) showed that in the radiance reflectance (upward radiance/downward irradiance) spectra of productive British Columbian coastal waters obtained at the Brewster angle from a low-flying aircraft, a distinct peak at 685 nm was present, the height of which was proportional to chlorophyll concentration in the upper few metres of water. In a coastal inlet the height of the fluorescence peak above the baseline (measured above the water, but from a boat) was linearly correlated ($r^2=0.85$) with a weighted average (allowing for vertical distribution) chlorophyll *a* concentration over the range 1 to 20 mg m^{-3}.[330] On the basis of the observed height of the peak it seems unlikely that the results would be of acceptable accuracy below about 1 mg chlorophyll *a* m^{-3}. The spectral curvature and radiance ratio algorithms described in the previous section can also be applied to the spectral distribution of emergent flux in the region of the fluorescence peak to estimate phytoplankton chlorophyll.[376,378]

Phytoplankton chlorophyll can be estimated using the fluorescence excited by light from an airborne laser rather than by sunlight, and the method has certain advantages. A key problem in the use of remotely sensed chlorophyll fluorescence as an indication of phytoplankton biomass is that the proportion of the fluorescent light which succeeds in passing up to the sensor, as well as the proportion of the exciting light (solar or laser) which succeeds in reaching the algae, depends on the optical properties of the water. Two different water bodies with the same phytoplankton content, but different attenuation properties, could give quite different fluorescence signals. To correct for the effects of attenuation by the aquatic medium, Hoge & Swift (1981) and Bristow *et al.* (1981) have made use of the laser-induced Raman emission of water.

When water molecules scatter light, most of the scattered light undergoes no change in wavelength. A small proportion of the scattered photons, however, when they interact with the scattering molecule, lose or gain a small amount of energy corresponding to a vibrational or rotational energy transition within the molecule, and so after scattering are shifted in wavelength. These appear in the scattered light as emission bands at wavelengths other than that of the exciting light, and are referred to as Raman emission lines, after the Indian physicist who discovered this phenomenon. A particularly strong Raman emission in the case of liquid water arises from the O–H vibrational stretching mode: this shows up as an emission band roughly 100 nm on the long-wavelength side of the exciting

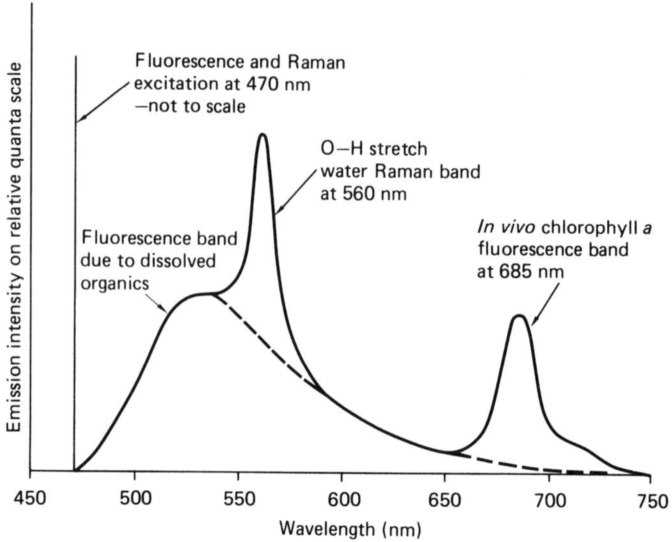

Fig. 7.9. Idealized emission spectrum of natural waters resulting from excitation with a laser at 470 nm (by permission from Bristow *et al.* (1981), *Applied Optics*, **20**, 2889–906).

wavelength (Fig. 7.9). Since the water content of the aquatic medium is essentially constant, the intensity of this Raman emission, when remotely sensed from above the water is, for a given exciting light source, determined entirely by the light-attenuating properties of the water, low Raman signals indicating high attenuation, and *vice versa.*

Thus to correct for variation in the optical character of the water, the measured chlorophyll fluorescence intensity at each station is divided by the corresponding water Raman emission intensity. Chlorophyll fluorescence values measured from the air, and normalized in this way, have been found to correlate closely with chlorophyll contents determined on water samples.[100]

Bristow *et al.* (1981) used a pumped dye laser, emitting at 470 nm, and exciting Raman emission at 560 nm, operated from a helicopter at 300 m above the water (Fig. 7.10*a*). The light returning from the water was collected by a telescope with a 30 cm diameter Fresnel lens; a beam splitter, interference filters, and separate photomultipliers (Fig. 7.10*b*) being used to separately detect the chlorophyll fluorescence and Raman emission. Hoge & Swift (1981) used a Nd:YAG laser emitting at 532 nm, and exciting a water Raman emission at 645 nm: the equipment was flown in a P-3A aircraft at 150 m above the water surface. In a further development[373,378] this

(*a*)

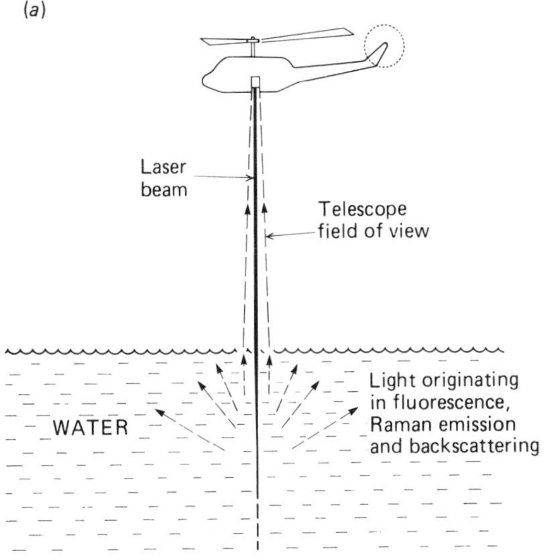

(*b*)

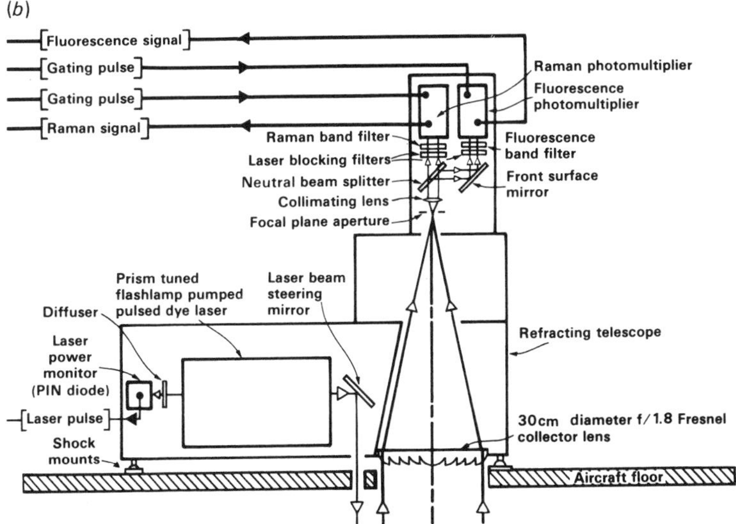

Fig. 7.10. Mode of operation of airborne laser system for detection of chlorophyll fluorescence and water Raman emission. (*a*) Schematic diagram of light fluxes. (*b*) Diagram of laser and optical receiver system (by permission, from Bristow *et al.* (1981), *Applied Optics*, **20**, 2889–906).

laser system (the NASA airborne oceanographic lidar, or AOL) has been combined with a passive spectroradiometer for concurrent measurement of the spectral distribution of the solar-induced upwelling radiant flux from the sea beneath the aircraft. The instrument has 32 contiguous 11.25-nm spectral channels forming a 360-nm bandwidth which can be placed between 350 and 800 nm, and is used for spectral curvature and radiance ratio measurements as described in the previous section. Separation of the signal originating in the backscattered solar flux from that due to the laser pulse is achieved electronically.

In waters in which the phytoplankton population includes a significant proportion of cyanophyte and/or cryptophyte algae, the laser-induced fluorescence spectrum includes a substantial peak at ~580 nm due to emission from the biliprotein photosynthetic pigment, phycoerythrin,[236,393] and thus can provide some information about the types of algae present. Gilvin, the dissolved yellow humic constituent of natural waters also emits fluorescence when excited by the laser beam: its emission spectrum extends through the visible region with a broad peak at ~530 nm (Fig. 7.9). It has been monitored in river water in terms of the combined emission at 531 and 603 nm,[101] and in coastal water in terms of the emission at 500 nm.[799] It is worth noting that the Raman signal itself can be used for mapping the attenuation properties of the water: the reciprocal of the signal is proportional to an attenuation coefficient intermediate in value between K_d and c.

Recognition of water optical type

In marine waters away from the influence of land, the major factor determining the optical character of the medium, apart from water itself, is the content of phytoplankton and their associated degradation products. This means that the reflectance spectra of oceanic waters vary in a roughly systematic way. A family of curves,[29] of progressively changing shape, determined mainly by phytoplankton concentration, is observed. Thus, for any given oceanic water, specification of the ratio of radiances or radiance reflectances at any two wavelengths should, in effect, specify the whole radiance reflectance curve, and therefore the optical character of the water. Austin & Petzold (1981), using measurements of the spectral distribution of the upwelling flux obtained from several different regions of the ocean, found that the ratios of upwelling radiance for one pair of wavelengths correlated well with the ratios for another pair of wavelengths (e.g.

log[L_u(443)/L_u(670)] plotted against log[L_u(443)/L_u(550)] was a straight line), indicating, in agreement with the above, that if one ratio is known, the shape of the spectral radiance curve is specified.

Concomitant with the systematic change in the spectral radiance curve, there is a corresponding progressive change in the curve of vertical attenuation coefficient for downwelling irradiance (K_d) as a function of wavelength; thus, the specification of the value of the upwelling radiance ratio at two wavelengths should also approximately specify K_d at any given wavelength. Austin & Petzold derived empirical relations by means of which K_d(490) or K_d(520) can be obtained from the ratio of radiances in the CZCS 443 and 550 nm bands. The calculated vertical attenuation coefficient at 490 or 520 nm can thus be used as a quantitative parameter in terms of which variation in oceanic water optical type can be mapped on the basis of satellite measurements. Agreement between the value of K_d(490) estimated from CZCS radiance data and that determined by direct measurement in the water, was found to be good for two stations in the Gulf of Mexico.[31] It is not, however, to be expected that this method would work satisfactorily for waters, such as estuarine and many coastal waters, in which the optical properties are significantly influenced by components other than phytoplankton.

On the basis of a large amount of field data, Højerslev (1981) established an empirical relationship between the Jerlov colour index (L_u[450]/L_u[520]), referred to earlier, and the depth of the euphotic zone (z_{eu}),

$$\frac{L_u(450)}{L_u(520)} = 0.00013\, z_{eu}^2 + 0.017\, z_{eu} + 0.14 \qquad (7.10)$$

which he considered to be applicable to most marine waters. Since the colour index can in principle be determined by remote sensing, then the Højerslev relationship could be used to map the distribution of z_{eu}, and of K_d(PAR) (since K_d[PAR]$=4.606/z_{eu}$).

In the Delaware River estuary/Delaware Bay, Stumpf & Pennock (1991) found quite a strong relationship between sea-level reflectances calculated from remotely sensed AVHRR radiances in the red and near-infrared bands, and the *in situ* values of K_d(PAR). However, in this turbid system attenuation is apparently dominated by resuspended sediments, and so the existence of a strong link with reflectance (which responds directly to suspended sediments) is not surprising. The relationship broke down in the presence of high levels of phytoplankton. Algorithms for remote sensing of K_d(PAR) in estuarine and coastal waters are likely to be locality-specific, and even then may vary with time as seasonal changes in river flow, humic colour and phytoplankton growth change the ratio of absorption to scattering, and thus disturb the nexus between reflectance and attenuation.

We saw earlier (§6.4) that reflectance is proportional to b_b/a. If, by remote sensing, in-water reflectance is determined in a band in the far red or near infrared, spectral regions in which absorption can be assumed to be almost entirely due to water itself, then a realistic value for a can be assigned, and so b can be calculated, and the distribution of backscattering, and by implication total scattering, within the water body can be mapped. This is, of course, only feasible for turbid waters: in clear (i.e. most marine) waters, far-red/near-infrared reflectance is too low to be measured by remote sensing.

In view of the direct dependence of reflectance on backscattering, and thus on total scattering, it is not surprising that a number of studies have shown good correlations between reflectances measured with Landsat and the *in situ* values for nephelometric turbidity.[136,349,464,540,562] Correlation with turbidity tends in fact to be better than correlation with suspended solids because, as we have noted earlier, the concentration of TSS corresponding to a given amount of scattering is a function of particle size.

In some studies, correlations between remotely sensed Landsat radiances or combinations of radiances in one or more bands, and the measured values of Secchi depth have been observed,[349,465,540,562,563,954] and used for monitoring, and mapping the distribution of, this optical property. Since reflected radiance in any waveband is a function of b_b/a, and since Secchi depth is proportional to $1/(K_d+c)$, where K_d and c are each separate functions of a and b, it is not surprising that for any given locality some connection may be observed, but no generally applicable algorithm should be expected. Indeed, in Lake Michigan, Lathrop & Lillesand (1986) found that quite different algorithms applied in the turbid Green Bay and in the clear central region of the lake.

7.5 Specific applications of remote sensing to the aquatic environment

Since our concern in this book is with light and its availability for aquatic photosynthesis, we shall consider the remote sensing only of those qualities of the aquatic environment which relate to its optical properties and/or to photosynthetic biomass.

Total suspended solids and turbidity

Remote sensing, particularly using Landsat, has been quite commonly used for mapping the distribution of total suspended solids in inland and coastal waters. We shall examine just a few examples.

Klemas *et al.* (1974) used Landsat MSS 600–700 nm radiance values, together with their previously determined empirical relation to study the distribution of TSS in the Delaware River estuary and Delaware Bay. Concentrations varying through six intervals, from ~5.6 p.p.m. to >211 p.p.m., were mapped. Turbidity in the system decreased with distance towards the sea: within the bay it was highest around the sides and lowest in the deeper central part (Fig. 7.11). Landsat 600–700 nm radiance values have also been used as a broad indication of the distribution of TSS in western Lake Ontario,[112] in Lake Superior,[797] and off the Georgia coast.[670]

Rouse & Coleman (1976) used Landsat 600–700 nm radiance values to study the distribution of turbidity from the Mississippi River in the Louisiana Bight. They mapped the band 5 radiance (for which they had a calibration curve relating it to TSS) at six different levels of intensity. They were able to show that the size and shape of the discharge plume were a function not only of the amount of fresh water discharged, but also of the speed and direction of the wind. Brakel (1984) used Landsat imagery of Kenyan coastal waters in the 500–600 nm, 600–700 nm and 700–800 nm bands to study changes in the movement of the sediments discharged by the Tana and Sabaki rivers, as determined by the East African Coastal Current and the seasonally changing monsoon winds. Landsat imagery has also been used to monitor ocean dump plumes in the New York Bight (US Atlantic Coast).[88]

The limnology of the large impoundment Lake Kainji, Nigeria, is dominated by the alternation of the 'white flood' (sediment-laden water) from the Sokoto River and the 'black flood' (clear water) from the Niger River, which occur at different times of the year. Abiodun & Adeniji (1978) used Landsat 600–700 nm radiance values to follow the movement and mixing of the two types of water in the 1972–4 period. Landsat radiance data, combined with empirical radiance/TSS relations have been used to map the distribution of suspended solids in the Bay of Fundy.[12] Radiance values in bands 4, 5 and 6 of Landsat have been used, in conjunction with appropriate empirical relations, to map the distribution of nephelometric turbidity in an Australian impoundment.[136]

The Landsat Thematic Mapper scanner has also been used to map turbidity: radiance in the red band was used for Green Bay and adjoining

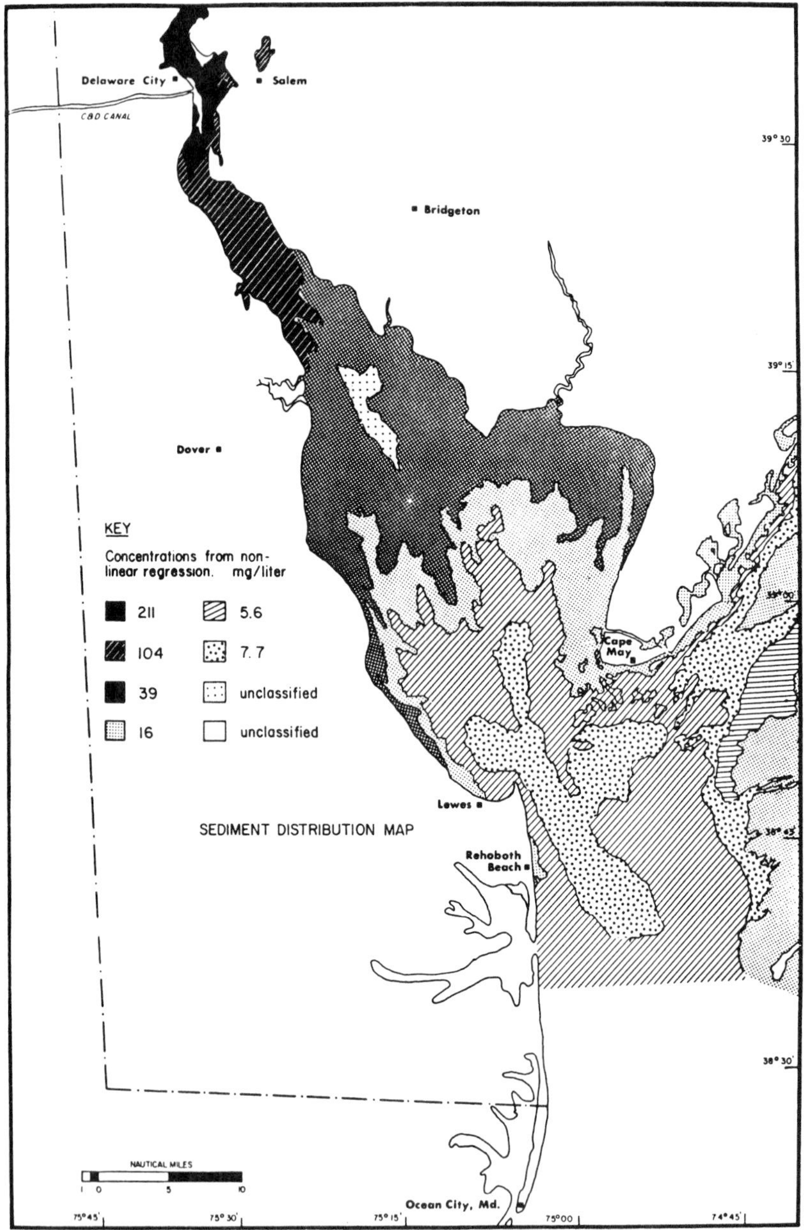

Fig. 7.11. Distribution of suspended solids concentration in Delaware River estuary and Delaware Bay derived from Landsat 600–700 nm radiance measurements (by permission, from Klemas *et al.* (1974), *Remote Sensing of Environment*, **3**, 153–74).

regions of Lake Michigan,[540] whereas radiance in the green band was used for the nearshore seawater in and around Augusta Bay, Sicily.[465] In a higher resolution study on Green Bay,[541] radiances in the green, red and near-infrared bands of the HRV sensor on the SPOT-1 satellite were used to map the distribution of suspended sediments from the Fox River plume.

Because its channels are saturated at quite low radiance values the CZCS has not been very much used for mapping suspended sediments at the high levels often found in turbid nearshore waters. It has, however, been used to provide interesting information on the distribution of suspended solids and its seasonal variation in coastal shelf seas, such as the North Sea,[387] the Irish Sea,[979] and the northern Adriatic Sea.[799]

Airborne multispectral scanners have been used to map the distribution of suspended solids in Swansea Bay and the Bristol Channel (UK),[757] and around a sewage sludge dump site in the New York Bight.[433]

A special category of suspended solids is responsible for the phenomenon known as 'whitings'. This is produced by the precipitation of a fine cloud of calcium carbonate particles, possibly resulting from the uptake of CO_2 by photosynthesizing phytoplankton. Strong (1978) used Landsat to observe the distribution of 'whitings' in the Great Lakes. The 'whitings' are formed some metres below the surface. They showed up as a strongly increased radiance in the more highly penetrating green (500–600 nm) band, but were hardly detectable in the strongly absorbed red (600–700 nm) band.

Phytoplankton – aircraft studies

Detection of phytoplankton from aircraft is generally based on the decreased reflectance in the blue, relative to that in the green. Studies carried out from low altitude have used photometers of the non-(spatially) scanning type so that distribution along the aircraft's linear track, rather than over an area, has been determined. Arvesen *et al.* (1973) used their differential radiometer, operating at 443 and 525 nm from a plane at 150 m altitude, to determine the distribution of phytoplankton within and outside San Francisco Bay and along the Pacific coastline north of San Francisco. Clarke & Ewing (1974) used the ratio of reflectance at 540 nm to that at 460 nm (which they call the colour ratio) measured with an airborne spectrometer, as a broad indication of the phytoplankton content, and hence productivity, of waters in the Caribbean region and Gulf of California.

Viollier, Deschamps & Lecomte (1978), using the difference between irradiance reflectance in the green (525 nm) and that in the blue (466 nm) observed from 150 m altitude as an approximate indication of phytoplankton concentration, mapped the changing distribution of phytoplankton during the development of a coastal upwelling in the Gulf of Guinea (tropical East Atlantic). Borstad *et al.* (1980) used both the ratio of radiance reflectance at 560 nm to that at 440 nm, and the chlorophyll fluorescence emission at 685 nm, to map the distribution of phytoplankton along the coast of British Columbia from a plane at 100 m altitude. They concluded that airborne remote sensing is a useful and powerful technique for the detection and measurement of phytoplankton abundance, and can moreover, unlike satellite-based measurements, be used under cloudy conditions.

Moderate- to high-altitude aircraft studies have used spatially scanning multispectral photometers to observe phytoplankton distribution within swaths some kilometres in width. Such mapping has been carried out, for example, in the James River estuary, Virginia;[432] the New York Bight;[432] Monterey Bay, California;[472] San Francisco Bay;[463] for a large diatom patch within a Gulf Stream frontal eddy;[1013] in the eutrophic Loosdrecht Lakes in the Netherlands;[185] and (with the new AVIRIS scanning radiometer) in Lake Tahoe (California–Nevada).[337] Use of a scanner at moderate altitude, as in the James River study (2.4 km), provides better spatial resolution, and so is better suited for the study of aquatic systems with complex shapes, such as estuaries.

An alternative to the blue/green reflectance ratio, as a retrieval variable is, as we saw earlier, the spectral curvature of the emergent flux from the water body. Harding *et al.* (1992) used the NASA ODAS radiometer system operating at 460, 490 and 520 nm, and the spectral curvature algorithm (eqn 7.8) to map the distribution of, and follow the seasonal changes in, phytoplankton populations in Chesapeake Bay (USA), from a plane at 150 m altitude.

Laser excitation of chlorophyll fluorescence, using the NASA airborne oceanographic lidar (AOL), from low-altitude aircraft has been used to map phytoplankton distribution in the German Bight (North Sea),[374] in a Gulf Stream warm core ring,[375] and in the New York Bight.[377] Bristow *et al.* (1985) used an airborne laser fluorosensor to measure the longitudinal distribution of chlorophyll, dissolved organic matter and beam attenuation coefficient (approximately estimated from reciprocal of Raman emission) along a 734 km segment of the lower Snake and Columbia rivers (Oregon/Washington, USA). Kondratyev & Pozdniakov (1990) used a helicopter-borne laser fluorosensor to map phytoplankton distribution in a number of Russian lakes: in the Rybinsk reservoir it was possible to observe by this means major changes in phytoplankton distribution, in response to changing weather conditions, over a few days.

Airborne multispectral scanners have been used to map the distribution of benthic macrophytes (seagrass, macroalgae) in St Joseph Bay, Florida,[793] and in the Venice Lagoon (Italy).[1020]

Phytoplankton – satellite studies

The Landsat MSS optical system can, as we have noted previously, be used to detect high concentrations of phytoplankton by taking advantage of the increased reflectance in the near infrared (700–800 nm) band without a corresponding increase in the red (600–700 nm) band. Bukata & Bruton (1974) used the radiance in the 700–800 nm band as a broad indication of the distribution of phytoplankton in western Lake Ontario, Canada. Strong (1974) used Landsat radiances to map algal blooms in Utah Lake, USA, and in Lake Erie, and also subsequently in Lake Ontario.[881] Stumpf & Tyler (1988) used the difference between the red and near-infrared reflectance, as measured with the AVHRR, to map phytoplankton blooms in the estuarine waters of Chesapeake Bay (USA).

Apart from some preliminary studies,[215,473] little use appears so far to have been made of the Landsat Thematic Mapper for observing phytoplankton distributions, despite its having a channel in the blue, as well as in the green and red, regions.

The major breakthrough in the remote sensing of phytoplankton, which led in turn to major advances in our understanding of the marine ecosystem, was the advent of the Coastal Zone Colour Scanner on the Nimbus-7 satellite, which makes use, as we saw earlier, of the decreased ratio of blue to green light in the emergent flux as the algal concentration increases, and which can remotely sense phytoplankton at the low concentrations found in the ocean. During its operational life from 1978 to 1986 it accumulated data covering most of the world ocean. The results of many CZCS studies on phytoplankton distribution have been published, all providing information of great interest and value, and brief accounts of just some of these are given below. The data are generally presented as colour-coded maps covering many increments of concentration in the ranges <0.05 to >1.0 mg (chlorophyll + phaeopigments) m^{-3} or 0.1 to >10.0 mg m^{-3}, depending on whether the 443/550 nm or 520/550 nm radiance ratios are used.

Hovis *et al.* (1980) and Gordon *et al.* (1980) used CZCS data to characterize the distribution of phytoplankton in the Gulf of Mexico. Smith & Baker (1982) were able to map phytoplankton distribution in the California Bight from CZCS measurements: one such map, for a day in March 1979, is shown in Fig. 7.12. Phytoplankton in shelf and slope waters

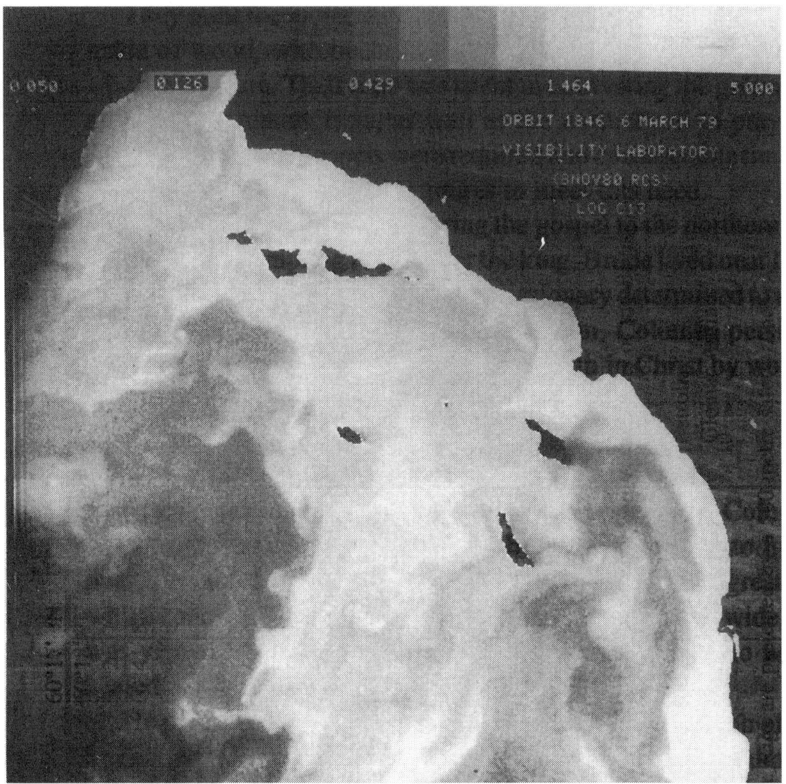

Fig. 7.12. Phytoplankton distribution in the California Bight on 6th March 1979, derived from Coastal Zone Colour Scanner (Nimbus-7 satellite) data (by permission, from Smith & Baker (1982), *Marine Biology*, **66**, 269–79). The relation between chl *a* conc. (mg m^{-3}) and the density of the image is indicated on the scale at the top of the figure.

of the Middle Atlantic Bight, and in the Sargasso Sea, were mapped by Gordon *et al.* (1983). Bricaud & Morel (1986) show CZCS-derived phytoplankton distributions for the Mediterranean between the Riviera and Corsica, and for an upwelling area off the coast of Portugal.

Brown *et al.* (1985), combining CZCS information with sea-surface temperature data from the AVHRR, were able to observe the spatial and temporal development of the spring phytoplankton bloom in the western Atlantic Ocean off the US East Coast. Pelaez & McGowan (1986) observed recurring annual patterns of phytoplankton in the California Current over a three-year period. Of particular interest was a remarkably close correspondence between phytoplankton distribution as revealed by the CZCS and sea-surface temperature as revealed by the AVHRR: the colder the water, the higher the phytoplankton concentration.

Holligan *et al.* (1989) have published the results of a very comprehensive CZCS study of the North Sea over the period 1979–86. A series of CZCS images is presented showing *inter alia* the annual coccolithophore blooms (readily detectable by their high reflectance at 550 nm), the development of the phytoplankton spring bloom in 1980 in the eastern and northwestern parts of the North Sea, and the summer phytoplankton distribution in the German Bight over a number of seasons. Fiedler & Laurs (1990) were able to follow the seasonal changes in the plume of the Columbia River (Oregon/Washington) – the largest point source of freshwater flow into the eastern Pacific Ocean – over the 1979–85 period, by means of its associated high phytoplankton levels as detected with the CZCS. Sathyendranath *et al.* (1991) used CZCS data to show the major seasonal increase in phytoplankton population in the Arabian Sea resulting from the upwelling caused by the southwest monsoon, and went on to show that the resulting distribution of phytoplankton exerts a controlling influence on the seasonal evolution of sea-surface temperature.

The information provided by the CZCS shows aspects of large-scale phytoplankton distribution previously quite unknown. The regions of transition from highly productive coastal waters to less productive oceanic waters are sharply delineated, and within the oceanic regions a complex array of whorls and eddies of varying productivity is revealed.

7.6 Further outlook

The development of remote sensing scanners such as the CZCS for mapping phytoplankton distribution is one of the most important developments in the aquatic sciences in recent years. It promises to provide detailed synoptic information on the productivity of the oceans on a scale that could never be envisaged previously. The waters on which the procedures developed so far work best are those oceanic waters in which the phytoplankton are, apart from the water itself, the major influence on the upwelling light flux. Extending the method to marine ecosystems, such as those bordering northern Europe, with much higher levels of dissolved yellow colour from river discharge, will be more difficult. The phytoplankton influence on upwelling flux in the blue region will be hard to separate from the background absorption due to gilvin. Other wavebands in which there is greater selectivity are needed, and should be available on the new generation of spaceborne radiometers.

Background absorption by yellow substances will be an even greater problem in the remote sensing of phytoplankton in inland waters. Turbidity due to suspended sediments will also interfere. Inland waters are typically more productive than marine waters and so there are higher concentrations of phytoplankton available for detection. A finer spatial resolution than the 825 × 825 m provided by the CZCS will, however, be required for inland waters and should be available on some of the new instruments.

Even when the phytoplankton signal cannot be separated from the background, remote sensing data can provide useful information about natural waters. Maps of turbidity distribution such as those generated already from Landsat data are of great interest, and it seems possible that with the new generation of scanners which can measure radiance in the blue region, the distribution of total yellow colour could also be mapped.

The big challenge, however, especially in the context of the role of the ocean in the global carbon cycle and climate change is how, once we have satisfactorily mapped oceanic phytoplankton distribution, we then proceed to determine the corresponding primary productivity of the global ecosystem. This is a topic to which we return in a later chapter.

Part II

Photosynthesis in the aquatic environment

8

The photosynthetic apparatus of aquatic plants

In Part I, we considered the underwater light climate: the particular characteristics that it has in different types of natural water bodies, the scattering and absorption processes that take place in the aquatic medium, and the ways that these operate upon the incident light stream to produce the kinds of underwater light field that we observe. Now, in Part II, we turn our attention to the utilization of this underwater light for photosynthesis by aquatic plants. We begin, in this chapter, by asking: with what intracellular structures, from the level of organelles down to that of molecules, do aquatic plants harvest radiant energy from the underwater light field and convert it to chemical energy?

8.1 Chloroplasts

In eukaryotic plants, photosynthesis is carried out by the organelles known as *chloroplasts*, the best-known members of the great class of related and interconvertible organelles known as *plastids*. Detailed accounts of these organelles may be found in Kirk & Tilney-Bassett (1978) and Staehelin (1986): we shall here content ourselves with a rather brief treatment.

The chloroplasts contain the pigments which capture the light, the electron carriers which use the absorbed energy to generate reducing power in the form of $NADPH_2$ and biochemical energy in the form of ATP, and the enzymes which use the $NADPH_2$ and the ATP to convert CO_2 and water to carbohydrate. The pigments and electron carriers are contained in a specialized type of membrane known as the *thylakoid*. The enzymes of CO_2 fixation are distributed throughout the rest of the inner volume – the *stroma* – of the chloroplast, although in certain algal chloroplasts some of the enzymes may be located in a specific structure within the chloroplast known as the *pyrenoid*.

Chloroplasts occupy a substantial proportion of the cytoplasmic volume outside the vacuole. There is usually enough chloroplast material per cell to ensure that most of the light incident on the cell passes through one or more chloroplasts. As we shall see in a later chapter, however, the disposition of the chloroplast(s) within the cell, and hence the proportion of the cellular cross-section occupied by chloroplast substance, can vary with light intensity.

In higher plants, including aquatic species, chloroplasts are typically lens-shaped bodies, 4–8 μm in diameter and about 2 μm thick in the centre. There are commonly 20-400 chloroplasts per mature leaf cell, depending partly on the size of the cell. In terrestrial plants, the chloroplasts occur mainly in the palisade and mesophyll cells of the leaves, whereas the plastids in the epidermal cells usually do not develop into chloroplasts. In submerged aquatic plants, however (the only kind with which we are concerned in this book), the epidermal cells often have well-developed chloroplasts. The palisade layer within the leaf in some cases does not develop at all, being replaced by spongy parenchyma.[399] Stomata are scarce or absent in submerged leaves, the CO_2 presumably being taken up directly through the epidermal cell walls.

Amongst the eukaryotic algae, chloroplasts vary greatly in size, number and morphology. Comprehensive accounts may be found in Fritsch (1948), Bold & Wynne (1978) and Raymont (1980). In both unicellular and multicellular algae, chloroplasts can occur singly, in pairs, or in numbers of up to 100 or more per cell. In many algal species the chloroplasts contain dense bodies, 1–5 μm in diameter – the pyrenoids. These are often the site of polysaccharide deposition and are tightly packed with granular or crystalline aggregates of what appears to be ribulose bisphosphate carboxylase.

The chloroplasts may be lens-shaped (*Euglena*, *Coscinodiscus*), helical (*Spirogyra*), star-shaped (*Zygnema*), plate-like (*Pinnularia*), in the form of an irregular network (*Oedogonium*), or various other shapes (Fig. 8.1). Where they occur singly they are usually large as, for example, the flat, approximately rectangular chloroplast of *Mougeotia* which is commonly 100–150 μm long. Where they occur in large numbers they are usually lens-shaped and of similar size (a few μm in diameter) to higher plant chloroplasts.

If we direct our attention to individual groups of algae, some useful generalizations are possible. Most of the green algae (Chlorophyta), major components of both the attached and the planktonic flora, contain one chloroplast per cell. This may be a cup-shaped structure as in *Chlamydomonas*, *Chlorella* and other unicellular planktonic species, or it may have a spiral, flat or other shape in the large-celled filamentous forms. Some

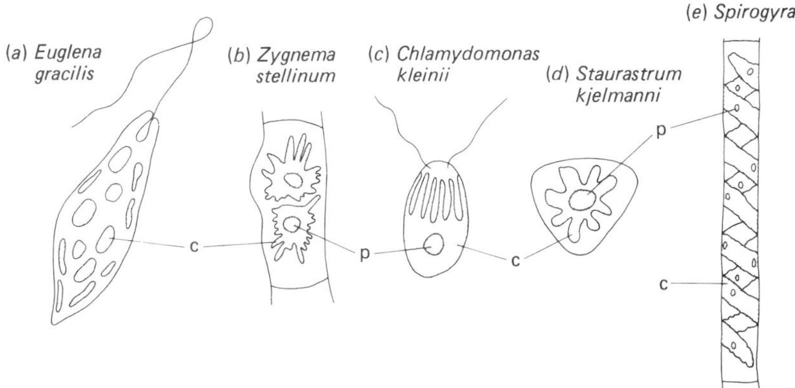

Fig. 8.1. Some algal chloroplasts. c = chloroplast, p = pyrenoid. (*b*)–(*e*) after West, G. S. (1904), *The British freshwater algae*, Cambridge University Press.

desmids contain two chloroplasts per cell. Three of the 15 or so orders of green algae, the Caulerpales, Siphonocladales and Dasycladales, have a *coenocytic* construction. As well as having some cells of normal construction they have tubular filaments lacking cross-walls. These tubular structures contain large numbers of, usually lens-shaped, chloroplasts.

The brown algae (Phaeophyta) may contain one, a few, or many chloroplasts per cell. The Laminariales and Fucales, two orders of particular importance for coastal primary production, both contain numerous chloroplasts per cell in their large thalli. A characteristic of these brown-algal chloroplasts is that when they contain a pyrenoid, this forms a bulbous structure protruding out from the otherwise lens-shaped chloroplast.

In that very diverse division the Chrysophyta (yellow, yellow-green or yellow-brown single cells), we shall consider, on the grounds of their major contribution to planktonic primary production, two of the six classes. The class Prymnesiophyceae (formerly Haptophyceae) includes the coccolithophorids, major constituents of the marine phytoplankton, particularly in the warmer regions. Members of this class typically have two, but sometimes four or more, lens-shaped chloroplasts. The class Bacillariophyceae, or diatoms, often forms the major component of the phytoplankton in both marine and fresh waters. In the centric (circular) diatoms the chloroplasts are commonly small, lens-shaped and numerous. In the pennate (elongated) diatoms, fewer (often only two) and larger (lobed or plate-like) chloroplasts are present.

The division Pyrrophyta, or dinoflagellates, is a major contributor to aquatic, especially marine, primary production; being comparable in importance to the diatoms. The Pyrrophyta are made up of two classes, the Dinophyceae with numerous small, lens-shaped chloroplasts and the Desmophyceae with few, plate-like chloroplasts.

Members of the Euglenophyta, unicellular flagellates of common occurrence in ponds, have a variety of types of chloroplast structure, ranging from large numbers of lens-shaped plastids (Fig. 8.1) to small numbers of large plate-like or ribbon-like bodies. Another group of unicellular flagellates, the Cryptophyta, occasionally significant components of the phytoplankton in coastal waters, contain usually two, but sometimes one, chloroplast(s) per cell.

The Rhodophyta, or red algae, are predominantly thalloid in form and occur mainly in marine habitats. They are major components of the benthic (growing, attached, on the bottom) flora of coastal waters. They can be divided into two groups: the Bangiophycidae which usually have a single, stellate, axial chloroplast; and the Florideophycidae which normally contain numerous small lens-shaped chloroplasts around the periphery of the cell.

8.2 Membranes and particles

In electron micrographs of sections, chloroplasts are seen to be bounded by a double- or, in some algae, triple-membraned envelope and to be filled with a granular matrix. This matrix is referred to as the *stroma* and is composed of a concentrated solution or gel of proteins, consisting mainly of the enzymes used in carbon dioxide fixation. Embedded within the stroma there is an array of flattened, membrane-bounded sacs known as *thylakoids*. These lie approximately parallel to the main plane of the chloroplast and when seen in electron microscope sections at right angles to that plane, they present the appearance of membranes in pairs. There are large numbers of thylakoids in each chloroplast. In the algae most of the thylakoids extend from one end of the chloroplast to the other.

The different classes of algae differ from one another in the way in which the thylakoids are grouped together. The simplest arrangement is that found in the red algae (Fig. 8.2*b*). These have single thylakoids, about 20 nm thick, lying separately in the stroma. Attached to the outer surface of each thylakoid is an array of particles 30–40 nm in diameter, known as *phycobilisomes* and consisting mainly of biliprotein molecules.

The next simplest type of thylakoid arrangement is that found in the Cryptophyta, which have thylakoids loosely associated in pairs (Fig. 8.3*a*).

(*a*)

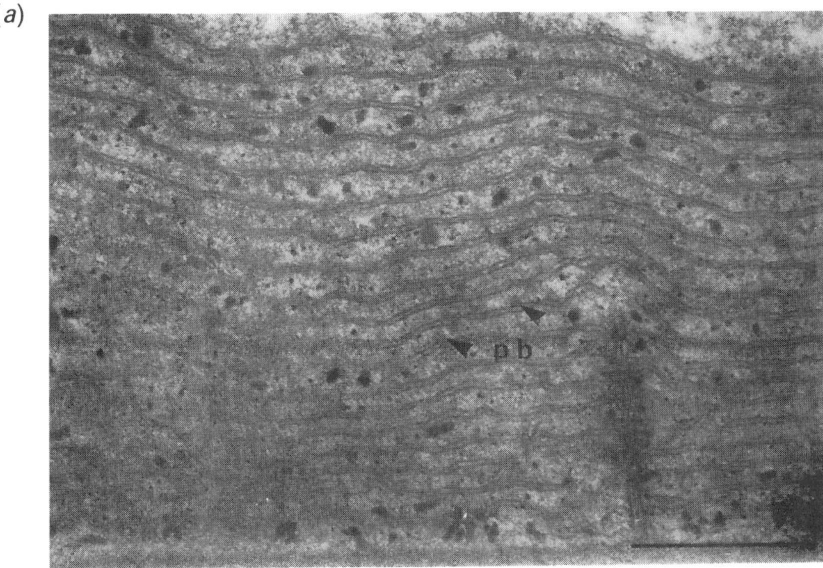

(*b*)

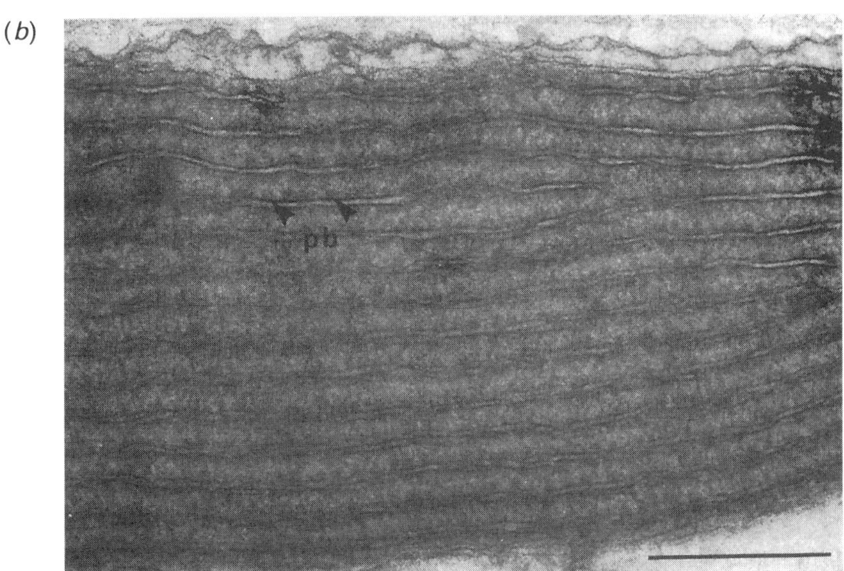

Fig. 8.2. Parallel single thylakoids with adherent phycobilisomes in blue-green and red algae (by courtesy of Dr M. Vesk). pb – phycobilisomes. Bar = 0.5 μm. (*a*) Thylakoids lying free within the cytoplasm of the blue-green alga *Oscillatoria brevis*. (*b*) Part of a chloroplast in the red alga *Haliptilon cuvieri*, showing the regularly spaced single thylakoids.

(*a*)

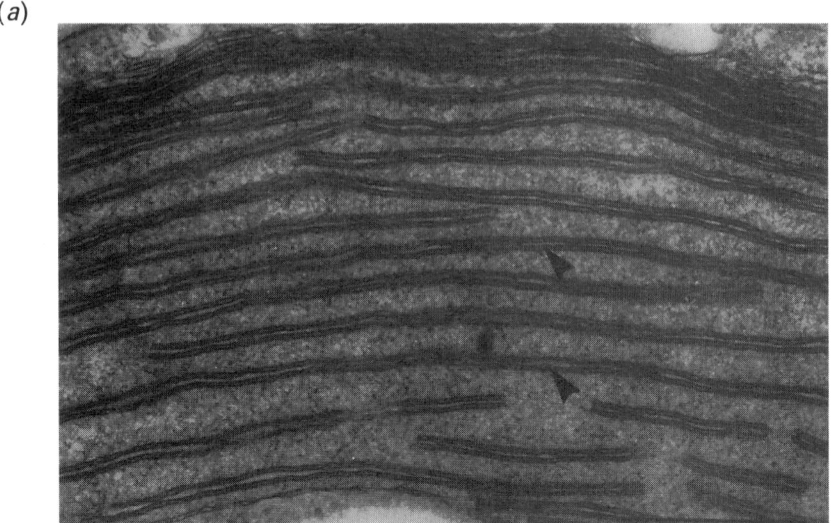

(*b*)

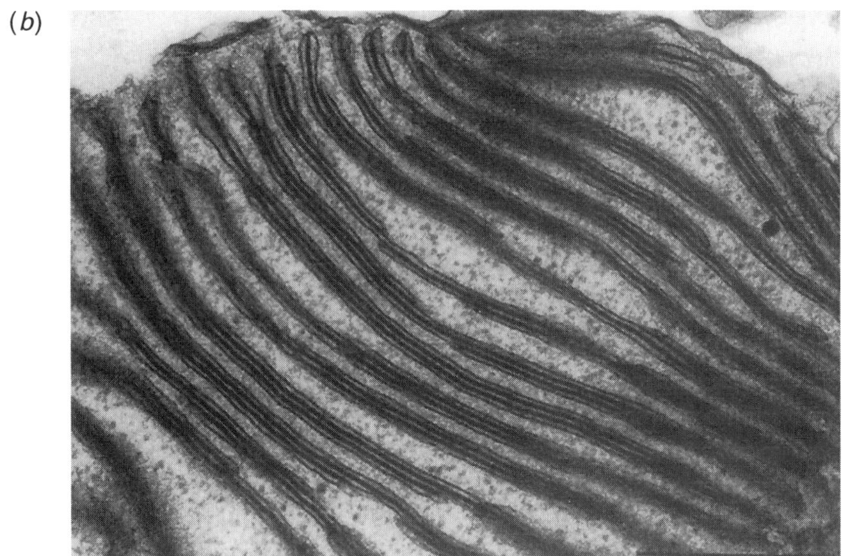

Fig. 8.3. Algal chloroplasts with thylakoids grouped in pairs and triplets (by courtesy of Dr M. Vesk). Bar = 0.5 μm. (*a*) Part of a chloroplast of the cryptophyte alga *Chroomonas* sp., showing the paired thylakoids (arrows) filled with a phycobiliprotein matrix. (*b*) Part of a chloroplast of the dinoflagellate *Gymnodinium splendens*, showing the compound lamellae, each consisting of three closely appressed thylakoids.

They are thicker than the thylakoids of other algae, being 19–36 nm across, and their interior space is filled with a finely granular, electron-dense material which is thought to consist of the biliprotein pigments.

In all other classes of algae, with the exception of some of the Chlorophyta, the thylakoids are grouped, or stacked, in threes (Fig. 8.3*b*). These stacks of thylakoids are referred to as compound lamellae. In aquatic higher plants and in some green algae a more complex arrangement of the thylakoids is found. They occur in much bigger stacks consisting typically of 5–20 thylakoids, but the individual thylakoids in the stacks are of much smaller diameter than a typical algal thylakoid. Each such stack is referred to as a *granum*. There may be 40–60 *grana* in a typical chloroplast. There are numerous interconnections between the different grana and between the different thylakoids within a stack (Fig. 8.4). In this type of chloroplast the thylakoids can best be regarded as individual compartments in a complex, ramifying system of interconnected compartments that constitutes the photosynthetic membrane system of the chloroplast. The thylakoid system in the chloroplast of a submerged aquatic higher plant may be seen in Fig. 8.4*b*.

Blue-green algae (Cyanophyta) are prokaryotes and so the various cellular functions are not compartmented within separate membrane-bounded organelles. These algae also have thylakoids but these lie free within the cytoplasm (Fig. 8.2*a*). They are similar in type to those of the red-algal chloroplasts, occurring as separate, single thylakoids with phycobilisomes on the surface. A type of prokaryotic alga, *Prochloron*, which occurs in symbiotic association with certain marine ascidians, does not contain biliproteins and so lacks phycobilisomes. It has been proposed by Lewin (1976) that these algae should be placed in a new algal division, the Prochlorophyta, but the counterproposal has been put by Antia (1977) that the criteria defining the Cyanophyta be widened so that the group could include algae such as *Prochloron*. Different isolates of *Prochloron* have thylakoids lying within the cytoplasm singly, in pairs, or in stacks of up to 12 thylakoids.[918,989] Free-living prochlorophyte species are now known to exist: a filamentous form has been found in a lake, and coccoid forms are abundant in the sea.[120,142]

The thylakoid membrane in all plants consists of a lipid bilayer with protein particles embedded within it. The thylakoids of higher plant chloroplasts are made up of about 32% colourless lipid, 9% chlorophyll, 2% carotenoid and 57% protein, and it is likely that algal thylakoids have a broadly similar average chemical composition. The nature of the polar lipids constituting the bilayer varies from one class of alga to another.

(*a*)

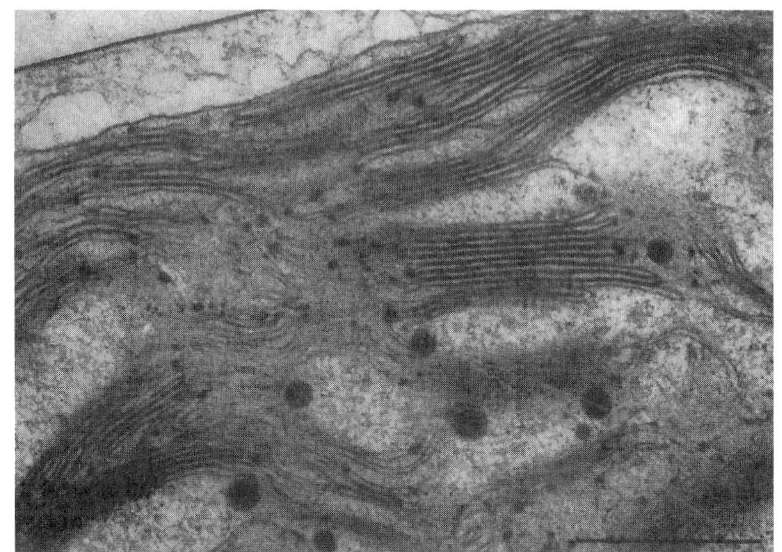

(*b*)

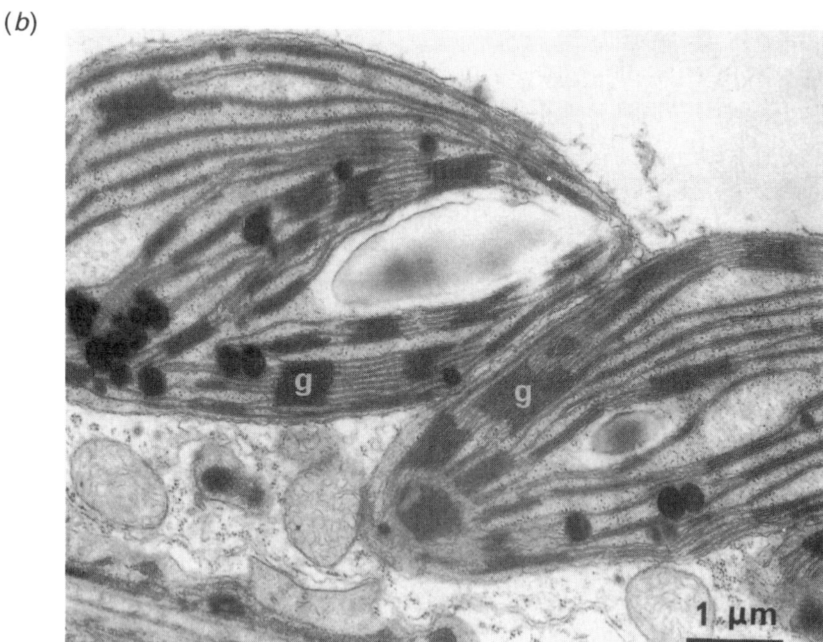

Fig. 8.4. Chloroplast structure in green aquatic plants. (*a*) Part of a chloroplast of the unicellular green flagellate *Dunaliella tertiolecta* (Chlorophyta), showing the somewhat irregular stacking of thylakoids to form rudimentary grana (by courtesy of Dr M. Vesk). Bar = 0.5 µm. (*b*)

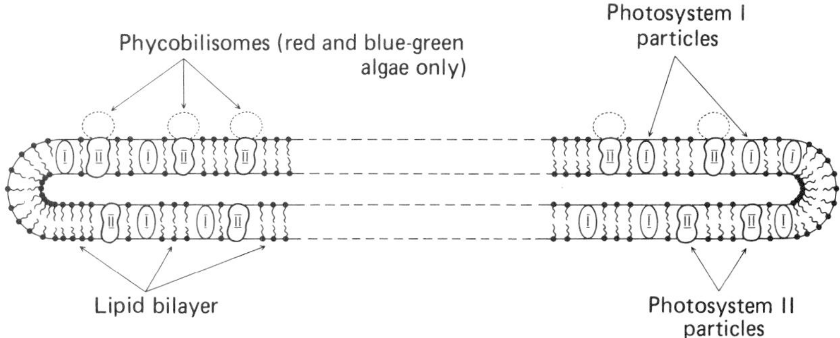

Fig. 8.5. Possible mode of organization of thylakoid membrane in algae and aquatic higher plants.

The light absorption/electron transport system of photosynthesis consists of two subsystems, referred to as photosystem I and photosystem II (see §8.5). It is thought that these exist within the thylakoid membrane as two distinct, but interacting, types of particle, each containing a reaction centre and light-harvesting pigment–proteins, and having specific types of electron transfer components associated with it. The molecular structure of the photosynthetic membrane is likely to vary markedly from one algal class to another, and is not known in any detail for any type of plant. There are, however, grounds for supposing that the mode of organization is something along the lines indicated in Fig. 8.5.

8.3 Photosynthetic pigment composition

The task of collecting light energy from the underwater light field is carried out by the photosynthetic pigments – molecules whose structures are such that they efficiently absorb light somewhere in the 400–700 nm range. There are three chemically distinct types of photosynthetic pigment: the chlorophylls, the carotenoids and the biliproteins. All photosynthetic plants contain chlorophyll and carotenoids; the red algae, the blue-green algae and the cryptophytes contain biliproteins as well. A comprehensive account of the photosynthetic pigments of algae has been given by Rowan (1989).

Chloroplasts in the leaf of the submerged aquatic plant *Vallisneria spiralis* (by courtesy of Mr D. Price and Dr S. Craig), showing the stacking of thylakoids to form grana (g), and the stroma thylakoids connecting the grana. A large starch grain is evident in the left-hand chloroplast.

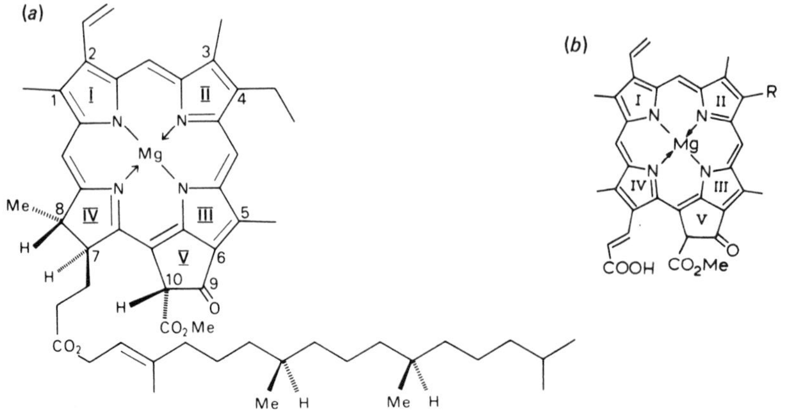

Fig. 8.6. The chlorophylls. (*a*) *a*/*b*. In chlorophyll *b*. the $-CH_3$ on ring II is replaced by a $-CHO$. In chlorophylls a_2 and b_2 the ethyl on ring II is replaced by a vinyl. (*b*) c_1/c_2. In c_1 R is $-C_2H_5$; in c_2 R is $-CH{=}CH_2$.

The chlorophylls

The chlorophylls are cyclic tetrapyrrole compounds with a magnesium atom chelated at the centre of the ring system. Chlorophylls *a* and *b* are derivatives of dihydroporphyrin: their structures are shown in Fig. 8.6*a*. In the coccoid marine prochlorophyte, *Prochlorococcus marinus*, chlorophylls *a* and *b* are replaced by the divinyl forms, a_2 and b_2,[140,308] in which the ethyl group on ring II is replaced by a vinyl group. Chlorophyll *d*, which is found only in small amounts in certain red algae, has a structure similar to that of chlorophyll *a*, except that the vinyl group on ring I is replaced by a formyl ($-CHO$) group. Chlorophylls *a*, *b* and *d* are rendered hydrophobic by the presence of a C_{20} isoprenoid alcohol, phytol, esterified to the propionic acid residue on ring IV. Chlorophylls c_1 and c_2 are porphyrins rather than dihydroporphyrins and lack the phytol group: their structures are shown in Fig. 8.6*b*.

All photosynthetic plants contain chlorophyll *a* (or a_2), and most classes of plant contain in addition, either chlorophyll *b* (or b_2), or one or more of the chlorophyll *c*s, or (rarely) chlorophyll *d*. The distribution of the chlorophylls (other than chlorophyll *d*, which may have no photosynthetic function) amongst the different plant groups is summarized in Table 8.1. Chlorophyll *a* normally constitutes most of the chlorophyll present.

Amongst the algae the chlorophyll *a* content varies widely. The ranges of concentrations found among the three major pigment classes of the littoral multicellular marine algae off Helgoland were, as a percentage of dry mass:

Table 8.1. *Distribution of chlorophylls amongst different groups of plants. (After Kirk & Tilney-Bassett, 1978)*

Plant group	Chlorophyll *a*	*b*	c_1	c_2
Angiosperms, Gymnosperms Pteridophytes, Bryophytes	+	+	−	−
Algae				
Chlorophyta	+	+	−	−
Euglenophyta	+	+	−	−
Chrysophyta				
Chrysophyceae/Haptophyceae	+	−	+	+
Xanthophyceae	+	−	−	−
Eustigmatophyceae	+	−	−	−
Bacillariophyceae	+	−	+	+
Pyrrophyta	+	−	−(+)	+
Phaeophyta	+	−	+	+
Cryptophyta	+	−	−	+
Rhodophyta	+	−	−	−
Cyanophyta	+	−	−	−
Prochlorophyta[a]	+	+	−	−

Notes:
The majority of dinoflagellates (Pyrrophyta), those containing peridinin as the major carotenoid, with one known exception contain only chlorophyll c_2: those dinoflagellates in which peridinin is replaced by fucoxanthin contain both c_1 and c_2.[415] Some c_2 containing chrysophytes also contain the newly described chl c_3.[418,869]
[a] Some coccoid oceanic prochlorophytes contain the divinyl forms, a_2 and b_2, instead of the normal forms, of chls *a* and *b*.[142,308]

red, 0.09–0.44; brown, predominantly 0.17–0.55; green, 0.28–1.53.[219] The amounts per unit area of thallus, in mg dm^{-2}, were: red, 0.5–2.8; brown, predominantly 4.3–7.6; green, 0.5–1.4. So far as the phytoplankton in natural waters are concerned, the chlorophyll *a* content is at its highest in nutrient-rich waters favouring rapid growth. Steele & Baird (1965) found that in the northern North Sea the ratio of carbon to chlorophyll *a* in the mixed phytoplankton population was at its lowest value of 20:1 in the spring, after which it decreased to a value of about 100:1 in late summer. Assuming the phytoplankton to contain about 37% carbon, and disregarding the silica of the frustules of the diatoms present, these ratios correspond to chlorophyll *a* contents of about 1.8% and 0.37% of the dry mass, respectively. In the upwelling area off Peru, the average carbon/chlorophyll *a* ratio of the phytoplankton throughout the euphotic zone was 40

(≡chlorophyll ~0.9% non-SiO_2 dry mass): this population consisted almost entirely of diatoms.[570] In the eastern Pacific ocean, off La Jolla, California, USA, the carbon/chlorophyll *a* ratio of the phytoplankton average about 90 (≡chlorophyll ~0.4% non-SiO_2 dry mass) in nutrient-depleted surface waters, and about 30 (~1.2% non-SiO_2 dry mass) in deeper, nutrient-rich waters.[230]

When grown in culture under conditions favouring high pigment content – low light intensity, high nitrogen concentration in the medium – algae usually have chlorophyll levels higher than those observed under natural conditions. Unicellular green algae such as *Chlorella* commonly have chlorophyll contents in the range 2 to 5% of the dry mass and *Euglena gracilis* has been observed to contain 3.5% chlorophyll, when grown under these conditions.

The molar ratio of chlorophyll *a* to chlorophyll *b* is about 3 in higher plants and in fresh-water green algae. Marine species of green algae, both multicellular and unicellular, are characterized by low *a*/*b* ratios, in the range 1.0–2.3.[411,653,1017] Outside the Chlorophyta, chlorophyll *b* occurs only in the Euglenophyta and in the prokaryotic Prochlorophyta. In *Euglena gracilis*, *a*/*b* is commonly about 6, and in the *Prochlorophyta* values of 1–12.0 have been reported.[918,120,142]

In those algae which contain chlorophyll *c*, this pigment constitutes (on a molar basis) about the same proportion of the total chlorophyll as chlorophyll *b* does in the green algae and higher plants. In surveys by Jeffrey (1972, 1976) and Jeffrey *et al.* (1975), the range of values of the molar ratio of *a* to *c* (where $c = c_1 + c_2$) found for different classes of marine algae were: diatoms, 1.5–4.0 (mean, 3.0); peridinin-containing dinoflagellates, 1.6–4.4 (mean, 2.3); fucoxanthin-containing dinoflagellates, 2.6–5.7 (mean, 4.2); chrysomonads, 1.7–3.6 (mean, 2.7); cryptophytes, 2.5; and brown algae, 2.0–5.5 (mean, 3.6). In the majority of those algae which contained both c_1 and c_2 they were present in approximately equal amounts: however, ratios (c_1:c_2) from 2:1 to 1:5 were found. Most of the dinoflagellates and cryptomonads were found to lack chlorophyll c_1. The newly defined algal class, Synurophyceae,[14]contains c_1 but not c_2.

The absorption spectra of chlorophylls *a* and *b* in organic solvent are shown in Fig. 8.7. They each have a strong absorption band (α) in the red and another stronger band (the Soret band) in the blue region, together with a number of satellite bands. Absorption is very low, but not zero, in the middle, green, region of the spectrum, hence the green colour of these pigments. As chlorophylls c_1 and c_2 are metalloporphyrins rather than metallodihydroporphyrins, they have spectra (Fig. 8.8) in which the Soret

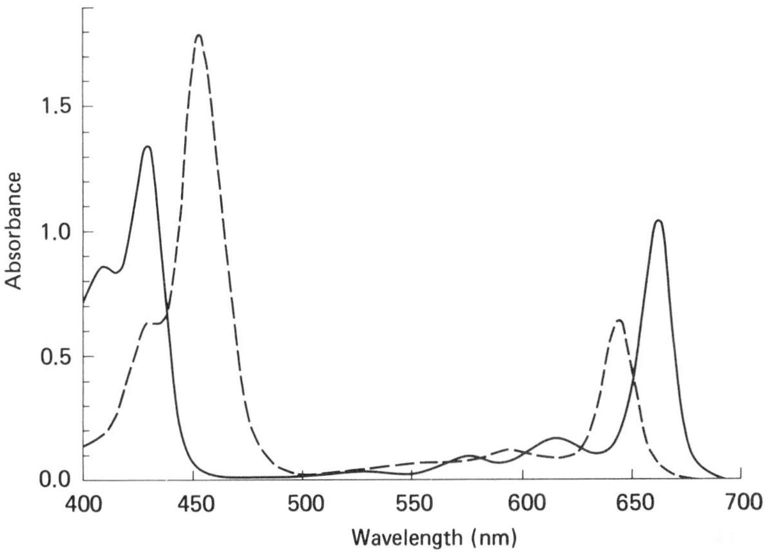

Fig. 8.7. Absorption spectra of chlorophylls *a* and *b* in diethyl ether at a concentration of 10 μg ml^{-1} and 1 cm pathlength. Calculated from data given by French (1960). (Chlorophyll *a* ——. Chlorophyll *b* ------.)

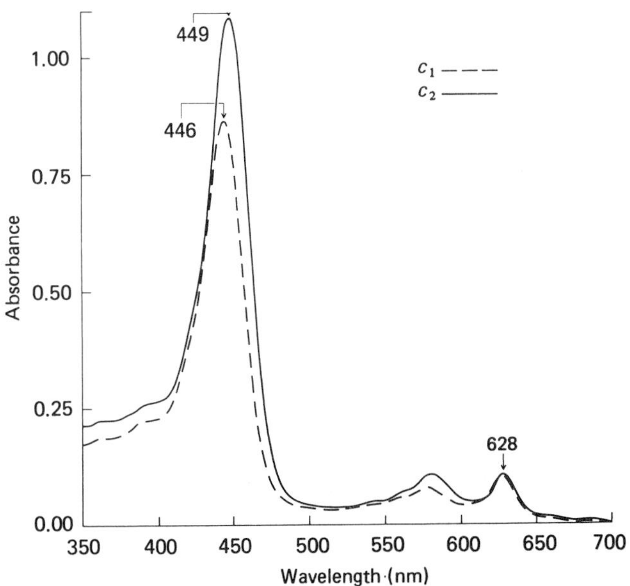

Fig. 8.8. Absorption spectra of chlorophyll c_1 and c_2 in acetone containing 2% pyridine with a pathlength of 1 cm (Jeffrey, S. W., unpublished data). (Chlorophyll c_1 (2.68 μg ml^{-1}) ------; chlorophyll c_2 (2.74 μg ml^{-1}) ——.)

band is more intense, and the α band less intense, than the corresponding bands in chlorophylls *a* and *b*. Also, the β band (~580 nm) in the *c* chlorophylls is comparable in intensity to the α band (~630 nm).

It can be seen from Figs 8.7 and 8.8 that chlorophyll *a* absorbs only weakly between 450 and 650 nm, and that chlorophylls *b* or *c*, when present, have the effect of increasing absorption within this window, at both the long- and the short-wavelength ends.

The carotenoids

The carotenoids are another class of photosynthetic pigment, which extend absorption still farther into the 'window', at the short-wavelength end. Chemically they are quite distinct from the chlorophylls, being C_{40} isoprenoid compounds. There are far more known carotenoids than there are chlorophylls. The distribution of the various chloroplast carotenoids amongst different plant groups is shown in Table 8.2. β-Carotene is present in all except the Cryptophyta. In higher plants and green algae, which rely mainly on chlorophylls for light harvesting, the molar ratio of carotenoid to chlorophyll $(a+b)$ is about 1:3. Other classes of algae, apart from those which contain biliproteins, depend to a greater extent on their carotenoids to capture light, and this is evident in the pigment composition. The molar ratio of carotenoid to chlorophyll $(a+c)$ is about 1:0.5 in the diatom *Phaeodactylum*,[333] 1:1.4 in the dinoflagellate *Gyrodinium*,[415] and amongst the brown algae is about 1:2 in *Hormosira*[480] and 1:0.5 in *Laminaria*.[9] Even amongst the red algae, which have biliproteins, the carotenoid/chlorophyll molar ratio varies from 1:2.6 to 1:1.[72]

The structures of some of the most important carotenoids in higher plants and algae are shown in Fig. 8.9. Their ability to absorb visible quanta is due to their possessing systems of up to 11 conjugated double bonds. They absorb at the short-wavelength end of the visible range, which is why their characteristic colours are yellow, orange or red. The absorption spectrum of β-carotene is shown in Fig. 8.10.

Chlorophyll/carotenoid–protein complexes

Essentially all the chlorophyll and most of the carotenoid in chloroplasts occur complexed to protein. There is good reason to suppose that these pigments can function in photosynthesis only as components of specific complexes with protein. All such photosynthetic complexes that have been examined so far, contain chlorophyll *a*: some also contain chlorophyll *b* or

Table 8.2. *Major chloroplast carotenoids in various algal classes. (After Kirk & Tilney-Bassett, 1978)*

		Chrysophyta											
Carotenoid	Chlorophyta[a]	Xanthophyceae	Eustigmatophyceae	Bacillariophyceae	Chrysophyceae[b]	Chloromonadophyta	Euglenophyta	Phaeophyta	Pyrrophyta	Cryptophyta	Rhodophyta	Cyanophyta	Prochlorophyta
α-Carotene	+	−	−	−	+	−	−	−	−	+	+	−	+[g]
β-Carotene	+	+	+	+	+	+	+	+	+	−	+	+[c]	+[g]
Echinenone	−	−	−	−	−	−	−	−	−	−	−	+	−
Lutein	+[c]	−	−	−	−	−	−	−	−	−	+[c]	−	−
Zeaxanthin	+	−	−	−	−	−	−	−	−	−	+	+	+
Neoxanthin	+	−	−	−	−	−	+	−	−	−	−	−	−
Vaucheriaxanthin	−	+	+[c,d]	−	−	−	−	−	−	−	−	−	−
Violaxanthin	+	−	+[c]	−	−	−	−	+	−	−	−	−	−
Heteroxanthin	−	+	−	−	−	−	−	−	−	−	−	−	−
Fucoxanthin	−	−	−	+[c]	+[c]	−	−	+[c]	(+)[e]	−	−	−	−
Diatoxanthin	−	+	−	+	+	−	+	−	−	−	−	−	−
Diadinoxanthin	−	+[c]	−	+	+	+[c,f]	+[c]	−	+	−	−	−	−
Peridinin	−	−	−	−	−	−	−	−	+[c]	−	−	−	−
Alloxanthin	−	−	−	−	−	−	−	−	−	+[c]	−	−	−
Myxoxanthophyll	−	−	−	−	−	−	−	−	−	−	−	+	−

Notes:

[a] These data for the Chlorophyta apply to higher plants also. Sublittoral species of green algae usually possess siphonaxanthin or siphonein, instead of, or as well as, lutein.[676,1016] Some species in the Prasinophyceae contain prasinoxanthin as their major carotenoid.[261,775]

[b] δ-Carotene frequently present; Prymnesiophyceae (Haptophyceae) have similar carotenoid composition to Chrysophyceae.

[c] Predominant carotenoid(s) present.

[d] Predominantly esterified.

[e] In some dinoflagellates fucoxanthin (and/or fucoxanthin derivatives such as 19′-hexanoyloxyfucoxanthin) replaces peridinin as the major carotenoid.[415]

[f] Identity of xanthophylls still in some doubt.

[g] Carotene predominantly α in *Prochlorococcus* and β in *Prochloron* and *Prochlorothrix*.[120,142,308]

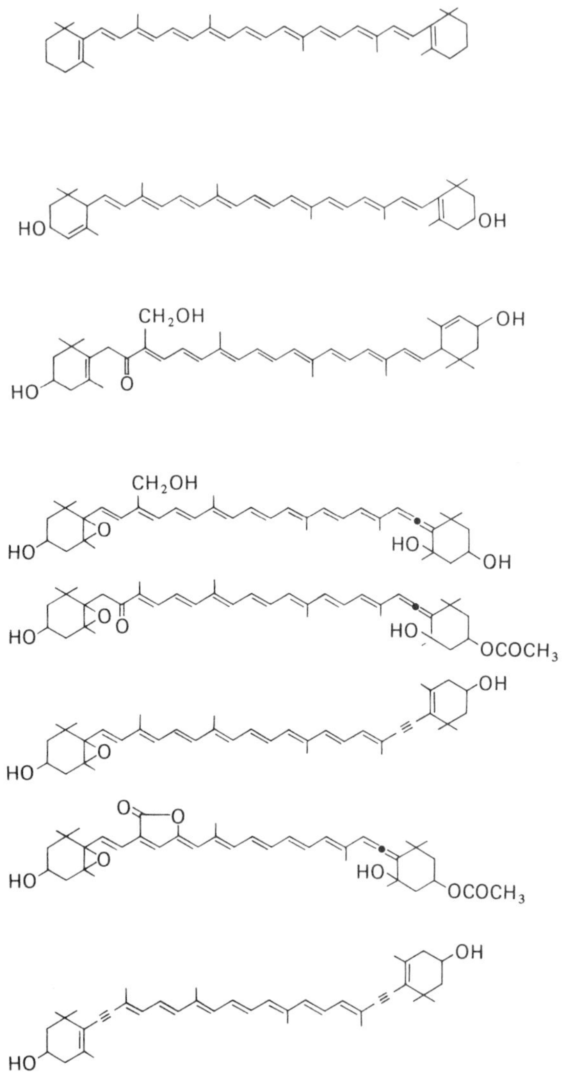

Fig. 8.9. Structures of chloroplast carotenoids. (*a*) β-Carotene, (*b*) Lutein. (*c*) Siphonaxanthin. (*d*) Vaucheriaxanthin. (*e*) Fucoxanthin. (*f*) Diadinoxanthin. (*g*) Peridinin. (*h*) Alloxanthin.

Each of carotenoids (*b*)–(*h*) is the major carotenoid in at least one algal group (see Table 8.2).

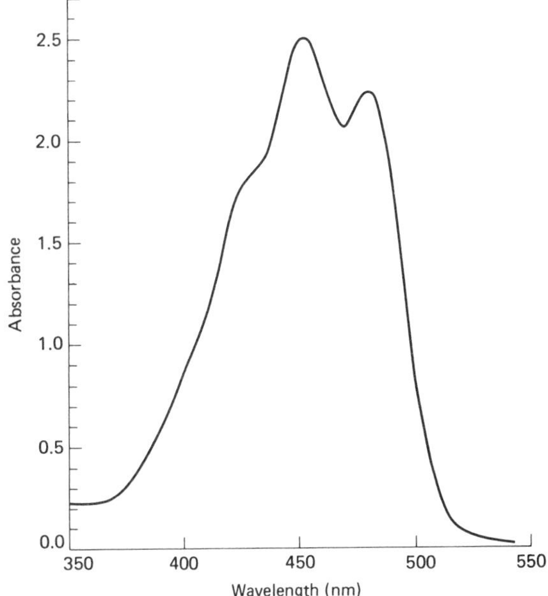

Fig. 8.10. Absorption spectrum of β-carotene in petrol at a concentration of 10 μg ml^{-1} and pathlength 1 cm.

c. The great majority, but possibly not all, of these complexes, contain one or more carotenoids as well as chlorophyll. There are usually several chlorophyll molecules per polypeptide together with one, or several, carotenoid molecules. Some of the complexes occur in the photosynthetic reaction centres and may participate in the light reactions of photosystem I or photosystem II. However, the bulk of the chlorophyll/carotenoid–protein consists of various light-harvesting pigment–proteins whose role is to collect light from the prevailing field and transfer the absorbed energy to the reaction centres.

Although the molecular structures of these pigment–protein complexes are not well understood, it seems likely that the chlorophylls and probably also the carotenoids are enclosed within the protein, rather than being attached at its surface. The bonding between the pigments and the protein, although no doubt highly specific, is not covalent. Some, perhaps all, of the chlorophyll molecules, are likely to have a ligand from the polypeptide chain to the central magnesium atom. Amino acid side chains such as histidine imidazole are believed to be the source of the ligands. In addition there must be hydrogen bonding between oxygen-containing ring substituents in the chlorophyll molecules, and appropriate groups in the polypeptide chains, as well as hydrophobic associations between the phytyl chains and the non-polar regions of the protein.

In the case of the carotenoids there must be hydrophobic associations between the central hydrocarbon chain region of the carotenoid and non-polar amino acid side chains of the polypeptide. The polar groups at each end of the carotenoid molecule presumably participate in hydrogen bonding with polar groups in the polypeptide. In the case of β-carotene, which has no polar groups, all interactions with the protein are presumably hydrophobic.

A very important consequence of the interaction between the pigments and the protein is that the absorption spectra of the former are modified. The absorption peaks are, to varying degrees, shifted to longer wavelength, with, in the case of the chlorophylls, some increase in complexity as well. Chlorophyll *a* in living cells or chloroplasts has a rather broad absorption peak in the red at about 676 nm, a shift of 9–15 nm relative to its position in organic solvents (usually 661–667 nm). Computer deconvolution of the absorption curve indicates that the peak is made up of four major forms of chlorophyll *a* with peaks at 662, 670, 677 and 684 nm, together with two minor forms at 692 and 703 nm.[266] Isolated chlorophyll–protein complexes show similar complexity: it may possibly be due to exciton interaction between the chlorophyll molecules within each pigment–protein complex, leading to splitting of the absorption bands and/or to the presence of different types of chlorophyll–protein binding.

In the case of some chloroplast carotenoids, such as β-carotene, the absorption peaks *in vivo* are shifted, relative to their positions in organic solvents, to longer wavelengths to about the same extent as the chlorophyll peak. However, certain major light-harvesting carotenoids in the algae – fucoxanthin, peridinin, siphonaxanthin – have *in vivo* spectra shifted by about 40 nm to longer wavelength relative to their spectra in organic solvents. This has the effect of substantially increasing absorption in the green window between the chlorophyll blue and red peaks. The diatoms and the Phaeophyta, for example, are brown in colour rather than green precisely because of their higher absorption in the green waveband. The shift in absorption to longer wavelengths is a consequence of the specific association between the carotenoid and the protein: if this association is disrupted by, say, heat denaturation, then the carotenoid spectrum reverts to the *in vitro* type. When a piece of brown-algal thallus is dipped in hot water it rapidly turns green. One possibility is that the binding between protein and carotenoid is such that the polyene chain of the latter is twisted: apparently this could account for the shift of the absorption band to longer wavelengths.[111]

Fucoxanthin and peridinin are the dominant carotenoids in those algae which contain them and show up in the *in vivo* spectra as a major shoulder in the 500–560 nm region (see Figs 9.3, 9.4*b*). Siphonaxanthin does not to the same extent dominate the carotenoid make-up of those algae which possess it, and shows up as a subsidiary peak at about 540 nm in the *in vivo* spectrum.[1016]

Many different chlorophyll/carotenoid–proteins from higher plants and algae have been described in recent years. They have in most cases been liberated from the thylakoid membranes, in which they are normally embedded, with the help of detergents. It seems likely that these are variants on a relatively small number of fundamental pigment–protein types, and we shall treat them accordingly. The reviews by Larkum & Barrett (1983), Thornber (1986), Anderson & Barrett (1986), Rowan (1989), Wilhelm (1990), Hiller, Anderson & Larkum (1991), and Thornber *et al.* (1991) have been extensively used for the following account, and should be consulted for further details. Where possible we shall use the nomenclature for these pigment-proteins that has been agreed upon by workers in this field.[916]

All algae and higher plants contain an essential chlorophyll *a*/β-carotene protein, Core Complex I, which functions in photosystem I, its absorbed light energy being transferred to the reaction centre known as P_{700} (see §8.4), which is contained within Core Complex I. There are thought to be two 84 kilodalton (kDa) protein subunits per P_{700}, each subunit binding 20–45 chlorophyll *a* and 5–7 β-carotene molecules.

While the Core Complex I does itself absorb light (in barley chloroplasts, for example, it contains about 20% of the total chlorophyll), it nevertheless has associated with it a light-harvesting pigment–protein complex to supply it with additional energy. This is referred to as Light-Harvesting Complex I (LHC I). In higher plants it contains chlorophylls *a* and *b* ($a/b \sim 3.5$) and xanthophylls (mainly lutein), bound to at least four protein subunits (24, 21, 17 and 11 kDa): in barley chloroplasts it accounts for about 18% of the total chlorophyll. The green flagellate *Chlamydomonas reinhardtii* also has an LHC I, rather similar to the higher plant complex.[1002] The siphonalean green alga *Codium* has an LHC I, containing chlorophyll *a* and *b*, in which siphonaxanthin is the main carotenoid,[145] and there is some evidence for the presence of an LHC I containing diadinoxanthin in *Euglena gracilis*.[170] The presence of light-harvesting pigment–proteins supplying additional energy specifically to Core Complex I, in other algal groups has yet to be demonstrated, although it seems likely that they exist.

Algae and higher plants also contain another essential chlorophyll *a*/β-carotene protein complex, Core Complex II, which functions in photosystem II, the absorbed light energy being transferred to the reaction centre known as P_{680}, which is part of the total complex. It is estimated that there are about 40 chlorophyll *a* molecules and one P_{680} reaction centre for each photosystem II unit. Most of the chlorophyll and β-carotene is present in two non-identical but similarly sized (apoproteins 43–50 kDa) pigment proteins, each accounting (in barley) for about 5% of the total thylakoid chlorophyll. The actual reaction centre part of the complex appears to contain four to five chlorophyll *a*, two phaeophytin and one β-carotene molecules associated in a manner yet to be determined, with four polypeptides (32, 30, 9 and 4 kDa): about 1% of the total thylakoid chlorophyll is contained in these reaction centres.

Like Core Complex I, Core Complex II, as well as absorbing light directly, is supplied with additional energy by an associated light-harvesting pigment–protein complex, which is referred to as Light-Harvesting Complex II (LHC II). In higher plants LHC II is made up of at least four individual pigment proteins – LHC IIa, b, c and d – all containing chlorophylls *a* and *b*, and the three chloroplast xanthophylls, lutein, violaxanthin and neoxanthin. LHC IIb is the major component containing as it does 40–45% of the total chloroplast chlorophyll: the other three LHC II pigment–proteins between them account for 10–15% of the total chlorophyll. The aggregate LHC II has a chlorophyll *a*/*b* ratio of about 1.4; for LHC IIb the ratio is about 1.33.

LHC IIb preparations contain three slightly different apoproteins – 28, 27 and 25 kDa – possibly resulting from post-translational modification of a single polypeptide. Each apoprotein molecule is thought to bind about 15 chlorophyll molecules (eight *a*, six to seven *b*), and about three xanthophyll molecules.

Amongst the various classes of algae, a large number of different light-harvesting pigment–proteins has been found in recent years. It seems likely that most, perhaps all, of these are of the LHC II type, feeding their energy to Photosystem II. They all contain chlorophyll *a* together with, in most cases, whatever accessory chlorophyll – *b* or *c* – is present in that particular alga. In addition they contain the major light-harvesting xanthophyll carotenoid(s) characteristic of that algal class. While they all have peaks in the red region of the spectrum, due to chlorophyll(s), their main light absorption, due to a combination of chlorophyll Soret and carotenoid bands, is in the 400–550 nm, blue-green waveband. Table 8.3 lists a selection of these putative LHC II algal proteins, with their pigment and polypeptide composition.

Katoh & Ehara (1990) have presented evidence that *in vivo* these light-harvesting proteins are organized into supramolecular assemblies. Using

Table 8.3. *Algal light-harvesting pigment–protein complexes (presumptive LHC IIs)*

Algal group	Number of pigment molecules per 100 molecules of Chl *a*				Apoprotein M. wt. (kDa)	Reference
	Chl *a*	Chl *b*	Chl *c*	Carotenoid		
Chlorophyta						
Chlorophyceae						
Chlamydomonas	100	106	—	Lutein 12 Violaxanthin 11 Neoxanthin 6 β-Carotene 5	29, 25	454
Bryopsis	100	116	—	Siphonaxanthin 51 Siphonein 15 Neoxanthin 28		654
Codium	100	58	—		25–19	145
Euglenophyta						
Euglena	100	49	—	Diadinoxanthin 31 Neoxanthin 9 β-Carotene 1	26.5, 28, 26	170
Chrysophyta						
Xanthophyceae						
Pleurochloris	100	—	22	Diadinoxanthin 26 Heteroxanthin 15	23	998
Eustigmatophyceae						
Nannochloropsis	100	—	—	Violaxanthin 26–44 Vaucheriaxanthin 22–25 Neoxanthin 2 β-Carotene 4	26	105, 889
Polyedriella	100	—	—	Violaxanthin 120 (predominant)	22	144

Table 8.3. (*cont.*)

Algal group	Number of pigment molecules per 100 molecules of Chl *a*				Apoprotein M. wt. (kDa)	Reference
	Chl *a*	Chl *b*	Chl *c*	Carotenoid		
Bacillariophyceae						
Phaeodactylum	100	—	40	Fucoxanthin 192	18	267
Nitzschia	100	—	—	Fucoxanthin 212	18.5	77
Prymnesiophyceae						
Pavlova	100	—	21	Fucoxanthin	21	369
Pyrrophyta						
Amphidinium[a]	100	—	—	Peridinin 450	32	362
Amphidinium	100	—	57	Peridinin 171 Diadinoxanthin 29	19	370
Phaeophyta						
Acrocarpia	100	—	50	Fucoxanthin 100		46
Fucus	100	—	16	Fucoxanthin 70 (predominant)	21.5, 23	134, 135
Cryptophyta						
Chroomonas	100	—	65	Unidentified (Alloxanthin?)	20, 24	401
Prochlorophyta						
Prochloron	100	42	—		34	368
Prochlorothrix	100	25	—		30–33	118

Note:
[a] This is the water-soluble peridinin–chlorophyll *a* complex.

the mild detergent, octyl sucrose, they were able to isolate, from chloroplasts of the brown algae *Petalonia fascia* and *Dictyota dichotoma*, pigment–protein complexes of molecular weight about 700 kDa, containing in each complex about 128 molecules of chlorophyll *a*, 27 of chlorophyll *c*, 69 of fucoxanthin and 8 of violaxanthin. Electron microscopy showed that each complex was discoidal in shape, being 11.2 nm in diameter and 10.2 nm in height, with a small pit at the centre of the disc.

Biliproteins

The biliprotein chloroplast pigments are found only in certain algae: the Rhodophyta, Cryptophyta and Cyanophyta. They are either red (phycoerythrins, phycoerythrocyanin) or blue (phycocyanins, allophycocyanins) in colour. They have been reviewed by Bogorad (1975), Gantt (1975, 1977), O'Carra & O'h Eocha (1976), Glazer (1981, 1985), MacColl & Guard-Friar (1987) and Rowan (1989).

The biliproteins of the red and blue-green algae are closely related and we shall consider them together. They are divided into three classes on the basis of the position of their absorption bands: these are, in order of increasing wavelength, the phycoerythrins (and phycoerythrocyanin), the phycocyanins and the allophycocyanins. The position of the main absorption peaks are listed in Table 8.4, and absorption spectra of a red-algal phycoerythrin and phycocyanin are shown in Fig. 8.11.

The general principle governing the distribution of these pigments appears to be that allophycocyanin and one or other of the phycocyanins occur in all red- and blue-green algal species: a phycoerythrin (or phycoerythrocyanin in some blue-green algae) may or may not, in addition, be present. Phycoerythrin is rarely absent in the red algae but is frequently absent in the blue-green algae. The typical situation is that in the red algae, phycoerythrin constitutes most of the biliprotein present, whereas in the blue-green algae, phycocyanin or (less commonly) phycoerythrin (or phycoerythrocyanin) is the major component: allophycocyanin is nearly always a minor component.

The chromophores to which biliproteins owe their colour are open-chain tetrapyrrole compounds known as phycobilins. There are three main chromophores – phycocyanobilin, phycoerythrobilin and phycourobilin. Unlike chlorophylls and carotenoids, they are covalently bound to their proteins. Their structures and modes of linkage to the apoproteins are shown in Fig. 8.12. Another phycobilin chromophore (provisionally cryptoviolin) occurs in phycoerythrocyanin.

Table 8.4. *Spectroscopic properties of the different biliproteins and the distribution of phycobilin chromophores amongst their subunits* [300,302,668,673,10,680,976,301 579,580,582]

Biliprotein	Algal type[a]	Absorption maxima in photosynthetic range[b] (nm)	α Subunit[c]	β Subunit[c]	γ Subunit[c]
Red and blue-green algae					
Allophycocyanin	R, B	650, (620)	1 PCB	1 PCB	—
Allophycocyanin B	R, B	671, 618	1 PCB	1 PCB	—
C-Phycocyanin	R, B	620	1 PCB	2 PCB	—
R-Phycocyanin	R	618, 553	1 PCB	1 PCB 1 PEB	—
Phycoerythrocyanin	B	(590), 568	1 CV	2 PCB	—
C-Phycoerythrin	B	562–565	2 PEB	3 PEB	—
CU-Phycoerythrin	B	490–500, 545–560	2 PEB	3 PEB 1 PUB	—
b-Phycoerythrin	R	(563), 545	2 PEB	4 PEB	—
B-Phycoerythrin	R	(565), 546, (498)	2 PEB	3 PEB	2 PEB 2 PUB
R-Phycoerythrin	R	568, 540, 598	2 PEB	2 PEB 1 PUB	1 PEB 3 PUB
Cryptophytes					
Phycocyanin-615		615, 585	1 PCB	2 PCB 1CV	—
Phycocyanin-630		630, 585	n.d.	n.d.	—
Phycocyanin-645		645, (620), 585	1 U	2 PCB 1 CV	—
Phycoerythrin-544		(565), 544	2 PEB 2 CV	PEB	—
Phycoerythrin-555		555	PEB	PEB	—
Phycoerythrin-568		568	PEB	PEB	

Notes:

[a] R: red algae; B: blue-green algae.

[b] Minor maxima or shoulders are in brackets.

[c] PCB: phycocyanobilin; PEB: phycoerythrobilin; PUB: phycourobilin; CV: cryptoviolin; U: unknown phycobilin chromophore, distinct from the previous four with peak at 697 nm. The absence of a number means that the number of chromophores is unknown. n.d.: no data.

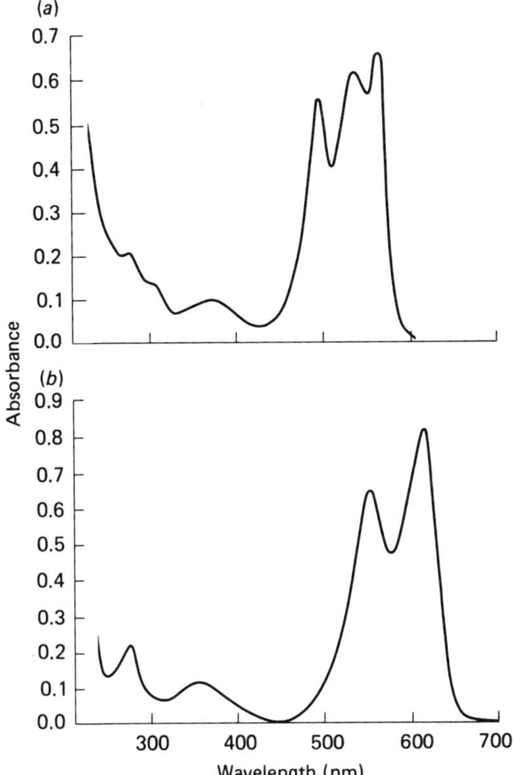

Fig. 8.11. Absorption spectra of biliproteins (after O'h Eocha, 1965). (*a*) R-phycoerythrin from *Ceramium rubrum*. (*b*) R-phycocyanin from *Porphyra laciniata*.

All the algal biliproteins contain equimolar amounts of two kinds of subunit, referred to as α and β: some of the phycoerythrins contain in addition much lesser amounts (one molecule for every six of α or β) of a third class of subunit, γ. The α and β subunits have mol wts of about 17–22 kDa but the γ subunit may be larger. Each α or β subunit in a biliprotein has at least one phycobilin chromophore attached. The α subunit can have one or two chromophores; the β subunit has one, two or three, and the γ subunit has four. The distribution of chromophores amongst the subunits in the different proteins is shown in Table 8.4. This distribution may not be immutable. In phycoerythrin from the red alga *Callithamnion roseum* it seems that the proportion of phycoerythrobilin to phycourobilin varies with light intensity during growth:[1018] it is possible that at some sites within the protein, either of these two chromophores may be attached.

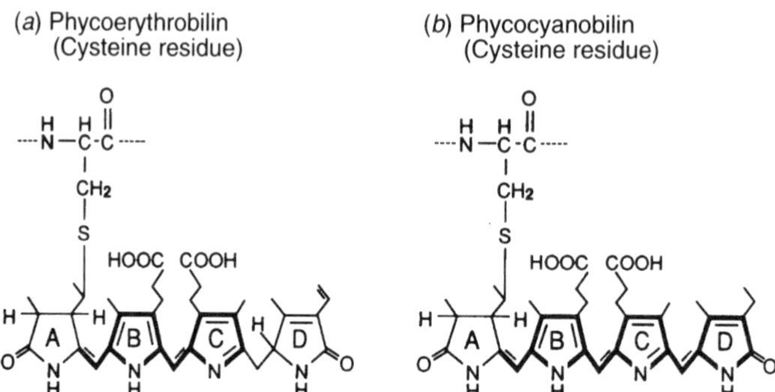

Fig. 8.12. Structures, and probable mode of binding to protein, of the phycobilin chromophores of algal biliproteins.[301,470] The systems of conjugated double bonds, upon which the spectral properties of the chromophores depend, are emphasized by heavy lines. Phycourobilin is likely to have CH_2 rather than CH groups joining rings A to B and C to D. Linkage to protein cysteine can also occur through the vinyl/ethyl group on ring D.[301]

Biliproteins occur as aggregates of the basic $\alpha\beta$ structure. In the isolated, soluble state, various sizes are found of which the most common appear to be the hexamer $\alpha_6\beta_6$ and the trimer $\alpha_3\beta_3$. Those phycoerythrins which have a γ subunit are likely to occur as $\alpha_6\beta_6\gamma$ aggregates in the isolated state. *In vivo*, biliproteins occur as much larger aggregates, the phycobilisomes, particles of diameter 30–40 nm attached to the outer surface of the thylakoids (Fig. 8.2). These are ordered, specific structures rather than random aggregates, and it seems likely that the three or so different biliproteins present in the alga are all present in each phycobilisome, together with small amounts of some colourless proteins. Allophycocyanin, which as we shall see later transfers the energy harvested by the phycobilisome to chlorophyll, is believed to be located at the base of the particle, adjoining the membrane: the other, more plentiful proteins form a shell around the allophycocyanin, with the phycocyanin being next to the allophycocyanin and the phycoerythrin or phycoerythrocyanin forming the periphery of the particle. Electron microscope and biochemical studies have given rise to a model of phycobilisome structure along the lines of that shown schematically in Fig. 8.13.

Turning to the Cryptophyta we find some similarities to, as well as marked differences from, the situation in the Rhodophyta and Cyanophyta. The cryptophytan pigments are either red or blue and accordingly are classified either as phycoerythrins or phycocyanins. Each cryptophyte species appears to contain only one biliprotein. No allophycocyanin-like

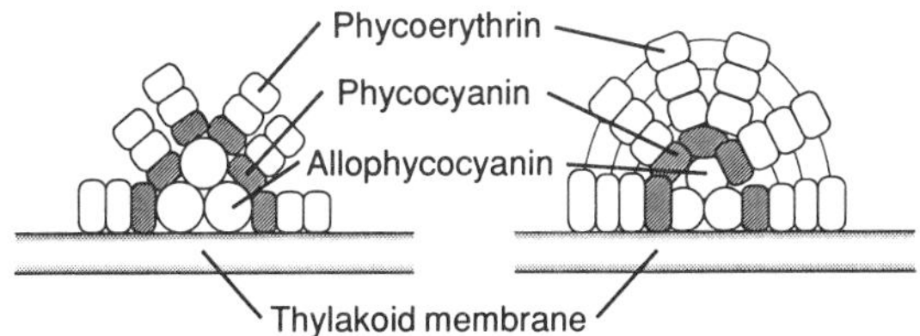

Fig. 8.13. Schematic representation of phycobilisome structure (after Glazer, 1985; Gantt, 1986; Glazer & Melis, 1987).

pigments have yet been found in this group. The main absorption peaks are in the 612–645 nm range in the case of the phycocyanins, and the 544–568 nm range in the case of the phycoerythrins. The chromophores so far detected in these proteins are phycoerythrobilin, phycocyanobilin, cryptoviolin and an unknown phycobilin with an absorption peak (in acid urea) at 697 nm.[580]

The cryptophyte biliproteins appear to be organized on the same $\alpha\beta$ subunit principle as the red- and blue-green algal proteins: in the isolated state they exist predominantly as $\alpha_2\beta_2$ dimers. The subunits, at 10 and 17.5 kDa respectively appear, however, to be smaller than those in the other algae, and there are two types of α subunit so that the dimer composition is better represented as $\alpha\alpha'\beta_2$.[332] Where they differ most strikingly from the corresponding proteins in the other algae is their location in the photosynthetic apparatus: instead of being present as phycobilisomes on the surface of the thylakoids, the cryptophyte biliproteins occur as a dense granular matrix filling the interior of the thylakoid (Fig. 8.3).

8.4 Reaction centres and energy transfer

The crucial step within photosystems I and II is the use of the absorbed light energy to transfer an electron from a donor molecule to an acceptor molecule. The particular donor and acceptor molecules are different in the two photosystems. The site in a photosystem at which this event occurs is known as the *reaction centre*. The central role of using the excitation energy to extract an electron from one molecule and transfer it to another is, in each photosystem, carried out by a special form of chlorophyll *a* complexed with a specific protein. When reaction centre chlorophyll receives excitation energy, it is raised to an excited electronic state. In this excited state it can reduce (transfer an electron to) the acceptor molecule. The oxidized chlorophyll then withdraws an electron from the donor molecule, and so returns to its original state. The loss of an electron is accompanied by a fall in the absorption spectrum of the reaction centre chlorophyll, in the region of its red peak. The spectral change is maximal at about 700 nm in photosystem I and at about 680 nm in photosystem II: accordingly, the specialized forms of chlorophyll are referred to as P_{700} and P_{680}, respectively. Since the reaction centre chlorophylls constitute only a very small proportion of the total chlorophyll *a* – one P_{700} and one P_{680} per 500 total chlorophyll molecules in green plants – their spectral changes are insignificant with respect to the total absorption spectrum of the system.

Excitation of reaction centre chlorophyll can be brought about by direct absorption of a photon. Since P_{700} and P_{680} constitute a tiny proportion of the total pigment, however, this does not take place very often. In fact, virtually all the excitation energy received by the reaction centre is energy initially captured by the vastly more numerous light-harvesting or antenna pigment molecules of that photosystem and transferred to the reaction centre. The mechanism by which this takes place is known as inductive resonance transfer, first postulated by the theoretical chemist T. Förster in 1947.

Absorption of a photon by one molecule is followed by vibrational energy dissipation, bringing the excited electron to its lowest excited state. This can be in a state of resonance with one of the upper vibrational levels of the excited state of another molecule (not, initially, in an excited state). The energy is transferred from the first molecule to the second, i.e. the first molecule reverts to the ground state and the second molecule is raised to an excited electronic state. For efficient transfer the fluorescence emission peak of the donating molecule must overlap the absorption spectrum of the receiving molecule. Since the fluorescence emission spectrum of any molecule is a mirror image of the absorption spectrum (on the long-wavelength side), with the peak shifted to longer wavelength, it follows that for efficient energy transfer the absorption peak of the donating molecule should be at a shorter wavelength than that of the receiving molecule. In addition the molecules must not be too far apart: efficient transfer can take place at distances up to about 5 nm.

The light-harvesting, as opposed to reaction centre, pigment molecules are made up of the great majority of the chlorophyll *a* molecules, chlorophylls *b*, c_1 and c_2, the various carotenoids, and the biliproteins. All these (except carotenoids) fluoresce actively *in vitro* and so the assumption that energy transfer amongst them takes place by inductive resonance presents no problem. Carotenoids, however, do not fluoresce measurably *in vitro*, and so the view is sometimes expressed that energy transfer from carotenoids to other pigments must involve some other mechanism. This need not be so. The lack of detectable fluorescence by the carotenoids simply means that their excitation energy decays quickly: it is, however, entirely possible that energy transfer by inductive resonance is even faster, so that transfer can still take place. The carotenoid molecules might, however, need to be rather close to the other pigment molecules to transfer energy efficiently to them.

In higher plants and green algae, energy absorbed by chlorophyll *b* is transferred to chlorophyll *a* with about 100% efficiency: energy absorbed by carotenoids is transferred (with lower efficiency), probably first to chlorophyll *b* and then to chlorophyll *a*. In all those algae containing chlorophyll *c*, the energy absorbed by this pigment is transferred efficiently to chlorophyll *a*. In those algae which contain major light-harvesting carotenoids – fucoxanthin, peridinin or siphonaxanthin – with substantial absorption in the 500–560 nm region, there is efficient energy transfer directly from the carotenoid to chlorophyll *a*. In red and blue-green algae, the sequence of transfer, 80–90% efficient overall, is phycoerythrin (or phycoerythrocyanin)→phycocyanin→allophycocyanin→chlorophyll *a*. In cryptophytes, which usually have only one biliprotein, this sequence is not possible: direct energy transfer from the biliprotein to chlorophyll *a* may occur, with quite high efficiency from phycocyanin, but lower efficiency from phycoerythrin.

In every case, no matter which pigment first captures the light, the absorbed energy always ends up in chlorophyll *a*. This is to be expected since chlorophyll *a* has its absorption peak at a longer wavelength than any of the other pigments and, as noted earlier, energy migration by this mechanism is in the direction of the molecules absorbing at the greatest wavelength. Amongst the bulk chlorophyll *a*-protein complexes, the excitation energy moves at random until it reaches a reaction centre, where it is immediately trapped and used for electron transfer.

8.5 The overall photosynthetic process

The photosynthetic process can be divided into two parts, the light reactions and the dark reactions. In the light reactions, which take place in the thylakoid membrane system, hydrogen is withdrawn from water and passed along a series of hydrogen carriers to NADP, so that $NADPH_2$ is formed and oxygen is liberated. Associated with this hydrogen (or electron) transport there is a conversion of ADP and inorganic phosphate to ATP, probably two (or, on average, some fractional number between one and two) ATP molecules being formed for every two electrons transferred or molecule of NADP reduced. These chemical changes are associated with a considerable increase in free energy: this is made possible by the light energy absorbed by the chloroplast pigments. Thus we may summarize the light reactions by the equation

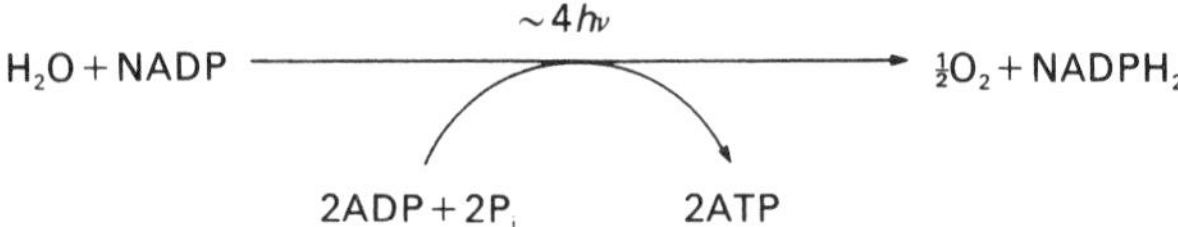

In the dark reactions, which take place in the stroma of the chloroplast, the $NADPH_2$ produced in the light reactions is used to reduce CO_2 to the level of carbohydrate. This too is associated with an increase in free energy, the energy being supplied by the concomitant breakdown of ATP produced in the light reactions. The dark reactions can be summarized by the equation

$$CO_2 + 2NADPH_2 \xrightarrow[3ATP \;\rightarrow\; 3ADP + 3P_i]{} (CH_2O) + H_2O + 2NADP$$

Thus, the overall photosynthetic process can be represented by

$$CO_2 + 2H_2O \xrightarrow{\sim 8h\nu} (CH_2O) + H_2O + O_2$$

The light reactions

It is now generally believed that there are two photochemical reactions, occurring in series in photosynthesis. The subject is comprehensively reviewed in the multi-author work, edited by Staehelin & Arntzen (1986): see also Mathis & Paillotin (1981) and the recent account by Parson (1991). Light reaction 1 is associated with the reduction of NADP; light reaction 2 brings about the liberation of oxygen from water. Each of the photoreactions takes place in a reaction centre in association with specialized light-harvesting pigment–proteins and electron transfer agents. The set of specific functional components associated with light reaction 1 is referred to as *photosystem I*; the set associated with light reaction 2 is referred to as *photosystem II*. The functional unit consisting of a single photosystem I and a single photosystem II working together, plus a set of light-harvesting pigment–proteins, is commonly referred to as a *photosynthetic unit*.

Each photosystem has a chlorophyll *a*/carotenoid–protein which both harvests light and is intimately associated with the reaction centre. This pigment–protein represents a funnel through which all the excitation energy collected by other pigment–proteins must pass before it is delivered to the reaction centre. Core Complex I, and Core Complex II chlorophyll *a*/β-carotene–proteins present in all plants (see §8.3) perform this role for photosystems I and II, respectively.

The extent to which each of the other light-harvesting pigment–proteins of algae and higher plants, described in §8.3, transfers energy specifically to one photosystem or the other, or to both, is uncertain. There is some evidence that the major light-harvesting pigment–proteins, such as the various chlorophyll carotenoid-containing LHC IIs of higher plants and green algae, and the biliproteins of red algae, transfer most, but not all, of their energy initially to photosystem II. Transfer of some of this energy from photosystem II to photosystem I (referred to as 'spillover') then takes place, thus making it possible for the two photoreactions to continue at the same rate. The review by Butler (1978) may be consulted for a detailed discussion of this topic.

Turning now to what actually happens in the light phase of photosynthesis, the current view of the sequence of events is summarized in Fig. 8.14. In photosystem I when the reaction centre chlorophyll, P_{700}, acquires excitation energy it loses an electron to a primary acceptor, A_0, which may be another molecule of chlorophyll *a*. From the primary acceptor, electrons are transferred, probably via one or more other electron carriers, to ferredoxin and from there, via the flavoprotein ferredoxin–NADP reductase, to NADP. The oxidized reaction centre chlorophyll, P_{700}^+, is restored to its original state by an electron from the copper–protein, plastocyanin, which is then in turn reduced by cytochrome *f*.

When the reaction centre chlorophyll in photosystem II, P_{680}, acquires excitation energy it loses an electron first to an associated phaeophytin molecule, which then rapidly reduces a bound form of plastoquinone (Q_A). The electron is then transferred possibly via another specialized plastoquinone molecule (Q_B), to the major pool of plastoquinone, from which in turn electrons are transferred to reduce the cytochrome *f* molecules which were oxidized by photosystem I. The oxidized reaction centre chlorophyll, P_{680}^+, is reduced by transfer of an electron from a donor, *Z*, which may be a tyrosyl residue in a reaction centre protein, and the oxidized form of *Z*, in association with the Mn-containing water-splitting complex, removes electrons from water, liberating oxygen.

Putting all these processes together, the transfer of hydrogen from water

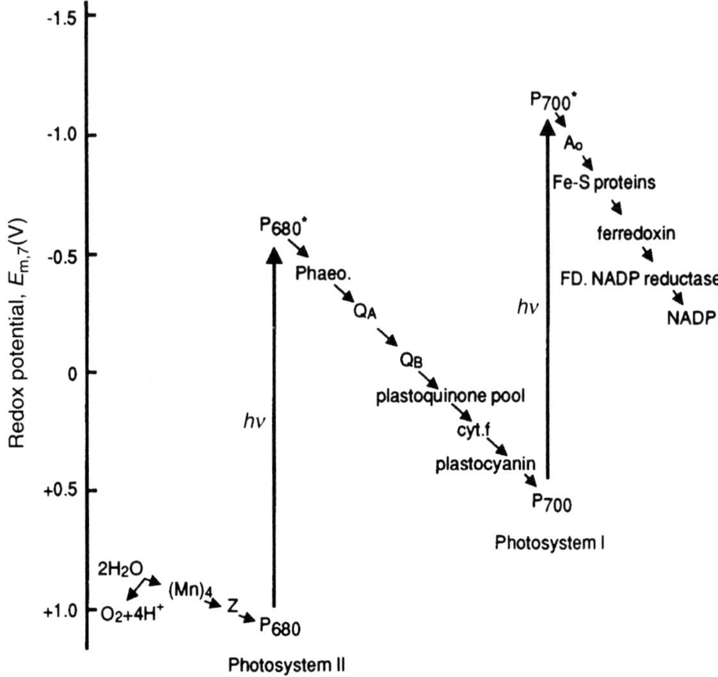

Fig. 8.14. The light phase of photosynthesis. The details are discussed in the text.

to NADP, giving rise to oxygen and $NADPH_2$, is now complete. As a consequence of the particular location of electron carriers within the membrane, and the particular reactions that occur, there is a movement of eight protons from outside the thylakoid to the intrathylakoid space for every oxygen molecule liberated. It is believed that the pH gradient and electric potential set up in this way operate, by means of a reversible ATPase in the membrane, working backwards, to bring about ATP synthesis, i.e. photophosphorylation.

The dark reactions

The pathway by which, in the stroma of the chloroplast, the $NADPH_2$ and ATP produced in the light reactions are used to convert CO_2 to carbohydrate was elucidated largely by the work of Calvin & Benson, and is outlined in Fig. 8.15. The cycle is somewhat involved and needs several turns to produce one molecule of hexose. To understand the overall effect of the dark reactions let us consider the fate of six molecules of CO_2. These react

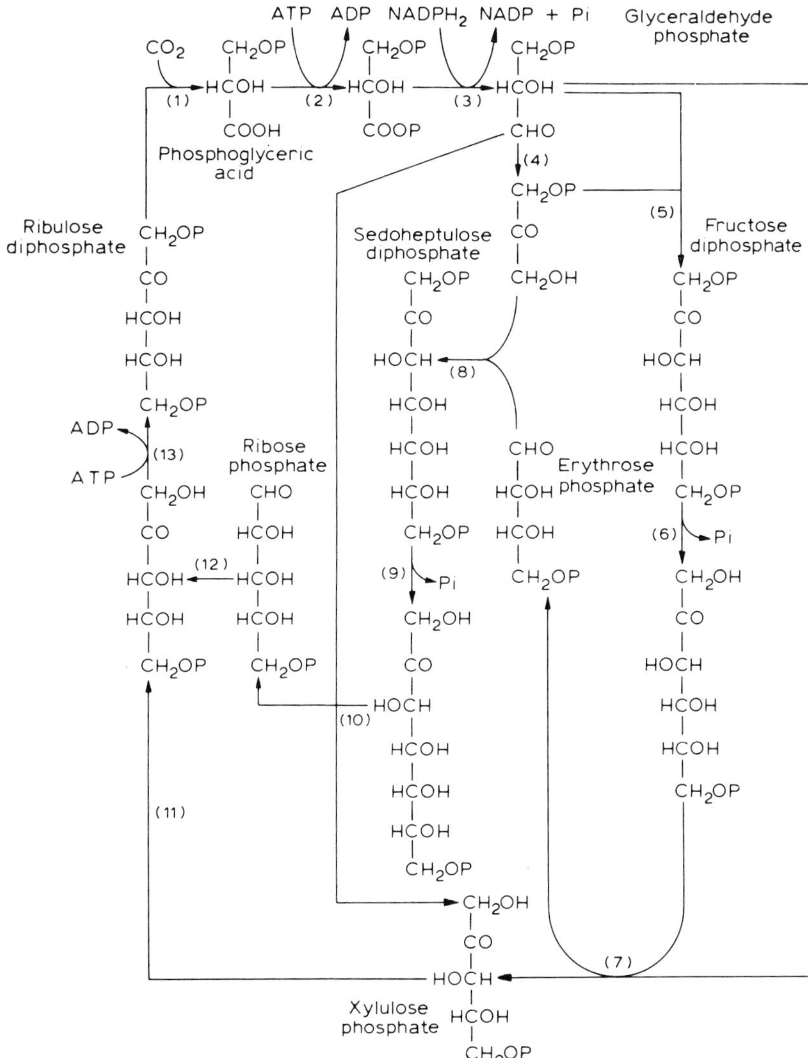

Fig. 8.15. The photosynthetic CO_2 fixation cycle. P = phosphate group; Pi = inorganic phosphate. The enzymes involved in each step are: (1) Rubisco; (2) 3-phosphoglyceric acid kinase; (3) glyceraldehyde-3-phosphate dehydrogenase; (4) triose-phosphate isomerase; (5) fructose-diphosphate aldolase; (6) fructose 1,6-diphosphatase; (7) transketolase; (8) aldolase; (9) sedoheptulose 1,7-diphosphatase; (10) transketolase; (11) xylulose-5-phosphate epimerase; (12) ribose-5-phosphate isomerase; (13) ribulose-5-phosphate kinase.

with six molecules of ribulose bisphosphate (C_5) to give 12 molecules of phosphoglyceric acid (C_3). The phosphoglyceric acid is, with the help of ATP and $NADPH_2$, reduced to triose phosphate. For simplicity we can regard these as being converted to six molecules of hexose phosphate (C_6). One of these molecules of hexose can be removed to form starch or other reserve carbohydrate, leaving five hexose phosphate molecules. These are now rearranged to form six pentose phosphate (C_5) molecules which are, with the help of ATP converted to six molecules of ribulose bisphosphate. Thus the cycle is completed with six molecules of CO_2 being converted to one of hexose. The overall process in terms of carbon atoms may be summarized

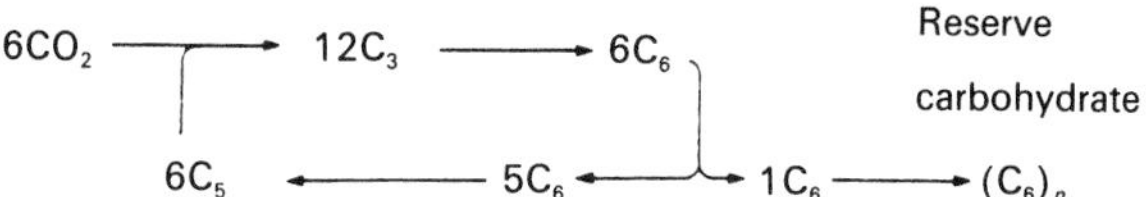

Some higher plants, known as C_4 plants, possess a variant of this cycle in which CO_2 is first fixed in the form of a C_4 acid, such as malic acid. The C_4 acid is then translocated to another tissue where CO_2 is liberated from it again and converted to carbohydrate by the normal cycle. It seems very unlikely that this C_4 pathway occurs in any alga, and it does not appear to exist in submerged aquatic higher plants, so we may disregard it as a contributor to aquatic photosynthesis within the water column. Some aquatic higher plants do, however, have a simplified form of C_4 pathway in which a substantial proportion of the CO_2 fixation is indeed initially into C_4 acids, from which the CO_2 is subsequently liberated and used for photosynthesis, but within the same cell.[87,461,746] The effect of this is to increase the CO_2 supply to ribulose bisphosphate carboxylase, this being a significant limiting factor in aquatic photosynthesis, as we shall discuss in more detail in a later chapter (see §11.3).

In bright sunlight, photosynthesis produces carbohydrates faster than they can be used in respiration or growth. Accordingly the plant must store the fixed carbon in some form that it can utilize later, most commonly as grains of polysaccharide. The chemistry of algal storage products was reviewed by Craigie (1974). In higher plants and green algae, fixed carbon is accumulated inside the chloroplast as starch grains in the stroma, or forming a shell around the pyrenoid in those algae which have them. In algal phyla other than the Chlorophyta, the photosynthetic end products accumulate outside the chloroplast. Where a pyrenoid is present the polysaccharide grains are usually found in the cytoplasm but in close contact with the pyrenoid region of the chloroplast.

The Floridean starch of the Rhodophyta is an α-D-(1→4)-linked glucan with α-D-(1→6) branch points, and can thus be regarded as an amylopectin, like the major component of higher plant starch. Cryptophytes accumulate a starch of the higher plant type, containing both amylose and amylopectin. Members of the Pyrrophyta accumulate grains of a storage polysaccharide with staining properties similar to starch. Brown algae (Phaeophyta) accumulate the polysaccharide laminaran, a β-D-(1→3)-linked glucan with 16–31 residues per molecule and a mannitol residue at the reducing end of each chain: some of the laminaran molecules are unbranched, some have two to three β-D-(1→6) branch points per molecule. In addition, these algae accumulate large amounts of free mannitol. In the Euglenophyta, the photosynthetic storage product accumulates as granules of paramylon, a β-D-(1→3) glucan, with 50–150 residues per molecule. The photosynthetic storage product of the Chrysophyceae is chrysolaminaran (also known as leucosin), a β-D-(1→3)-linked glucan with possibly two 1→6 branch points per 34-residue molecule. Diatoms also contain polysaccharides of the chrysolaminaran type.

9

Light capture by aquatic plants

The collection of light energy for photosynthesis by aquatic plants is, as we have seen, carried out by the photosynthetic pigments. We have already examined the spectral absorption capabilities of each of the different classes of pigment. We shall now consider the light-harvesting properties of the complete photosynthetic system, with particular reference to the dependence of these on the particular combination of pigments present and, in the case of phytoplankton, on the size and shape of the cells or colonies.

9.1 Absorption spectra of photosynthetic systems

We might measure the absorption spectrum of, say, phytoplankton or a multicellular algal thallus, for a number of different reasons. We might seek information on what pigments are present. We might wish to compare the spectral position and shape of an *in vivo* absorption peak with those of the same peak in the isolated pigment with a view to assessing the extent to which the absorption properties are modified by binding to protein. We might want to know to what extent an alga is equipped to efficiently harvest light from the underwater radiation field in which it lives.

An absorption spectrum is the variation of some measure of light absorption by a system with wavelength. Light absorption might be expressed in terms of the absorptance, A, the percent absorption (100 A), the absorption coefficient, a, the absorbance, D (where $D = -log_{10}(1-A)$), or some other function such as the first derivative of the absorbance. The particular light absorption parameter chosen will depend on the purpose of the absorption spectrum. If, for example, we wish to be able to estimate the contribution of a planktonic alga to the total absorption coefficient of the aquatic medium at any wavelength, then we would measure the absorbance of a suspension of known concentration of the algae, as a function of wavelength, and (since absorbance is proportional to the absorption coefficient) calculate the specific absorption coefficient per unit algal biomass or pigment at the wavelength of interest. If, on the other hand, we require information on the rate at which an algal thallus is absorbing quanta from a particular incident light field, then the absorptance, or percent absorption, spectrum of that thallus is what is needed.

Before returning to the absorptance spectra of algae we shall in some detail consider absorbance spectra: these are, in effect, absorption coefficient spectra, linked as they are by simple proportionality (see §3.2). At this point it should be noted that since living cells scatter, as well as absorb, light then all measurements, of whatever absorption parameter, must be carried out with procedures which eliminate the effect of scattering on the absorption spectrum, as described in §3.2. Except where otherwise specified, it should be assumed in the remainder of this chapter that all absorption spectra referred to have been corrected for scattering.

We saw in the previous chapter that what actually carries out the light absorption in aquatic plants, the fundamental light-absorbing system, is the thylakoid. The absorbance spectrum of the thylakoid is determined by the particular kind and quantity of chlorophyll/carotenoid–protein and, in some cases, biliprotein complexes present within, or attached to, the membrane. It might therefore be thought that two species of planktonic algae which have the same array of pigment–proteins, and consequently the same absorbance spectrum at the thylakoid level, must also have the same *in vivo* absorbance spectrum in suspensions of cells or colonies at the same total pigment concentration. This is not the case. While two such algal species would certainly have rather similar spectra with the peaks and the troughs in the same positions, the extent to which the peaks rise above the troughs, and the specific absorption coefficient per unit pigment at any wavelength, can differ markedly between the two species. This is because the *in vivo* absorbance spectrum is, as we shall see, determined by the size and shape of the chloroplasts, cells or colonies, as well as by the pigment composition.

It is not feasible to determine the absorbance spectrum of a single thylakoid directly. However, in many cases it is possible to disperse the chloroplasts, by physical disruption and/or detergents, into particles so small that the effects of size and shape on the spectrum are eliminated, but with the pigment–protein interaction, so important for spectral characteristics, being unaffected. Spectra of such preparations may be regarded as

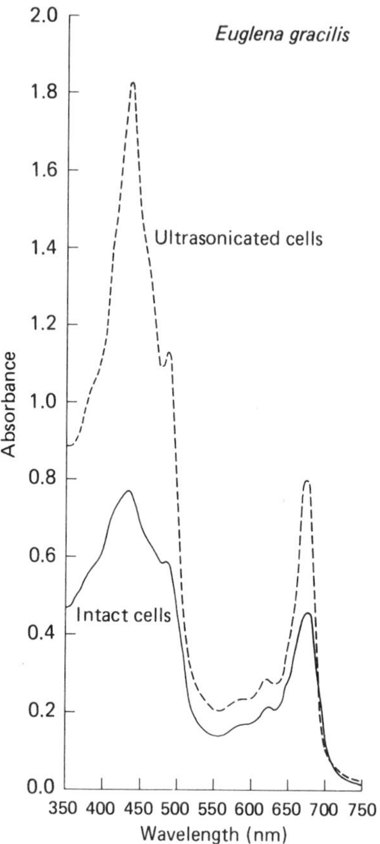

Fig. 9.1. Absorbance spectrum of whole cells of *Euglena gracilis* compared with that of disrupted cells in which absorption is essentially due to thylakoid fragments (Kirk, unpublished). The spectra of intact cells (——) and of cells fragmented by ultrasonication (------) have been corrected for scattering and in both cases correspond to 12 μg chlorophyll *a* ml^{-1} and 1 cm pathlength.

being reasonable approximations to the true *in situ* absorbance spectra of the thylakoids. Fig. 9.1 shows a spectrum of this type for thylakoid fragments of the green planktonic alga *Euglena gracilis*.

9.2 The package effect

The absorbance spectrum of a cell or colony suspension (in the case of unicellular algae) or of a segment of thallus or leaf (in the case of multicellular aquatic plants) will be found to differ noticeably from that of dispersed thylakoid fragments. The *in vivo* spectra (e.g. *Euglena*, Fig. 9.1) will be found to have peaks which are less pronounced with respect to the valleys, and to have, at all wavelengths, a lower specific absorption per unit pigment. These changes in the spectra are due to what we shall refer to as the *package* effect – sometimes, inappropriately, called the *sieve* effect. It is a consequence of the fact that the pigment molecules, instead of being uniformly distributed, are contained within discrete packages: within chloroplasts, within cells, within cell colonies. This, as one might intuitively suspect, lessens the effectiveness with which they collect light from the prevailing field – hence the lowered specific absorption. It is in the nature of the package effect to be proportionately greatest when absorption is strongest[214] – hence the flattening of the peaks. The influence of the package effect on the absorption spectra of algal suspensions was first studied, experimentally and theoretically, by Duysens (1956), and its implications for the absorption of light by phytoplankton populations have been analysed by Kirk (1975*a*, *b*, 1976*a*) and Morel & Bricaud (1981).

To better understand the package effect, let us compare the absorption properties of a suspension of pigmented particles with those of the same amount of pigment uniformly dispersed – in effect, in solution. We shall for simplicity ignore the effects of scattering, and absorption by the medium. Consider a suspension containing N particles per m^3, illuminated by a parallel beam of monochromatic light through a pathlength of 1 m. The jth particle in the light beam would, in the absence of the other particles, absorb a proportion s_jA_j of the light in unit area of beam, where s_j is the projected area of the particle (m^2) in the direction of the beam, and A_j is the particle absorptance – the fraction of the light incident on it which it absorbs. s_jA_j has the dimensions of area and is the absorption cross-section of the particle (see §4.1). The absorptance of unit area of the suspension due to the jth particle is thus s_jA_j, and so (from eqn 1.32) the absorption coefficient of the suspension due to the jth particle is $\ln 1/(1-s_jA_j)$, which, since s_jA_j is small, is approximately equal to s_jA_j. Assuming that Beer's Law applies to a suspension of particles, i.e. that the absorbance (or absorption coefficient) of the suspension is equal to the sum of the absorbances (or absorption coefficients) due to all the individual particles,[214,878] then the absorption coefficient of the suspension is

$$a_{sus} = \sum_{j=1}^{N} s_jA_j = N\overline{sA} \tag{9.1}$$

where $\overline{sA}$ is the mean value, for all the particles in the suspension, of the product of the projected area and the particle absorptance, i.e. the average absorption cross-section. The absorbance (0.434ar, see §3.2) of the suspension for a 1 m pathlength, is given by

$$D_{sus} = 0.434N\overline{sA} \tag{9.2}$$

If the particles have an average volume of $\bar{v}$ m^3, and the pigment concentration within the particle is C mg m^{-3}, then if the pigment was to be uniformly dispersed throughout the medium, it would be present at a concentration of $NC\bar{v}$ mg m^{-3}. If the specific absorption coefficient of the pigment (the absorption coefficient due to pigment at a concentration of 1 mg m^{-3}) at the wavelength in question is γ m^2 mg^{-1}, then the absorption coefficient due to the dispersed pigment is given by

$$a_{sol} = NC\bar{v}\gamma \tag{9.3}$$

and the absorbance (1 m pathlength) by $D_{sol} = 0.434NC\bar{v}\gamma$.

The extent of the package effect can be characterized by the ratio of the absorption coefficient or absorbance of the suspension to that of the solution. From eqns 9.1 and 9.3, it follows that this ratio is given by

$$\frac{a_{sus}}{a_{sol}} = \frac{D_{sus}}{D_{sol}} = \frac{\overline{sA}}{C\bar{v}\gamma} \tag{9.4}$$

It can be shown[475] that for particles of any shape or orientation, a_{sus}/a_{sol} is always less than 1.0. However, it can also be shown[477] that if the individual particles absorb only weakly, e.g. because they are small or because the pigment concentration within the particles is low, then

$$\overline{sA} \simeq C\bar{v}\gamma$$

and so, for such particles $a_{sus} \simeq a_{sol}$.

We shall now consider what happens if we make the particles absorb progressively more strongly. For simplicity we shall keep the size and shape of the particles constant, i.e. we shall keep s and $\bar{v}$ constant. Both C and γ can be increased in value almost indefinitely by increasing the pigment concentration within the particles, and by changing to a more intensely absorbed wavelength, respectively: thus the denominator in eqn 9.4 can be increased in value many-fold. As C or γ increases, so A in the numerator also increases, but since this is the *fraction* of the light incident on the particle which is absorbed, it can never exceed 1.0, and cannot increase in proportion to the increase in C or γ. When A is very low (weakly absorbing particles) a doubling in C or γ can bring about an almost commensurate increase in A, but the closer A gets to 1.0 the less leeway it has to increase in response to a given increase in C or γ. This is why the package effect, the discrepancy between the spectrum of the particle suspension, and the corresponding solution, becomes more marked as absorption by the individual particles increases: this in turn explains why the peaks of the spectrum are affected more than the troughs. In short, it is the dependence of the absorption spectrum of a suspension on the *fractional absorption per particle* (particle absorptance) which accounts for the flattening of the spectrum and the lowered specific absorption per unit pigment concentration. A more extensive treatment of this phenomenon can be found elsewhere.[475,476,477] It should be noted that the package effect still affects the absorption spectrum even in suspensions so dense that no photon can avoid traversing a chloroplast while passing through: this is one reason why the term 'sieve effect' is inappropriate.

In the special case of spherical cells or colonies, which present the same cross-section to the light, whatever their orientation, an explicit expression for the particle absorptance was derived by Duysens (1956)

$$A = 1 - \frac{2[1-(1+\gamma Cd)e^{-\gamma Cd}]}{(\gamma Cd)^2}$$

where d is the diameter of the particle. Morel & Bricaud (1981) and Kirk (1975*b*) have used this relation to carry out an analysis of the package effect and its implications for light absorption for spherical phytoplankton cells.

It is apparent from eqn 9.4 that the extent to which the package effect influences absorbance depends only on the absorption properties of the individual particles. It is therefore just as true for a suspension of particles as it is for a solution, that the *shape* of the absorbance spectrum, and the specific absorbance per unit pigment, are independent of concentration. Measured spectra of suspensions of algal cells might sometimes seem not to conform to this rule: concentrated suspensions may have higher absorbances than anticipated, even when spurious absorbance due to scattering of light away from the detector is eliminated. This is because there is a residual scattering artifact in dense suspensions which cannot be overcome by instrumental means; namely, that as a result of the multiple scattering that takes place within such suspensions, the pathlength of the photons and hence the number absorbed within the suspension is increased.

We noted at the beginning of this section that the package effect can be observed in the absorbance spectra of multicellular photosynthetic tissues as well as in those of suspensions. This is because within the tissue the pigments are also segregated into packages – the chloroplasts. However, since the chloroplasts are not randomly distributed in space and since, moreover, in many cases they change their position within the cells in response to changes in light intensity, no simple mathematical treatment of the light-absorption properties of multicellular systems is possible.

9.3 Effects of variation in cell/colony size and shape

We have seen in the previous section that the absorption coefficients and hence the light-harvesting efficiency of the photosynthetic pigments are lower when they are segregated into packages than when they are uniformly dispersed. The kind of packages within which the photosynthetic pigments occur in the aquatic biosphere vary greatly in size, shape and internal pigment concentration, however, and so we need some more general rules to assist our understanding of the light-intercepting capabilities of these different forms.

We already have two rules:

(i) $a_{sus} < a_{sol}$

(ii) At constant cell/colony size and shape (constant s_p), when A_p is increased (by raising the intracellular pigment concentration or altering the wavelength), a_{sus}/a_{sol} decreases.

It is possible also to deduce the following:

(iii) At constant total pigment and biomass in the system, when both s_p and A_p are increased (by decreasing the number and thus increasing the size of the cells or colonies, without changing their shape), a_{sus}/a_{sol} decreases.

(iv) At constant total pigment in the system, when biomass is increased (by increasing cell/colony number or volume, without changing the shape), a_{sus}/a_{sol} increases.

(v) At constant total pigment and biomass in the system, and constant cell/colony volume, as the shape becomes more extended, e.g. more elongated, so a_{sus}/a_{sol} increases.

In both (iv) and (v) the diminution in absorptance associated with the dilution of the pigment into more biomass, or the stretching-out of the cells/colonies, is proportionately less than the increase in projected area and so, overall, the absorption cross-section s_pA_p increases.

Having examined the underlying mechanism and the general rules governing the expression of the package effect, we shall now consider, in quantitative terms, just how much it can influence the light-harvesting capacity of planktonic algae. A convenient parameter to look at is the absorption cross-section of a given amount of algal biomass organized into packages of different sizes and shapes.[477] Fig. 9.2*a* shows the absorption cross-section at wavelengths from 350 to 700 nm calculated for model randomly oriented blue-green algal colonies of various geometrical forms: the volume of biomass is 100000 μm³ in each case, and the pigment concentration is constant at 2% (of the dry mass) chlorophyll *a*. It can be seen that the most inefficient arrangement for light collection is the large, spherical colony, 58 μm in diameter (lowest curve). Matters are improved somewhat if the spheres are elongated into prolate spheroids, 230 × 29 μm. A much greater increase in efficiency is achieved if the spheres are transformed into long, thin, cylindrical filaments, 3500 μm long and 6 μm in diameter. Only a marginal further increase in light-harvesting capacity is brought about if the filaments are chopped up into about 900 pieces, each piece being rounded up into a sphere of 6 μm diameter (uppermost curve). The advantages of the more extended package are much more evident at strongly absorbed, than weakly absorbed, wavelengths. For example, the ratio of the absorption cross-section of the thin cylinder to that of the large sphere is 3.82 at 435 nm but only 1.16 at 695 nm. A given change in shape cannot increase the absorption cross-section by more (proportionately) than it increases the average projected area. As the light absorption by the cell is intensified (i.e. as A_p approaches 1) so the effect of a given geometrical change on absorption cross-section tends to become identical to the effect on average projected area.

As another quantitative illustration of the significance of the package effect, Fig. 9.2*b* shows how the specific absorption coefficient per mg chlorophyll *a* present in the form of a suspension of spherical cells or colonies, at its red peak, decreases as the diameter of the cells/colonies increases.

That the package effect really does have a major influence on the light-harvesting capability of algal cells has now been demonstrated in numerous experiments comparing different phytoplankton species varying in cellular size and pigment content, or comparing cells of a given species having a range of pigment contents due to variation in growth irradiance.[61,176,241,335,403,639,642,657,789,850] A striking example in the field has been described by Robarts & Zohary (1984): in Hartbeespoort Dam (South Africa), as the colony size of the dominant blue-green alga *Microcystis aeruginosa* increased, there was a corresponding increase in euphotic depth resulting from the less efficient light interception of biomass distributed in larger packages. Phytoplankton populations in the coastal waters of the Antarctic Peninsula appeared, on the basis of chlorophyll-specific vertical attenuation coefficients measured at wavelengths through the photosynthetic

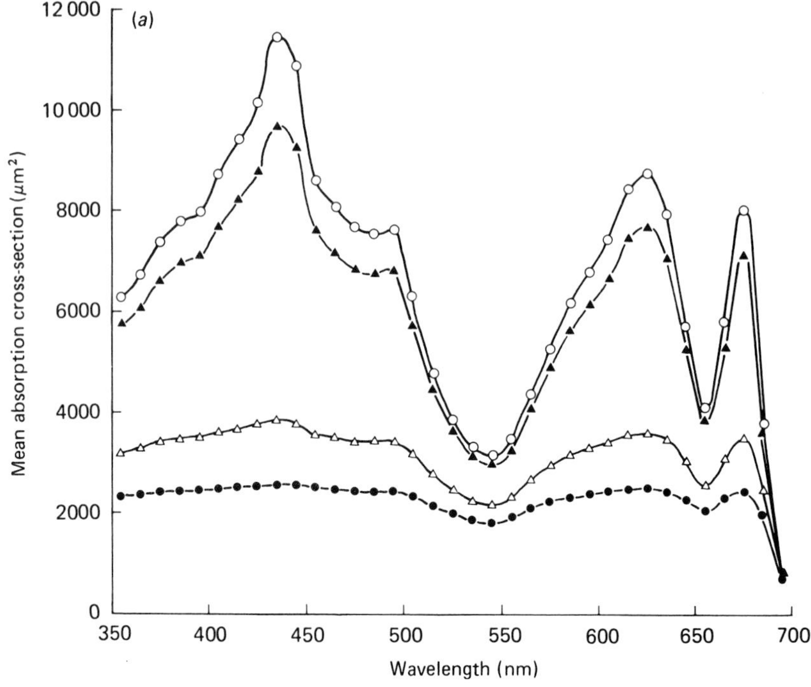

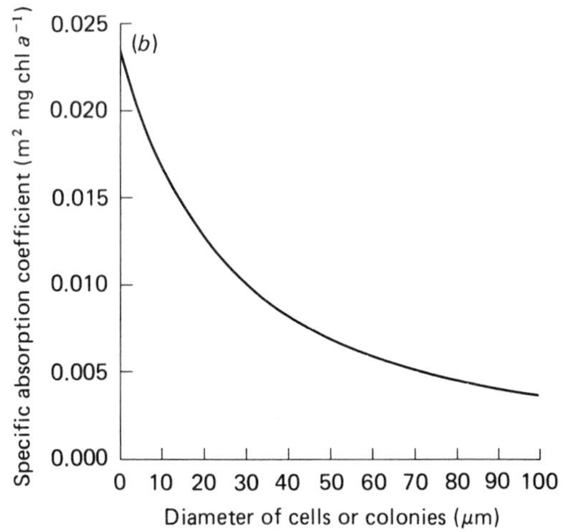

Fig. 9.2. Effect of size and shape on light absorption properties of phytoplankton. (*a*) The absorption cross-section spectra of randomly oriented blue-green algal colonies of various shapes and sizes (calculated by Kirk (1976*a*) for idealized colonies containing 2%, dry mass, chlorophyll *a*). In every case the data apply to 100000 μm^3 of algal volume. This corresponds to one particle in the cases of the 57.6 μm diameter spheres

spectrum, to have a markedly greater package effect than populations from temperate oceans: Mitchell & Holm-Hansen (1991*a*) attributed this to the presence of larger cells with high cellular pigment concentration resulting from chronic low-light adaptation in nutrient-rich waters.

Geider & Osborne (1987) showed that for the relatively small diatom *Thalassiosira* (~5 μm diameter) grown in culture, the package effect reduced light absorption efficiency by 50% at the blue absorption maximum (435 nm), and by 30% at the chlorophyll *a* red maximum (670 nm), but had no significant effect at the absorption minimum (600 nm): larger diatoms would show an even greater package effect. In the picoplankton (<2 μm diameter), such as the unicellular cyanophytes and prochlorophytes of the ocean, however, the package effect should be of no significance.[490]

9.4 Rate of light absorption by aquatic plants

The rate of photosynthesis by an aquatic plant must ultimately be limited by (although it is not always simply proportional to) the rate at which the higher plant leaf, or multicellular algal thallus, or individual phytoplankton cell or colony, is absorbing quanta from the underwater light field. In the case of the leaf or the algal thallus, the rate of absorption of quanta of a given wavelength incident at a particular angle on a particular element of tissue surface is equal to $E(\lambda, \theta, \phi).\ \delta s.A(\lambda, \theta, \phi)$, the product of the irradiance at that angle, the area of the element, and the absorptance for that wavelength and that angle. The total rate of absorption of light of that wavelength by the whole leaf or thallus is the sum of this product over all angles of incident light for each element of surface, and over all the elements which constitute the total area of the leaf or thallus.

It is clear that the crucial optical property of the tissue is the absorptance (the fraction of incident light absorbed) rather than the absorbance.

(●), the 230.4 × 28.8 μm prolate spheroids (△), and the 3537 × 6 μm cylinders (▲); and to 884 particles in the case of the 6 μm diameter spheres (○). (*b*) Specific absorption coefficient of phytoplankton chlorophyll at the red maximum (670–680 nm) as a function of cell or colony size (Kirk, unpublished). The values were obtained from the particle absorptance values calculated using the equation of Duysens (1956), assuming that the cells/colonies contained 2%, dry mass, (~4 g l^{-1} cell volume) of chlorophyll *a* and that the (natural logarithm) specific absorption coefficient of chlorophyll *a* in solution at its red peak is about 0.0233 $m^2\ mg^{-1}$, as it is, for example, in diethyl ether.[265] Similar calculations have been carried out by Morel & Bricaud (1981).

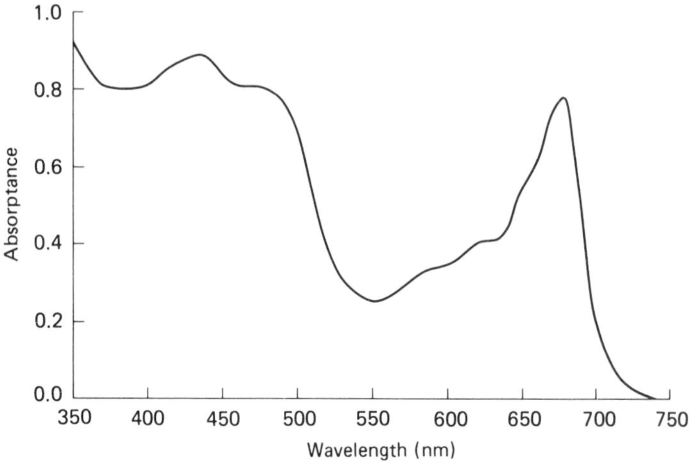

Fig. 9.3. Absorption spectrum of the leaf of a fresh-water macrophyte (Kirk, unpublished). The spectrum was measured on a piece of leaf of *Vallisneria spiralis* (Hydrocharitaceae), free of epiphytic growth, from Lake Ginninderra, ACT, Australia, with the sample cell close to the photomultiplier: the spectrum has been corrected for scattering.

Because of the variation in optical path through the tissue with angle, the absorptance of the leaf or thallus at any point will in fact vary somewhat with the angle of incidence of the light. The effective absorptance of the tissue at a given point for the total light of a given wavelength incident at all angles is a function of the angular distribution of the light field, and so it is not, strictly speaking, possible to attribute a particular absorptance to the tissue for underwater light of a given wavelength independent of the radiance distribution of that light. The absorptance is in fact usually determined with the measuring beam at right angles to the plane of the photosynthetic tissue. This is considered to provide at least an approximate measure of the absorptance that the tissue will present to the light incident on it within the water. Due to variations in thickness, and/or chloroplast number and pigment composition, there can be variation in absorptance from place to place within a leaf or thallus. This is particularly likely to be the case with long algal thalli, different parts of which are normally exposed to different light environments, e.g. because the lower parts are shaded. The increase in photon pathlength resulting from multiple scattering within the tissue can accentuate absorption by multicellular plants.

Fig. 9.3 shows the 90° absorptance spectrum for an aquatic higher plant; corresponding spectra for various kinds of multicellular algae are shown in Fig. 10.6*a*, *b* and *c*. The differences between the spectra are in part due to differences in concentration of chloroplast pigments per unit area, and this can of course vary markedly within any algal class as well as from one class to another. Some of the differences in the shape of the spectra are, however, attributable to the different types of pigment present. The relatively greater absorption in the 500–560 nm region in the brown alga compared to the green alga and higher plant, is due to the presence of fucoxanthin in the former. The broad peak at 520–570 nm in the spectrum of the red alga is due to the presence of the biliprotein, phycoerythrin.

In the case of phytoplankton, the rate of absorption by an individual cell or colony of light quanta of a given wavelength coming from a given direction, is equal to $E(\lambda, \theta, \phi).s_p(\theta, \phi).A_p(\lambda, \theta, \phi)$, the product of the irradiance in that direction, the cross-sectional area (in the specified direction) of the cell or colony and the absorptance of the particle in its particular orientation with respect to the light stream. Since the cells or colonies are randomly oriented, each will present a somewhat different projected area, and absorptance, to light flowing in the specified direction. The average rate of absorption of this light per particle is $E(\lambda, \theta, \phi).\overline{s_p A_p}$. It will be recalled that $\overline{s_p A_p}$ is the average absorption cross-section of the particles (§9.2). It follows from the random orientation of the cells or colonies that they have the same average absorption cross-section for light at all directions. We can therefore validly attribute an average absorption cross-section to a phytoplankton population, regardless of the angular distribution of the underwater light field. It is in fact possible to attribute an average absorptance to the individual particles in a plankton population, but this is not a useful thing to do, since it is the product of absorptance and cross-sectional area (which vary together, with orientation) rather than absorptance alone, which determines the rate of collection of quanta from a particular light stream.

What, in the present context, we wish to know about a phytoplankton population is the average absorption cross-section of the cells or colonies composing the population, at all wavelengths in the photosynthetic range. To carry out measurements on individual cells or colonies, although possible, is technically difficult and does not give results of high accuracy. We must therefore rely on spectroscopic measurements carried out on suspensions of phytoplankton. The concentrations at which phytoplankton normally occur are too low for accurate absorption measurements. It is therefore necessary to prepare more concentrated suspensions by filtration or centrifugation, followed by resuspension in a smaller volume.

Given a reasonably concentrated suspension of phytoplankton, by what sort of measurement can we determine the average absorption cross-section? Despite the importance of absorptance of the individual cell or colony in determining the absorption cross-section, measurements of the absorptance of the whole suspension tell us relatively little about the absorption properties of the individual particles in the suspension. The absorptance spectrum of the suspension changes shape as the phytoplankton concentration changes, and at very high concentrations tends to become a straight line, with $A_{sus} \simeq 1.0$ throughout the spectrum (total absorption of light at all wavelengths). The absorptance spectrum of the suspension, however, is of relevance if, for some reason, we need to know the rate of light absorption by the whole suspension – for example, in laboratory studies of photosynthetic efficiency.

To determine the average cross-section per individual free-floating particle, whether cell or colony, in the suspension, what we in fact measure is the absorbance of the suspension, D_{sus}. Fig. 9.4 shows the absorbance spectra of suspensions of three planktonic algae: *Chlorella* (green), *Navicula* (a diatom) and *Synechocystis* (a blue-green). The absorbance, in a 1-cm pathlength, of a suspension of particles, as we saw earlier (eqn 9.2, §9.2), is equal to 0.434 $n\overline{s_p A_p}$, where n is the number of particles per ml, and $\overline{s_p A_p}$ is the average absorption cross-section per particle. Thus we may obtain the values of $\overline{s_p A_p}$ throughout the photosynthetic range for any phytoplankton population by preparing a suitably concentrated suspension, measuring the absorbance spectrum and n, and applying the relation

$$\overline{s_p A_p} = \frac{D_{sus}}{0.434\, n} \qquad (9.5)$$

at a series of wavelengths from 400 to 700 nm. Since the pathlength is 1 cm then n, as well as being the number of cells per ml, is also the number of particles per cm^2 in the path of the measuring beam. Thus, in eqns 9.2 and 9.5 N or n can be taken to mean the number of particles per unit area, provided only that the units in which area is expressed are the same as those in which absorption cross-section ($\overline{s_p A_p}$) is expressed, i.e. with a given number of particles per unit area, the absorbance is the same no matter through what pathlength they are distributed.

The average rate of absorption of light of wavelength λ, per individual phytoplankton cell or colony at depth z m, is $E_0(\lambda, z).\overline{s_p A_p(\lambda)}$, where $E_0(\lambda, z)$ is the scalar irradiance of light of that wavelength at that depth and $\overline{s_p A_p(\lambda)}$ is the average absorption cross-section in m^2. The rate of light absorption (in W, or quanta s^{-1}) per horizontal m^2 by all the phytoplankton within a thin layer of thickness Δz m, at depth z m is given by

$$E_p = E_0(\lambda, z).\overline{s_p A_p(\lambda)}.\Delta z N \qquad (9.6)$$

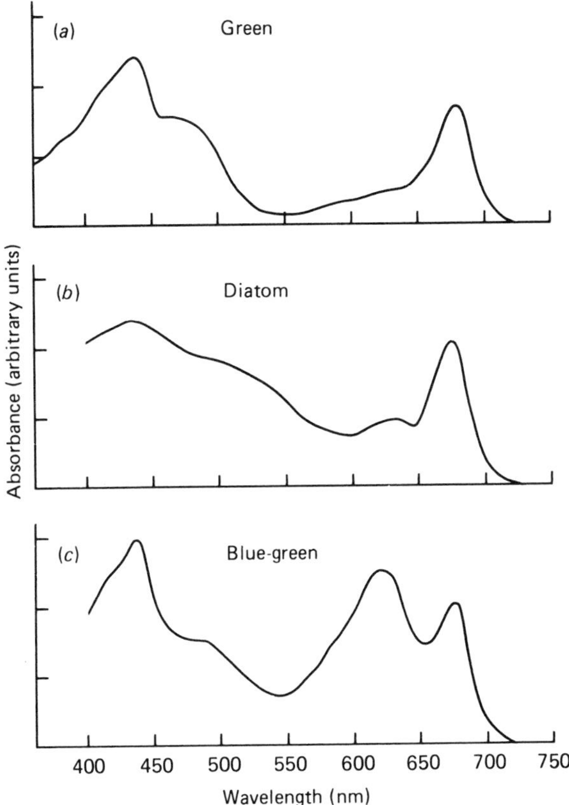

Fig. 9.4. Absorbance spectra of cultured cells of three species of planktonic algae measured using an integrating sphere (after Latimer & Rabinowitch, 1959). (*a*) *Chlorella pyrenoidosa* (green). (*b*) *Navicula minima* (diatom). (*c*) *Synechocystis* sp. (blue-green).

where N is the number of phytoplankton cells or colonies per m^3.

In the field of aquatic primary production, phytoplankton concentration is more commonly expressed in terms of mg chlorophyll a m^{-3}, than cells or colonies m^{-3}. Since, from eqn 9.1

$$N\overline{s_p A_p(\lambda)} = a_p(\lambda)$$

where $a_p(\lambda)$ is the absorption coefficient due to phytoplankton, we can write

$$E_p = E_0(\lambda,z).a_p(\lambda).\Delta z \tag{9.7}$$

and

$$E_p = E_0(\lambda,z).Chl.a_c(\lambda).\Delta z \tag{9.8}$$

where *Chl* is the concentration of phytoplankton chlorophyll *a* in mg m^{-3}, and $a_c(\lambda)$ is the specific absorption coefficient of the phytoplankton per mg chlorophyll *a* m^{-3}: $a_c(\lambda)$ has the units m^2 mg chlorophyll a^{-1}.

In any given waveband, if the total absorption coefficient due to all components of the medium is $a_T(\lambda)$, then the proportion of the total absorbed energy which is captured by phytoplankton is $a_p(\lambda)/a_T(\lambda)$. A set of values for the specific absorption coefficient of marine phytoplankton across the photosynthesis spectral region may be found in Fig. 3.9 (Chapter 3).

A useful concept when considering light capture by phytoplankton is the effective absorption coefficient of the phytoplankton population existing at a given depth for the light field at that depth, across the whole photosynthetic spectrum.[633] It may be thought of as a weighted average absorption coefficient of the phytoplankton for PAR, taking into account the actual spectral distribution of PAR at the depth in question, and is defined by

$$\bar{a}_p(z) = \frac{\int_{400}^{700} a_p(\lambda) E_0(\lambda, z) d\lambda}{\int_{400}^{700} E_0(\lambda, z) d\lambda} \tag{9.9}$$

where $E_0(\lambda,z)$ is the scalar irradiance per unit bandwidth (nm^{-1}) at wavelength λ and depth z m. We can also define a specific effective absorption coefficient of the phytoplankton, $\bar{a}_c(z)$, for PAR as the value of $\bar{a}_p(z)$ for phytoplankton at 1 mg chl *a* m^{-3}, and having the units m^2 mg chl a^{-1}.

Even if the concentration and nature of the phytoplankton remain the same, the values of the effective, and specific effective, absorption coefficients for PAR do vary with depth, in accordance with the change in spectral distribution of PAR. In the ocean, where there is generally little dissolved yellow colour, the light field with increasing depth becomes increasingly confined to the blue-green (400–550 nm) spectral region (§6.2, Fig. 6.4). Since this is where phytoplankton have their major absorption peak (Fig. 3.9), the value of $\bar{a}_c(z)$ in such waters increases with depth, by up to 50 or even 100% within the euphotic zone.[502,633] In the Pacific Ocean southeast of Japan, Kishino *et al.* (1986) found $\bar{a}_c(z)$ to increase rapidly with depth from 0.022 m^2 mg chl a^{-1} at the surface to 0.044 at 30 m. In the green waters of highly productive upwelling regions, $\bar{a}_c(z)$ can decrease with depth.[633]

9.5 Effect of aquatic plants on the underwater light field

We have concerned ourselves so far in this chapter with the ability of phytoplankton and macrophytes to make use of the underwater light field.

By harvesting light from the field, however, the plants in turn modify the light climate for any other plants below them in the water column. Within any substantial stand of aquatic macrophytes, such as a kelp forest, or a bed of sea grass or fresh-water aquatic higher plants, the intensity of PAR is greatly reduced. Phytoplankton also increase the rapidity of attenuation of light with depth, and in productive waters may do so to such an extent that by self-shading they become a significant factor limiting their own population growth. The contribution of phytoplankton to vertical attenuation of PAR must therefore be taken into account in any consideration of the extent to which light availability limits primary production in the aquatic biosphere.

We saw in §6.8 that the total vertical attenuation coefficient for monochromatic irradiance at a given depth can be regarded as the sum of a set of partial attenuation coefficients, each corresponding to a different component of the medium. This proposition does not, strictly speaking, hold for the average vertical attenuation coefficient throughout the euphotic zone, and even less does it hold for the whole photosynthetic waveband, 400–700 nm. Nevertheless, the assumption that the average K_d(PAR) for the euphotic zone can be partitioned in this way is so useful that, despite its only approximate truth, it is commonly made, and may be given approximate expression through the equation

$$K_d(\mathrm{PAR}) = K_W + K_G + K_{TR} + K_{PH} \qquad (9.10)$$

where K_W, K_G, K_{TR} and K_{PH} are the partial attenuation coefficients (for PAR) due to water, gilvin, tripton and phytoplankton, respectively. The contribution of phytoplankton to vertical attenuation of PAR can therefore be expressed in terms of the contribution of K_{PH} to the value of K_d(PAR) in eqn 9.10.

To arrive at a quantitative estimate of the contribution of phytoplankton to K_d(PAR) we must determine K_{PH}. To do this we make use of a further commonly made assumption, which as we saw in §6.7 is also only approximately valid, namely that the contribution of any component of the medium to K_d(PAR) is linearly related to the concentration of that component. Applying this to the phytoplankton we assume that

$$K_{PH} = B_c k_c \qquad (9.11)$$

where B_c is the phytoplankton biomass concentration expressed in terms of mg chlorophyll *a* m^{-3} (mg chl *a* m^{-3}), and k_c is the specific vertical attenuation coefficient (units m^2 mg chl a^{-1}) per unit phytoplankton concentration. We may now write

$$K_d(\mathrm{PAR}) = K_W + K_G + K_{TR} + B_c k_c \qquad (9.12)$$

If, for any given water body, an estimate of the specific attenuation coefficient, k_c, is available, then from a measurement of the phytoplankton concentration, B_c, we obtain a value for K_{PH}.

The value of k_c has been calculated for each of the four kinds of blue-green-algal colonies to which the data in Fig. 9.2 apply: k_c was 0.0063 $m^2\ mg^{-1}$ for the 58 μm spheres, 0.0084 $m^2\ mg^{-1}$ for the 230 × 29 μm prolate spheroids, 0.0133 $m^2\ mg^{-1}$ for the 3500 × 6 μm cylinders, and 0.0142 $m^2\ mg^{-1}$ for the 6 μm spheres.[477] This marked variation of k_c amongst the different types of algae confirms the importance of the package effect in light capture by phytoplankton. These values were all obtained by calculation for idealized algae: k_c can also be determined experimentally for real phytoplankton populations by measuring K_d for PAR in natural water bodies at different times as the algal population waxes and wanes, and determining the linear regression of K_d with respect to phytoplankton chlorophyll *a* concentration. Not many measurements of k_c covering the whole photosynthetic waveband have in fact been made. There are, however, other data in the literature, on the effect of varying phytoplankton concentration on the value of K_d in particular spectral bands within the 400–700 nm range, obtained with irradiance meters fitted with broad-band filters. Talling (1957*b*) found that for various natural waters an approximate value for K_d(PAR) could be obtained by multiplying the minimum value of $K_d(\lambda)$ for the water body concerned (usually that in the green waveband in inland waters) by 1.33; for Lough Neagh, N. Ireland, a factor of 1.15 was found to be more suitable.[428] Using this sort of relation, measurements of K_d in wavebands within the photosynthetic range can be used to provide estimates of k_c. Table 9.1 lists some values of k_c determined in various natural water bodies, from measurements either of irradiance for total PAR or for a particular waveband. It can be seen that k_c varies widely – by a factor of four between the lowest and the highest value – from one alga, one water body, to another.

There are several possible reasons for this variability. One is the influence of cell size and geometry. We noted above that the package effect – within the range that might occur in nature – can vary k_c by a factor of more than two in blue-green algae of identical pigment composition. We may reasonably attribute the low k_c values for the large dinoflagellates in Table 9.1 to the low efficiency of light collection by large pigmented particles. The package effect also, as outlined earlier (§9.2), increases the more strongly the particles absorb (at constant size and shape). Thus, even for algae of similar size, shape and pigment type, k_c will decrease as total pigment content increases. In addition, since k_c is expressed per unit chlorophyll *a*,

Table 9.1. *Values of specific vertical attenuation coefficient for PAR per mg phytoplankton chlorophyll* a, *obtained from* in situ *measurements of irradiance*

Water body	Phytoplankton type	k_c (m^2 mg^{-1})	Reference
L. Windermere, England	*Asterionella* (diatom)	0.027[a]	899
Esthwaite Water, England	*Ceratium* (large dinoflagellate)	~0.01[a]	901
L. George, Uganda	*Microcystis* (blue-green)	0.016–0.021[a]	283
Loch Leven, Scotland	*Synechococcus* (blue-green)	0.011[a]	67
L. Vombsjon, Sweden	*Microcystis* (blue-green)	0.021[a]	293
Lough Neagh,Ireland	*Melosira* (diatom)	0.014[b]	428
	Stephanodiscus (diatom)	0.008[b]	
	Oscillatoria (blue-green)	0.012–0.013[b]	
L. Minnetonka, Minn., USA	Mixed blue-green (*Aphanizomenon* etc.)	0.022[c]	608
L. Tahoe Calif.–Nevada, USA	Small diatoms (mainly *Cyclotella*)	0.029	923
Irondequoit Bay L. Ontario, USA	Mixed blue-green	0.019	980
L. Constance Germany	Mixed	0.015[c]	921
L. Zurich (0–5 m depth) Switzerland	Mixed	0.012	795
Sea of Galilee, Israel	*Peridinium* (large dinoflagellate)	0.0067	205
Various oceanic and coastal waters	Mixed	0.016	829

Notes:
[a] Obtained by multiplying $K_d(\lambda)_{min}$ by 1.33.
[b] Obtained by multiplying $K_d(\lambda)_{min}$ by 1.15.
[c] Derived from measurements of scalar, rather than downward, irradiance of PAR.

there can be marked variation in specific attenuation from one alga to another, due to differences in the type of other photosynthetic pigments present and their ratio to chlorophyll *a*. Calculations for model cells having the same chlorophyll *a* content indicated that k_c for diatoms would be about 70% higher than that for green algae because of the increased absorption in the 500–560 nm region due to fucoxanthin; k_c for blue-green algae with substantial levels of the biliprotein phycocyanin, absorbing in the 550–650 nm region, was calculated to be about twice that for diatoms.[477]

The colour of the aquatic medium in which the cells are suspended can also have a marked influence on the values of k_c. Green algal cells, for example, absorb strongly in the blue region (Fig. 9.4*a*). In typical inland waters, however, with high levels of yellow substances, the contribution of the blue spectral region to the underwater light field is greatly diminished, and so in such waters, green cells have a low value of k_c.[476] The k_c values we have considered so far have been the average values over some considerable optical depth, i.e. a depth in which the downward irradiance falls to some small fraction of that at the surface. In fact, since the spectral distribution of the light changes progressively with depth, so the value of k_c calculated over a small increment of depth also changes.[23] In the case of blue-green algae which, due to the presence of ample levels of biliprotein, absorb strongly in the green, as well as in the blue and red regions, k_c does not vary markedly with depth in any water type. In the case of green algae, however, in an inland water absorbing strongly in the blue, calculations by Atlas & Bannister (1980) indicate that k_c diminishes from about 0.012 $m^2\ mg^{-1}$ in the surface layer to about 0.005 $m^2\ mg^{-1}$ at the bottom of the euphotic zone. In the ocean we would expect the same sort of changes in k_c with depth as are found for the effective specific absorption coefficient, $\bar{a}_c(z)$ (see previous section).

One further possible cause of variation in the value of k_c from one kind of phytoplankton to another is variation in their light-scattering properties. Scattering, as we saw earlier (§6.7), contributes in various ways to the vertical attenuation of irradiance. In dense algal blooms the contribution of the algal population to total scattering could significantly increase k_c. The amount of scattering – especially per unit chlorophyll *a* – can vary markedly from one species to another (see Table 4.2). Coccolithophores and diatoms, for example, scatter light more intensely than algae enclosed within less refractile integuments.

The effects of macrophytes on the underwater light field vary so much with the growth habit of the plants and the morphology of the leaves or thallus that a general theoretical account is not feasible. According to Westlake (1980*c*) the specific vertical attenuation coefficient per mg chlorophyll is lower for macrophytes than for phytoplankton. Dense stands of emergent and floating macrophytes can make the whole water column virtually aphotic. Within submerged weed beds the spectral distribution of irradiance is predominantly green.[988]

10

Photosynthesis as a function of the incident light

The rate of photosynthesis achieved by a phytoplankton cell or aquatic macrophyte depends on the rate of capture of quanta from the light field. This is determined by the light absorption properties of the photosynthetic biomass, which we have considered in some detail, and by the intensity and spectral quality of the field. The rate of photosynthesis is not, however, simply proportional to the rate of capture of photons. The efficiency with which the photosynthetic apparatus can make use of the absorbed energy to fix CO_2 varies from one plant cell to another and within a given cell as its physiological state changes. Light quanta may be collected by the pigments faster than the electron carriers and enzymes can make use of them. In particularly high light intensities the excess absorbed energy can inactivate the photosynthetic system. The relation between the rate of photosynthesis and the characteristics of the incident light is thus not a simple one: we shall examine it in this chapter.

In order to study the effects of light intensity and spectral quality on photosynthesis, suitable quantitative procedures for determining the photosynthetic rate per unit biomass must be used. Detailed descriptions of such methods for use in the field or the laboratory may be found elsewhere,[786,962] and so they will only be briefly mentioned here. Photosynthesis can be measured in terms of either carbon dioxide fixed or oxygen released. Because of the stoichiometry of the overall photosynthetic process (§8.5), approximately one O_2 molecule is liberated for every molecule of CO_2 fixed. However, since the average composition of plant biomass differs somewhat from CH_2O due to the presence of protein, lipid and nucleic acid as well as carbohydrate, the O_2/CO_2 ratio (known as the *photosynthetic quotient*) is usually in the range 1.1 to 1.2, rather than exactly 1.0. In the case of the more active photosynthetic systems, it is convenient to measure O_2 liberation – using chemical analysis, an oxygen electrode, or manometry.

For field measurements of phytoplankton photosynthesis, however, in all except highly productive waters, the much more sensitive procedure of measuring fixation of $^{14}CO_2$ is used: this method was introduced by Steemann Nielsen in 1952. Bottles containing water samples with the natural phytoplankton population present and with small amounts of [^{14}C]-bicarbonate (hydrogen carbonate) added, are suspended at a series of depths throughout the euphotic zone, generally for periods of a few hours in the middle of the day. The amount of radioactivity fixed in cells, collected on a filter and treated with acid, is determined.

Alternatively, the incubations of the phytoplankton samples with [^{14}C]-HCO_3 can be carried out in the laboratory at the same temperature as that in the water body, and at a series of irradiance values designed to correspond to different depths. For marine phytoplankton, Jitts (1963) introduced the technique of carrying out the laboratory incubations under a series of thicknesses of blue glass, to simulate the variation of spectral distribution, as well as total irradiance, with depth. Failure to reproduce, in the laboratory, spectral distributions similar to those found underwater can lead to substantial errors in estimates of primary production.[543]

The rate may be expressed as either *gross* or *net* photosynthesis. Gross photosynthesis is the total rate of carbon dioxide fixation, making no allowance for the fact that some CO_2 is simultaneously lost in respiration. Net photosynthesis is the total rate of photosynthetic CO_2 fixation minus the rate of loss of CO_2 in respiration. The rate of increase in oxygen concentration within an illuminated bottle containing phytoplankton is a measure of net photosynthesis; a value for gross photosynthesis may be obtained by adding to this the rate of respiratory oxygen consumption measured in a darkened bottle incubated in parallel. Whether the $^{14}CO_2$ fixation method measures net or gross photosynthesis, or something between the two, remains a matter of controversy. In short-term incubations, such as are possible in the more productive waters, there are *a priori* as well as experimental grounds for considering $^{14}CO_2$ fixation as an approximate measure of gross photosynthesis.

Rates of photosynthesis, whether net or gross, can be expressed per unit biomass (specific photosynthetic rate) or per unit area or volume of the water. Typical units of specific photosynthetic rate (P) for a phytoplankton population are μmoles CO_2 (or O_2) or mg C (carbon), per mg chl *a* per h. It is convenient to indicate it as $P(CO_2)$, $P(O_2)$ or $P(C)$ in accordance with the units being used. If the rate per m^3 of water (P_v) is summed for every 1-m depth interval from the surface to the bottom of the euphotic zone, then the areal (or integral) photosynthetic rate, P_A, i.e. the rate of photosynthesis in the whole water column beneath 1 m^2 of surface, is obtained: its units are moles (or μmoles) CO_2 (or O_2), or g (or mg) C, m^{-2}

$$P_A = \int_{z_{eu}}^{surface} P_v dz$$

It may be useful to indicate areal photosynthetic rate as $P_A(CO_2)$, $P_A(O_2)$ or $P_A(C)$ in accordance with the units used. Rates of photosynthesis of benthic macrophytes may be expressed per g dry mass of tissue, or per m^2 of the substrate on which they are growing.

10.1 Photosynthesis and light intensity

The variation of the photosynthetic rate of a phytoplankton population with incident light intensity can be studied using bottles suspended at a series of depths: in this case the attenuation of light with depth provides the necessary range of irradiance values. Alternatively the measurements can be carried out under artificial light in the laboratory.

P *versus* E_d *curves*

In the dark there is of course no photosynthesis and aquatic plants exhibit a net consumption of O_2 and liberation of CO_2, due to cellular respiration. As the light intensity is gradually increased from zero, some photosynthetic O_2 production and CO_2 consumption takes place, but at very low intensities shows up as a diminution in the rate of O_2 consumption rather than a net liberation of O_2, i.e. there is significant gross photosynthesis but no net photosynthesis. An irradiance value (E_c) is eventually reached at which photosynthetic oxygen liberation just equals respiratory oxygen consumption: this is the *light compensation point*. Beyond this point liberation exceeds consumption and net photosynthesis is achieved. The typical pattern of behaviour from here on is that P increases linearly with E_d up to a certain value. The graph then begins to curve over and eventually levels off. In the range of irradiance values where P does not vary with E_d, photosynthesis is said to be light-saturated, P now having the value P_m, the maximum specific photosynthetic rate, sometimes referred to as the *photosynthetic capacity*. With further increase in irradiance, P begins to decrease again, a phenomenon referred to as *photoinhibition*. Fig. 10.1 shows two typical P versus E_d curves, one for marine phytoplankton,[778] the other for a mixed population of fresh-water diatoms.[56] P versus E_d curves for macrophytes have a similar shape but often do not show photoinhibition in full sunlight. Fig. 10.2 shows the variation of P with irradiance in four fresh-water

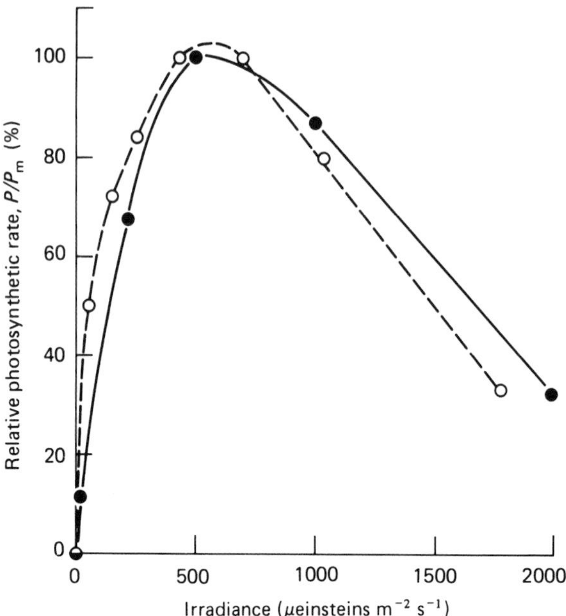

Fig. 10.1. Relative specific photosynthetic rate (P/P_m) of phytoplankton as a function of irradiance (E_d). Marine phytoplankton from the mid-point of the euphotic zone in the Sargasso Sea (redrawn from data of Ryther & Menzel, 1959) —●—. Fresh-water diatoms (mainly *Asterionella formosa* and *Fragilaria crotonensis*) in Lake Windermere, England (redrawn from data of Belay, 1981) -----○-----. Appropriate conversion factors have been used to convert the authors' original E_d values to μeinsteins (PAR) m^{-2} s^{-1}.

Fig. 10.2. Photosynthetic rates of aquatic macrophytes as a function of irradiance of PAR (*a*) Fresh-water macrophytes from lakes in Florida, USA (by permission, from Van, Haller & Bowes (1976) *Plant Physiology*, **58**, 761–8. The rates are limited by low, but typical *in situ*, CO_2 concentration (0.42 mg l^{-1}). Temperature is 30 °C. L.S. = irradiance required for light saturation. L.C.P. = irradiance at the light compensation point. $\frac{1}{2}V_{max}$ = irradiance required for half-maximal photosynthetic rate. (*b*) Green, brown and red multicellular algae from the western Baltic Sea (plotted from data of King & Schramm, 1976). The samples of *Ulothrix speciosa* (eulittoral green), *Scytosiphon lomentaria* (eulittoral brown) and *Phycodrys rubens* (sublittoral red) were collected in the spring and the measurements were carried out at 10 °C. The published irradiance values have been converted from mW cm^{-2} to μeinsteins (PAR) m^{-2} s^{-1}.

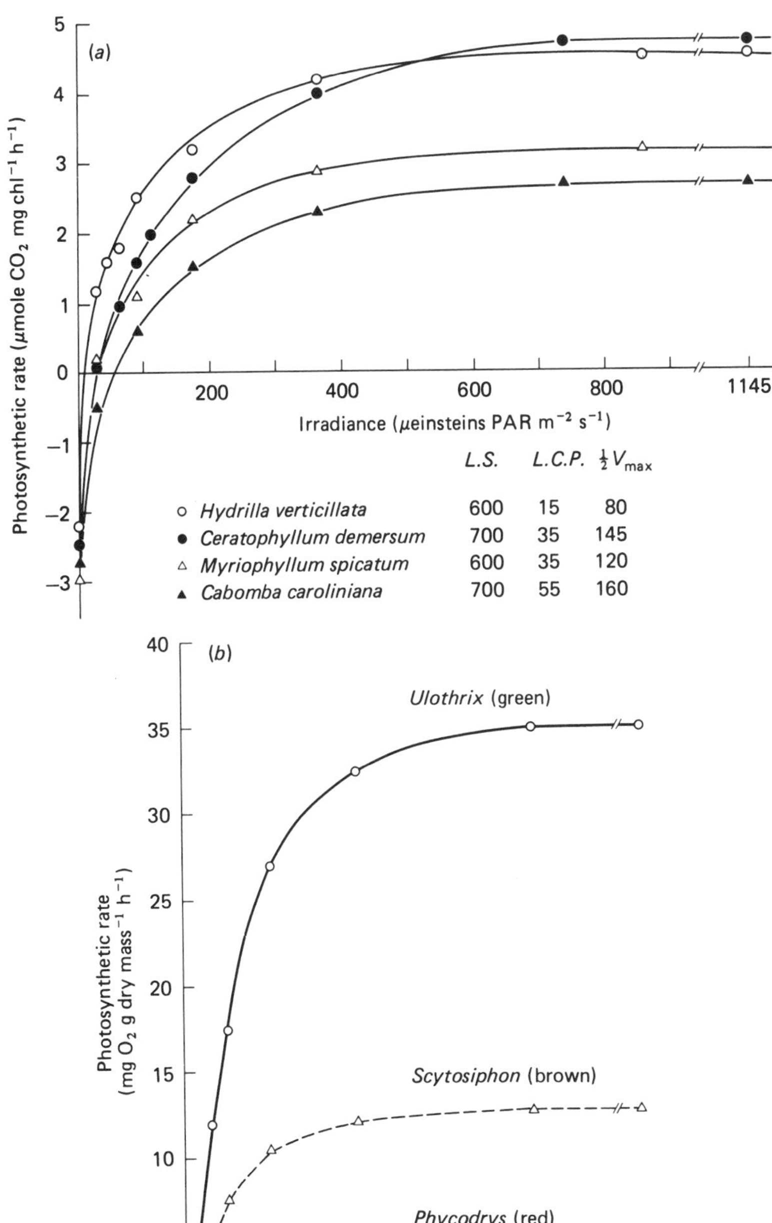
(a)
Photosynthetic rate (μmole CO_2 mg chl^{-1} h^{-1})
Irradiance (μeinsteins PAR m^{-2} s^{-1})
L.S. L.C.P. ½ V_{max}
Hydrilla verticillata 600 15 80
Ceratophyllum demersum 700 35 145
Myriophyllum spicatum 600 35 120
Cabomba caroliniana 700 55 160
(b)
Ulothrix (green)
Scytosiphon (brown)
Phycodrys (red)
Photosynthetic rate (mg O_2 g dry mass^{-1} h^{-1})
Irradiance (μeinsteins m^{-2} s^{-1})

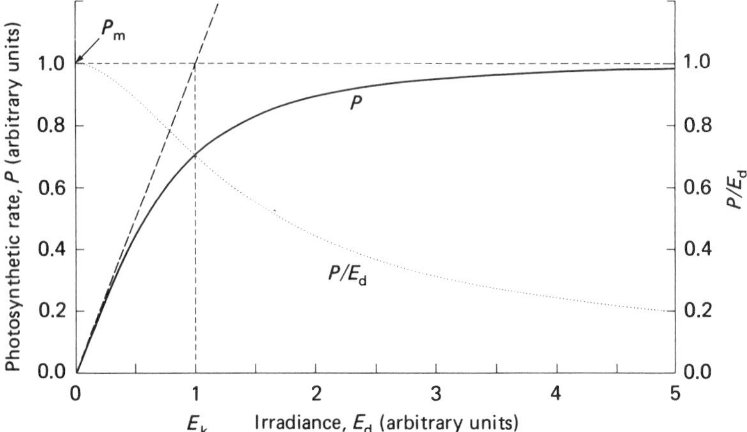

Fig. 10.3. Idealized curve of specific photosynthetic rate (P) as a function of irradiance (E_d), illustrating the maximum photosynthetic rate, P_m, and the saturation onset parameter, E_k. The variation of P/E_d (a measure of the efficiency of utilization of incident light) with irradiance value is also indicated (······).

macrophyte species,[946] and in green, brown and red marine algal species.[474]

Because of the gradual onset of saturation as E_d increases, it is difficult to pinpoint the irradiance value at which photosynthesis is just saturated. A more easily measured parameter by means of which the onset of saturation may be characterized[898] is the irradiance (E_k) at which the maximum rate, P_m, would be reached if P were to continue to increase linearly with E_d. The value of E_k is that value of E_d corresponding to the point of intersection between the extrapolated linear part of the curve and the horizontal line at P_m. This is illustrated on the idealized P versus E_d curve in Fig. 10.3. It can be seen that the slope, α, of the linear part of the curve is equal to P_m/E_k: α is the rate of photosynthesis per unit biomass per unit of incident irradiance, and so is a measure of the efficiency with which the biomass utilizes light, at low intensities, to fix CO_2.

The light intensity required to saturate photosynthesis, and the compensation point, vary markedly from one species to another. Furthermore, as we shall see in a later section, these parameters also depend on the CO_2 concentration and the temperature. Thus, if measurements of the photosynthetic response as a function of light are to be ecologically meaningful, they must be measured under conditions approximating those in the aquatic ecosystem. In the case of measurements on phytoplankton the effect of photoinhibition, which develops to a greater extent during incubation in stationary bottles than in freely circulating cells (see later), is to lead to underestimation, both of the irradiance required for light saturation and of the maximum photosynthetic rate.[594] Thus, for phytoplankton, P versus E_d data obtained in relatively short incubations are to be preferred: there is inevitably considerable uncertainty associated with results obtained from incubations of many hours duration, such as are required for unproductive oceanic waters. The ability of aquatic plants, including phytoplankton, to utilize light of any given intensity can be highly dependent on the light climate to which they were exposed during growth: it is therefore preferable, for ecological interpretation, if P versus E_d curves are measured on naturally occurring, rather than laboratory-grown, plant material. Table 10.1 presents a selection of published data on the irradiance values which have been observed to correspond to light compensation, onset of light saturation and saturation, in a range of naturally occurring aquatic plants. The data do not permit any firm generalization with respect to differences in light saturation values between one algal class and another. This is partly because of the inherent difficulty in identifying the saturation irradiance in a P versus E_d curve. Comparisons would be facilitated if the practice of always recording the more easily identified parameter, E_k, corresponding to onset of saturation, was generally adopted. One permissible generalization is that for any given algal species the light compensation point is lower in the winter or spring than in the summer or autumn: whether this is due simply to the differences in temperature or whether other factors are involved remains uncertain.[474] Within some species the greater the depth from which the sample was taken, the lower the irradiance required to saturate photosynthesis:[738] this is true of phytoplankton as well as macroalgae.[874] Adaptation to different ambient light levels is discussed more fully in Chapter 12. Within the phytoplankton there is evidence that the dinoflagellates have higher respiration rates, and therefore higher light compensation points, than the diatoms:[244,845] this may be due to the energy required to sustain motility in the former group.

A number of attempts have been made to find mathematical expressions which give a reasonable fit to the empirical curves relating P to E_d. Since it is a fact of observation that for any given phytoplankton population the curve will exhibit a fairly well-defined initial slope and a maximum, asymptotically approached, value of P, the values of α and P_m being characteristic of that population, then we may reasonably anticipate that the relationship we are seeking will express P as a function of α and P_m, as well as E_d. Furthermore the relationship will be such that $P = f(\alpha, P_m, E_d)$ reduces to $P = \alpha E_d$ as E_d tends to zero, and approaches $P = P_m$ as E_d tends to infinity. Jassby & Platt (1976) tested eight different expressions which have

Table 10.1. *Irradiance values required for saturation and light compensation of photosynthesis in various aquatic plants. Only data obtained for naturally occurring plant material, measured in natural water or its equivalent, have been used. Irradiance values published in other units have where necessary been converted to μeinsteins of PAR* m^{-2} s^{-1} *(where 1 μeinstein* $= 6.02 \times 10^{17}$ *quanta) with the use of appropriate conversion factors. In many cases saturation,* E_k, *or compensation irradiance values where not published explicitly have had to be estimated from authors' data; the uncertainty is particularly great in estimates of the irradiance required to saturate photosynthesis*

Species or plant type	Location, season	Temperature (°C)	Irradiance, μeinstein (PAR) m^{-2} s^{-1}			Reference
			At saturation	At onset of saturation (E_k)	At compensation point	
Freshwater algae						
Diatoms						
Asterionella formosa	L. Windermere,	5	—	28	—	897
	England, spring	10	—	50	—	897
Melosira italica	L. Windermere,	5	—	16	—	897
	England, winter					
Blue-green algae						
Microcystis etc.	L. George, Uganda	27–34	—	135–323	—	284
Oscillatoria sp.	L. Neagh, N. Ireland					
	spring	9	145	49	—	427
	summer	15	203	64	—	427
Green						
Cladophora glomerata	Green Bay, L. Michigan USA, July–Aug	25–27	345–1125	—	44–104	546
Freshwater macrophytes						
Hydrilla verticillata	Lakes, Fla, USA	30	600	—	15	946
Ceratophyllum demersum	Lakes, Fla. USA	30	700	—	35	946

Myriophyllum spicatum	Lakes, Fla, USA	30	600	—	35	946
Cabomba caroliniana	Lakes, Fla, USA	30	700	—	55	946
Myriophyllum brasiliense	Orange L., Florida, USA	30	250–300	—	42–45	783
Vallisneria americana	Lake, Wisc., USA summer	25	140	—	—	930
Nuphar japonicum	Japan					
floating leaf		20	400–600	—	3	400
submerged leaf		21	75	—	3	400
Marine microalgae						
Oceanic phytoplankton (0 m)	Pacific (3° S)	20–25	600	—	—	431
Oceanic phytoplankton	Pacific Ocean off Japan,					
(10 m)	summer	23	> 700	240	—	896
(80 m)		23	~ 140	50	—	896
Oceanic picoplankton	Coral Sea Oct–Nov					
(10 m)		~ 27	—	344–818	—	277
(100 m)			—	72–245	—	277
Continental shelf phytoplanklton	Bransfield Strait, Antarctica Dec–March	0–1	50	18	0.5–1.0	389,616
Coastal phytoplankton (1–10 m)	Nova Scotia, Canada, all year	0–15	~ 300	105 (av.)	4 (av.)	707
Coastal phytoplankton (0 m)	Baltic, Denmark					
Feb. 3rd		1	400	200	—	875
July 15th		17	1200	500	—	875
Oct. 31st		12	800	300	—	875
Coastal phytoplankton	S. California Bight, USA mid-day, 15 July					
(1 m)		16	—	254	—	839
(32 m)		12	—	42	—	839

Table 10.1. (*cont.*)

Species or plant type	Location, season	Temperature (°C)	Irradiance, μeinstein (PAR) $m^{-2} s^{-1}$ At saturation	At onset of saturation (E_k)	At compensation point	Reference
Surf zone phytoplankton	Algoa Bay	15	700	300	—	128
	S. Africa	22.5	1000	450	—	128
Estuarine phytoplankton	Chesapeake Bay					
Prorocentrum mariae-lebouriae	USA, June					
Surface mixed layer (0.5 m)		24	—	412	—	342
Subpycnocline (15 m)		21	—	67	—	342
Intertidal benthic diatoms	Cape Cod, USA, summer	27	57	—	—	911
Sea-ice algae (pennate diatoms)	Canadian Arctic spring	−1	3–25	0.5–4.3	0.18	160
Sea-ice algae (pennate diatoms)	McMurdo Sound Antarctica summer	−2	25	5.4	—	686
Seagrasses						
Cymodocea nodosa	Malta	25	—	158	17	199
Halophila stipulacea	Malta	25	—	83	8	199
Posidonia oceanica	Malta	17	—	108	17	199
Phyllospadix torreyi	Calif., USA	15	—	149	21	199
Zostera marina	Calif., USA	15	—	208	25	199
Zostera angustifolia	Scotland	10	—	133	12	199
Marine macroalgae						
Chlorophyta						
Enteromorpha intestinalis	W. Baltic spring	10	450	—	6	474

	summer	20	—	—	15	474
	Calif., USA	21	245	56	6	18
Cladophora glomerata	W. Baltic, summer	20	700	—	8	474
Acrosiphonia centralis	W. Baltic, spring	10	200	—	7	474
Ulothrix speciosa	W. Baltic, spring	10	700	—	6	474
Monostroma grevillei	N. Baltic	5	120	—	—	970
Ulva lactuca	Woods Hole, Mass., USA summer	23	250	160	—	738
Ulva lobata	Calif., USA	16	245	76	11	18
Ulva rigida	Calif., USA	21	412	50	9	18
Codium fragile	Calif., USA	21	346	50	9	18
Chaetomorpha linum	Calif., USA	21	418	82	10	18
Phaeophyta						
Fucus serratus	W. Baltic					
	winter	5	350	—	5	474
	autumn	15	200	—	12	474
Laminaria saccharina	W. Baltic					
	winter	5	25	—	4	474
	autumn	15	700	—	18	474
L. solidungula	Alaskan High Arctic, summer	2	—	38–46	—	212
Scytosiphon lomentaria	W. Baltic, spring	10	700	—	8	474
Ectocarpus confervoides	W. Baltic, spring	10	200	—	5	474
Dictyosiphon foeniculaceus	N. Baltic	14	300	—	18	970
Pilayella littoralis	N. Baltic	14	200	—	20	970
		4	100	—	—	970
Macrocystis integrifolia	Vancouver I., B.C., Canada, September	13	80	50	—	844
M. pyrifera	S. Calif., USA March–August	16–21	300	140–300	—	294
Nereocystis luetkeana	Vancouver I., B.C., Canada February	9.5	—	22	—	993

Table 10.1. (*cont.*)

Species or plant type	Location, season	Temperature (°C)	Irradiance, μeinstein (PAR) $m^{-2} s^{-1}$			Reference
			At saturation	At onset of saturation (E_k)	At compensation point	
	September	14	—	64	—	993
Rhodophyta						
Dumontia incrassata	W. Baltic					
	winter	5	100	—	5	474
	spring	10	500	—	8	474
Phycodrys rubens	W. Baltic					
	spring	10	200	—	5	474
	summer	20	200	—	14	474
Polysiphonia nigrescens	W. Baltic					
	spring	10	400	—	7	474
	autumn	15	300	—	24	474
Ceramium tenuicorne	N. Baltic	11	100	—	—	970
Rhodomela confervoides	N. Baltic	4, 10	40	—	—	970
Chondrus crispus	Woods Hole, Mass., USA summer	23	120	60	—	738
Porphyra umbilicalis	Woods Hole, Mass., USA summer	23	250	90	—	738
Coral						
Stylophora pistillata	Sinai, Egypt					
High-light form		28	600–2000	—	350	240
Low-light form		28	200	—	40	240
Coral reef algal turf	Virgin Islands Caribbean Sea Jul, Oct, Nov, Dec	28	1400–1800	780–1060	60–105	138

at various times been proposed, against 188 P versus E_d curves measured for marine phytoplankton in coastal Nova Scotia waters. The two which fitted best were

$$P = \frac{P_m \alpha E_d}{(P_m{}^2 + \alpha^2 E_d{}^2)^{\frac{1}{2}}} \tag{10.1}$$

originally (in a somewhat different form) proposed by Smith (1936), and

$$P = P_m \tanh(\alpha E_d / P_m) \tag{10.2}$$

proposed by Jassby & Platt (1976), the latter expression giving somewhat the better fit.

These two expressions for P have been chosen simply on the basis of goodness of fit to the observations: they are not based on assumptions about the mechanisms of photosynthesis. There is a third, equally simple, equation

$$P = P_m[1 - e^{-\frac{\alpha E_d}{P_m}}] \tag{10.3}$$

originally proposed by Webb, Newton & Starr (1974) to describe photosynthesis in the tree species, *Alnus rubra*, and which Peterson *et al.* (1987) have found to satisfactorily describe P versus E_d curves in a wide range of phytoplankton systems, that can be given a plausible rationale in terms of the mechanism of photosynthesis. By application of simple Poisson distribution statistics to the capture of photons by the photosynthetic unit per unit of time t, where t is the turnover time, and assuming that excess photons are not utilized, it can be shown that the rate of photosynthesis is proportional to $(1 - e^{-m})$ where m is the mean number of incident photons captured by the photosynthetic unit in time t.[698] Since m must be proportional to the incident flux, E_d, it is clear that eqn 10.3, or its alternative version

$$P = P_m[1 - e^{-\frac{E_d}{E_k}}] \tag{10.4}$$

(since $P_m/\alpha = E_k$), is in accordance with this simple mechanistic model.

Eqns 10.1 to 10.4 describe the variation of P with E_d only up to the establishment of saturation; they do not encompass the decline in P at higher values of E_d. Platt, Gallegos & Harrison (1980) have obtained an empirical equation which describes the photosynthetic rate of phytoplankton as a single continuous function of available light from the initial linear response up to and including photoinhibition.

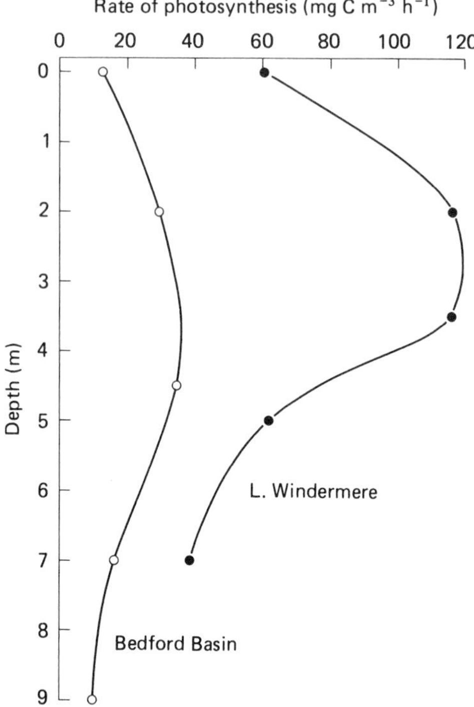

Fig. 10.4. Depth profiles of phytoplankton photosynthetic rate per unit volume of water. The curves are for an inland water (Lake Windermere, England; plotted from data of Belay, 1981, assuming a photosynthetic quotient of 1.15), and a coastal water (Bedford Basin, NS, Canada; plotted from data of Marra, 1978).

Photoinhibition

The inhibition of photosynthesis at high light intensities must be taken into account in ecological studies, since the intensities typically experienced in the surface layer of natural waters in sunny weather are in the range that can produce photoinhibition. Indeed if the depth profile of phytoplankton photosynthetic activity is measured by the suspended bottle method in inland or marine waters, a noticeable diminution in the specific photosynthetic rate or the rate per unit volume is commonly, although not invariably, observed near the surface. Fig. 10.4 shows examples of this surface inhibition of photosynthesis in a coastal and an inland water. With increasing depth and diminishing light intensity, photoinhibition lessens and the maximum, light-saturated but not inhibited, photosynthetic rate is achieved. With further increase in depth, irradiance falls to the point at which light intensity becomes limiting and from here on the photosynthetic rate diminishes roughly exponentially with depth approximately in parallel with irradiance.

Many, but not all, macrophyte species also show inhibition of photosynthesis at light intensities in the range of full sunlight. Ecologically, however, this phenomenon is of less significance since any given macrophyte species is generally to be found growing at a depth where the light intensity is one to which it is well adapted (Chapter 12), whereas phytoplankton are circulated within a range of depths by water movement. Marine macrophytic algae of the intertidal zone are intermittently exposed to very high light intensities. At irradiance values equivalent to full sunlight some of these species show no inhibition, but others are partially inhibited.[474]

Inhibition of photosynthesis by high light intensities takes time to develop. In the case of phytoplankton populations from Lake Ontario, Canada, the decline in photosynthetic activity began after about 10 min exposure.[353] Measurements of the time course of photosynthesis by populations of the diatom *Asterionella* in bottles suspended at the surface of a Welsh lake indicated that the inhibitory effect was small during the first hour but became significant during the second hour.[56] The higher the temperature at a given light intensity, the more rapidly inhibition ensues.[353] In the case of laboratory cultures of *Asterionella* grown at 18 °C and 200 μeinsteins $m^{-2} s^{-1}$, exposure to 2000 μeinsteins $m^{-2} s^{-1}$ (~ full sunlight) for 1 h at 18 and 25 °C reduced the subsequently measured photosynthetic rate by about 10 and 50%, respectively.[57]

Phytoplankton can recover from the inhibitory effects of intense light if they are transferred to a lower light intensity.[309] The longer the exposure to bright light, the longer the recovery takes. In the case of *Asterionella* populations from a Welsh lake, recovery from 2 h exposure to bright sunlight was complete after 4 h in low light intensity: after 6 h bright sunlight, which reduced photosynthetic rate by 70%, recovery took 20 h.[56]

The mechanism of photoinhibition has been studied in most detail in higher plants. Jones & Kok (1966) measured the action spectrum of photoinhibition of electron transport in spinach chloroplasts. The spectrum showed its main activity in the ultraviolet (UV) region with a peak at 250–260 nm. Photoinhibition also occurred in the visible region but with a very much lower quantum efficiency. Between 400 and 700 nm, the action spectrum followed the absorption spectrum of chloroplast pigments, with a distinct chlorophyll peak at 670–680 nm. The lesion appears primarily to affect the light reactions of photosynthesis by damaging the reaction centre of photosystem II.[164,165,143]

The shape of the action spectrum in the UV region suggests that plastoquinone or some other quinone functional in the reaction centre may be the sensitive molecule so far as UV inhibition is concerned. The shape of the action spectrum in the visible region indicates that at very high light intensities some of the energy absorbed by the photosynthetic pigments themselves is transferred to a sensitive site – not necessarily the same site as that affected by UV – where it causes damage.

Although detailed studies on the basis of photoinhibition of algal photosynthesis have not been carried out, the most plausible and economical hypothesis is that the mechanism is the same as in higher plants. For oceanic phytoplankton, the observed photoinhibition was found to vary linearly with the daily biological dose calculated using the Jones & Kok action spectrum.[835] In the surface layer of clear ocean waters, 50% of the photoinhibitory dose is at wavelengths less than 390 nm; in moderately productive waters (0.5 mg chl *a* m^{-3}) at 10 m depth, 50% of the photoinhibitory dose is at wavelengths less than 430 nm.[831] Thus, for oceanic waters we may attribute about 50% of the photoinhibition to UV and about 50% to visible light. Field measurements by Smith *et al.* (1992) in the Bellingshausen Sea in the austral spring of 1990 indicated that primary production in the Antarctic marginal ice zone was 6–12% inhibited by the increased UV flux resulting from ozone depletion.

Aquatic yellow substances absorb strongly in the UV. We may therefore expect photoinhibition to be less apparent in the more coloured waters: this has been observed to be the case in highly productive tropical oceanic waters with high levels of gilvin.[509] By the same token it seems likely that photoinhibition in the more highly coloured (i.e. most inland) waters is caused mainly by the visible (400–700 nm) component of the solar radiation.

Depth profiles of phytoplankton photosynthesis, such as those in Fig. 10.4, determined by the suspended bottle method, tend to overestimate the extent to which photoinhibition diminishes primary production.[353,594] In nature the phytoplankton are not forced to remain at the same depth for prolonged periods. Some, such as dinoflagellates and blue-green algae, can migrate to a depth where the light intensity is more suitable (see §12.6). Even the non-motile algae will only remain at the same depth for extended periods under rather still conditions. Wind blowing across a water surface induces circulatory currents known as Langmuir cells, after the eminent physical chemist, Irving Langmuir, who first studied them.[536] Langmuir cells are horizontal tubes (roll vortices) of rotating water, their axes aligned approximately parallel to the wind direction (Fig. 10.5). Adjacent tubes

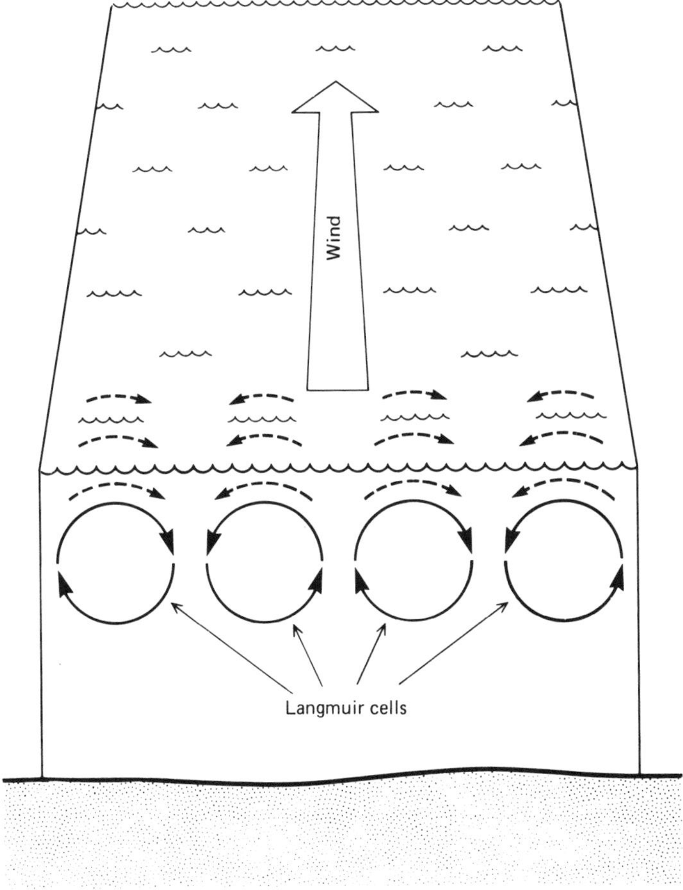

Fig. 10.5. Wind-induced circulatory currents (Langmuir cells) in a water body.

rotate in opposite directions and tubes of varying diameter can be present at the same time. The simultaneous occurrence of both wind and waves is necessary for the generation of these roll vortices,[247] but even a light wind over small-amplitude waves can set them going. Cells can have diameters ranging from a few centimetres to hundreds of metres: with a wind speed of 5 m s^{-1}, a typical cell might have a diameter of 10 m and a surface speed of 1.5 cm s^{-1}.[235] Measurements by Weller *et al.* (1985) from the research platform FLIP, drifting off the coast of southern California, showed that with quite moderate windspeeds (mainly 1–8 m s^{-1}), downwelling flows typically between 0.05 and 0.1 m s^{-1} were generated. The mixed layer above the seasonal thermocline was at the time about 50 m deep, and the strongest downwelling flows were observed between 10 and 35 m depth, corresponding to the middle region of the mixed layer. Above and below that region, downwelling flows were generally less than 0.05 m s^{-1}, and there appeared to be no downwelling flow in or below the seasonal thermocline.

Thus it will very commonly be the case that phytoplankton are not held in the intensely illuminated surface layer but are slowly circulating throughout the mixed layer. Harris & Piccinin (1977) point out that on the Great Lakes of North America the average monthly wind speed throughout all the winter, and most of the summer period, is sufficient to generate Langmuir cells, and that the residence time at the surface under such conditions will not be long enough for photoinhibition to set in. Within any given month of course, although the average wind speed may be enough to ensure Langmuir circulation, there will be calm periods in which it does not occur. Phytoplankton sampled in winter from the waters of Vineyard Sound (Massachusetts, USA) showed marked photoinhibition in bottles held at surface light intensities, but were in fact well adapted to the average light intensity that they would actually encounter in this well mixed shallow coastal water.[305]

Photoinhibition is only likely to be of frequent significance in water bodies in which high solar irradiance commonly occurs together with weak wind activity, leading to the formation of transient shallow temperature/density gradients in the surface layer which impede mixing and thus trap phytoplankton for part of the day in the intense near-surface light field. A well-documented example is the high-altitude (3803 m), low-latitude (16° S) Lake Titicaca (Peru–Bolivia). Vincent, Neale & Richerson (1984) found the typical pattern of thermal behaviour to be that a near-surface thermocline began to form each morning, persisted during the middle part of the day, and was then dissipated by wind mixing and convective cooling towards evening and through the night. While the near-surface stratification persisted, phytoplankton photosynthesis in the upper layer was strongly depressed. That this was not an artifact resulting from phytoplankton immobilization in bottles was shown by the observation that the cellular fluorescence capacity (believed to correlate with the number of functional photosystem II complexes) of phytoplankton samples taken from the water was also greatly reduced. Neale (1987) estimated the diminution of total water column photosynthesis in L. Titicaca on such days to be at least 20%. Elser & Kimmel (1985) have also used measurements of cellular fluorescence capacity to show that in reservoirs in temperate regions (southeastern USA) photoinhibition does occur in the surface layer under calm sunny conditions.

On balance we may reasonably conclude that photoinhibition of photosynthesis in the surface layer, although it exists, is not as frequent a phenomenon as was originally thought. It can significantly reduce areal photosynthesis under sunny, still, conditions, but is likely to be of small or no significance when there is even a light wind. Underestimates of primary production resulting from the use of stationary bottles are likely to be more serious in oligotrophic waters requiring long incubation times, than in productive waters. It should be noted, however, that circulation does not by any means always increase primary production: as we shall discuss more fully in the next chapter, circulation through too great a depth can diminish total photosynthesis by keeping the cells for significant periods in light intensities too low for photosynthesis.

10.2 Efficiency of utilization of incident light energy

Of the light energy incident on the water surface, only a small fraction is converted to chemical energy in the form of aquatic plant biomass. We shall now consider the reasons why this is so.

The first mode by which energy is lost is reflection at the surface. As we saw earlier, however, (§2.5, Table 2.1) such losses are small. For that range of solar angles at which most aquatic primary production takes place, only 2–6% of the incident light is lost by surface reflection. Thus the main causes of inefficiency of light utilization are to be found beneath the water surface.

In shallow water bodies (very shallow in coloured and/or turbid waters; moderately shallow in clear waters) substantial amounts of light reach the bottom. Some is absorbed, some reflected, the proportion depending on the optical characteristics of the substrate. Of the bottom-reflected light, a fraction will succeed in passing up through the water column again and escaping through the surface. Thus in shallow waters, bottom absorption and bottom reflection, followed by surface escape, are mechanisms preventing utilization of some of the light in photosynthesis. The light lost can be anything from a trivial proportion up to nearly 100% in, say, very shallow clear water over a white sandy bottom. Our main concern here, however, is with optically deep waters in which the fraction of the incident light which penetrates to the bottom is negligible. In such waters, most of the light which penetrates the surface is absorbed within the aquatic medium. A fraction of the light, however, usually small, is backscattered upwards within the water (see §6.4) and some of this succeeds in passing up to, and out through, the surface. Combining data for the irradiance reflectance just beneath the surface (§6.4) with the fact that about half the upwelling flux is reflected down again at the water/air boundary,[26] we may conclude that the amount of incident PAR lost in this way is 1–2.5% in oceanic waters, 1–10% in inland waters of low to high turbidity, and as little as 0.1-0.6% in waters with intense colour but low scattering.

Proportion of incident light captured by phytoplankton

A major factor limiting conversion of solar energy to chemical energy by phytoplankton is, as was pointed out by Clarke (1939), the competition for radiant energy by all the non-living components of the water. We saw in Chapter 3 that the different components of the aquatic medium – water, soluble colour, tripton and phytoplankton – each account for a proportion of the total light absorbed by the water body. We also saw that to obtain an accurate estimate of the amount of PAR captured by each component separately, calculations should be carried out using the absorption coefficients for a series of narrow wavebands followed by summation across the photosynthetic spectrum. A useful approximation, however, is to first consider the total photosynthetically available radiation and then assume that the relative rates of absorption of light by the different components of the aquatic medium are in proportion to their individual contributions to the total vertical attenuation coefficient for downward irradiance of PAR, in accordance with eqn 9.10

$$K_d(\mathrm{PAR}) = K_W + K_G + K_{TR} + K_{PH}$$

For many, perhaps most, waters of interest to limnologists and marine biologists, this assumption will not be intolerably far from the truth but it does presuppose that absorption rather than scattering is the dominant contributor to each of the partial attenuation coefficients on the right-hand side of eqn 9.10. For K_W (water), K_G (gilvin) and K_{PH} (phytoplankton) this will be true, and when the tripton fraction is strongly coloured (e.g. by insoluble humic material) it will also be true for K_{TR}. If, however, the tripton fraction is high in concentration so that K_{TR} is high, but consists of mineral particles low in intrinsic colour so that K_{TR} is made up mainly of the scattering contribution (see §6.8, eqn 6.16), then the assumption that relative rates of absorption are proportional to the partial vertical attenuation coefficients will be seriously in error.

Nevertheless, for waters other than the kind we have just described, the

Table 10.2. *Proportion of total absorbed PAR captured by phytoplankton in idealized water bodies of different types. Calculations carried out using eqn 10.5 and assuming* $k_c = 0.014$ *m*^{2} *mg chl* a^{-1}. K_{NP} *is vertical attenuation coefficient due to all non-phytoplankton material*

Type of water body	K_{NP} (m^{-1})	Phytoplankton (mg chl *a* m^{-3})	Proportion of absorbed PAR captured by phytoplankton (%)	Proportion of absorbed PAR captured by non-phytoplankton material (%)
Clear oceanic	0.08	0.2	3.4	96.6
		0.5	8.0	92.0
		1.0	14.9	85.1
Coastal	0.15	1.0	8.5	91.5
		2.0	15.7	84.3
		4.0	27.2	72.8
Clear lake, limestone catchment	0.4	4.0	12.3	87.7
		8.0	21.9	78.1
		12.0	29.6	70.4
Productive lake, coloured water	1.0	8.0	10.1	89.9
		16.0	18.3	81.7
		32.0	30.9	69.1
		64.0	47.3	52.7
Oligotrophic lake, coloured water	2.0	1.0	0.7	99.3
		2.0	1.4	98.6
		4.0	2.7	97.3

fraction of the total absorbed light which is captured by the phytoplankton is, with this approximate treatment, given by

$$\frac{K_{PH}}{K_d(PAR)} = \frac{B_c k_c}{K_d(PAR)} = \frac{B_c k_c}{K_W + K_G + K_{TR} + B_c k_c} = \frac{B_c k_c}{K_{NP} + B_c k_c} \qquad (10.5)$$

where K_{NP} (which is equal to $K_W + K_G + K_{TR}$) is the vertical attenuation coefficient due to all non-phytoplankton components, B_c is the phytoplankton concentration (mg chl *a* m^{-3}) and k_c is the specific vertical attenuation coefficient per unit phytoplankton concentration. Thus the extent to which the phytoplankton succeed in competing with other components of the medium for the available quanta depends on the relative size of $B_c k_c$ and K_{NP}. The range of possibilities is limitless but we shall consider a few specific examples. We shall take 0.014 m^2 mg^{-1} as a typical mid-range value of k_c (see Table 9.1). Table 10.2 lists some values for the proportion of absorbed

PAR captured by phytoplankton in a number of hypothetical (but typical) water bodies ranging from very pure oceanic water with K_{NP} not much greater than that due to pure water ($K_W = 0.03$–$0.06\ m^{-1}$), to a quite highly coloured, but productive, inland water. Very approximate though these calculations are, they do show that the share of the available quanta collected by the phytoplankton can vary from a few per cent in the less productive waters, to well over 50% in highly productive systems. They also emphasize the point that quite dilute algal populations can collect a substantial proportion of the quanta, provided the background absorption is low.

Some calculations of this type have been carried out for real water bodies. Dubinsky & Berman (1981) estimated that in the eutrophic Lake Kinneret (Sea of Galilee) the proportion of the absorbed quanta captured by phytoplankton (mainly the dinoflagellate *Ceratium*) varied from about 4 to 60% as the algal concentration rose from about 5 to 100 mg chl a m^{-3}. For the eutrophic, blue-green-algal-dominated Halsted Bay, Lake Minnetonka, USA, Megard *et al.* (1979) calculated that the proportion of absorbed PAR collected by the algae rose from about 8 to 80% as the phytoplankton concentration increased from about 3 to 100 mg chl a m^{-3}. From the data of Talling (1960) these workers estimated that for Lake Windermere (*Asterionella*-dominated), England, which has relatively low background absorption, the proportion of absorbed PAR collected by the algae rose from about 5 to 25% over the population density range of 1 to 7 mg chl a m^{-3}. For the eutrophic, shallow (and therefore turbid) Lough Neagh, Ireland, Jewson (1977) estimated that phytoplankton accounted for about 20% of the absorbed light at the lowest population level (26.5 mg chl a m^{-3}) and 50% at the highest (92 mg chl a m^{-3}). In mesotrophic Lake Constance (Germany), Tilzer (1983) calculated fractional light absorption by phytoplankton to vary between about 4 and 70% over a two-year period in which chlorophyll a levels varied between about 1 and 30 mg m^{-3} (Fig. 11.8).

As we saw in the previous chapter, in the sea, where the waters are deep and usually with little dissolved yellow colour, and the light field with increasing depth becomes increasingly confined to the blue-green (400–550 nm) spectral region, the effective specific absorption coefficient of the phytoplankton for PAR, $\bar{a}_c(z)$ (defined below), also increases with depth. Where, as is normally the case in the ocean, there is a layer of increased phytoplankton concentration (the deep chlorophyll maximum, §11.1) near the bottom of the euphotic zone, the combination of increased pigment concentration and enhanced light-harvesting efficiency leads to a great increase in the proportion of the total light absorption which is carried out by the phytoplankton. In the Pacific Ocean, off southeastern Japan, Kishino *et al.* (1986) found the fractional light absorption by phytoplankton to increase from 1.7% at the surface to 40% in the middle of the deep chlorophyll maximum at 75 m.

As we have just discussed, the usefulness of a given light field for photosynthesis is not simply a function of the total intensity of PAR, but is very much determined by how well the spectral distribution of the PAR matches the absorption spectrum of the phytoplankton or other aquatic plants. Morel (1978, 1991) has introduced the concept of *photosynthetically usable radiation*, or PUR, which may be thought of as a modified PAR obtained by weighting the actual PAR, across its spectrum for absorbability by the phytoplankton. This can be achieved by multiplying the PAR in each narrow waveband by some dimensionless quantity proportional to phytoplankton absorption in that waveband, and then summing across the spectrum. As a suitable dimensionless quantity, Morel in fact chose the ratio of the phytoplankton absorption coefficient in any given waveband to the maximum absorption coefficient – which is commonly the value it has at about 440 nm. PUR can thus be defined by

$$\mathrm{PUR}(z) = \int_{400}^{700} E_0(\lambda, z)\,\frac{a_\mathrm{p}(\lambda, z)}{a_\mathrm{p}(\lambda_\mathrm{max}, z)}\,\mathrm{d}\lambda \qquad (10.6)$$

where $a_p(\lambda,z)$ and $a_p(\lambda_{max},z)$ are the absorption coefficients at wavelengths λ and λ_{max} (wavelength of maximum phytoplankton absorption), respectively, of the phytoplankton population existing at depth z m, and $E_0(\lambda,z)$ is the scalar irradiance per unit bandwidth (nm^{-1}) at wavelength λ and depth z m. From the definition of the effective absorption coefficient, $\bar{a}_p(z)$, of the phytoplankton for the whole PAR waveband (eqn 10.9, see below) it follows that we can also write

$$\mathrm{PUR}(z) = \mathrm{PAR}(z)\,\frac{\bar{a}_\mathrm{p}(z)}{a_\mathrm{p}(\lambda_\mathrm{max}, z)} \qquad (10.7)$$

Efficiency of conversion of absorbed light

Once the light energy is absorbed by the chloroplast pigments of the phytoplankton or aquatic macrophytes it is used, by means of the photosynthetic fixation of CO_2, to generate useful chemical energy in the form of carbohydrate. We shall now consider the efficiency of this conversion of excitation energy to chemical energy.

An upper limit to the efficiency is imposed by the nature of the physical and chemical processes that go on within photosynthesis. We saw in Chapter 8 that the transfer of each hydrogen atom from water down the electron transport chain to NADP requires two photons, each driving a distinct photochemical step. The reduction of one molecule of CO_2 to the level of carbohydrate uses four hydrogen atoms ($2 \times NADPH_2$) and so requires eight photons. Putting it another way, to convert one mole of CO_2 to its carbohydrate equivalent (one-sixth of a mole of a glucose unit incorporated in starch) requires not less than eight molar equivalents, i.e. 8 einsteins, of light, where an einstein is Avogadro's number ($\sim 6 \times 10^{23}$) of photons. The energy in a photon varies with wavelength ($\epsilon = hc/\lambda$) and so we shall obtain an average value by taking advantage of the observation by Morel & Smith (1974) that for a wide range of water types 2.5×10^{18} quanta of underwater PAR $\equiv$ 1 J, with an accuracy of better than 10%. We may thus regard typical underwater light as containing 0.24 MJ (megajoules) of energy per einstein and so 8 einsteins is equivalent to 1.92 MJ. The increase in chemical energy associated with the photosynthetic conversion of one mole of CO_2 to its starch equivalent is 0.472 MJ. Thus, of the light energy absorbed and delivered to the reaction centres, about 25% is converted to chemical energy as carbohydrate and this is the maximum possible efficiency.

To equate the plant biomass to carbohydrate is an oversimplification, since the aquatic plants also contain protein, lipids and nucleic acids, none of which conform closely in their overall composition to CH_2O. The biosynthesis of these substances requires additional photosynthetically generated reducing power and chemical energy in the form of $NADPH_2$ and ATP, and so requires additional light quanta per CO_2 incorporated. The true minimum quantum requirement per CO_2 for growing cells is likely to be about 10–12 rather than 8[737] which brings the maximum efficiency down to 16–20%. Thus the best efficiency we can hope for in the conversion of absorbed light energy to chemical energy in the form of new aquatic plant biomass is about 18%.

The conversion efficiency or quantum yield actually achieved by a given phytoplankton population or macrophyte can be determined from measurements of the photosynthetic rate and the irradiance, provided that information on the light absorption properties of the plant material is available. Using the absorption spectrum, 400–700 nm, rates of light absorption for a series of wavebands can be calculated and summed. For example, the rate of absorption of PAR by phytoplankton per unit volume of medium at any given depth, z, is

$$\frac{d\Phi_p(z)}{dv} = \int_{400}^{700} a_p(\lambda, z) E_0(\lambda, z)\, d\lambda \tag{10.8}$$

where $a_p(\lambda,z)$ is the absorption coefficient at wavelength λ of the phytoplankton population existing at depth z m, and $E_0(\lambda,z)$ is the scalar irradiance per unit bandwidth (nm^{-1}) at wavelength λ and depth z m.

A useful concept here is that of the effective absorption coefficient of the phytoplankton for the whole PAR waveband. This is defined by

$$\bar{a}_p(z) = \frac{\int_{400}^{700} a_p(\lambda, z) E_0(\lambda, z)\, d\lambda}{\int_{400}^{700} E_0(\lambda, z)\, d\lambda} \tag{10.9}$$

and as an alternative to eqn 10.8 we can therefore write

$$\frac{d\Phi_p(z)}{dv} = \bar{a}_p(z) E_0(\text{PAR}, z) \tag{10.10}$$

The specific absorption coefficient of the phytoplankton for PAR, $\bar{a}_c(z)$, is defined by substituting $a_c(\lambda,z)$ for $a_p(\lambda,z)$ in eqn 10.9: also $\bar{a}_p(z) = B_c\bar{a}_c(z)$, so that

$$\frac{d\Phi_p(z)}{dv} = B_c \bar{a}_c(z) E_0(\text{PAR}, z) \tag{10.11}$$

In any attempt to calculate the rate of energy absorption by phytoplankton (and, by implication, quantum yield – see below) with eqn 10.11, using estimated values of $\bar{a}_c(z)$, the fact that, as we saw earlier, the effective specific absorption coefficient of the phytoplankton for PAR can vary markedly with depth, must be taken into account.

It is often more convenient to work in terms of downward, rather than scalar, irradiance, but E_o is always greater than E_d (see Fig. 6.10, §6.5) by a factor which depends on the angular structure of the light field at that particular depth. Following Morel (1991) we shall indicate this 'geometrical' correction factor by g. This correction is not trivial: for wavelengths in the photosynthetically important 400–570 nm region, and for phytoplankton concentrations in the 0.1–1.0 mg chl *a* m^{-3} range, Morel (1991) calculated values of g varying between 1.1 and 1.5. Using the geometrical correction factor we can now write another expression for the rate of absorption of PAR by phytoplankton per unit volume of medium at depth z m, namely

$$\frac{d\Phi_p(z)}{dv} = \int_{400}^{700} a_p(\lambda,z) E_d(\lambda,z) g(\lambda,z)\, d\lambda \tag{10.12}$$

where $E_d(\lambda,z)$ is the downward irradiance per unit bandwidth (nm^{-1}) at wavelength λ and depth z m.

An alternative approach starts from the fact that the rate of absorption of radiant energy per unit volume at depth z m is given by

$$\frac{d\Phi(z)}{dv} = K_E \vec{E}(z) \tag{10.13}$$

where $\vec{E}(z)$ is the *net* downward irradiance at depth z m and K_E is the vertical attenuation coefficient for net downward irradiance. From this it can readily be shown that

$$\frac{d\Phi(z)}{dv} = K_d E_d(z)\left[1 - R(z)\left(\frac{K_u}{K_d}\right)\right] \simeq K_d E_d(z)[1 - R(z)] \tag{10.14}$$

where K_u is the vertical attenuation coefficient for upward irradiance and (as is usually the case) $K_u \simeq K_d$, and $R(z)$ is irradiance reflectance ($E_u[z]/E_d[z]$). If we choose to ignore the contribution of the upwelling flux – a reasonable approximation in most marine waters, with reflectance values of only a few percent, but not in turbid waters – then we can write

$$\frac{d\Phi(z)}{dv} \simeq K_d E_d(z) \tag{10.15}$$

for the total rate of absorption of energy per unit volume, as a function of downward irradiance. To calculate the rate of absorption of energy by phytoplankton we make use of the fact that in any given waveband the proportion of the absorbed energy which is captured by the phytoplankton is $a_p(\lambda,z)/a_T(\lambda,z)$, the ratio of the absorption coefficient due to phytoplankton to the total absorption coefficient, at that wavelength. The rate of absorption of PAR by phytoplankton per unit volume of medium, as a function of downward irradiance is therefore given by

$$\frac{d\Phi_p(z)}{dv} = \int_{400}^{700} [a_p(\lambda, z)/a_T(\lambda, z)] K_d(\lambda, z) E_d(\lambda, z)\, d\lambda \tag{10.16}$$

where $K_d(\lambda,z)$ is the vertical attenuation coefficient for downward irradiance at wavelength λ and depth z m. In eqns 10.8, 10.9 and 10.12 we can, of course, replace $a_p(\lambda,z)$ with $B_c(z)a_c(\lambda,z)$, the product of the phytoplankton concentration (mg chl a m^{-3}) and the specific absorption coefficient (m^2 mg chl a^{-1}) at wavelength λ, of the phytoplankton population present at depth z m.

To arrive at an accurate determination of photosynthetic efficiency, the spectral variation both of the light field and of absorption by the biomass should be taken into account, along the lines indicated above, and in recent years increasing numbers of workers have sought to do this.[65,151,238,500,502,543,633,801] Useful information can, however, still be obtained from photosynthesis measurements combined with broad-band irradiance data. Since, as we saw earlier, the fraction of the total absorbed PAR which is captured by the phytoplankton is approximately $B_c k_c / K_d(\mathrm{PAR})$, then from eqn 10.15 we can write

$$\frac{d\Phi_p(z)}{dv} = B_c k_c E_d(\mathrm{PAR}, z) \tag{10.17}$$

for the rate of absorption of light energy by phytoplankton per unit volume, at depth z m. Estimates of energy absorption by phytoplankton (and consequently of quantum yield – see below) made with eqn 10.17 can, however, be grossly inaccurate if the variation of k_c with the type of phytoplankton present, the background colour of the water, and the depth ($k_c[z]$ varying with depth in a similar manner to $\bar{a}_c[z]$) are not taken into account.

It is useful in dealing with the present topic to have a specific symbol for the rate of absorption of PAR by phytoplankton per unit volume of medium at a given depth. We here introduce the symbol χ, defined by

$$\chi(z) = \frac{d\Phi_p(z)}{dv} \tag{10.18}$$

where $\chi(z)$ can have the units W m^{-3}, or MJ m^{-3} h^{-1}, or quanta (or μeinsteins, or μmoles photons) m^{-3} s^{-1}. We can, in addition, define $\chi_c(z)$ to be the *specific* rate of absorption of PAR by phytoplankton per unit volume at depth z m, i.e. the rate per unit phytoplankton concentration, expressed in terms of mg chl *a* m^{-3}. Thus $\chi(z) = B_c \chi_c(z)$, and $\chi_c(z)$ has the units W mg chl a^{-1}, or quanta (or μeinsteins, or μmoles photons) s^{-1} mg chl a^{-1}. For any given aquatic system, $\chi(z)$ is determined by using one or other of eqns 10.8, 10.10, 10.16 and 10.17, or some equivalent procedure.

To obtain the energy conversion efficiency at a given depth we divide the rate of accumulation of chemical energy per unit volume at that depth by the rate of absorption of light energy by phytoplankton per unit volume. Given a specific photosynthetic rate of $P(CO_2)$ moles CO_2 fixed mg chl a^{-1} h^{-1}, and an increase in chemical energy of 0.472 MJ associated with the fixation of each mole of CO_2, then the rate of accumulation of chemical energy per unit volume is 0.472 $B_c P(CO_2)$ MJ m^{-3} h^{-1}. Dividing by $\chi(z)$ we obtain the conversion efficiency

$$\epsilon_c = \frac{0.472\, B_c P(CO_2)}{\chi(z)} = \frac{0.472\, P(CO_2)}{\chi_c(z)} \tag{10.19}$$

$\chi(z)$ being expressed in MJ m^{-3} h^{-1} (quanta measurements can be converted using 2.5×10^{24} quanta $\equiv$ 1 MJ, see above). Alternatively, if P is expressed in terms of mg carbon fixed mg chl a^{-1} h^{-1} (P(C)), then, since there is an increase in chemical energy of 3.93×10^{-5} MJ associated with the fixation of 1 mg C, the conversion efficiency is given by

$$\epsilon_c = \frac{3.93 \times 10^{-5} B_c P(\mathrm{C})}{\chi(z)} = \frac{3.93 \times 10^{-5} P(\mathrm{C})}{\chi_c(z)} \quad (10.20)$$

Another way of expressing the efficiency of conversion of absorbed light energy to chemical energy by aquatic plants is the quantum yield, ϕ. This is defined to be the number of CO_2 molecules fixed in biomass per quantum of light absorbed by the plant. Given the quantum requirement per CO_2 fixed, imposed by the mechanism of photosynthesis (see above), it follows that the quantum yield could never be greater than 0.125, and for growing cells, even under ideal conditions, is unlikely to exceed about 0.1. Quantum yield and per cent conversion efficiency are, of course, linearly related. Allowing 0.472 MJ chemical energy per CO_2 fixed and 0.24 MJ per einstein of underwater PAR, we arrive at

$$\epsilon_c = 1.97\phi \quad (10.21)$$

Equations corresponding to 10.19 and 10.20 can be written for the calculation of quantum yield from $\chi(z)$ or $\chi_c(z)$ and specific photosynthetic rate

$$\phi = \frac{B_c P(\mathrm{CO_2})}{\chi(z)} = \frac{P(\mathrm{CO_2})}{\chi_c(z)} \quad (10.22)$$

$$\phi = \frac{B_c P(\mathrm{C})}{12000.\chi(z)} = \frac{P(\mathrm{C})}{12000.\chi_c(z)} \quad (10.23)$$

$\chi(z)$ in these equations has the units einsteins m^{-3} h^{-1}.

The quantum yield attained by an aquatic plant is a function of the light intensity to which it is exposed. That this is so is apparent from the variation of specific photosynthetic rate with irradiance (Fig. 10.3). Ignoring for the moment any changes in chloroplast shape or position with light intensity, we may assume that the rate of absorption of quanta is proportional to the incident irradiance. Thus, at any point in the photosynthesis versus irradiance curve the value of P/E_d (see Fig. 10.3) is proportional to the quantum yield.

For a plant to be able to make efficient use of light quanta being absorbed at a given rate, the activity of the electron transfer components in the thylakoid membranes and of the enzymes of CO_2 fixation in the stroma must both be high enough to ensure that the excitation energy collected by the light-harvesting pigments is utilized as fast as it arrives at the reaction centres. If this situation exists, then a moderate increase in light intensity, causing a proportionate increase in quantum absorption rate, leads to a corresponding increase in the specific rate of photosynthesis. In the initial, linear, region of the P versus E_d curve this is what is happening. Over this range of intensity, P/E_d is constant and has its highest value, indicating that the plants are achieving their highest conversion efficiency and quantum yield. If the absorption characteristics of the plant are known, then this maximum value of P/E_d can be used to calculate ϕ_m, the maximum quantum yield.

As the incident light intensity is further increased, the rate of absorption of quanta reaches the point at which excitation energy begins to arrive at the reaction centres faster than it can be made use of by the electron transfer components and/or the CO_2 fixation enzymes. At this stage some of the additional absorbed quanta (over and above those the system can readily handle) are utilized for photosynthesis, and some are not, the energy of the latter being eventually dissipated, mainly as heat. For this reason, in this range of light intensity, increments in E_d are accompanied by less than commensurate increases in P, i.e. the slope of the curve progressively diminishes, until eventually the point is reached at which $\Delta P/\Delta E_d$ becomes zero. In this light-saturated state, the electron transfer and/or CO_2 fixation enzymes (most likely, the latter) are working as fast as they are capable, and so any additional absorbed quanta are not used for photosynthesis at all. From the end of the linear region through to the light-saturated region, since photosynthetic rate does not increase in proportion to irradiance (P/E_d steadily falls – Fig. 10.3) the quantum yield and conversion efficiency necessarily undergo a progressive fall in value. This is accentuated further if, at even higher light intensities, photoinhibition sets in.

The characteristic manner in which P varies with E_d can, as we saw earlier, be represented mathematically in a number of different ways (eqns 10.1, 10.2, 10.3). Since quantum yield and P/E_d are linearly related, then for each particular form of the function, $P = \mathrm{f}(E_d)$, there will be a corresponding expression for quantum yield as a function of E_d, i.e. $\phi = \text{Constant}.E_d^{-1}.\mathrm{f}(E_d)$. Using the tanh form (eqn 10.2) Bidigare, Prézelin & Smith (1992) arrived at

$$\phi = \phi_m \frac{E_k}{E(\mathrm{PAR}, z)} \tanh\left[\frac{E(\mathrm{PAR}, z)}{E_k}\right] \qquad (10.24)$$

and using the exponential form (eqn 10.3) we obtain

$$\phi = \phi_m \frac{E_k}{E(\text{PAR}, z)} [1 - e^{-\frac{E_d(\text{PAR}, z)}{E_k}}] \qquad (10.25)$$

The consequence in natural water bodies of the decrease in quantum yield with increasing light intensity is that quantum yield and conversion efficiency vary markedly with depth, the general tendency being, as might be expected, for ϕ and ϵ_c to increase with depth.[204,633] Quantum yields in the surface layer in the middle period of the day, when irradiance values are generally above the range corresponding to the linear part of the P versus E_d curve, are usually below ϕ_m.

Morel (1978) calculated the quantum yield at a series of depths from $^{14}CO_2$ fixation, chlorophyll and light data for a variety of oceanic waters from the highly oligotrophic Sargasso Sea to the productive waters of the Mauritanian upwelling area. In most cases ϕ increased with depth, i.e. with decreasing irradiance. On average, the ϕ values for green eutrophic waters were higher than those observed in blue, oligotrophic waters. In the surface layers the ϕ values were mainly in the range 0.003–0.012 (equivalent to ϵ_c values of 0.6–2.4%). Kishino *et al.* (1986) found, in the Pacific Ocean southeast of Japan, that quantum yields of photosynthesis were 0.005–0.013 at the surface, 0.013–0.033 at that depth (10–20 m) where photosynthesis reached its maximum rate in the surface mixed layer, and 0.033–0.094 in the deep chlorophyll maximum ($\sim$70 m). If, as these authors suggest, we reduce all these values by 20%, as an approximate correction factor for the fact that they are based on measured irradiance, rather than scalar irradiance, values then we obtain quantum yields of 0.004–0.01, 0.01–0.026, and 0.026–0.075, respectively.

In oligotrophic Lake Superior, Fahnenstiel, *et al.* (1984) found quantum yields to be very low, $\sim$0.003, at the surface, and to increase with depth, reaching maximum values of 0.031–0.052 (corrected for scalar irradiance) at 15–25 m. Dubinsky & Berman (1981) estimated that during the spring *Peridinium* bloom in Lake Kinneret (Sea of Galilee), ϵ_c rose from about 5% at the surface to 8.5% at 3 m. In late summer, with a much lower population of different (green) algae, ϵ_c was 2.5% at the surface but rose to about 12% at 5–7 m.

The maximum quantum yield that a phytoplankton population, or aquatic macrophyte, can exhibit, ϕ_m, is a parameter of considerable theoretical and ecological interest. On the assumption (see above) that in the linear region of the P versus E curve the cells are achieving their maximum efficiency, estimates of ϕ_m are normally obtained from the observed slope, P/E, commonly referred to as α, in this region of the curve.

From the equations for ϕ it follows that ϕ_m should be proportional to α. For example, if E is E_0, then from eqns 10.22, 10.18 and 10.11 it follows that $\phi_m = \alpha/\bar{a}_c$. If E is E_d, then from eqns 10.22, 10.18 and 10.17 it follows that $\phi_m = \alpha/k_c$. The photosynthetic process imposes by its essential nature a maximum value for ϕ_m of ~ 0.1, for all photosynthetic systems. On the basis of a critical analysis of literature data, Bannister & Weidemann (1984) have concluded that published values of *in situ* ϕ_m in excess of 0.10 are almost certainly in error. The fact that a ϕ_m value ~ 0.1 is one which any plant species might in principle achieve, and also that α is linearly related to ϕ_m (since $\alpha = \bar{a}_c\phi_m$ or $k_c\phi_m$), has aroused expectations that α should not vary markedly for a given species in different environments, or from one species to another. Such expectations, however, ignore the extent to which the proportionality factor, $\bar{a}_c$ or k_c, can vary. We saw earlier (Chapter 9) how markedly $\bar{a}_c$ varies with the size and shape of the cells or colonies, for example decreasing as the absorbing units become larger or more intensely pigmented. Taguchi (1976), from studies on seven species of marine diatom, found, as might be predicted, that α decreased with increasing size of the cells. Also, since $\bar{a}_c$ and k_c are expressed per unit chl *a*, then they can vary markedly in value in accordance with variation in the type of accessory photosynthetic pigments present, and their ratio to chl *a* (§9.5): the resulting changes in shape of the absorption spectrum markedly influence the rate of energy capture from the white light fields normally used in the determination of α. Welschmeyer & Lorenzen (1981), in a comparison of six species of marine phytoplankton growing exponentially under identical conditions, found that there were no significant differences in ϕ, but there were significant differences in α: these they attributed to differences in the light absorption efficiency per unit of chlorophyll.

Another problem in the determination of α values, particularly when we wish to compare different data sets from the literature, is that there is no generally accepted standard for the light source to be used in its measurement. Some workers use natural sunlight, others use tungsten–halogen lamps, while some will use lamps fitted with blue filters. The spectral distribution of the incident light will be quite markedly different in all three cases, with the result that for any given phytoplankton population, with a given absorption spectrum, α will have a different value for each light source because the effective specific absorption coefficient, $\bar{a}_c$, for the incident PAR, will be different for each light source.

Thus, even if we did not expect ϕ_m to vary dramatically with species or environment, we should not expect the same constancy of α, because of the variability of $\bar{a}_c$ and k_c. Unfortunately this variability in $\bar{a}_c$ and k_c, and the consequent uncertainty in their values, make the determination of ϕ_m in natural populations, from α, difficult. The best estimates are undoubtedly those based on full spectral data for phytoplankton absorption and the underwater light field, and combining eqn 10.22 with $\chi(z)$ values from eqns 10.8 10.12 or 10.16.

Some typical values of α for natural phytoplankton populations, in the units mg C mg chl a^{-1} h^{-1} (μeinsteins m^{-2} s^{-1})$^{-1}$, are: 0.05 (range 0.007–0.15) for Nova Scotia coastal waters,[707] 0.06 for nanoplankton (< 22 μm) in the lower Hudson estuary, USA,[589] 0.033–0.056 for picoplankton in the Celtic Sea,[435] and 0.024 for diatoms and 0.034 for blue-green algae in eutrophic Lough Neagh, N. Ireland.[427]

For reasons that are not well understood, the maximum quantum yield of phytoplankton populations in the ocean is extremely variable. For example, Prézelin *et al.* (1991) found that in a 200 km transect of a hydrographically variable region of the Southern California Bight, the value of ϕ_m varied over the range ~0.01–0.06, the variation being found both with distance along the transect through different water masses, and with depth at any given station. Some of this spatial variation may have a genetic base, i.e. it may be due to the presence of taxonomically different phytoplankton populations: diatom-dominated communities in the Southern California Bight had ϕ_m values twice as high as the cyanobacterial picoplankton in the deep chlorophyll maximum.[801] Some of it may have a physiological basis, due to populations differing in nutritional status, or recent light exposure history. For phytoplankton photosynthesis in the Sargasso Sea, Cleveland *et al.* (1989) found an inverse relationship between ϕ_m – which varied from 0.033 to 0.102 – and distance from the top of the nitracline. Similar results were obtained by Kolber, Wyman & Falkowski (1990) in the Gulf of Maine, and in both cases the conclusion was reached that quantum yield was related to nitrogen flux. As well as the spatial variation, there is at any given location in the ocean, a diurnal variation in ϕ_m,[502,723,728] often, but not invariably, involving a decrease in the afternoon. In Lake Constance (Germany), Tilzer (1984) found maximum quantum yield at mid-day to vary over the range 0.022–0.092 throughout the year, but on any given day ϕ_m could vary by up to three-fold, the tendency again being for a diminution in the afternoon.

Areal and volumetric efficiencies

We have seen that the efficiency of utilization of the light incident on the aquatic ecosystem for primary production is determined by two main factors: the extent to which the aquatic plants succeed in competing with the other components of the system for the quanta in the underwater light field, and the efficiency with which the absorbed light energy is converted to chemical energy. We shall now consider the overall efficiency which results from the simultaneous operation of these two factors.

The most common and generally useful way of expressing the overall efficiency is as the proportion of the light energy (400–700 nm) incident per m^2 of water surface which is photosynthetically stored as chemical energy in plant biomass throughout the water column. This we shall refer to as the areal efficiency, ϵ_A: it is obtained by dividing the areal (integral) photosynthetic rate expressed in energy units (MJ equivalent of photosynthetic assimilate) per m^2 per unit of time (hour or day), by the total PAR (in MJ) incident per m^2 of water surface in the same time

$$\epsilon_A = 0.472 P_A(CO_2)/E_d(\text{on surface}) \tag{10.26}$$

For oceanic waters, ϵ_A values vary by a factor of 100–300 from the least to the most productive. Koblents-Mishke (1979) reviewing Russian and other work reported that marine ϵ_A values worldwide ranged from 0.02 to 5%. Morel (1978) reported ϵ_A values (calculated on a daily basis) of about 0.02% for the very oligotrophic Sargasso Sea, 0.02–0.07% for the Caribbean Sea (oligotrophic), 0.06–0.25% for the moderately productive eastern equatorial Pacific, and 0.4–1.66% for the eutrophic Mauritanian upwelling area in the eastern tropical Atlantic. Smith *et al.* (1987) studying primary productivity across a coastal front in the Southern California Bight, found ϵ_A to vary from 1.57% on the cold side of the front where phytoplankton concentration was high ($\sim$2.5 mg chl *a* m^{-3}), down to 0.11% on the warm side of the front where phytoplankton levels were much lower (0.1–0.5 mg chl *a* m^{-3}).

Brylinsky (1980) analysed the collected International Biological Programme data for lakes and reservoirs throughout the world. The calculated ϵ_A values for the whole growing season ranged from about 0.002 to 1.0%, most of the values being in the range 0.1 to 1.0%. Talling *et al.* (1973) found high efficiencies in two very productive Ethiopian soda lakes dominated by blue-green algae: ϵ_A based on half-hour incubations at high incident irradiances ranged from 0.5 to 1.6% in Lake Kilotes and from 1.2 to 3.3% in Lake Aranguadi. In the Sea of Galilee (Lake Kinneret), Dubinsky & Berman (1981) found ϵ_A (measured 09.00–12.00 h) to vary from 0.3% in August when the phytoplankton consisted mainly of small chlorophytes, to a maximum of 4% in April during the *Peridinium* (dinoflagellate) bloom.

From literature data for eight lakes covering a wide range of latitude and trophic status, Tilzer, Goldman & De Amezaga (1975), calculated values of ϵ_A ranging from 0.035% in the very oligotrophic, high-altitude Lake Tahoe (California, USA) up to 1.76% in the eutrophic Loch Leven (Scotland). Areal efficiency was strongly correlated with the concentration of algal biomass per unit volume: Tilzer *et al.* considered that the key factor responsible for the variation in ϵ_A in these lakes was the proportion of the total incident light captured by the phytoplankton. Areal efficiency tends to decrease with increasing surface irradiance.[610,728]

A way of expressing the overall efficiency of light utilization which can provide information about its variation with depth through the water column is the volumetric efficiency, ϵ_V. We may define this as the proportion of the downwelling light energy (400–700 nm) incident upon the upper surface of any unit volume within a water body which is photosynthetically stored as chemical energy in plant biomass within that volume. A more all-embracing definition of ϵ_V should take into account the light incident upon the unit volume from all directions, i.e. scalar irradiance, E_0, rather than downward irradiance, E_d. The definition based on E_d is, however, more convenient and serves the purpose well enough: ϵ_V is given by

$$\epsilon_V = \frac{0.472 B_c P(CO_2)}{E_d(z)} \tag{10.27}$$

$P(CO_2)$ being expressed in moles CO_2 mg chl a^{-1} h^{-1}, and $E_d(z)$ in MJ m^{-2} h^{-1}. Unlike ϵ_C and ϵ_A, ϵ_V is not dimensionless, since it has the units m^{-1}. Platt (1969) pointed out that volumetric efficiency at a specific depth has the same dimensions, m^{-1}, as the vertical attenuation coefficient for irradiance, and that it is in fact equivalent to that part of the total vertical attenuation coefficient for downward irradiance which is due to removal of light by photosynthetic conversion to chemical energy

$$K_d(\text{total}) = K_d(\text{photosynthetic}) + K_d(\text{physical})$$

K_d(physical) being that part of the total vertical attenuation coefficient which is due to removal of light by all processes other than photosynthetic conversion to chemical energy: it thus includes that part of the light absorption by phytoplankton which fails to result in photosynthesis.

K_d(photosynthetic) is equivalent to ϵ_V. Platt's data for St Margaret's Bay, Nova Scotia, Canada, indicate a general tendency for ϵ_V to increase with depth (average ϵ_V was 0.07% m^{-1} at 1 m and 0.21% m^{-1} at 10 m). This is to be expected since, as we have already noted, conversion efficiency, ϵ_c, increases with depth, and $\epsilon_V = B_c k_c \epsilon_c$.

Having noted the equivalence of ϵ_V with K_d(photosynthetic), we can now go on to define one more efficiency parameter, namely, the proportion of the total light energy *absorbed* within unit volume of medium which is photosynthetically stored as biomass chemical energy. This is equal to K_d(photosynthetic)/K_d(total) and, following Morel (1978), we shall refer to it as the radiation utilization efficiency, and give it the symbol, ϵ

$$\epsilon = \epsilon_V / K_d \qquad (10.28)$$

From the definitions of ϵ_c and ϵ_V it follows that

$$\epsilon = \frac{B_c k_c}{K_d} . \epsilon_c \qquad (10.29)$$

i.e. ϵ is the (dimensionless) product of the fraction of the total absorbed light which is captured by phytoplankton and the energy conversion efficiency of the phytoplankton. Thus, ϵ combines directly the two factors controlling the efficiency with which the aquatic ecosystem utilizes incident light energy for photosynthesis: ϵ varies with depth and the integral of $\epsilon(z)$ with respect to depth over the whole euphotic zone gives ϵ_A, the areal efficiency.

Given the tendency of the conversion efficiency, ϵ_c, to increase with depth we might expect ϵ also to increase with depth, and this generally seems to be the case.[633] However, variation in ϵ with depth depends also on the variation of k_c with depth resulting from changes in the spectral distribution. From one water body to another ϵ will increase as the phytoplankton biomass increases (B_c in eqn 10.29) but decrease as the background attenuation due to dissolved colour etc. (increasing K_d in eqn 10.29) rises. For the oligotrophic Sargasso Sea and Caribbean Sea, Morel (1978) found ϵ to increase with depth from about 0.01 to 0.1%. In the somewhat more productive waters of the equatorial eastern Pacific the range was from about 0.01% near the surface to approaching 1% at low light levels. In the productive waters of the Mauritanian upwelling, ϵ rose from 0.1–0.4% near the surface to 2–7% near the bottom of the euphotic zone. In Lake Kinneret (Sea of Galilee), Dubinsky & Berman (1981) found that during the *Peridinium* bloom (428 mg chl *a* m^{-2}), ϵ rose from just under 1% at the surface to 6.5% at 3 m. After the *Peridinium* bloom had collapsed, to be replaced with a much lower biomass (50 mg chl *a* m^{-2}, chlorophytes), ϵ was ~0.3% at the surface, and rose to about 2% at 10 m.

10.3 Photosynthesis and wavelength of incident light

The spectral composition of underwater light in a given water body varies markedly with depth and at any specified depth it varies with the optical properties of the water (Chapter 6). Thus, in order to assess the suitability of a given underwater light field for photosynthesis by different kinds of aquatic plant, we need to know in what ways the photosynthetic rates of the various plant types depend upon the wavelengths of the light to which they are exposed. Some of the information we need can be provided by an action spectrum of photosynthesis: this is the curve obtained by plotting the photosynthetic response of a plant per unit incident irradiance (quanta m^{-2} s^{-1}) of incident light at a series of wavelengths across the photosynthetic range. Action spectra are normally measured at light intensities low enough to ensure that the response is proportional to the incident irradiance, i.e. they are measured in the initial, linear, region of the P versus E_d curve. The rate of photosynthesis at any wavelength under these conditions will be equal to the rate of photon absorption times the quantum yield, ϕ. In the case of a macrophyte thallus or leaf, the rate of light absorption at a given wavelength is proportional to the absorptance at that wavelength. Thus the action spectrum will have a shape similar to that of the absorptance spectrum, modified by any variation in quantum yield with wavelength that may occur (see later).

The action spectrum of macrophytes can usually be measured on a single leaf or thallus or piece thereof, and can reasonably be taken as representative of the ability of the original plant *in situ* to make use of light of different wavelengths. In the case of phytoplankton, however, action spectra must for practical reasons be measured on suspensions containing many cells or colonies: sometimes, if reasonable rates are to be achieved, on concentrated suspensions. How does an action spectrum determined on a suspension relate to the action spectrum we wish to know, namely that of the individual cell or colony? The action spectrum of a cell or colony would be identical to the spectral variation in rate of absorption of quanta, modified as usual by any variation in quantum yield. The average rate of absorption of quanta per individual cell or colony, from a light beam, is the product of the irradiance and the mean absorption cross-section ($\overline{s_p A_p}$) of the cells or colonies (§9.4). Thus the action spectrum of a single cell or colony would have the same shape (allowing for variations in ϕ) as the spectral variation in $\overline{s_p A_p}$, and this is fixed, being determined only by the absorption characteristics of the individual cells or colonies.

The action spectrum of the phytoplankton suspension measured in the laboratory will, by contrast, have the same shape (allowing for variations in ϕ), as the absorptance spectrum of the whole suspension. The shape of this spectrum is not fixed, being a function of the number of cells/colonies per unit area of the illuminating beam. Since absorptance can only approach

1.0 and can never exceed it, the absorptance at weakly absorbed wavelengths (e.g. in the green in the case of chlorophytes, diatoms etc.) progressively approaches absorptance at strongly absorbed wavelengths (e.g. at the chlorophyll red peak) as the concentration of cells or colonies increases.

From its definition (§3.2) it follows that the *absorbance* of the suspension, D_{sus}, is equal to $-0.434 \ln (1-A_{sus})$ where A_{sus} is the absorptance of the suspension. From this it in turn follows (since $\ln (1+x) \simeq x$ when $x \ll 1$) that for low values of A_{sus}, $D_{sus} \simeq 0.434\ A_{sus}$. We saw earlier (§9.2) that $D_{sus} = 0.434\ n\overline{s_p A_p}$, and therefore at low values of A_{sus}, $A_{sus} \simeq n\overline{s_p A_p}$, where n is the number of cells or colonies per ml. The true action spectrum of the individual cells or colonies follows (allowing for variations in ϕ) the spectral variation in $\overline{s_p A_p}$ (see above). Therefore, the action spectrum of the suspension will have approximately the same shape as the action spectrum of the individual cells or colonies provided that it is measured on suspensions with low values of absorptance. The discrepancies are about 5, 11 and 19% when A_{sus} is 0.1, 0.2 and 0.3, respectively. The general effect of the error is that the action spectrum of the suspension will be flattened (i.e. peaks lowered relative to the valleys) compared to the true action spectrum of the cells or colonies.

The earliest measurements of photosynthetic action spectra were those of Engelmann (1884). He illuminated pieces of algal tissue with a spectrum and then observed the migration of oxygen-requiring bacteria to the different parts of the tissue. He used the concentration of bacteria around different parts of the tissue as an indication of the relative ability of different colours of light to elicit O_2 evolution. Approximate though this method was, the results did indicate that other pigments in addition to chlorophyll participated in photosynthesis. Nowadays, laboratory determination of action spectra is most commonly carried out using a high-intensity monochromator as a light source, and a platinum electrode to measure the rate of O_2 evolution.[361] Manometry has also been used,[224,225] as has $^{14}CO_2$ fixation.[406,556]

When the action spectrum is measured by determining the rate at a series of different wavelengths, then the curve of photosynthetic rate per unit irradiance obtained is generally found to be somewhat similar, but not identical, to the curve of absorptance for the tissue or cell suspension. Fig. 10.6 shows the action and corresponding absorptance spectra measured on multicellular green, brown and red algae,[361] and on the unicellular green alga *Chlorella*,[225] together with an action spectrum for the marine diatom *Skeletonema*.[406]

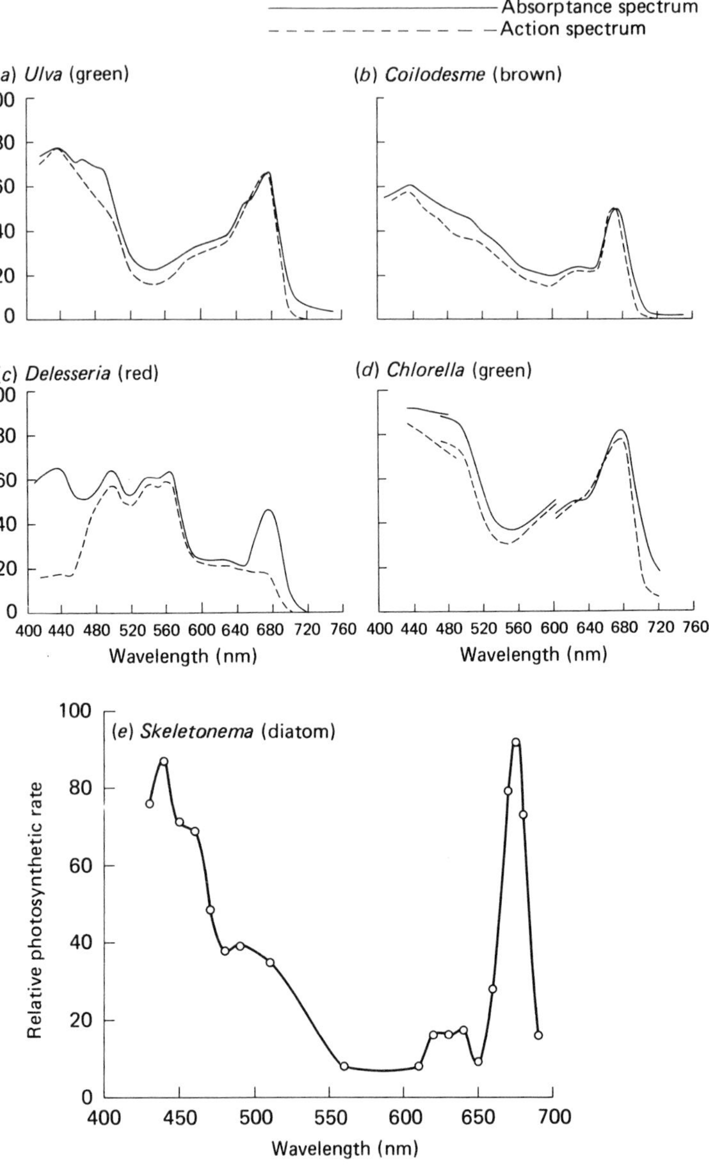

Fig. 10.6. Action spectra of photosynthesis in multicellular and unicellular algae. In (*a*)–(*d*) the action spectrum (photosynthetic rate per unit incident irradiance) has been plotted so that it coincides with the absorptance spectrum at some appropriate wavelength. (*a*), (*b*) and (*c*) – bare platinum electrode method (after Haxo & Blinks, 1950). (*d*) – manometric method (after Emerson & Lewis, 1943); the spectra in this case were obtained with dense suspensions and so do not accurately correspond to the absorption

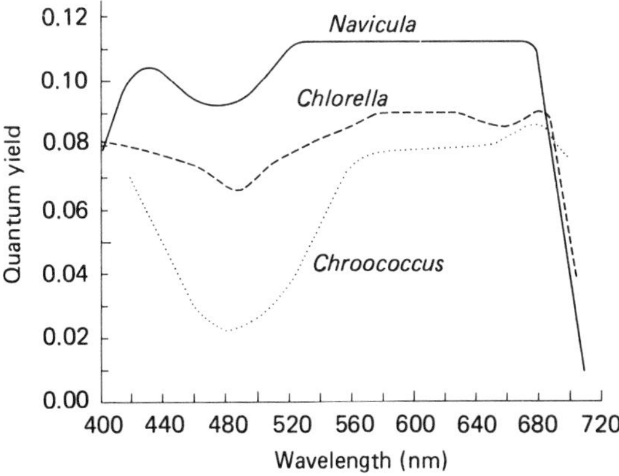

Fig. 10.7. Quantum yield as a function of wavelength in three unicellular algae. (—— *Navicula minima*, diatom (after Tanada, 1951); ----- *Chlorella pyrenoidosa*, green (after Emerson & Lewis, 1943); ...... *Chroococcus* sp., blue-green (after Emerson & Lewis, 1942).)

It will be observed that the action spectrum falls below the absorption spectrum in certain spectral regions. This means that the amount of photosynthesis achieved per unit light absorbed is lower, i.e. the quantum yield is lower, at these wavelengths than at others. Fig. 10.7 shows the quantum yield as a function of wavelength in the blue-green alga *Chroococcus*, the green alga *Chlorella*, and the diatom *Navicula*.[224,225,907] A fall in quantum yield is observed on the long-wavelength side (680–710 nm) of the chlorophyll *a* red absorption peak, and also in the region of carotenoid absorption (440–520 nm), this latter fall being particularly marked in the blue-green alga and significant in the green alga but only slight in the diatom. Such observations led to the belief, initially, that while light absorbed by the carotenoid fucoxanthin in diatoms is efficiently used for photosynthesis, light absorbed by carotenoids in blue-green and green algae is used with much lower efficiency. While there can indeed be some variation in the efficiency with which light absorbed by different pigments is utilized for photosynthesis, it is now thought that this is not the major factor involved in the fall in quantum yield. Furthermore, the explanation

and action spectra of individual cells (see text). (*e*) – $^{14}CO_2$ method (plotted from data of Iverson & Curl, 1973). (*a*) *Ulva taeniata* (multicellular green). (*b*) *Coilodesme californica* (multicellular brown). (*c*) *Delesseria decipiens* (multicellular red, from shaded habitat). (*d*) *Chlorella pyrenoidosa* (unicellular fresh-water chlorophyte). (*e*) *Skeletonema costatum* (marine diatom).

originally suggested for the decrease in ϕ in the far-red – namely, that the quanta are insufficiently energetic to bring about photosynthesis – is also incorrect.

The variation in quantum yield with wavelength is now considered to be primarily due to the fact that there are two light reactions, both of which must be energized for photosynthesis to occur, and which have – to an extent varying with the algal type – different light-harvesting pigment arrays with differing absorption spectra. Thus, during the measurement of an action spectrum, when the wavelength of the monochromatic light provided is strongly absorbed by one photosystem but weakly by the other, then the two light reactions do not operate at the same rate and so photosynthetic efficiency is low. For example, 680–710 nm light is absorbed quite well by photosystem I but poorly by photosystem II. Consequently when cells are illuminated with far-red light, the rate of photosynthesis is low because photosystem II cannot keep up with photosystem I.

Emerson (1958) showed that if *Chlorella* cells illuminated with far-red light were simultaneously exposed to light absorbed by chlorophyll *b* (which transfers most, perhaps all, its energy to photosystem II), then the quantum yield increased above that expected on the basis of the additive contributions of the two kinds of light alone. This phenomenon, of a synergistic effect of two wavebands of light, is referred to as *enhancement* and can plausibly be explained in terms of the existence of the two photosystems. Invoking the photosystems to explain the dependence of quantum yield on wavelength immediately raises problems of its own however. The virtually constant and high value of ϕ between 560 and 680 nm (Fig. 10.7) indicates that both photosystems are working at the same rate throughout this wavelength range, which is rather surprising in view of the differing pigment complements of the two photosystems. The generally accepted explanation of this phenomenon, originally proposed by Myers & Graham (1963), is that when photosystem II is absorbing light faster than photosystem I, some of its excitation energy is transferred to photosystem I, so that both systems can operate at the same rate. This is known as the 'spillover' hypothesis, and there is now good evidence for its validity:[123] transfer of energy from photosystem I to II does not appear to take place. Put simply, light absorbed by photosystem II can be used to drive both light reactions; light absorbed by photosystem I drives only light reaction 1. We may therefore take the fall-off in quantum yield in certain parts of the spectrum as indicating that in those spectral regions absorption is mainly due to photosystem I pigments.

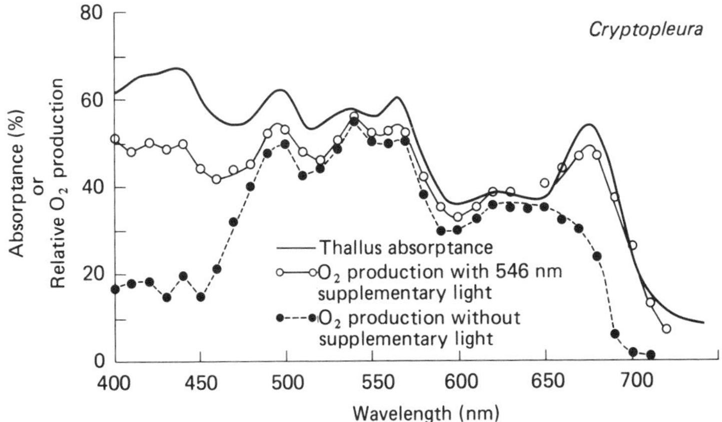

Fig. 10.8. Photosynthetic action spectra of the red alga *Cryptopleura crispa* measured with and without background green (546 nm) light, together with the absorptance spectrum (from Fork, 1963). Reproduced from *Photosynthetic mechanisms in green plants*, National Academy Press, Washington DC, 1963.

It follows from all this that action spectra measured with monochromatic light are going to be biased heavily in their shape towards the absorption spectrum of photosystem II, and may therefore be a poor guide to the ability of a particular plant to make use of the mixture of wavelengths (however enriched this may be in one spectral region or another) that is invariably present in underwater light fields. A more meaningful action spectrum, for ecological purposes, may be obtained by carrying out the measurements with two simultaneously applied light sources; one with wavelength changing progressively to cover the whole photosynthetic range, the other constant at a wavelength absorbed by photosystem II. In this way, even when the variable source is set to spectral regions where absorption is mainly due to photosystem I, the fixed-wavelength background source ensures that photosystem II continues to operate. Fig. 10.8 shows, for a red alga, the difference between action spectra measured[260] in the absence or the presence of background green (546 nm) light, absorbed by phycoerythrin, which is known to supply its energy directly to photosystem II. When photosystem II is energized in this way, there is a great increase in the photosynthetic response to light in the 400–480 nm and 600–700 nm regions. It appears likely that most of the chlorophyll *a* (400–450 nm, 650–700 nm) and carotenoids (400–500 nm) in this alga are part of photosystem I. The discrepancy between action spectra measured with and without background photosystem II light is particularly great in biliprotein-containing algae, because in such algae it seems to be generally true that the biliproteins, which constitute such a major but spectrally confined part of the light-harvesting apparatus, transfer all their energy directly to photosystem II, whereas the chlorophyll *a* directs most of its energy to photosystem I. In algae such as the diatoms, brown algae and green algae, where the pigment arrays of the two photosystems are not so different, the difference between action spectra measured with and without background photosystem II light is not so marked.

It can be observed in Fig. 10.8 that the action spectrum measured with background photosystem II light is fairly close to the absorptance spectrum of the thallus. A similar observation has been made for other red algae.[754] In the absence of action spectra data measured with background photosystem II light, the absorptance spectrum may be used as an approximate guide to the ability of an aquatic plant not only to capture but also to use light of different wavelengths. The discrepancies between the absorptance spectra and the photosystem II-supplemented action spectra of aquatic plants are nevertheless significant and of interest. It has been observed for a number of red algae,[260,754] and can be seen in Fig. 10.8, that the green-supplemented action spectrum falls significantly below the absorptance spectrum in the 400–520 nm region, indicating that some of the pigment molecules absorbing in this region are not transferring their excitation energy efficiently to the reaction centres. The molecules in question are probably carotenoids. It seems unlikely that the carotenoids in the red algae are completely inactive: if they were inactive we would expect a much greater fall in the green-supplemented action spectrum between 450 and 500 nm. The photosystem II-supplemented action spectrum of the blue-green alga *Anacystis* also indicates that carotenoids contribute excitation energy for photosynthesis.[439] The photosynthetically inactive carotenoid is in some cases likely to consist mainly of zeaxanthin, which appears to have the function of protecting the photosynthetic system against excessive light levels.

Thus, in the red and blue-green, and possibly also green and xanthophycean algae[307] the absorption spectrum is likely, because of the inactivity of some of the xanthophyll molecules, partly to overestimate the ability of the cells to utilize light in the blue region of the spectrum. In those algal groups (diatoms, phaeophytes, dinoflagellates, siphonous green algae) which rely heavily on specialized carotenoids (fucoxanthin, peridinin, siphonaxanthin) for light harvesting, however, the major carotenoid species transfer their absorbed energy to the reaction centres as efficiently as does chlorophyll. In these algae we may reasonably assume that the absorption spectrum (absorptance or absorption cross-section, as appropriate) is quite an accurate guide to the ability of the cells to utilize light in all regions of the photosynthetic spectrum, although even in these cases there are minor carotenoids – such as diadinoxanthin in the coccolithophorid, *Emiliania huxleyi*[360] – which are photosynthetically ineffective.

11
Photosynthesis in the aquatic environment

Having considered the photosynthetic response of aquatic plants to light of different intensities and spectral qualities, we shall now examine how the availability of light influences where, when and how much photosynthesis takes place in aquatic ecosystems, and also the extent to which other parameters of the environment can limit photosynthesis. Aquatic production ecology is an enormous field: a comprehensive account will not therefore be attempted. Rather, the broad principles governing the controlling influence of light and other parameters will be outlined and illustrated by examples. More detailed accounts and extensive bibliographies can be found in the books on phytoplankton ecology by Reynolds (1984), Harris (1986) and Fogg & Thake (1987), and the symposium proceedings edited by Platt & Li (1986) and Falkowski & Woodhead (1992). The essay by Fogg (1991) on 'The phytoplanktonic way of life' provides a particularly valuable overview of the multifarious interactions between the phytoplankton and its environment.

11.1 Circulation and depth

We saw in the previous chapter that except under very still conditions with virtually no wind or waves, there is always circulation of water in the upper layer. We also saw that this can be an advantage to the phytoplankton insofar as, by ensuring that they are not exposed to the intense light just below the surface for very long, they avoid photoinhibition. This circulation can, however, also be a disadvantage to the phytoplankton if, in the lower reaches of the mixed layer, the light intensity is too low for net photosynthesis to be achieved. As the depth of the mixed layer through which the phytoplankton is circulating increases, so the average light intensity to which the cells are exposed decreases, and consequently the total rate of photosynthesis by the whole phytoplankton population throughout the water column decreases. The rate of respiration of the whole phytoplankton population, on the other hand, will be essentially constant whatever the mixing depth. Thus, as was first pointed out by Braarud & Klem (1931), there exists a mixed-layer depth – the *critical depth*, z_c – beyond which respiratory carbon loss by the whole population exceeds photosynthetic carbon gain, and so net phytoplankton growth cannot occur. Even when the critical depth is not exceeded, increases in mixing depth tend to reduce total photosynthesis.

The depth through which circulation can occur is limited either by the depth of the bottom or (from spring to autumn in temperate latitudes and throughout the year in tropical oceans) by the presence of a shallow thermocline. Thus, one answer to the question, where does aquatic photosynthesis take place is that it takes place best in shallow waters or in waters in which circulation is confined to a shallow layer by thermal stratification. The converse is that net photosynthesis occurs poorly, or not at all, in waters in which the phytoplankton is circulated through great depths. For a series of Japanese lakes covering a wide range of depths, Sakamoto (1966) found a general tendency for deeper lakes to be less productive than shallow ones. Even when differences in nutrient supply were allowed for, the inhibiting effect of depth remained. Comparing two South African impoundments, of comparable optical and chemical character, Grobbelaar (1989) found the shallow Wuras Dam ($z_{max} = 3.4$ m) to be eight times as productive as the deep Hendrik Verwoerd Dam ($z_{max} > 60$ m). In the marine environment too, we may plausibly attribute the greater productivity of shallow coastal waters (relative to the deeper waters) in part (i.e. in addition to nutrient differences) to their lesser depth. Isolated shallow areas within the oceans such as the Faroebank (100 m depth) west of the Faroe Islands are also more productive than the surrounding oceanic waters.[873] In the turbid waters of the San Francisco Bay estuary, phytoplankton biomass is generally higher in the lateral shallows than in the channel.[154]

The higher productivity of shallow water applies to the benthic flora also. Any surface within the euphotic zone of a water body is usually found to support a productive plant community. This applies not only to the more obvious macrophyte communities such as kelp forests, seagrass beds, brown and red algal associations on underwater rocks, but also to seemingly bare sand and mud surfaces which usually harbour a dense microflora amongst the grains. This microflora is likely to consist of specialized benthic diatoms in temperate mud flats and sand, or of symbiotic algae within Foraminifera in the sands of tropical lagoons.[852] Moving in a seawards direction, as depth increases so the standing crop of biomass (macrophytes and microflora) per unit area, and the rate of primary production per unit area, both decrease until eventually a depth is reached beyond which the light intensity is too low to support any plant growth. For the important group of large sublittoral brown algae, loosely referred to as the kelps, the lower limit occurs at depths where the downward irradiance is 0.7–1.4% of that penetrating the surface: this depth limit ranges from as little as 8 m in the turbid water around the island of Helgoland (North Sea) to about 95 m in the very clear waters of the central Mediterranean.[574] For the coralline crustose red algae – commonly the most deeply occurring algal type – the lower depth limit occurs where the irradiance is 0.01–0.1% of the subsurface value: this corresponds to about 15 m depth off Helgoland but to as much as 175 m in the Caribbean.[574] In the ultra-clear waters east of the Bahamas, Littler *et al.* (1986) using a submersible, have observed a zone of crustose coralline red algae between 189 and 268 m depth, on a seamount. Lüning & Dring (1979) found for the sublittoral region off Helgoland that the total downwelling light (400–700 nm) received *per year* was about 15 MJ m^{-2} or 70 einsteins m^{-2} at the lower kelp limit (8 m) and 1 MJ m^{-2} or 6 einsteins m^{-2} at the lower red-algal limit (15 m): these may be regarded as the approximate minimum yearly requirements of PAR for growth of these algal types. Much higher annual light requirements (einsteins m^{-2} yr^{-1}) have been estimated for some freshwater macrophytes by Sand-Jensen & Madsen (1991): 40–200 for charophytes (*Nitella* and *Chara*), 416 for a bryophyte (*Fontinalis antipyretica*), 455 for an isoetid angiosperm (*Isoetes lacustris*) and 1760 for the shallow-water angiosperm *Littorella uniflora*.

The depth distribution of aquatic macrophytes is controlled by a number of environmental factors, but the underwater light climate is frequently the primary one.[854] An important ecological parameter for any water body is z_{col}, the maximum depth to which it is colonized by macrophytes, and this tends to be approximately proportional to the reciprocal of the vertical attenuation coefficient for PAR in the water body, although the constant of proportionality appears to be a function of latitude.[952] For nine New Zealand (North Island) lakes at ~38° S, Vant *et al.* (1986) found macrophyte depth limits to vary with the attenuating character of the water in accordance with $z_{col} = 4.34/\bar{K}_d$, where $\bar{K}_d$ was the vertical attenuation coefficient of the lake for PAR averaged over a year. For lakes in Scotland and the English Lake District (54–57° N), Spence (1976) found the approximate relationship $z_{col} = 1.7/K_d(\text{PAR})$. For seagrasses Duarte

(1991), from a survey of literature data, arrived at the relationship $z_{col} = 1.86/K_d$.

It is instructive to consider under what conditions phytoplankton primary production may be prevented altogether as a result of the mixed depth's exceeding the critical depth. Using Talling's (1957*b*) model for calculating integral photosynthesis (see §11.5) we can derive an approximate expression for the critical depth

$$z_c = \frac{N}{24\, K_d \rho} \ln \left(\frac{\bar{E}_d(0)}{0.5\, E_k} \right) \tag{11.1}$$

where ρ is the ratio of respiration rate to light-saturated photosynthetic rate in the phytoplankton, K_d is the vertical attenuation coefficient for downwards irradiance of PAR, N is daylength in hours (readily calculated from eqn 2.11), $\bar{E}_d(0)$ is the average value of downward irradiance just below the surface during daylight hours, and E_k is the irradiance value defining the onset of saturation (§10.1). We can see from eqn 11.1 that the likelihood of the mixed layer depth exceeding the critical depth increases (i.e. z_c decreases) as respiration rate (relative to photosynthetic rate), attenuation by the water, or the light-saturation parameter, increase: an increase in incident irradiance or in daylength has the opposite effect.

As an example of a temperate-zone lake with relatively low attenuation we shall take Lake Windermere, England, and use the data of Talling (1957*a*). For a day in late spring (1st May, 1953), with $\bar{E}_d(0) \simeq 170$ W m^{-2}, daylength $\simeq 15$ h, $K_d \simeq 0.43$ m^{-1}, and the phytoplankton (predominantly the diatom, *Asterionella*) having $E_k = 6.3$ W m^{-2}, and $\rho = 0.033$, we may calculate that the critical depth was about 177 m. Since the lake itself (North basin) has a maximum depth of only about 63 m, there is clearly no possibility of net primary production being prevented altogether. Even if the lake were much deeper, once the spring–summer thermal stratification had set in, the comparatively shallow depth of the epilimnion (the layer, commonly 5–20 m deep, above the thermocline, within which wind-induced circulation occurs) would prevent the critical depth's being exceeded. For Lake Windermere in midwinter, on the other hand, with daily incident light down to a small fraction of the summer value and circulation through the full depth of the lake, Talling (1971) estimated that phytoplankton growth was impossible. In the much shallower Esthwaite Water ($z_{max} \simeq 15$ m), in contrast, he concluded that phytoplankton growth should be possible throughout the year.

Even in shallow water bodies, if the attenuation is high enough the circulation depth can exceed the critical depth. For turbid Lough

Neagh (N. Ireland), with a mean depth of 8.6 m and K_d values commonly 1.5–3.5 m^{-1}, Jewson (1976) estimated that the critical depth would be exceeded in winter, thus precluding growth. In spring, despite the lack of stratification in this well-mixed lake, the increase in illumination was sufficient to ensure that z_c was not exceeded.

When high attenuation is combined with a requirement for high light intensity for saturation, then the critical depth can be very shallow. In Lake George, Uganda, on the equator, Ganf (1975) observed K_d values in the region of 9 m^{-1}, and the blue-green algal population had a light-saturation onset parameter (E_k) of about 55 W m^{-2}, on a day (12th April, 1968) when the average incident irradiance ($\bar{E}_d(0)$) was 360 W m^{-2}. Assuming ρ to be about 0.1 and the daylength to be 12 h, eqn 11.1 gives a critical depth of 1.4 m. Thus, despite the high irradiance incident on this tropical lake, the mixed depth could often exceed the critical depth. In productive lakes of this type a large part of the attenuation is due to the phytoplankton itself and so the situation is self-correcting. Cessation of growth is likely to be followed by breakdown of the algae, following which attenuation falls and photosynthesis and growth can recommence.

In the marine environment, the relation between the mixed depth and the critical depth is of particular importance, and indeed the first attempt to characterize this relation quantitatively was carried out by Sverdrup (1953) for the Norwegian Sea. In middle and high latitudes the mixed layer is deep at the end of the winter (varying from 100–400 m in March at 66° N 2° E), but in the spring a shallow mixed layer (25–100 m) develops due to thermal stratification. For the 66° N station that Sverdrup studied, he estimated that until the last week of April the mixed layer was much deeper than the critical depth, thus precluding phytoplankton growth, but that after the middle of May the mixed-layer depth was smaller than the critical depth so that growth was possible. In locations where shallow coastal waters are permanently well mixed by tidal action, such as the coast of Brittany (France) in the western English Channel, a stratification-induced spring bloom does not occur, and the phytoplankton population reaches its maximum only in the summer, when the light intensity is such that the euphotic depth becomes comparable to the water depth.[1001]

Density stratification brought about by vertical variation in salt concentration, rather than by solar heating, can also induce a phytoplankton bloom. This has been observed in the case of lower-salinity upper layers resulting from freshwater inflow in estuarine and coastal waters,[153,693,933] and from melting ice in the marginal ice zones around Antarctica,[616,846] or from intrusion of dense continental slope water beneath less-saline coastal water.[933]

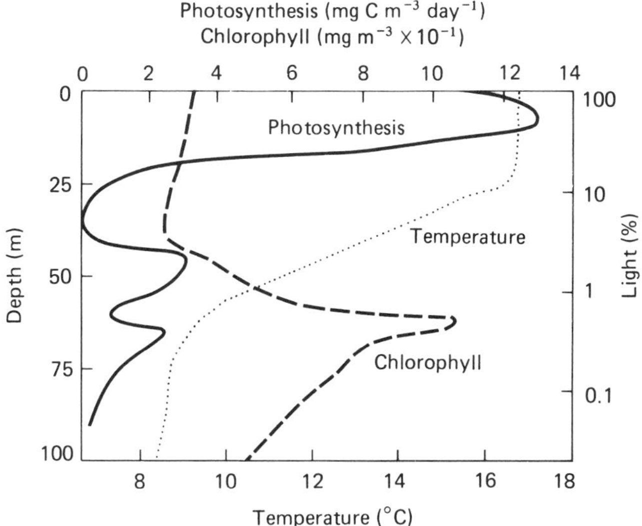

Fig. 11.1. The deep chlorophyll maximum in the Northeast Pacific Ocean (after Anderson, 1969). The temperature profile shows that the mixed layer is about 25 m deep.

Strictly speaking, these various interactions between mixed depth and critical depth apply to non-motile forms such as diatoms. Motile algae such as dinoflagellates can to some degree escape these effects by migrating again up to regions of higher irradiance if the circulating water moves them down. Nevertheless the evidence is clear that for the total phytoplankton population, circulation of water within the mixed layer is a major factor influencing primary production, and frequently determines whether such production takes place at all.

In the sea, the mixed layer, once thermal stratification has been established (spring to autumn in moderate and high latitudes, all year in the tropics), is typically 20–100 m deep. Within and below the thermocline there is comparatively little circulation of the water. Since the euphotic zone will frequently be deeper than the mixed layer (for marine waters with K_d values of 0.03, 0.05, 0.11, 0.16 m^{-1} (Table 6.1), $z_{eu} \simeq 153$, 92, 42 and 29 m, respectively), there is, throughout much of the oceans and coastal seas, a substantial layer of water which combines sufficient light intensity to support photosynthesis, with a stable water column.

When the vertical profile of phytoplankton chlorophyll is measured in the (stratified) ocean, a distinct peak of chlorophyll concentration is normally found close to the bottom of the (nominal) euphotic zone. An example of this *deep chlorophyll maximum* as it is often called, in the northeast Pacific Ocean is shown in Fig. 11.1. The peak of the chlorophyll distribution is either a little above or a little below (as in Fig. 11.1) the depth at which E_d is 1% of the subsurface value. It will be noted that there is a layer of increased photosynthetic activity in this region indicating that the algal cells are active, not moribund. This layer of deep phytoplankton, commonly about 10 m thick, is very widespread in the world's oceans: it appears, for example, to extend right across the Pacific.[816] It occurs only where the water column is stabilized by a pycnocline (density gradient).[15,953] It seems likely that the increase in chlorophyll concentration in this deep layer, relative to that in the mixed layer, is due to an increase in chlorophyll concentration within the cells as well as to an increased population of phytoplankton.[15,167,468,681,] Kiefer, Olson & Holm-Hansen (1976) found the chlorophyll per unit biomass in the deep phytoplankton layer to be about twice the value near the surface.

The mechanism (apart from shade adaptation) by which the deep chlorophyll maximum becomes established is uncertain. A commonly held view is that it is related to the distribution of nutrients. In the well-illuminated mixed layer the phytoplankton rapidly consumes the nutrients in the spring, and the population then decreases. Below the thermocline nutrient concentrations are much higher. Furthermore, additional nutrients can diffuse upwards from the deep water below. The continued availability of nutrients in this region, possibly combined with a diminution in the sinking rate of the phytoplankton once it reaches these nutrient-rich waters, may account for the establishment and maintenance of this phytoplankton layer.[15,407,871] The deep phytoplankton layer makes a significant contribution to total primary production in those waters where it occurs.

In most inland waters the euphotic depth is less than the mixed depth, and so the conditions for the development of a deep chlorophyll maximum do not occur. However, Fee (1976) found a very narrow layer with very high phytoplankton chlorophyll concentration in the region of, or below, the thermocline in a series of lakes in northwestern Ontario (Canada), which were both clear and thermally stratified. This layer was found between 4 and 10 m depth (depending on the lake) where the irradiance was 0.3–3.5% of the subsurface value. Phytoplankton samples from this layer were always dominated by large colonial chrysophycean flagellates of the genera *Dinobryon*, *Synura*, *Uroglena* and *Chrysosphaerella*, just one species normally being dominant in the deep chlorophyll layer of any given lake. A curious and interesting feature of these chrysophycean phytoplankton layers is that, as well as photosynthesizing, they derive a substantial proportion of their carbon by phagotrophic ingestion of bacteria.[69] In the very clear, colourless waters of Crater Lake and Lake Tahoe, comparable optically to ocean water, a deep chlorophyll maximum develops at considerable depth (~75 m) in the summer.[539,924] In Lake Michigan a deep chlorophyll layer develops below the thermocline, after seasonal stratification sets in. This appears to be due to *in situ* growth but also, to a lesser extent, to sedimentation and shade adaptation.[237]

11.2 Optical characteristics of the water

We saw in the previous chapter that a major factor in limiting efficiency of utilization of incident light in aquatic ecosystems is the removal of a large proportion of the light energy by the aquatic medium. This occurs in waters in which the vertical attenuation of PAR by non-phytoplanktonic material is high (see Table 10.1). We would therefore expect, for example, brown-water lakes with a high concentration of gilvin to be on the average less productive than lakes with low background colour. Observation suggests that this is indeed the case although no wide-ranging survey appears yet to have been carried out. However, with the ready availability nowadays of spectrophotometric techniques for determining the absorption properties of natural waters, it is to be hoped that the relation between productivity and background water colour will be studied quantitatively for a variety of water-body types. It should be noted that a high total vertical attenuation coefficient need not necessarily be associated with low productivity since that high coefficient may be due to a high concentration of phytoplankton: it is high attenuation by non-algal material that we expect to lower the productivity. So far as seagrasses and fresh-water macrophytes are concerned, we have already seen (§11.1) that the depth of colonization is inversely proportional to the vertical attenuation coefficient for PAR, indicating that the more highly attenuating the water, the lower the macrophyte productivity of the water body.

Estuaries are often highly attenuating water bodies, due to the presence of high concentrations of suspended sediment particles, which both absorb and scatter light, and to dissolved humic colour in the river inflow. Turbidity, and consequently attenuation, varies longitudinally down the estuary, rising to a peak at the so-called 'turbidity maximum', the position of which varies with the tide, and then diminishing seawards. Phytoplankton productivity within estuaries is found to vary inversely with attenuation and turbidity.[152,156,344,436,586,693]

Cole & Cloern (1984, 1987) have found, for a number of estuaries that phytoplankton daily productivity can be expressed as a linear function of

$B_c E_d(0,+)/K_d(\text{PAR})$ (or, equivalently, of $B_c E_d[0,+]z_{eu}$). To interpret this, we note that in optically deep waters all the daily surface-incident light, $E_d(0,+)$ (apart from the small fraction which is reflected at, or backscattered through, the surface), is absorbed in the water column, and also that the proportion of this light which is absorbed by the phytoplankton is a_p/a_T (see §9.4). Given that a_p (the phytoplankton absorption coefficient) is linearly related to the phytoplankton concentration, B_c, and that $K_d(\text{PAR})$ is very approximately proportional to a_T (where this is understood as a depth- and spectrally-averaged total absorption coefficient for the estuarine water), we can see that $B_c E_d(0,+)/K_d(\text{PAR})$ is an approximate measure of the daily absorption of light by the phytoplankton population.

Turbid waters in which the turbidity is due to large numbers of mineral particles which are themselves of low intrinsic colour, and in which dissolved colour is also low, represent a special case. Because of the intense scattering, the total vertical attenuation coefficient ($K_d(\text{PAR})$) may be quite high so that the euphotic zone is shallow. This fact alone does limit production by reducing the volume of medium which the phytoplankton can exploit photosynthetically. Nevertheless, within the euphotic zone the relative ability of the phytoplankton to collect light is dependent on its absorption properties compared to the other constituents of the medium (a_p/a_T), and since in the present case the rest of the medium does not absorb the light strongly despite its intense scattering, the phytoplankton is well placed to compete for the available photons. In a medium of this type, the assumption, contained in eqn 10.5, that the relative amounts of the absorbed light captured by phytoplankton and by the rest of the medium are in proportion to $B_c k_c$ and K_{NP}, is no longer even approximately valid. We saw in §10.2 that this assumption can in fact only be made with reasonable accuracy for waters in which vertical attenuation is absorption dominated.

The kind of mechanism we have envisaged, namely effective competition by phytoplankton for the photons within the shallow euphotic zone, may in part account for the surprisingly high productivity of some waters with high inorganic turbidity.[670] The shallowness of such waters (the turbidity often being due to wind resuspension of bottom sediments) will further promote their productivity by ensuring that the circulating cells do not for long remain below the illuminated layer. In addition, the ratio of scalar to downward irradiance is high in turbid waters with a high ratio of scattering to absorption (see Fig. 6.10), and so there is substantially more light available for photosynthesis than the values of downward irradiance alone would indicate.

11.3 Other limiting factors

To fully understand the extent to which light availability limits primary production it is necessary to be aware of the limitations which are imposed at the same time by other environmental factors. Consider, for example, a diatom population photosynthesizing in lake water at a saturating light intensity of ~400 μeinsteins m^{-2} s^{-1}, as illustrated in Fig. 10.1. A small increase in light intensity will bring about no change in photosynthetic rate and a large increase will even lead to photoinhibition. What is happening within the cells is that the rate of formation of the products of the light reactions of photosynthesis, namely $NADPH_2$ and ATP, is so high that the enzyme system responsible for the dark reactions is saturated so far as $NADPH_2$ and ATP are concerned. Any increase in their steady-state concentration resulting from an increase in light intensity is therefore not accompanied by an increase in the rate of CO_2 fixation. This does not mean, however, that the system is necessarily working at its maximum possible rate. The CO_2 concentration may be too low to saturate the first enzyme, ribulose bisphosphate carboxylase (Rubisco), in the dark reaction sequence so that it is not functioning at its maximum capacity. Alternatively, or in addition, the temperature may be too low for maximum activity: at a higher temperature the enzymes of the whole dark reaction system might operate more rapidly and thus be able to consume the $NADPH_2$ and ATP at a faster rate.

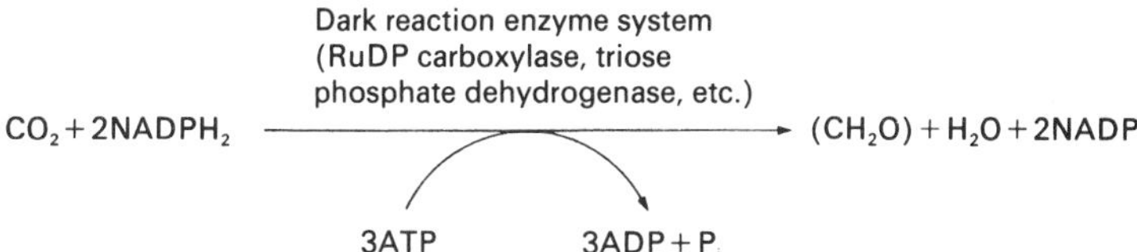

Carbon dioxide

We shall first consider the extent to which CO_2 availability limits the overall rate. Given that CO_2 is a substrate which is used by an enzyme (system), then we may plausibly suppose that the photosynthetic rate at any given light intensity will vary with CO_2 concentration approximately in accordance with the well-known Michaelis–Menten equation for enzyme kinetics

$$v = \frac{Vs}{K_m + s}$$

where v is the rate of enzyme reaction at substrate concentration s, V is the maximum rate obtainable at saturating substrate concentrations, and K_m is

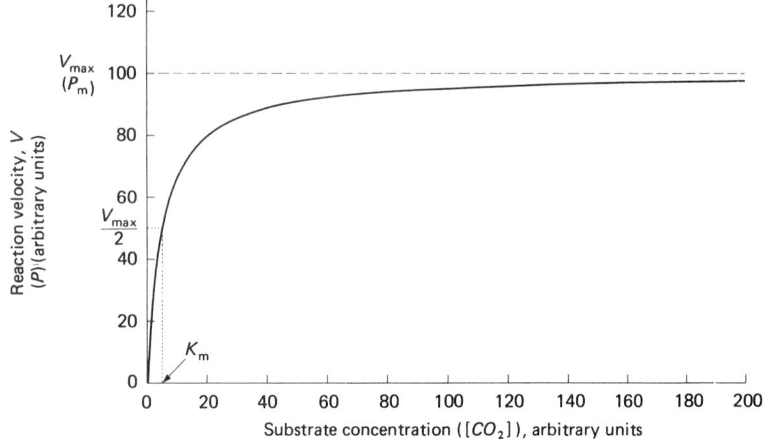

Fig. 11.2. Idealized curve of rate of an enzymic reaction as a function of substrate concentration. Calculated from Michaelis–Menten equation.

the dissociation constant of the enzyme–substrate complex, but is also equivalent to the substrate concentration which gives half the maximum rate. Re-expressing the equation in photosynthetic terms we obtain

$$P = \frac{P_m[CO_2]}{K_m(CO_2) + [CO_2]} \tag{11.2}$$

Fig. 11.2 shows an idealized curve of rate versus substrate concentration for an enzymic reaction, in accordance with the Michaelis–Menten equation. Observed curves of photosynthetic rate versus CO_2 concentration for aquatic plants are approximately of this type, and so by obtaining the K_m values from such curves and comparing them with the *in situ* CO_2 concentrations we may be able to assess to what extent CO_2 availability limits photosynthesis in natural waters.

In the isolated state, Rubisco from aquatic plants has a K_m value for CO_2 of 30–70 μM.[1011] The $K_m(CO_2)$ values for living plants, however, can be higher because of diffusive resistance to entry of carbon dioxide into the plant, or lower due to active uptake of CO_2. Some values for $K_m(CO_2)$ determined for phytoplankton and macrophyte species, at pH values low enough (pH < 6) to ensure that essentially all the inorganic carbon exists as CO_2 (or its hydrated form, H_2CO_3), are listed in Table 11.1: they range from 4 to 185 μM. Fresh water in equilibrium with the atmosphere (~0.034 vol% CO_2) at 15 °C contains dissolved CO_2 at about 14 μM concentration. Calculations for sea water at 26 °C indicate a free CO_2 concentration of about 12 μM.[84] We might therefore expect it commonly to be the case that

Table 11.1. *Apparent half-rate constants* (K_m) *for* CO_2 *for photosynthesis in certain phytoplankton and macrophyte species*

Plant species	$K_m(CO_2)$ (μM)	Reference
Phytoplankton		
Pediastrum boryanum (Chlorococcales)	40	Allen & Spence (1981)
Cosmarium botrytis (desmid)	170	Allen & Spence (1981)
Anabaena cylindrica	60	Allen & Spence (1981)
Macrophytes		
Nitella flexilis	100	Allen & Spence (1981)
Eurhynchium rusciforme	80	Allen & Spence (1981)
Fontinalis antipyretica	170	Allen & Spence (1981)
Elodea canadensis	30	Allen & Spence (1981)
Potamogeton crispus	20	Allen & Spence (1981)
Hydrilla verticillata	170	Van, Haller & Bowes (1976)
Myriophyllum spicatum	150	Van, Haller & Bowes (1976)
Ceratophyllum demersum	165	Van, Haller & Bowes (1976)
Marine macrophytes		
Ulva sp.	30	Beer & Eshel (1983)
Ulva lactuca	185	Drechsler & Beer (1991)
Marine phytoplankton		
Stichococcus bacillaris (Chlorococcales)	4	Muñoz & Merrett (1988)

aquatic plant photosynthesis in natural waters is undersaturated with respect to CO_2 concentration, and should respond to an increase in CO_2 concentration with an increase in photosynthetic rate. This need not apply only at saturating light intensity: the assumption that photosynthetic rate depends on CO_2 concentration in accordance with the Michaelis–Menten relation implies that even at low light intensities an increase in CO_2 concentration above a typical starting value of 12–14 μM would lead to an increase in photosynthetic rate. This means that at subsaturating light intensities, photosynthesis by plants within a water body is likely to be simultaneously limited by availability of both light and CO_2. The aquatic moss, *Fontinalis antipyretica*, and the aquatic higher plant, *Cabomba caroliniana*, both show marked increase in photosynthetic rate with increasing CO_2 concentration in the range 10–25 μM at low, subsaturating, as well as at high, light intensities.[340,824] Fig. 11.3 shows Smith's (1938) data for *Cabomba*.

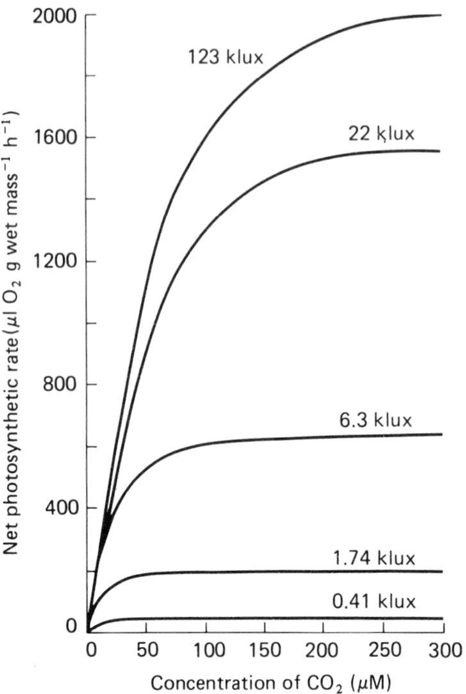

Fig. 11.3. Photosynthetic rate as a function of CO_2 concentration at different light intensities in the aquatic higher plant *Cabomba caroliniana* (after Rabinowitch, 1951, based on data of Smith, 1938).

Sea water and most inland waters contain much more inorganic carbon in the form of bicarbonate ion, HCO_3^-, than in the form of CO_2. The different forms of inorganic carbon are interconverted in accordance with

$$H_2O + CO_2 \rightleftharpoons H_2CO_3 \rightleftharpoons H^+ + HCO_3^- \rightleftharpoons 2H^+ + CO_3^{2-}$$

The higher the pH of the water, the more this equilibrium shifts to the right with HCO_3^- constituting more than 50% of the total from pH 6.2 to 9.3, and more than 80% between pH 6.7 and 8.8. Given that CO_2 tends, as we have seen, to be present at suboptimal concentration, it would clearly be an advantage in many waters for aquatic plants to be able to use bicarbonate as a carbon source for photosynthesis, and in fact many, but not all, species do. This topic has been reviewed by Raven *et al.* (1985), Prins & Elzenga (1989) and Madsen & Sand-Jensen (1991). The inorganic carbon species used by the carboxylase enzyme is always CO_2. In the plants which can utilize bicarbonate, the HCO_3^- ions are transported into the cell where they give rise to CO_2 by the reversal of the first two of the above interconversion reactions. The reversible dehydration of H_2CO_3 to give CO_2 is catalyzed by carbonic anhydrase, a chloroplast enzyme. The liberated CO_2 is then used in photosynthesis. The combination of HCO_3^- ions with H^+ to give H_2CO_3 leads to a corresponding accumulation of OH^- ions (from $H_2O \rightarrow H^+ + OH^-$) which are excreted from the cells to balance the uptake of HCO_3^-.

So far as marine ecosystems are concerned, photosynthetic utilization of bicarbonate has been found in most of the seaweed species – green, brown and red – which have been examined, but is not universally present. Maberley (1990) found that out of 35 species of marine macroalgae, six species – all rhodophytes – were unable to use HCO_3^-, and five of these occurred in relatively low-light habitats, beneath a canopy of larger Phaeophyta. He suggested that most species growing at depths where light is low will be unable to use HCO_3^-. The red macroalga *Chondrus crispus*, although it does not take up bicarbonate directly, nevertheless gains access to this plentiful inorganic carbon pool by dehydrating HCO_3^- to CO_2 with an external carbonic anhydrase.[842,843] Bicarbonate utilization is found in some,[53,584,732] but apparently not all,[1] seagrasses. Even in those macrophytes, such as *Ulva*, which can utilize bicarbonate it seems that photosynthesis in bright light can in some cases be limited by the level of inorganic carbon in sea water.[547] Amongst the marine phytoplankton, HCO_3^- utilization has been shown in the coccolithophorid *Coccolithus huxleyi*,[682] in the diatom *Phaeodactylum tricornutum*,[191] the unicellular red alga *Porphyridium cruentum*,[159] and in the cyanobacterium *Synechococcus*,[33] but the green plankter *Stichococcus bacillaris* has a high affinity for CO_2 (Table 11.1) and a low affinity for bicarbonate,[649] and a number of common marine diatoms appear to be unable to use bicarbonate.[184,753]

In fresh waters, according to Hutchinson's (1975) review, most aquatic higher plants can utilize bicarbonate for photosynthesis. The few species which cannot utilize bicarbonate typically occur in soft waters in which bicarbonate concentrations are low. Aquatic mosses mainly seem to lack the ability to utilize HCO_3^-. Most charophyte species (other than those from soft waters) can utilize HCO_3^- as can benthic filamentous algae such as *Cladophora*. Amongst the fresh-water phytoplankton, the ability to utilize HCO_3^- for photosynthesis varies widely. *Chlorella emersonii*, *Scenedesmus quadricauda*, *Chlamydomonas reinhardtii* (green), *Ceratium hirundinella* (dinoflagellate), *Fragilaria crotonensis* (diatom), *Microcystis aeruginosa* and *Anabaena cylindrica* (blue-green) can all effectively use HCO_3^-, but *Chlorella pyrenoidosa* (green), *Asterionella formosa* and *Melosira italica* (diatoms) cannot.[11,577,745,902,935]

Despite the wide occurrence of ability to utilize bicarbonate in the aquatic plant kingdom, it is generally true that, with the possible exception of phytoplankton species adapted to alkaline waters, free CO_2 if available is the preferred, i.e. more effectively used, carbon source. For example, the fresh-water macrophyte *Myriophyllum spicatum*, which can use bicarbonate, gives a much higher photosynthetic rate with free CO_2 at its optimum concentration than with HCO_3^- at its optimum concentration.[874] Steemann Nielsen (1975) suggests that one of the reasons for the less effective use of HCO_3^- is that energy must be used for its active transport into the plant, whereas CO_2 diffuses in freely. The data of Allen & Spence (1981) for fresh-water macrophytes in Table 11.1 show that in all cases, including the known bicarbonate users *Elodea canadensis* and *Potamogeton crispus*, the apparent K_m for HCO_3^- is 50- to 100-fold higher (indicating a much lower affinity) than the K_m for CO_2. Furthermore, to give the same photosynthetic rate as a certain concentration of CO_2 (at pH 5.5), HCO_3^- (pH 8.8) at 52 to 132 times the concentration was needed. In the case of the blue-green alga *Anabaena cylindrica*, on the other hand occurring typically in alkaline eutrophic waters, the K_m values for CO_2 and HCO_3^- are about the same, and the concentrations of CO_2 and HCO_3^- which produce a certain photosynthetic rate are also about the same.

On the basis of their studies, Allen & Spence conclude that despite the ability of most fresh-water macrophytes to utilize bicarbonate, they do not in fact obtain much of their carbon from HCO_3^- until the pH of the water exceeds 9.0, and at these high pH values their photosynthetic rates are greatly reduced anyway. Allen & Spence suggest that the natural rates of photosynthesis of fresh-water macrophytes and some planktonic algae are functions mainly of CO_2 (as opposed to HCO_3^-) concentration. This conclusion may not, however, apply to the more effective bicarbonate users amongst the macrophytes, such as *Myriophyllum spicatum*. Adams, Guilizzoni & Adams (1978) found that in a series of rather alkaline (pH values mainly 7.5–8.8) Italian lakes, all well supplied with phosphorus, the rate of photosynthesis of *M. spicatum* varied with the total dissolved inorganic carbon concentration in accordance with a Michaelis–Menten type of relation (Fig. 11.4). From calculations of the amounts of the different forms of inorganic carbon present, it appeared that photosynthetic rate was related primarily to the concentration of HCO_3^- rather than to the concentration of CO_2.

Algae such as *Anabaena cylindrica*, which photosynthesize optimally at pH values in the region of 9.0, make very efficient use of bicarbonate. For these planktonic species and the more effective bicarbonate users amongst

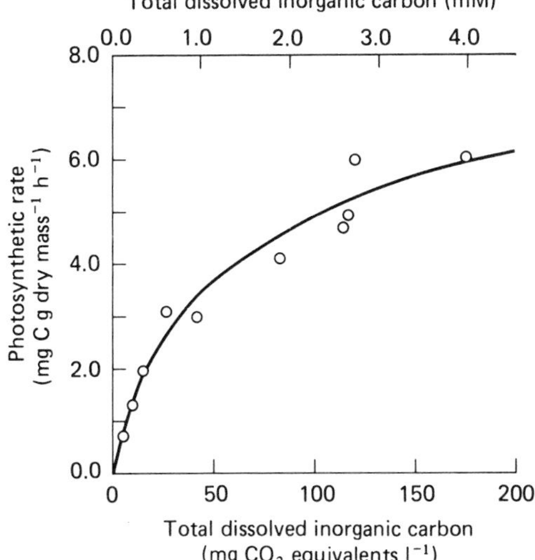

Fig. 11.4. Variation of photosynthetic rate of *Myriophyllum spicatum* with total dissolved inorganic carbon concentration in a series of Italian lakes (plotted from data of Adams *et al.*, 1978). Light intensities were at or near saturation in most cases. Using non-linear regression to the Michaelis–Menten equation, Adams *et al.* calculated a K_m (half-saturation constant) value of 1.06 mM total dissolved inorganic carbon (46.5 mg CO_2 equivalents l^{-1}) and a P_m value of 7.24 mg C g^{-1} dry mass h^{-1}.

the macrophytes, the natural rate of photosynthesis (as far as carbon supply is concerned) can be regarded as a function of the concentration of HCO_3^- plus CO_2. The ability of blue-green algae such as *Anabaena*, *Microcystis* and *Spirulina* to continue photosynthesizing in water of high pH, and effectively zero free CO_2 concentration could be one of the main factors contributing to their frequent domination of eutrophic lakes in late summer and their invariable domination of highly alkaline waters such as the African soda lakes.

Rattray, Howard-Williams & Brown (1991) found that two rooted freshwater macrophytes, *Lagarosiphon major* and *Myriophyllum triphyllum*, grew and photosynthesized twice as fast in the water of oligotrophic Lake Taupo (New Zealand) as in the nitrogen- and phosphorus-rich water of eutrophic Lake Rotorua. Rattray *et al.* attributed this to the two-fold higher level of CO_2 found to be present in the oligotrophic, than in the eutrophic, water. Soft-water oligotrophic lakes are often characterized by the presence of benthic macrophytes with the isoetid growth form: small plants with short stems and rosettes of leaves.[399] Isoetids have solved the problem presented by the low inorganic carbon concentration in these poorly mineralized waters by extracting the CO_2 from the sediments, where it is formed from decaying organic matter, through their roots.[86,584,746] This CO_2 diffuses from the roots to the leaves via longitudinal air-filled channels, and in some cases provides virtually all the photosynthetically fixed carbon.

The free CO_2 concentrations of 12–14 μM referred to earlier apply only to waters in equilibrium with the atmosphere. Such concentrations might normally be found in the surface layer of the oceans, or of unproductive inland waters, or of most inland waters in the winter. However, any biologically productive water body is likely not to be in equilibrium with the atmosphere so far as CO_2 is concerned. The free CO_2 concentration at any time is likely to differ from the equilibrium value and will vary with time and depth in accordance with where and how fast the processes of consumption (photosynthesis) and production (respiration, decomposition) are taking place. On the basis of model calculations of rates of CO_2 diffusion to the cell, and of laboratory measurements of growth rates as a function of CO_2 concentration, Riebesell, Wolf-Gladrow & Smetacek (1993) concluded that during phytoplankton blooms in the ocean, when CO_2 concentration in the surface layer falls drastically, the growth rate of marine diatoms can be limited by the CO_2 supply to the cell surface.

Free CO_2 can be even more affected by changes in pH caused by photosynthesis. Talling (1976, 1979) presents data for the changes in free CO_2 concentration in the eutrophic English lake Esthwaite Water from April to July, 1971. On April 19th, before thermal stratification, the CO_2 concentration was the same from the surface down to 10 m depth. By May 3rd when the phytoplankton population had begun to increase, and there was some warming of the surface layer, CO_2 concentration had fallen by about 33% in the upper 4 m but had increased greatly between 6 and 10 m. During July, thermal stratification became established, the phytoplankton population above the thermocline increased substantially and free CO_2 levels in the epilimnion were reduced virtually to zero, while rising to even higher values than before in the 6–10 m layer below the thermocline. The amount of inorganic carbon removed by photosynthesis represented about 50% of the total inorganic carbon in the surface layer. The remainder of the *c.* 1000-fold decrease in free CO_2 was due to the rise in pH from ~7.0 to above 9.0, resulting from the photosynthetic activity of the dense phytoplankton population. Similar changes are likely to take place in any productive water body. For one of the components of the phytoplankton, the diatom *Asterionella formosa* (not a bicarbonate user), Talling (1979) calculated that on July 12th its photosynthetic rate was virtually zero in the surface layer due to lack of CO_2, that it rose between 2 and 3 m due to an increase in CO_2 accompanied by reasonably high irradiance values, and that it declined again below this depth because of the fall in irradiance. Thus with increasing depth, its photosynthesis was limited first by a CO_2 supply, then by CO_2 and light simultaneously, and then mainly by light availability alone. These changes with depth did not apply to total phytoplankton photosynthesis, however, since when the pH rose and the CO_2 concentration fell, *Asterionella* was largely replaced by the bicarbonate users *Ceratium* and *Microcystis*.

The sinking rate of diatom cells increases when CO_2 uptake becomes limited: the CO_2 depletion associated with summer stratification might lead to an increased sinking rate of diatoms, and this may contribute to the paucity of diatoms frequently observed in productive lakes in the summer.[410]

Immediately next to the surface of any aquatic plant, there is an unstirred layer of water that CO_2 and/or HCO_3^- ions must diffuse across before they can enter the cells and be used for photosynthesis. The thickness of this layer diminishes if the plant is subjected to turbulence or rapid stirring but the layer never disappears completely. Under well-stirred conditions, the unstirred layer might be only about 5 μm thick around a small cell such as *Chlorella* but 30–150 μm thick around the surface of a macrophyte such as *Chara*:[745,825] in still or slowly moving water the unstirred layer will often be much more than 150 μm thick. Raven (1970) and Smith & Walker (1980), who have reviewed this topic, conclude that diffusion of CO_2 (and presumably HCO_3^-) across this unstirred layer can be an important rate-limiting step in photosynthesis by aquatic macrophytes. An increase in photosynthetic rate of river plants associated with increased flow velocities in natural waters has in fact been demonstrated by Westlake (1967). Wheeler (1980*b*) found that the photosynthetic rate of the blade of the great kelp, *Macrocystis pyrifera*, in saturating light, increased about four-fold when the current speed increased from 0 to 5 cm s^{-1}: in low light, however, the increase was only about 50%. Any morphological adaptation which increases the surface-to-volume ratio will help to overcome the diffusion problem and many aquatic macrophyte species have achieved this by evolving highly dissected leaves.[399] The more dissected the leaf form, the less the stimulation of photosynthesis brought about by increased turbulence.[297] Whether the existence of an unstirred layer around each cell has any significant limiting effect on photosynthesis by phytoplankton remains uncertain.

In the majority of terrestrial higher plant species, those which lack the C_4

pathway, photosynthesis is significantly inhibited by oxygen at the normal atmospheric level (21%), primarily as a result of the direct competition between O_2 and CO_2 at the active site of Rubisco. In the oxygenase reaction of Rubisco, the oxygen reacts with ribulose bisphosphate to produce phosphoglycollate and phosphoglycerate, and this is the first step in the metabolic pathway known as photorespiration. Photosynthesis by aquatic higher plants and algae, by contrast, in most cases shows relatively little inhibition by oxygen. This appears to be due to the ability of aquatic plants to increase, by one or other of a range of mechanisms, the concentration of CO_2 in the proximity of the Rubisco molecules within the cell, thus increasing the CO_2/O_2 competitive ratio.

There is evidence for active transport of CO_2, and in some cases of HCO_3^- as well, achieving a manyfold higher internal than external concentration, in cyanobacteria and unicellular green algae,[33,34,649,853,936] in the fresh-water diatom *Navicula pelliculosa*,[770] and in the marine green macroalga *Ulva fasciata*.[54] There is also evidence for such biophysical CO_2-concentrating mechanisms in brown algae[890] and, according to Raven *et al.* (1985), also in many fresh-water vascular macrophytes, most marine vascular macrophytes (seagrasses) and most seaweeds. Active transport of inorganic carbon is not, however, universally present in aquatic plants: Patel & Merrett (1986) found it not to occur in the marine diatom *Phaeodactylum tricornutum*, and Smith & Bidwell (1989) found no evidence for it in the marine red macroalga *Chondrus crispus*.

As well as the 'biophysical' strategy there are biochemical solutions to the problem of raising the internal CO_2 concentration. No submerged aquatic plant has yet been found which exhibits C_4 photosynthesis as it occurs in terrestrial C_4 species, i.e. with fixation of CO_2 into a C_4 acid such as malate in the mesophyll cells, followed by transport of this acid to specialized bundle sheath cells where CO_2 is released and then re-fixed into carbohydrate by the normal photosynthetic carbon reduction (PCR) cycle. What might, however, be regarded as a structurally abbreviated version of the C_4 pathway has been found by Salvucci & Bowes (1983) in the fresh-water macrophyte *Hydrilla verticillata*, where the whole biochemical sequence appears to take place within the same cell.

It seems likely that most of the initial CO_2 fixation is carried out by phosphoenolpyruvate (PEP) carboxylase in the cytoplasm. The oxaloacetic acid produced is reduced to malic acid. This is transported into the chloroplasts where it is decarboxylated by NADP malic enzyme, and the CO_2 is fixed into carbohydrate by the PCR cycle.[87] Evidence for a C_4 pathway of this type was found in one seagrass species *Cymodocea nodosa*, but not in ten others.[55] There is also evidence for this C_4 pathway, but using PEP carboxykinase as the carboxylating enzyme, in the marine green macroalga *Udotea flabellum*.[749]

Another biochemical CO_2-concentrating mechanism, crassulacean acid metabolism (CAM), occurs in some, but not all, of the isoetid macrophyte species of oligotrophic lakes.[86,461,746] These aquatic CAM plants take advantage of the higher levels of CO_2 produced by respiration of the biota in the water during the night to fix CO_2 by means of PEP carboxylase in the form of organic acids such as malate, which accumulate in the vacuole. During the day, when light is available, but CO_2 levels in the water fall due to photosynthesis by other plants, CO_2 is liberated by decarboxylation of the C_4 acids, and fixed into carbohydrate by the PCR cycle.

Temperature

In considering the effect of temperature on aquatic photosynthesis, it is necessary to distinguish between the effects immediately following a change in temperature and the effects obtained if the plants are allowed to adjust to the new temperature for one to several days. Considering the immediate effects of temperature change first, if the photosynthetic rate of a phytoplankton population or a marine or fresh-water macrophyte is measured under saturating light, at a series of temperatures covering the range from just above freezing to temperatures becoming unfavourable for life, say 5–40 °C, it will be found that photosynthetic rate per unit biomass at first increases with temperature, then levels off, and finally begins to decrease again, i.e. there is a temperature optimum. The temperature optimum determined in the laboratory varies in accordance with the temperature of the normal habitat of the aquatic species. For example, benthic marine algae from tidal pools exposed to high temperatures in the summer have somewhat higher optimum temperatures for photosynthesis than algae, sometimes even of the same species, obtained from the sub-tidal zone.[620] Fig. 11.5 compares the dependence of photosynthetic rate on temperature for specimens of the green alga *Ulva pertusa* isolated from these two habitats: the temperature optimum for the subtidal sample is about 5 °C lower than that for the pool sample. Yokohama (1973) found that for several species of green, brown and red benthic algae collected around the shore of the Izu peninsula, Japan (34°40′ N), the temperature optimum was 5–10 °C higher for material collected in summer than for samples obtained in winter. The reasons for the decrease in photosynthetic rate at temperatures above the optimum are not well understood: denaturation of

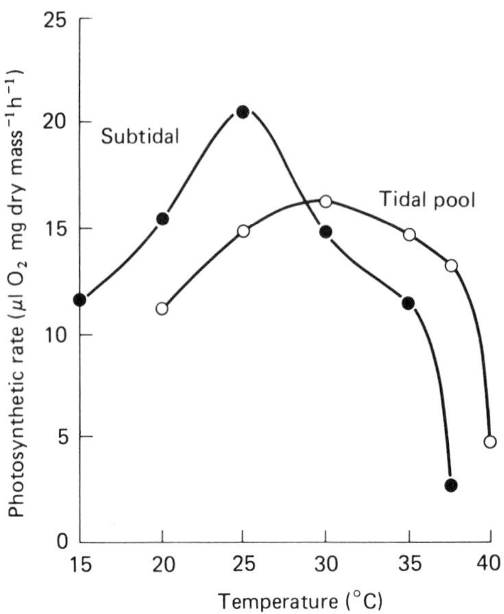

Fig. 11.5. Photosynthetic rate as a function of temperature in samples of the marine benthic alga *Ulva pertusa* (Chlorophyceae) collected from the subtidal zone and from a tidal pool (after Mizusawa *et al.*, 1978).

enzymes, a runaway increase in respiration and other forms of thermal damage are likely to be involved.

If photosynthesis as a function of irradiance is measured at a series of temperatures (not exceeding the optimum), then it is observed that the maximum photosynthetic rate at light saturation increases with temperature: this is shown for the marine diatom *Skeletonema costatum* and the lower-littoral-zone red alga *Gigartina stellata* in Fig. 11.6. It is, however, generally found in experiments of this type (i.e. in which the plants are not given time to adapt to each temperature) that at low light intensities, in the linear part of the P versus E_d curve, variation in temperature has little effect on the rate (i.e. the initial slope, α, is essentially constant over the physiological temperature range. We may reasonably attribute the fall-off in the light-saturated rate (P_m) with decreasing temperature to the enzymic reactions of the dark carboxylation system working progressively more slowly as temperature decreases. The irradiance value, E_k, corresponding to onset of light saturation also decreases together with the temperature: since $E_k = P_m/\alpha$, then since α is unaffected by temperature, E_k must decrease in parallel with P_m. When, due to low temperature, the photosynthetic dark reactions are working slowly, less light is required to produce $NADPH_2$ and

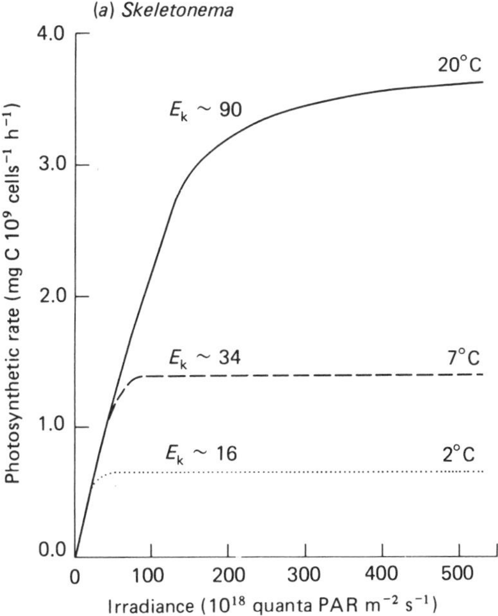

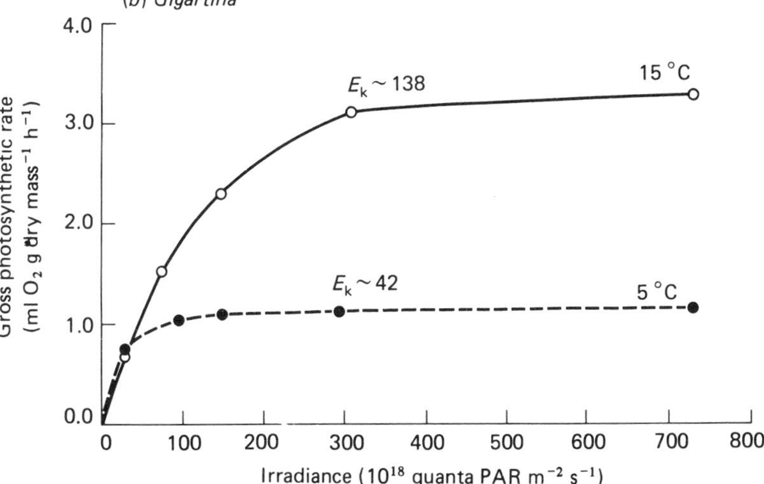

Fig. 11.6. Photosynthetic rate as a function of irradiance at different temperatures. (*a*) In the marine diatom *Skeletonema costatum* (after Steemann Nielsen & Jørgensen, 1968, and Steemann Nielsen, 1975). The cells were grown under an irradiance of 75×10^{18} quanta m^{-2} s^{-1}, and exposed to the experimental temperature for 30 min before measurements were carried out. (*b*) In the multicellular red alga *Gigartina stellata* (plotted from data of Mathieson & Burns, 1971). Plants were collected from the lower littoral zone of the coast of New Hampshire, USA. The original irradiance values (in foot-candles) have been converted to quantum units.

ATP fast enough to saturate the dark reaction system. The comparative insensitivity of the photosynthetic rate at low light intensity to temperature is commonly attributed to the fact that photochemical processes are not very sensitive to temperature, and it is mainly the light reactions which determine rate in the light-limited part of the curve. Respiration, however, does increase markedly with temperature. The failure of the light-limited photosynthetic rate to respond means that as temperature increases the light intensity required for photosynthesis to equal respiration, i.e. the light compensation point, also increases.

Although the photosynthetic rate of any individual aquatic plant or phytoplankton population sample is positively correlated with temperature (up to the optimum) in short-term experiments, in the case of the whole aquatic ecosystem undergoing slow seasonal temperature changes, there is time both for physiological adaptation by any given species, and for changes in the dominant species present. Thus the relation between primary production and temperature may not be so simple at the ecosystem level. Considering adaptation first, if samples of a phytoplankton population are exposed to each new temperature for some days before the P versus E_d curve is measured at that temperature, then in the case of some species the cells can adapt and the differences between the light-saturated rates of photosynthesis at the different temperatures is much smaller. For example, Steemann Nielsen & Jørgensen (1968) found that in cells of the diatom *Skeletonema costatum*, grown at 20 °C, an immediate transfer to 8 °C reduced the light-saturated photosynthetic rate by about two-thirds: in contrast, cells grown and measured at 8 °C had a light-saturated photosynthetic rate only about 10% lower than that of cells grown and measured at 20 °C.

Given that we are interpreting the reduction in light-saturated photosynthetic rate brought about by an immediate temperature lowering as being due to a lowered specific activity of the dark reaction enzyme system, then we might expect the adaptation to lower temperature to consist of an increase in the cellular content of this enzyme system.[874] Jørgensen (1968) found that the protein content of *S. costatum* per cell was twice as high in cells grown at 7 °C as in cells grown at 20 °C. In *Dunaliella*, Morris & Farrell (1971) observed that the level of photosynthetic enzymes increased as the growth temperature was lowered.

Skeletonema costatum is a ubiquitous diatom which in nature grows well over a wide range of temperature. It is therefore not surprising that marked adaptive abilities can be demonstrated in the laboratory in this species. Yentsch (1974), however, doubts if most phytoplankton species have this ability. Oceanic measurements from various parts of the world indicate that the photosynthetic rate per unit phytoplankton chlorophyll does not vary in any systematic manner with the water temperature:[1007] in all temperature regions, areas with high and low rates of total primary production are found.[874] This could be due to adaptation but more recent studies suggest that there is little temperature adaptation of most oceanic phytoplankton species in nature, and that the lack of a clear dependence of photosynthetic rate on temperature in the oceans, is due to the dominance of different (presumably genetically adapted) species in different temperature regimes.[1007]

Notwithstanding the apparent lack of a systematic effect of temperature on photosynthesis in the oceans, it is generally found for shallow, productive coastal waters and for inland waters, that both the *in situ* photosynthetic capacity (the maximum specific photosynthetic rate per unit phytoplankton or macrophyte) and the integral (areal) rate of photosynthesis, are positively correlated with the prevailing temperature. For example, in Nova Scotia (Canada) coastal waters, Platt & Jassby (1976) found that in the period July 1973 to March 1975 (temperature range 0–15 °C) the photosynthetic capacity of the phytoplankton was linearly dependent on ambient temperature, with a regression slope of 0.53 mg C mg chl a^{-1} h^{-1} $°C^{-1}$. For the waters of the Hudson estuary and the New York Bight, through an 11-month period, Malone (1977*a*, *b*) found that for both netplankton and nanoplankton, the photosynthetic capacity was an exponential function of ambient temperature from 8 to 20 °C, and 8 to 24 °C, respectively. The Q_{10} values (proportionate increase in rate per 10 °C rise in temperature) were 2.0–2.6 for the nanoplankton and 4.0 for the netplankton. Below 8 °C the rates for both fractions were higher than expected on the basis of the exponential dependence on temperature. Above 20 °C the photosynthetic capacity of the netplankton declined with temperature. In a number of temperate-zone eutrophic lakes, dominated by blue-green algae, the photosynthetic capacity of the phytoplankton has been found to vary throughout the year in an exponential manner with the prevailing temperature,[67,293,427] the Q_{10} being about 2 over the approximate range 4 to 20 °C. It should be noted that the changes in coastal and inland waters we have just been considering are seasonal changes which must include changes in the character (species composition, physiological adaptation) of the phytoplankton as well as in the temperature itself, and this may account for the very high seasonal Q_{10} of 4.0 observed by Malone (1977*b*) in the New York Bight. If a given phytoplankton population is exposed to a given series of temperatures (i.e. allowing no time for adaptation or species succession), photosynthetic capacity varies exponentially with temperature with a Q_{10} of about 2.3.[897]

We noted earlier that when a given phytoplankton population or macrophyte sample is exposed to different temperatures, E_k, the irradiance value corresponding to onset of saturation, increases as temperature increases (Fig. 11.6). This also appears to be true on a seasonal basis (i.e. including any changes in the phytoplankton population). In Lough Neagh, N. Ireland, Jewson (1976) found the *in situ* E_k to be correlated with prevailing temperature throughout a two-year period, rising 2.5- to 4.5-fold between the winter (temperature 3–4 °C) and late summer (18.5 °C). This rise in E_k is best seen as being part of the rise in photosynthetic capacity with temperature, than as a separate phenomenon. Given the increase in maximum specific photosynthetic rate, due to the increased rate of operation of the carboxylation system, it simply means that more light is required to produce $NADPH_2$ and ATP fast enough to saturate the dark reaction system.

In summary, there is no doubt that in coastal and inland waters temperature, operating mainly through its effect on photosynthetic capacity, is a major limiting factor for photosynthesis. In simple terms we may say that in any optically deep water body at a temperature below the photosynthetic optimum, photosynthesis per unit volume is likely to be limited by light in the lower part of the euphotic zone, by temperature in the upper part of the euphotic zone and by CO_2 everywhere. In water bodies so shallow that adequate light penetrates to the bottom, temperature (apart from CO_2) can become the dominant factor controlling photosynthesis in the whole system. For a number of stratified lakes with high light penetration in Ontario (Canada), Dale (1986) concluded that depth of colonization by macrophytes was limited by temperature. In the Beaufort Channel, North Carolina, USA – a shallow estuary of average depth 1 m – Williams & Murdoch (1966) observed a pronounced seasonal cycle in phytoplankton primary production which appeared to follow the water temperature cycle but not to be related to insolation. They point out that such temperature-driven annual cycles in the productivity of marine phytoplankton are likely to be characteristic of shallow embayments in temperate regions.

Indirect factors

There are environmental factors other than light, CO_2 and temperature which can have a major influence on the total amount of photosynthesis that takes place (i.e. on primary production in the ecosystem) by their effects on the amount of plant biomass present. Since our concern here is mainly with factors which influence photosynthesis directly we shall touch on these indirect factors only briefly.

Inorganic nutrition – particularly the concentrations of the key elements phosphorus and nitrogen – is the most important indirect factor. If the nutrient concentration is low, then although the existing phytoplankton may be photosynthesizing at a higher rate per unit biomass, they cannot increase their biomass and so the rate of photosynthesis per unit volume or area remains low. When nutrient levels increase, due to agricultural runoff or sewage input in inland waters, or an increase in river outflow in coastal waters caused by increased rainfall, or the commencement of a seasonal upwelling off the west coast of a continental landmass, or a breakdown of stratification in the sea or a lake in the autumn bringing up fresh nutrients from below the thermocline, then the phytoplankton population and therefore total photosynthesis, both increase as well.

When the phytoplankton population has depleted the level of any essential nutrient, then photosynthetic biomass ceases to increase. In the case of nitrogen depletion, however, certain filamentous species of blue-green algae are not prevented from continued growth since they have the ability to fix molecular nitrogen. This ability is present in common bloom-forming genera such as *Anabaena* and *Aphanizomenon* in inland water bodies. In the sea, the unicellular cyanobacterium *Synechococcus*, a major component of the picoplankton throughout the oceans, cannot fix nitrogen,[976] but the large colonial filamentous cyanobacterium *Trichodesmium*, which is often the major component of the phytoplankton in tropical ocean waters, does fix nitrogen. Carpenter & Romans (1991) have concluded that in the tropical North Atlantic Ocean *Trichodesmium* is the most important primary producer, and that its nitrogen fixation contributes about 30 mg of nitrogen per m^2 per day, a value exceeding the estimated upward flux of nitrate from below the thermocline.

For the diatoms, with their external skeleton (frustule) made of silica, silicon is an essential element. Depletion of silica can bring diatom blooms to an end in fresh water,[750] but silicon appears generally not to be a limiting element for diatom growth in the sea, except in some very productive coastal upwelling systems.[748]

The other major environmental factor which limits total phytoplankton photosynthesis by limiting biomass is grazing by the aquatic fauna. The spring bloom of phytoplankton is typically followed by a bloom of zooplankton which graze upon the phytoplankton cells. It is undoubtedly true that zooplankton grazing can have a major impact on phytoplankton populations and in some cases can be the cause of the decline in phytoplankton numbers following the bloom: the relations, however, are complex and fluctuations in phytoplankton populations cannot always be interpreted easily in terms of zooplankton grazing. The matter is discussed in more detail for inland waters by Hutchinson (1967) and Reynolds (1984) and for the oceans by Raymont (1980) and Frost (1980). Large colonial phytoplankton forms, such as occur in the Peruvian upwelling, can be grazed directly by herbivorous fish such as the anchovy.[777]

Benthic filter-feeding fauna, such as shellfish, in shallow coastal or estuarine waters can also make significant inroads into phytoplankton populations. Asmus & Asmus (1991) found that an intertidal mussel (*Mytilus edulis*) bed in the German Wadden Sea (eastern North Sea) reduced phytoplankton biomass by about 37% between the incoming and outgoing tide. Parallel with the uptake of phytoplankton there was, however, a substantial release of nitrogen, as ammonium, from the mussel bed, leading these authors to suggest that the shellfish were simultaneously reducing the standing stock of phytoplankton, and promoting phytoplankton primary production, or to put it another way, the mussel bed was accelerating phytoplankton turnover.

Although throughout most of the world ocean, phytoplankton biomass appears to be limited by the availability of the major nutrients, there are substantial regions – the Southern Ocean, the equatorial Pacific, and the northeast Pacific Ocean – where there is a surplus of phosphate and nitrate in the surface waters. The reasons for this are not yet understood. A number of plausible hypotheses have been put forward to explain the failure of the phytoplankton in these regions to fully utilize the major nutrients. It has been proposed, for example, notably by Martin and coworkers,[596] that phytoplankton growth in these waters is limited by iron deficiency. Another proposal, specifically for the equatorial Pacific, is that the standing crop of phytoplankton is controlled by closely coupled zooplankton grazing.[169,974] For the Antarctic Circumpolar Current in the Southern Ocean, Mitchell *et al.* (1991) argue that as a consequence of the high wind strength, low solar irradiance (persistent clouds) and weak stratification, deep circulation will lead to light limitation of phytoplankton growth to such an extent that they cannot possibly use more than a small fraction of the available nutrients. Another possibility, for the Southern Ocean, is the inhibitory effects of chronic low temperature on nitrate uptake by the phytoplankton.[209] Detailed discussions of these and other possible explanations of 'What controls phytoplankton production in nutrient-rich areas of the open sea?'

may be found in the special issue of *Limnology and Oceanography* published under this title and edited by Chisholm & Morel (1991).

The impact of zooplankton grazing on phytoplankton populations can be greatly affected by the extent to which the zooplankton themselves are subject to predation by larger animals. The addition of planktivorous fish to a pond in Minnesota was found to lead to an order-of-magnitude increase in total phytoplankton biomass.[575] In addition to the increased levels of those algal species already present, several new species appeared.

A limiting factor for one particular group amongst the phytoplankton, the dinoflagellates, is small-scale turbulence. It appears that turbulence in the water, at levels such as might result from quite moderate winds, can cause the cells to lose their longitudinal flagellum: cell division and growth are also inhibited.[59,915] The inhibitory effect of turbulence is thought to explain why dinoflagellate red tides only develop in calm weather.

In fresh-water bodies, infection of phytoplankton populations by parasitic fungi can, particularly in eutrophic lakes, significantly reduce their numbers, and even terminate blooms.[108,398,750] In marine waters, pathogenic viruses have been shown to infect a range of phytoplankton types including diatoms, cryptophytes, prasinophytes and cyanobacteria.[733,891] Although little quantitative evidence is yet available, preliminary indications are that viral mortality could be another significant factor regulating phytoplankton populations and primary production, in the sea.

Phytoplankton growth can itself limit primary production by the benthic flora. In eutrophic waters interception of light by dense phytoplankton populations can prevent the growth of benthic macrophytes.[446] From their studies on the Norfolk Broads (England), Phillips, Eminson & Moss (1978) consider that during progressive eutrophication of water bodies, overgrowth of the macrophyte leaves by epiphytes and mats of filamentous algae, with consequent reduction in light availability, initiates the macrophyte decline even before dense phytoplankton populations have developed. Overgrowth of leaves by epiphytes has also been implicated in the decline of seagrasses in eutrophicated Cockburn Sound, Western Australia.[127]

The density and therefore total photosynthetic rate of the benthic flora is also affected by grazing by invertebrate and vertebrate animals. In the case of macrophytes, snails and other gastropods feed on the photosynthetic tissue, as well as on the attached epiphytes, in both fresh and salt waters: in the sea, limpets and sea urchins are important invertebrate grazers, while in fresh waters insect larvae can be significant consumers. Some specialized species of herbivorous fish such as the grass carp (China) in fresh waters, and the parrot fish which feeds on seagrasses, eat macrophytes. Higher aquatic grazing animals such as sea turtles and dugongs, as well as ducks and geese, also eat macrophytes. In inland waters, grazing by invertebrates, fish and birds is a significant factor limiting biomass and photosynthesis in some macrophyte communities:[987] protection of *Potamogeton filiformis* in Loch Leven (Scotland) from grazing by waterfowl substantially increased its biomass throughout the growing season.[446] Herbivory on freshwater macrophytes has been reviewed by Lodge (1991). In shallow coastal sea waters, grazing by sea urchins can be a major factor limiting the growth of kelps and other seaweeds, and grazing by sea turtles can have a major impact on seagrass beds. The biotic relations are complex: whether or not a particular region of the sublittoral zone is colonized by brown algae, and the density at which they grow, are likely to be determined by the abundance of animals such as crabs, lobsters and sea otters (which prey on sea urchins), and the numbers of these animals may in turn be influenced by the activities of man.

The lower limit of *Laminaria hyperborea* at one site in the Isle of Man (UK) was found to be controlled by the sea urchin *Echinus esculentus*, which was present at a density of five animals m^{-2}.[448] In a study area in Nova Scotia (Canada), Breen & Mann (1976) found that the lobster *Homarus americanus*, a major predator of sea urchins, decreased by nearly 50% in 14 years: in the latter six years of this period the sea urchin *Strongylocentrotus* destroyed 70% of the *Laminaria* beds. Areas of bare rock protected from sea urchin grazing were rapidly recolonized by *Laminaria* plants. Similarly, off California, the mass mortality by disease of a localized population of sea urchins on the seaward side of a kelp forest was followed by the rapid seaward expansion of four species of brown algae.[689] On the basis of archaeological evidence from the Aleutian Islands, Simenstad, Estes & Kenyon (1978) concluded that when the Aleut people colonized this area and commenced hunting the sea otter, the diminution in otter numbers led to a population explosion of sea urchins, which was in turn followed by the decimation of the abundant kelp beds in the area: since fur hunting in the Aleutians ceased in 1911, both the sea otter and abundant kelp beds have become re-established. In Fiji, extinction of the dugongs and a reduction in sea turtle numbers has caused the seagrasses to increase to the point of becoming a nuisance.[606]

11.4 Temporal variation in photosynthesis

The short answer to the question 'When does aquatic photosynthesis take place?' is that it takes place when and to the extent that the various limiting constraints we have already discussed permit it to take place. Thus, to understand temporal variation in photosynthesis we need to know the manner in which these limiting factors vary with time.

Considering diurnal variation first, there is of course no photosynthesis at night. Photosynthesis begins at dawn and ends at dusk. The total photosynthesis of the whole water column roughly follows the approximately sinusoidal variation of solar irradiance during the day.[350,442,901] Under still conditions there may be a reduction of photosynthetic rate per unit volume near the surface due to photoinhibition. Also, where dinoflagellates are the dominant organisms, their downward migration to lower light intensities can lower photosynthetic rate per unit volume near the surface in the middle part of the day.[920] There have been many reports of variation in the photosynthetic capacity (i.e. photosynthesis in samples taken from the water body and exposed to saturating light) of phytoplankton populations and benthic algae during the day (see, for example, the reviews by Sournia (1974), Raymont (1980) and Harris (1980)). This is discussed later (§12.6).

In temperate or Arctic/Antarctic latitudes, there is marked seasonal variation in aquatic photosynthesis. This approximately follows, and is also a major contributing factor to, the seasonal variation in plant biomass. For the marine phytoplankton a detailed account has been given by Raymont (1980). In the winter, phytoplankton biomass and total photosynthesis are both very low, partly because of low solar irradiance[776] but even more because, as a consequence of lowered temperature and winter storms, thermal stratification is lost and the mixed layer through which the phytoplankton are circulated comes to exceed the critical depth for net column photosynthesis (see §11.1). In the spring, with increased heating of the surface water resulting from increased daily insolation, thermal stratification sets in, circulation below the critical depth is prevented, and this, combined with the increase in incident PAR and the availability of nutrients in the water, leads to a massive increase in phytoplankton biomass, typically of the order of 20-fold, measured as chlorophyll. Riley (1942) found that for a number of stations over the Georges Bank (Gulf of Maine) during the spring bloom the rate of phytoplankton increase (late March to mid-April) was approximately inversely proportional to the depth of the mixed zone. He concluded that the balance between the effects of vertical turbulence and the increasing vernal radiation determines the beginning of the spring bloom (see also §11.1). When vertical wind mixing is absent or weak, the spring phytoplankton bloom can begin even before thermal stratification is established.[932]

Typically, the spring bloom is followed by a decline to a lower level of

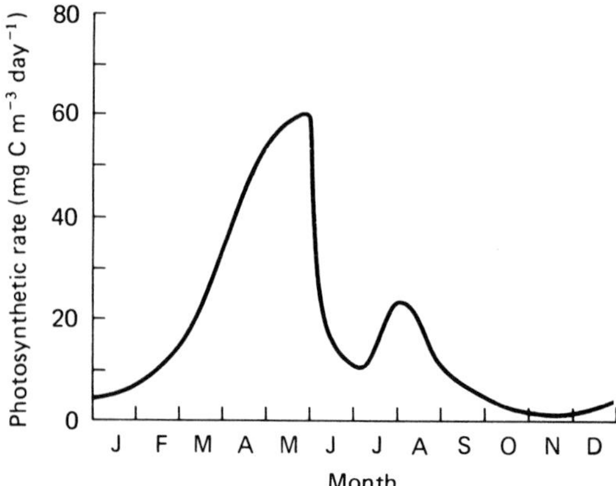

Fig. 11.7. Variation of photosynthetic rate per unit volume throughout the year in North Pacific coastal water (after an average curve derived by Koblents-Mishke, 1965, from literature data).

biomass and integral photosynthesis in the summer, caused by zooplankton grazing and possibly also by a loss of nutrients from the mixed layer resulting from the sedimentation of zooplankton faecal pellets down below the thermocline. In the autumn, the fall in temperature and increased wind strength lead to intermittent disruption of thermal stratification with consequent transport of nutrients up from below the thermocline. Thus, in oceanic and continental shelf waters in intermediate and high latitudes, there is typically a second phytoplankton bloom and associated increase in areal photosynthesis in the autumn: this is terminated when disruption of the thermocline becomes so severe that stratification disappears, circulation below the critical depth sets in and productivity declines to the low winter values. Fig. 11.7 shows the variation in photosynthetic rate throughout the year in North Pacific coastal water illustrating the large spring, and small autumn, peaks.[508] Broadly speaking, as latitude increases, the growing season becomes progressively shorter and the spring and autumn phytoplankton peaks tend to merge into one.

The pattern of temporal variation described above is by no means universal; it can be greatly modified by local conditions. In the Kattegat (Baltic), lowered salinity (and therefore lower density) in the upper layer maintains stratification throughout the year, and so productivity is substantial at all times of the year, diminishing somewhat in the winter due to reduced light availability.[874] In coastal waters so shallow that the mixed depth rarely or never exceeds the critical depth, production also tends to remain high throughout the year. Seasonal variation in production may follow the temperature cycle if the water is so shallow that light is non-limiting.[1000]

In the turbid estuarine waters of San Francisco Bay, Cloern (1991) found that the spring phytoplankton bloom each year was associated with the density stratification that resulted from the seasonal increased input of fresh water to the estuary. On a shorter time scale, rapid phytoplankton growth occurred during neap tides (low tidal energy, weak vertical mixing), but populations declined during spring tides (high tidal energy, intense vertical mixing).

In tropical oceanic waters where there is no seasonal variation in the hydrological regime, there is usually no pronounced variation in phytoplankton biomass and photosynthetic production. In regions where there is a seasonal upwelling of nutrient-rich deep water, such as parts of the west coast of Africa, there is an accompanying massive increase in phytoplankton and photosynthetic production. Seasonal variations in river outflow can also cause seasonal variations in primary production in tropical coastal waters.

Inland waters in intermediate and high latitudes show marked seasonal variation in biomass and total photosynthesis. Both are low in the winter due to a decrease in irradiance (sometimes accentuated by ice or snow cover) and temperature, and, in the deeper waters, circulation of the phytoplankton through a depth greater than the critical depth. Increased irradiance and temperature in the spring, together with the availability of nutrients and the onset of thermal stabilization, lead to a phytoplankton bloom and associated increase in areal photosynthesis. The behaviour during the rest of the seasons tends to be both complex and highly variable. Production may remain consistently high, or there may be wide fluctuations in populations and production caused by zooplankton grazing, fluctuations in nutrient content due to influxes of nutrients from the hypolimnion associated with temporary wind-induced disruption of the shallow (by marine standards) thermocline, or major changes in species composition resulting from the seasonal changes in water quality (e.g. higher pH values in late summer tend to favour blue-green algae rather than diatoms).

Discrete spring and autumn blooms separated by a summer minimum, are often observed in lakes,[398,750] but this temporal pattern is more commonly found in eutrophic, than in oligotrophic water bodies.[595] Fig. 11.8 shows the variation in phytoplankton biomass, and in daily phytoplankton photosynthesis through one and a half years in mesotrophic Lake Constance (Germany–Austria–Switzerland).

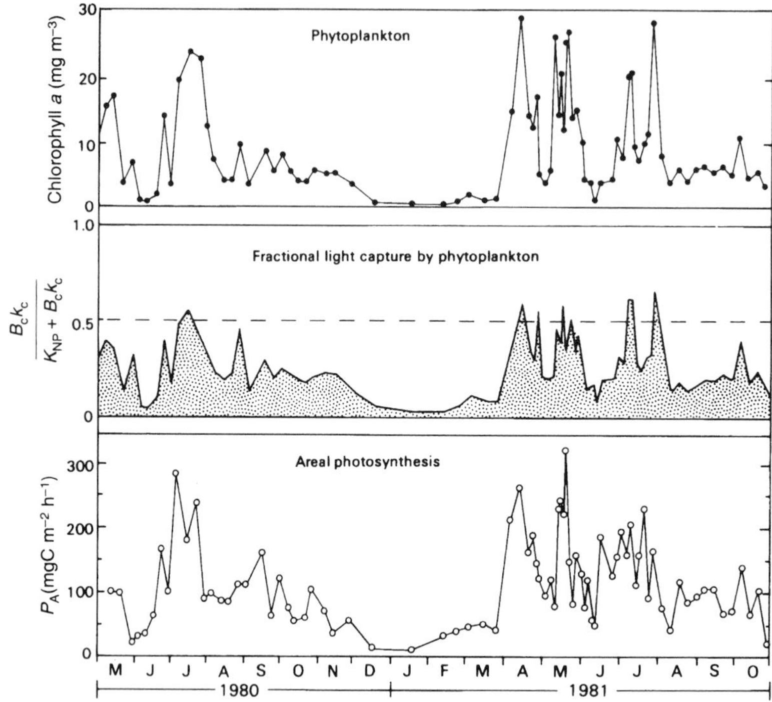

Fig. 11.8. Seasonal variation in phytoplankton biomass and photosynthetic production. Redrawn from data of Tilzer (1983) for the mesotrophic Lake Constance (Germany–Austria–Switzerland). The middle curve represents the estimated proportion of incident PAR that is captured by the phytoplankton. The areal (integral) photosynthetic rate is the hourly average for the middle four hours of the day.

In some tropical lakes such as Lake George, Uganda, there may be no seasonal variation and primary production and phytoplankton biomass remain high throughout the year.[284] On the other hand, where, as in the case of Lake Chad (Chad, Africa), there are alternating dry and rainy seasons, there can be seasonal variation in temperature, insolation and water level, with associated changes in areal photosynthesis.[548]

Benthic macrophytes, being fixed in position, are not, unlike the phytoplankton, directly affected by the seasonal onset or disappearance of thermal stratification. They do, however, show seasonal variation in photosynthetic production in response to the yearly cycles of irradiance and temperature. Fig. 11.9 shows the variation in net photosynthetic rate (per unit area of thallus) of the sublittoral brown alga *Laminaria longicruris* (a kelp) in Nova Scotia, Canada, coastal waters through a 12-month

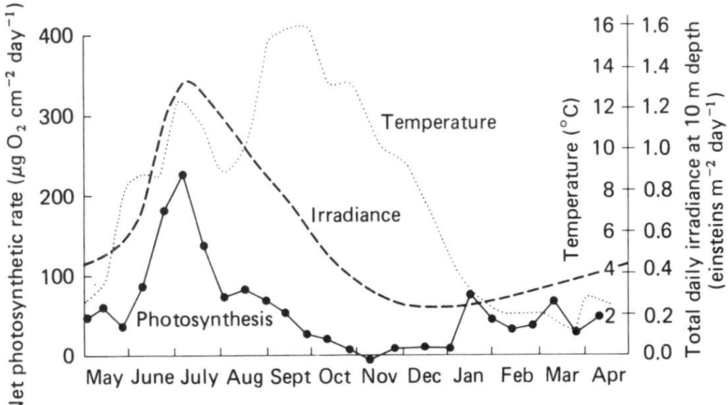

Fig. 11.9. Seasonal variation of the photosynthetic rate of the kelp *Laminaria longicruris* and of total daily irradiance (PAR) and water temperature at 10 m depth in a Nova Scotia, Canada, coastal water (St Margaret's Bay), from data of Hatcher *et al.*, 1977). Photosynthesis measurements were made *in situ* at the prevailing water temperature. The graph of total daily irradiance was obtained by drawing a smoothed curve through the data points of Hatcher *et al.*

period, together with values of the total amount of PAR received per day at the sea bottom (10 m depth) on which the algae were growing, and the prevailing water temperature.[358] Photosynthetic rate was highest in summer, declined through the autumn to approximately zero in November–December, became significant once more in late winter and spring, and then rose steeply in early summer. The temporal variation in photosynthesis corresponded roughly to the variation in irradiance. By multiple regression analysis, Hatcher, Chapman & Mann (1977) found that 61% of the variance in daily photosynthesis could be accounted for by irradiance. Variation in temperature (from 1.5 to 13 °C in these cold waters) did not account for any significant part of the variation in daily photosynthesis. Temperature did account for 56% of the observed variation in light-saturated photosynthetic rate (measured at ambient water temperature, back in the laboratory), but since for most of the time the plants *in situ* would be photosynthesizing at subsaturating irradiance values, the effect of temperature on the saturated rate is irrelevant.

Kirkman & Reid (1979) studied the seagrass, *Posidonia australis*, growing in shallow water – from low-water mark to a depth of 3 m below this – at Port Hacking, Australia (34° S). Fig. 11.10 shows the variation in relative growth rate (mg C g^{-1} day^{-1}, which should be closely related to photosynthetic rate), and water temperature during a 12-month period. The growth

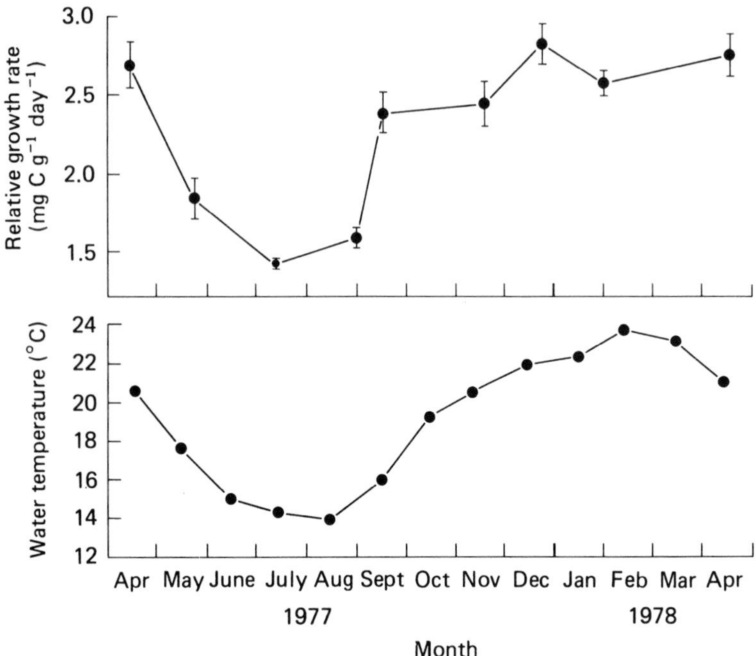

Fig. 11.10. Seasonal variation in relative growth rate of the seagrass *Posidonia australis*, and in surface water temperature, at Port Hacking, NSW, Australia (by permission, from Kirkman & Reid (1979), *Aquatic Botany*, **7**, 173–83).

rate was found to be strongly correlated ($r=0.79$) with water temperature. In shallow seagrass beds in the northern Gulf of Mexico, also, the seasonal growth cycle of the dominant species, *Thalassia testudinum*, correlates more closely with water temperature than with solar irradiance.[578] It is possible that in these shallow, well-illuminated environments, light intensities were at or near saturation for much of the time, so that temperature became a limiting factor.

Hanisak (1979) carried out a combined field and laboratory study of the growth pattern of the siphonous green alga *Codium fragile*, growing in shallow water (average depth 1.5–2.3 m below mean low-water mark) on the northeastern (41° N) coast of the USA. The main period of growth (as dry matter accumulation, therefore closely related to photosynthesis) was from late spring (April–May) through to early autumn, the maximum rate being in July. It appeared that the total daily irradiance levels were above saturation for growth from April to August/September so that temperature was the major limiting factor, but that in the autumn, when temperatures were still high, the reduced light levels were limiting for growth. In the winter it seemed that both temperature and light were limiting.

In a two-year study of seasonal change in the benthic macroalgae/seagrass community in Puget Sound (Washington, USA), Thom & Albright (1990) found that while water temperature appeared to correlate best overall with standing stock, changes in solar irradiance appeared to trigger the onset of biomass buildup in the spring, and die-back in the autumn. During periods when neither light nor temperature were limiting, nitrate level in the water became the limiting factor for growth.

In some coastal regions the diminution in underwater light availability in the winter due to the lowered solar elevation and shorter days is further aggravated by a marked increase in the vertical attenuation of light resulting from increased water turbidity caused by the stirring up of sediments in winter storms. At Banyuls-sur-Mer (western Mediterranean), K_d rose from a minimum value of about 0.075 m^{-1} in the summer to a maximum of about 0.19 m^{-1} in the winter,[982] and off Helgoland (North Sea) also, water transparency was much lower from October to March than in the rest of the year.[574]

These data suggest the tentative generalization that primary production of marine macrophytes is limited in deep water by light availability all the year, but in shallow water is usually limited by temperature except that in higher latitudes light becomes limiting in the winter, and that the temporal variation in primary production is in large part a function of the seasonal variation in these two physical parameters. It is likely that macrophyte photosynthesis (other than in deep water) is, in addition, always and everywhere limited by CO_2 availability, but this does not vary with the seasons. Growth, and therefore photosynthesis, are also determined by the availability of nutrients in the water, particularly nitrate, and in higher latitudes this varies on a seasonal basis being high in the winter but decreasing greatly during the spring phytoplankton bloom. This is less of a limiting factor for many macrophytes than it is for phytoplankton however. Seagrasses, for example, being rooted plants, can obtain nutrients from the sediments in which they are growing. In the summer, when nitrate concentration in the water is low, the kelps (Phaeophyta) continue to photosynthesize and accumulate carbohydrate reserves for later use, in the winter.[358] In the tropics we might *a priori* expect relatively little seasonal variation in macrophyte photosynthesis except where there is variation in the hydrological regime, associated for example with seasonal variation in river outflow.

The extent to which seasonal variation in marine macrophyte photosynthesis is influenced by internal, as well as external, changes is unclear. King & Schramm (1976) in their study of many green, brown and red algal species in the Baltic, found, for example, that light compensation points were lower in the winter than at other times of year, and that the light-saturated photosynthetic rates (per g dry mass) varied markedly within any species (but not in the same way in all species) according to the season. However, their measurements were in all cases carried out at the prevailing seasonal water temperature (which, of course, varied during the year) and so they were unable to conclude whether the seasonal adaptation was simply a direct response to temperature change or whether other factors were involved. An example of a seasonal change in photosynthesis caused by an internal physiological change is, however, provided by the seagrass *Posidonia oceanica*. Drew (1978, 1979) found that in plants of this species growing off Malta the photosynthetic rate (per unit area of leaf) declined markedly in the summer below the spring value, whereas in another seagrass species, *Cymodocea nodosa*, growing nearby, the photosynthetic rate remained high. The decline in *P. oceanica* photosynthesis appeared to be due to senescence (leaf chlorophyll content declined markedly in the summer), possibly triggered by daylength changes.

In fresh waters, the typical pattern of benthic macrophyte primary production in temperate regions is that biomass accumulation does not occur in the winter, that it begins in the spring as a consequence of the increase in solar irradiance and temperature, and rises to its maximum rate (g C or dry matter per m^2 of bottom per day) in the early summer. In the macrophyte-dominated Gryde River (Denmark), Kelly, Thyssen & Moeslund (1983) found that the daily primary productivity of the whole plant community throughout one year closely followed the daily insolation. As a consequence, presumably, of light saturation, there was a clear tendency for the *efficiency* of production within a given day to decrease with increasing irradiance during the day. The relative importance of light and temperature in determining the seasonal pattern of fresh-water macrophyte photosynthesis is not well understood. We may reasonably suspect that it is largely a matter of depth, temperature being the more important variable in shallow, and light in deeper, water. In the late summer and in the autumn, the rate of accumulation decreases for a number of reasons including a general deterioration in the vigour of the plant due to disease, grazing damage and excessive temperatures.[987] The rise in pH late in the summer, due to high phytoplankton photosynthesis, can also be inimical to some macrophytes. Since macrophytes can derive nutrients from the sediments in which they are growing, their productivity is not likely to be limited by the seasonal

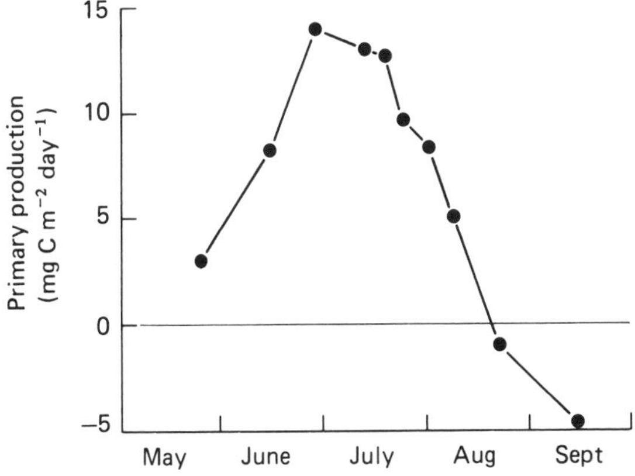

Fig. 11.11. Seasonal variation of the primary production rate of a mixed macrophyte community (predominantly *Chara vulgaris*, *Potamogeton Richardsonii* and *Myriophyllum alterniflorum*) in West Blue Lake, Manit., Canada (after Love & Robinson, 1977).

variations in levels of phosphorus and mineral nitrogen in the water. Fig. 11.11 shows the seasonal changes in primary production rate in a mixed macrophyte community in a Canadian lake.[571] Internal as well as external changes may influence the seasonal variation in photosynthesis: the photosynthetic capacity of fresh-water macrophytes is typically low in the winter and high in the spring.[986] Variations in photosynthetic capacity could influence the rate of photosynthesis in plants growing at saturating intensities in shallow water but would be of little consequence for plants growing at subsaturating intensities in deep water.

11.5 Photosynthetic yield per unit area

In our consideration of where and when aquatic photosynthesis takes place we have noted that some ecosystems are much more productive per unit area than are others. We shall now look more closely at the basis of this variation in productivity, particularly with respect to the role of light, and shall also examine some of the published data on the primary production rates actually achieved by aquatic ecosystems.

Integral photosynthetic rate

Comparisons of phytoplankton productivity in different waters are best carried out in terms of the total phytoplankton production per unit area, i.e.

the integral, or areal, photosynthetic rate (P_A) for the whole water column beneath 1 m^2 of surface. This can be determined, as we saw earlier, by measuring the depth profile of photosynthesis per unit volume at a given location, and summing through the euphotic zone. This is, however, a very expensive and time-consuming way of determining productivity, and is not feasible if it is wished to map the distribution of productivity within some short time frame, over substantial regions of the ocean. A variety of attempts have therefore been made over the years to find ways of estimating integral photosynthetic rates from smaller amounts of data or more easily measured parameters.[249,776,898,963] In recent times this has been given new impetus by the perceived need to understand the role of oceanic photosynthesis in the global carbon cycle and therefore the greenhouse effect, and also by the wealth of data on the distribution of phytoplankton biomass through the world's oceans which has come from remote sensing.

We shall here consider just some of the many different approaches that have been proposed for the determination of integral productivity. One useful generalization that can be made straight away is that expressions for areal production normally incorporate a ($B_c P_m / K_d$) term. Proportionality of production to the concentration, and to the photosynthetic capacity, of the phytoplankton in the water column is to be expected. Proportionality to $1/K_d$ is readily understood if we remember that the total amount of light instantaneously available in the water column is approximately proportional to $1/K_d$ (see §6.5).

Two of the earliest algorithms were those of Ryther (1956) and Talling (1957*b*); both of them for vertically homogeneous phytoplankton populations. Ryther arrived, on the basis of measurements of photosynthesis/light intensity curves on cultures of 14 marine phytoplankton species, at a notional average relative light response curve (ratio of photosynthetic rate to maximum photosynthetic rate, as a function of irradiance) for marine phytoplankton populations. Using this curve he was then able to calculate a dimensionless parameter R_s, which specified how the daily integral photosynthesis would vary with the daily insolation, and obtained the expression

$$P_A = R_s \frac{P_{sat}}{K_d} \tag{11.3}$$

where P_{sat} is the light-saturated photosynthetic rate of the population per unit volume of water (mg C m^{-3} hr^{-1}), and K_d is the vertical attenuation coefficient for downward irradiance of PAR. P_{sat} can be replaced by the product of phytoplankton concentration and the maximum specific photosynthetic rate.

$$P_A = R_s \frac{B_c P_m}{K_d} \tag{11.4}$$

If a plausible value for the maximum specific photosynthetic rate, or *assimilation number* (units mg C mg chl a^{-1} hr^{-1}), P_m, can be assumed for the phytoplankton in a particular region, and if K_d can be estimated as a function of phytoplankton concentration (B_c), then daily integral production can be estimated simply from daily insolation (giving R_s, using the Ryther curve) and phytoplankton concentration. Ryther & Yentsch (1957), on the basis of their own, and literature, data suggested that 3.7 mg C chl a^{-1} hr^{-1} was a good estimate of the assimilation number of oceanic phytoplankton, and found that when used in eqn 11.4 it gave estimates of integral production in good agreement with those from *in situ* measurement. Falkowski (1981), however, with the benefit of access to much more oceanographic productivity information than was available in 1957, concluded that phytoplankton assimilation numbers are affected by nutrient levels, temperature, cell size and light history, and in fact vary commonly between 2 and 10 mg C mg chl a^{-1} hr^{-1}.

It will be noted that Ryther (1956), by adopting a standard relative light curve implicitly assumed a fixed value for the saturation onset parameter, E_k, of marine phytoplankton. In the Talling (1957*b*) algorithm for calculating areal production, by contrast, E_k is included as one of the parameters whose value is selected at each location. On the basis that phytoplankton photosynthesis varies with light intensity in accordance with the equation of Smith (1936) (eqn 10.1), and that irradiance of PAR diminishes exponentially with depth, and ignoring photoinhibition, Talling calculated a standard curve for the photosynthetic rate per unit volume, as a function of depth. By measurement of the area under the curve, and relating it to various parameters of the system, he was able to show that the following relationship approximately holds

$$P_A = \frac{B_c P_m}{K_d} \ln \left[\frac{E_d(0)}{0.5 E_k} \right] \tag{11.5}$$

where $E_d(0)$ is the downward irradiance of PAR just below the surface. Talling found that eqn 11.5 gave reasonably accurate estimates of P_A over a wide range of values of $E_d(0)/E_k$, but significant discrepancies did arise for low surface light intensities ($E_d[0] < E_k$). To obtain the daily total of photosynthesis per unit area, Talling found the relationship

$$\text{Areal photosynthesis per day} = \frac{B_c P_m}{K_d} \left[0.9 N \ln \left(\frac{E_d(0)}{0.5 E_k} \right) \right] \tag{11.6}$$

where $\overline{E_d(0)}$ is the mean value of subsurface downward irradiance during daylight hours, N is daylength in hours, and 0.9 is an empirical correction factor, to be satisfactory. The expression in square brackets corresponds to the R_s term in the Ryther equation (11.4). As with the Ryther equation, the Talling equation can be used to estimate daily integral production from daily insolation and phytoplankton concentration, with the additional requirement that an appropriate value be assigned to the saturation onset parameter, E_k. Platt, Sathyendranath & Ravindran (1990) have, by a mathematically sophisticated approach, arrived at an exact but complex analytic solution for the daily integral of photosynthesis by phytoplankton in a vertically homogenous water column.

Conceptually, one of the simplest solutions to the problem of estimating oceanic primary production from remote sensing data makes use of the parameter ψ, sometimes known as the *water column light utilization index*,[239] which is the total net integral daily primary production in a 1 m^2 water column (g C m^{-2} day^{-1}), divided by the total amount of phytoplankton chlorophyll *a* in the water column (g chl *a* m^{-2}), and by the number of einsteins (mole photons) of PAR incident on 1 m^2 of surface during the day (einsteins m^{-2} day^{-1}). If it proved to be the case that ψ (units g C $[\text{g chl}]^{-1}$ $[\text{einstein m}^{-2}]^{-1}$) had, even if only very approximately, the same value everywhere in the ocean, then global marine primary production could be estimated from remotely sensed values of phytoplankton chlorophyll and surface-incident irradiance.

On the basis of a survey of literature data available up to that time, Platt (1986) arrived at the encouraging conclusion that ψ varied only over the range 0.31–0.66 g C $(\text{g chl})^{-1}$ $(\text{einstein m}^{-2})^{-1}$. More recent data, however, indicate that ψ is in fact much more variable than this. Campbell & O'Reilly (1988) found, for example, in a wide-ranging study over the oceanographically diverse northwest Atlantic continental shelf, that ψ varied about 100-fold, from about 0.1 to 10, and furthermore that its average value was 1.47 g C $(\text{g chl})^{-1}$ $(\text{einstein m}^{-2})^{-1}$, much higher than the values reported by Platt.

Even though it does not seem to be quite the 'biogeochemical constant'[637] that it was originally hoped to be, ψ is nevertheless a useful unifying concept, and consideration of why in fact it varies so much may be fruitful. Platt (1986) showed that in vertically homogeneous waters, $\psi = \alpha/4.6$, where α is the initial slope of the photosynthesis versus light intensity curve (§10.1) of the phytoplankton population. Platt *et al.* (1992) found the value of α to vary about 12-fold with time and place in cruises in the western North Atlantic. Measurements within one water mass during the spring bloom indicated a marked decrease in α as nitrate levels declined.

Ways of increasing the accuracy with which the geographical distribution of integral primary productivity can be calculated from remote sensing data are under active study in a number of centres. In order to discuss what parameters need to be estimated, it is useful to recall that algorithms for integral production are broadly of the form

$$P_A = \frac{B_c P_m}{K_d} f[E_d(0,+)] \qquad (11.7)$$

where $f[E_d(0,+)]$ is some function of the surface-incident solar irradiance. Values of B_c can be obtained by remote sensing, but apply only to the surface layer: in the ocean, phytoplankton concentration can vary markedly with depth. Morel & Berthon (1989) found that the profile of chlorophyll as a function of optical depth varied in a systematic manner from quasi-uniformity in eutrophic waters to a clearcut deep chlorophyll maximum in oligotrophic waters. Thus, once the trophic level of the water mass has been identified from the surface chlorophyll concentration, a plausible depth profile can be arrived at. Alternatively, total euphotic zone chlorophyll can be obtained from surface layer chlorophyll using empirically derived relationships[638,829] which have been found to apply, between these quantities. Balch, Eppley & Abbott (1989) used two standardized profiles, based on accumulated data, of relative chlorophyll concentration versus optical depth, one for nearshore and one for offshore waters.

Once the phytoplankton chlorophyll has been estimated, an approximate value of K_d(PAR) can be arrived at, using an empirical relationship.[39] In the case of the very comprehensive Morel (1991) algorithm, however, the chlorophyll values are used, together with a standard phytoplankton absorption spectrum, to give the actual absorption coefficients in a series of 5-nm wavebands over the 400–700 nm interval.

The daily surface-incident solar irradiance, $E_d(0,+)$, is determined by the date, latitude, and atmospheric conditions: the dependence of irradiance on the first two is readily calculable, but the third is the cause of enormous variation, particularly due to the influence of cloud cover. Kuring *et al.* (1990) have developed a procedure for estimating per cent cloud cover from remotely sensed satellite images and using these values to derive sea-surface irradiance.

There now remains the problem of finding a value for P_m, and it is here that the major uncertainty in the remote sensing of oceanic productivity lies. As we have seen, the photosynthetic characteristics of phytoplankton populations are highly variable both in space and in time, and the reasons are not well understood. One possible cause of variation is temperature.

The light-saturated maximum photosynthetic rate, P_m, increases with temperature (Fig. 11.6). Balch, Abbott & Eppley (1989) obtained, using data from the Southern California Bight, an empirical expression for the dependence of P_m on water temperature, and another empirical expression for mixed-layer depth, and the rate of decline of temperature with depth below that, as a function of surface temperature. Their PTL algorithm (pigments, temperature, light) combines the calculated chlorophyll and P_m depth profiles to calculate photosynthesis as a function of depth through the euphotic zone.[37,38,39] The Morel (1991) algorithm also takes temperature into account. The significance of algorithms incorporating temperature in this manner is that sea surface temperature can be mapped from space.

P_m is a linear function of maximum quantum yield, ϕ_m (since $P_m = \bar{a}_c E_k \phi_m$), and ϕ_m appears to be strongly correlated with nitrogen flux[151,512] (see §10.3). The decline of α as nitrate concentration decreased, in the western North Atlantic spring bloom referred to earlier, may be attributable to a decline in ϕ_m (since $\alpha = \bar{a}_c \phi_m$). Variation in ϕ_m due to variation in nitrogen availability is therefore another possible basis for variability in P_m, but unfortunately, nitrate concentration, or the depth of the nitracline, can not be measured by remote sensing. Yet another major possible cause of variation in ϕ_m, and hence of P_m, is community composition, which can vary either seasonally at a given location, or from one hydrographically distinct water mass to another along a transect in coastal waters.[64,839]

P_m is also linearly related to $\bar{a}_c$, the chlorophyll-specific absorption coefficient of the phytoplankton, and this can not only vary for a given phytoplankton species in accordance with its nutritional status, but will vary even more markedly with the taxonomic composition of the phytoplankton community, which as we have just seen is highly variable in both time and space.

All in all, therefore, there are good reasons why we must expect P_m to be a highly variable, and difficult-to-predict, quantity from one part of the ocean to another. The most feasible approach to the mapping of marine productivity may therefore be to divide the ocean up into regions, within each of which realistic estimates of the relevant physiological parameters can be made.[710]

The integral primary production we have discussed above is total primary production. A large part, usually most, of the plant biomass produced is eventually recycled within the euphotic zone as the result of grazing, fungal or viral infection etc., followed by microbial mineralization, but a proportion passes down through the thermocline, in the form of zooplankton faecal pellets and aggregated or dead phytoplankton cells at the end of a bloom, to the deep sea. This downward flux of organic matter from the euphotic zone is sometimes referred to as *export production*. It brings about a net transfer of carbon from the atmosphere to the deep sea, and is the reason why oceanic photosynthesis is of major importance in the global carbon cycle.

The downward transport of biomass removes not only carbon (which is continuously replenished by atmospheric CO_2), but also essential mineral nutrients, especially nitrogen, and if the phytoplankton ecosystem of the euphotic zone is to remain productive, levels of these nutrients must be restored. Nutrients are in fact supplied from outside the euphotic zone in a number of ways: from deep water during seasonal, or temporary wind-induced, breakdown of stratification, from the atmosphere, from nitrogen fixation, and from rivers. Such considerations led Dugdale & Goering (1967) to introduce the concept of *new production*, which is that primary production associated with these nutrient inputs from outside the euphotic zone, as opposed to the primary production associated with the nutrients recycled within the euphotic zone. Clearly, continued export production is only made possible by the existence of new production, and quantitatively the two are approximately equivalent. Eppley & Peterson (1979) defined the f ratio as the ratio of new production to total primary production at any given location, and they showed that this varied regionally: f increases as total productivity increases, but levels off at about 0.5 for the most productive oceanic regions, such as the Peruvian upwelling. For the oceanic ecosystem as a whole they estimated an f ratio of 0.18–0.20.

Estimating total primary production by remote sensing, in the manner we have discussed, can provide indirect information about new, or export, production if plausible values of the f ratio can be assigned, and so is of great potential value for understanding the role of the oceans in the global carbon cycle. Clearly, however, it would be an advantage if some information on the prevailing value of the f ratio in the region under study could be obtained from remote sensing measurements. Sathyendranath *et al.* (1991) have taken advantage of the fact that nitrate concentration is often negatively correlated with temperature, and that the f ratio is positively correlated with the nitrate concentration. They combined CZCS-derived phytoplankton estimates with AVHRR sea-surface temperature measurements to estimate not only total primary production, but also the f ratio and hence new production in the sea over and around Georges Bank, east of Cape Cod (USA).[790] This method can, of course, only be applied where, as in this case, there is a substantial body of accumulated field data, relating nitrate concentration to temperature.

Geographical variation of photosynthetic yield

Equations such as 11.4 or 11.6, or the more complex but equivalent expression derived by Platt *et al.* (1990), predict the amount of phytoplankton photosynthesis per unit area of the ecosystem per day for a specified set of values of the crucial parameters of the system: concentration (B_c) and photosynthetic characteristics (P_m,E_k) of the phytoplankton, penetration of light into the water (K_d), incident radiation ($E_d(0)$,N). All these parameters vary not only throughout the course of the year, but also, taken over the whole year, with geographical location – from one part of the world's oceans to another, from one inland water body to another – because of differences in the average values of the controlling physicochemical parameters of the system such as depth, insolation, nutrient concentration, temperature, optical properties of the water, stability of the water column, etc. The consequent geographical variation in annual photosynthetic yield can conveniently be characterized in terms of the total amount of carbon fixed by photosynthesis per m^2 of aquatic ecosystem per year. Data for phytoplankton primary production in various parts of the sea are listed in Table 11.2. Detailed discussions of productivity in various parts of the ocean may be found in Raymont (1980), Cushing (1988), and in the proceedings of the workshop on *Productivity of the ocean: present and past*, edited by Berger, Smetacek & Wefer (1989).

The least productive waters are the deep oceans in tropical latitudes. Such waters have a permanent thermocline which greatly impedes transport of nutrients up from the depths where they are regenerated by mineralization of sedimenting phytoplankton and zooplankton faecal pellets. In the temperate oceans, nutrient levels in the upper layer are restored each winter when thermal stabilization breaks down. The highest rates of production are achieved when the high insolation and temperatures of tropical regions are combined with a plentiful nutrient supply in the form of upwelling water from the depths: the Peruvian upwelling is probably the most productive region of the oceans. Phytoplankton productivity is higher in the shallower waters of the continental shelf, and higher still close inshore, than in the deep oceans, even in tropical latitudes. Nutrient supply from the land, and tidal mixing, are thought to be important factors. Close inshore, recycling of nutrients from the bottom and the impossibility of deep circulation of the phytoplankton (below the critical depth), are also likely to contribute to the enhanced productivity.

In the case of inland waters, oligotrophic (nutrient-poor) lakes have phytoplankton primary productivities usually in the 4–25 g C m^{-2} yr^{-1}

Table 11.2. *Annual phytoplankton primary production in the oceans. Data from literature surveys by* (a) *Platt & Subba Rao (1975)*, (b) *Sakshaug & Holm-Hansen (1984)*, (c) *Koblents-Mishke (1965)*, (d) *Walsh (1981)*, (e) *Sherman* et al. *(1988)*, (f) *Zijlstra (1988)*, (g) *Dragesund & Gjøsaeter (1988)*, (h) *Boynton, Kemp & Keefe (1982)*

	Primary production ($g\ C\ m^{-2}\ yr^{-1}$)		
	Continental shelf	Deep ocean	Reference
Ocean			
Indian	259	84	*a*
Atlantic	150	102	*a*
Pacific	190	55	*a*
Antarctic	25–130	16	*b*
Arctic	50–250	25–55	*b*
Regions within Pacific Ocean			
Tropical deep ocean	28		*c*
Tropical/temperate transition zone	49		*c*
Temperate deep ocean	91		*c*
Continental shelf, temperate	102		*c*
Inshore coastal, temperate and tropical	237		*c*
Peruvian upwelling	1350–1570		*d*
Various coastal and estuarine waters			
Northeast US coastal	260–505		*e*
North Sea	100–300		*f*
Barents Sea	60–80		*g*
45 estuaries	190		*h*

range whereas eutrophic (nutrient-rich) lakes typically fix 75–700 g C m^{-2} yr^{-1},[338,767] although, for the saline Red Rock Tarn in Australia, an annual production of 2200 g C m^{-2} has been reported.[338] Hammer presents an extensive compilation of productivity data for inland waters in the 1980 IBP report.[544] Brylinsky's survey of a large number of lakes all over the world for the IBP report indicated that phytoplankton productivity is negatively correlated with latitude. This can reasonably be explained in terms of the diminution of annual solar radiation input with increasing latitude.

12
Ecological strategies

Of the factors which limit the rate of primary production in aquatic ecosystems – light, nutrients, carbon dioxide, temperature – that which shows the most extreme variation within the aquatic medium is light. As we have seen (Chapter 6) the irradiance decreases with depth from intensities which are so high as to be damaging down to levels which cannot support photosynthesis, and the spectral distribution of the light also changes markedly. We have also seen that at any given depth the intensity and spectral quality of the light vary greatly in accordance with the optical properties of the water. Furthermore, to a much greater extent than the other limiting factors, light availability varies with time: both within the day – from darkness to the full noon Sun, and as clouds pass across the Sun – and with the seasons during the course of the year.

In this chapter we shall consider the ways in which the aquatic flora is adapted to this variability of the light climate.

12.1 Aquatic plant distribution in relation to light quality

As we saw in Chapters 8 and 9, there are major differences between the main taxonomic groups of aquatic plants with respect to the kinds of photosynthetic pigment present and, as a consequence, major differences in the absorption spectra. Given the variation in intensity and spectral quality of the light field in the aquatic environment, we may reasonably suppose that for any given location within a water body there will be certain species which are well equipped to exploit the particular prevailing light field and others which are not. It thus seems likely that photosynthetic pigment composition could be a major factor determining which species of aquatic plant grow where.

In the case of the benthic algae of marine coastal waters it has been observed that the different major algal groups are not mixed at random. In some parts of the benthic environment browns predominate while in others the reds or the greens predominate, although domination by any type is not usually complete. Furthermore, progressive changes in the proportions of the different algal groups along the depth gradient can often be discerned. It has been a commonly held belief in marine biology since the last century that the most important factor determining algal zonation is the variation of the light field with depth. This theory has taken two forms which have, perhaps needlessly, been seen as opposed rather than complementary. According to the *chromatic adaptation* theory of Engelmann (1883) it is the varying colour (spectral distribution) of the light with depth which determines the algal distribution, i.e. as the predominant colour changes due to selective absorption, those algae which have absorption bands corresponding best to the spectral distribution of the surviving light can photosynthesize most effectively and so predominate. Berthold (1882) and Oltmanns (1892), on the other hand, proposed that it is the varying intensity of the light with depth which determines the distribution of the different algal types.

In fact both the colour and the intensity of the light field change simultaneously with depth and the plants must adapt to both. For example, any plant growing near the bottom of the euphotic zone must be able to make use not only of a restricted spectral distribution, but also of a very low total irradiance. One possible form of adaptation to low irradiance is a lowering of the respiration rate; another would be an increase in the concentration of all the photosynthetic pigments present without alteration of the ratios. Harder (1923) suggested that both chromatic and intensity adaptation are involved: it is not in fact easy to disentangle the two. Nevertheless there is persuasive circumstantial evidence that chromatic adaptation plays an important role in determining the distribution of the different plant groups.

Before we examine this evidence, we must make clear the important distinction between *phylogenetic* and *ontogenetic*, chromatic adaptation.[735] Phylogenetic adaptation is adaptation that has taken place during phylogeny, i.e. during the evolution of the species. In the present context it refers to the genetically determined differences in pigment composition between the different taxonomic groups of aquatic plants. Within a given species, while the nature of the pigments formed is fixed, the proportion of the different pigments can alter, with significant effects on absorption properties, in accordance with the environmental conditions prevailing during growth and development, i.e. during ontogeny. This is ontogenetic chromatic adaptation. We shall consider both kinds of chromatic adaptation.

Evidence for phylogenetic chromatic adaptation of benthic plants

Changes in the pigment category of aquatic plants with depth are striking in sea water but hard to identify in fresh water, and so we shall here consider marine ecosystems. We shall begin by noting that in all marine waters there is ample light in all wavebands near the surface, and so the theory of chromatic adaptation, which is specifically concerned with light-limited situations, has nothing to tell us about the relative success of the different algal groups in this region: resistance to wave action, for example, or, in the case of upper sublittoral forms, to occasional exposure to the atmosphere, are likely to be more important factors.

While in all waters total irradiance diminishes exponentially with depth, the nature of the change of spectral distribution with depth varies in accordance with the absorption properties of the water. In all waters irradiance in the red waveband diminishes quite rapidly due to absorption by water itself. In very clear colourless waters attenuation is least in the blue region, and with increasing depth the underwater light becomes first blue-green and then predominantly blue in colour (Fig. 6.4*a*). In waters with significant amounts of yellow substances there is rapid attenuation in the blue region as well as the red, so that as depth increases the light becomes increasingly confined to the green waveband (Fig. 6.4*b*). On the basis of chromatic adaptation we might therefore expect somewhat different patterns of algal zonation in the different types of water, and this in fact turns out to be the case.

The depth distribution of the three major eukaryotic algal groups can be expressed in terms of either the biomass, or the number of species, in each group, at a series of depths. Both kinds of information are interesting, the first perhaps more so from our present point of view since it is more directly related to the competitive success through primary production of the different groups: species number, in contrast, can be regarded as a measure of the number of different solutions that the green, brown or red algae have found to the problems of growing at a given depth, or alternatively as a measure of the amount of green, brown or red algal genetic information to be found at that depth. Quantitative distribution data of either type seem to be rather scarce and unfortunately this is especially true in the case of biomass.

We shall first consider waters of the northern hemisphere temperate zone. For the west coast of Sweden (at the entrance to the Baltic Sea), Levring (1959) reported (although quantitative data were not given) that most members of the Chlorophyta occur in the littoral and upper sublittoral zones, the Phaeophyta occur mainly from the littoral zone down to the middle of the sublittoral zone ($\sim$15 m), and the Rhodophyta occur throughout the euphotic zone but are the predominant algal group in the lower sublittoral zone (15–30 m). Distributions described in other northern hemisphere temperate waters are broadly similar to this but clear-cut zonation appears to be absent. Fig. 12.1 shows the variation of species number in the three algal groups with depth at three sites in the British Isles and one from northeastern North America. The numbers of green and brown species decreased with depth. The number of red species at some places increased at first with depth but then began to decrease. At all sites green species were much less common than the other two, and brown species were less numerous than red. At the three sites where green algae were found (Chlorophyta apparently being absent or insignificant at the Scilly Isles site) they penetrated well down into the sublittoral – to about the middle or a bit further – but the brown algae penetrated deeper, and the red algae deeper still. In all cases the red algae dominated the lowest region of the sublittoral. Norton, Hiscock & Kitching (1977) measured the depth distribution of biomass of the more abundant seaweeds on a headland in southwest Ireland, and their data are plotted in Fig. 12.2. Green algae were apparently of no quantitative significance at this site. It can be seen that the brown algae constituted nearly all the underwater biomass throughout most of the sublittoral zone. Brown algal biomass consisted almost entirely of the large kelp *Laminaria hyperborea*: it reached its peak between about 6 and 10 m depth but decreased sharply below this, falling virtually to zero biomass at 18 m. Below 18 m the vegetation consisted mainly of a comparatively sparse cover of red algal species. On a deep-water rock pinnacle in the Gulf of Maine (USA), Vadas & Steneck (1988) observed that kelps (mainly *Laminaria* sp.) were dominant from 24 m (pinnacle summit) down to 33 m depth. Foliose red algae were present through the kelp zone but extended further: they were dominant at 37 m and reached a maximum depth of 50 m. Crustose red algae became a significant component at about 37 m, and were the dominant algal type in the lowest region, fleshy crusts extending to 55 m and coralline crusts to 63 m depth. Green algae appeared to be absent from the rock pinnacle.

Northern European waters contain sufficiently high levels of gilvin to ensure that the underwater light becomes predominantly green in col-

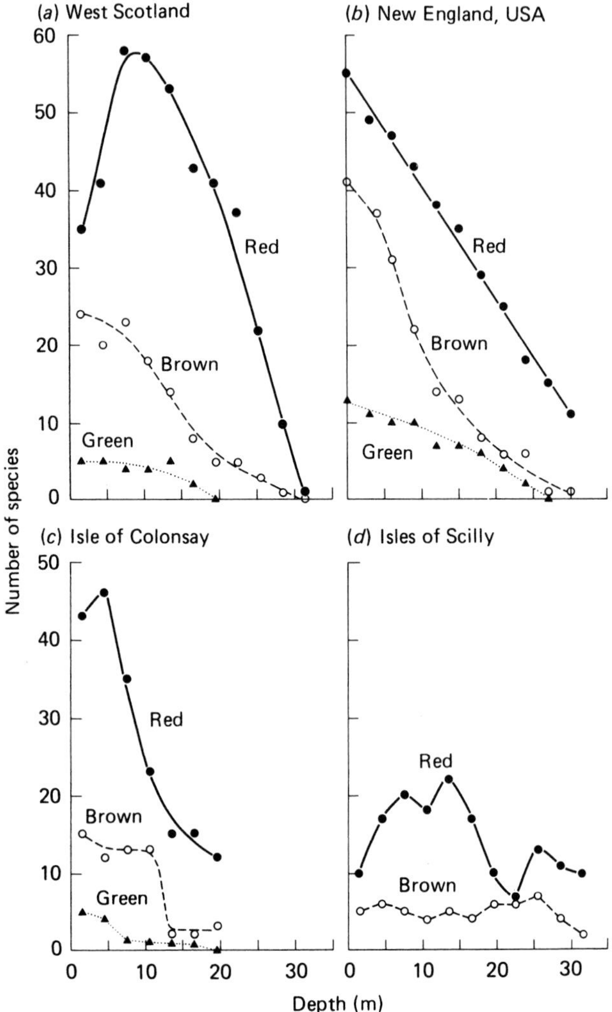

Fig. 12.1. Variation of taxonomic composition of benthic algal flora with depth in northern hemisphere temperate waters. (*a*) West of Scotland (Argyll and Ayrshire, 56–57° N). Derived from data of McAllister, Norton & Conway (1967). (*b*) New England, USA (Maine and New Hampshire, 42–43° N). Plotted from data of Mathieson (1979). (*c*) Isle of Colonsay, Inner Hebrides, Scotland (56° N). Derived from data of Norton *et al.* (1969). (*d*) Isles of Scilly, England (50° N). Derived from data of Norton (1968).

The curves show the number of red (—●—), brown (– –○– –) and green (... ▲ ...) algal species found at each depth. Curves (*a*), (*c*) and (*d*) were derived from the published depth distribution data for each algal species.

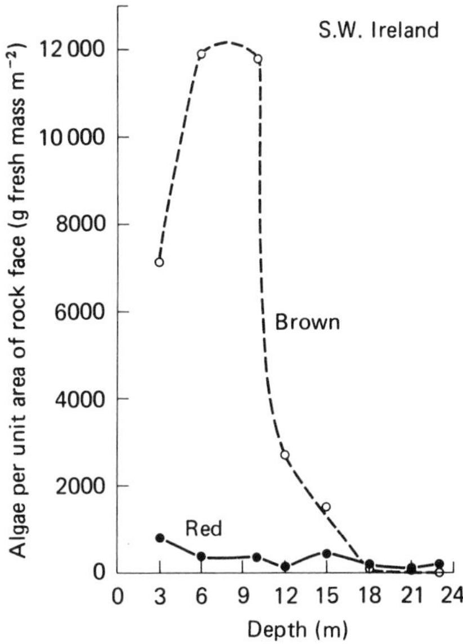

Fig. 12.2. Variation of biomass per unit area of brown and red algae with depth in a northern hemisphere temperate water (Carrigathorna, Lough Ine, southwest Ireland, 51° N). Plotted from data of Norton, Hiscock & Kitching (1977).

our,[336,422] and the same is likely to be true of the northeastern American coastal waters. In such waters the rate of assimilation achieved by an alga with increasing depth is going to depend on, amongst other things, how much absorption it has in the green (500–600 nm) band. The green algae, which show relatively low absorption in this spectral region, are the most disadvantaged, which would explain their small contribution – both in species number and biomass – to the algal community, and the fact that they penetrate least deeply. The brown algae, which show, due to the presence of fucoxanthin, substantial absorption in the 500–560 nm region, hold their own through most of the sublittoral euphotic zone. At the lower fringe of the euphotic zone, where the spectral distribution becomes quite narrow, the red algae, whose biliprotein pigments have their absorption peaks within the green region, can compete better for the available light and so come to dominate.

We shall now consider the very different kinds of algal distribution that are observed in coastal waters very low in colour. Taylor (1959) found in the Caribbean that while red, brown and green species were all well represented in shallow water there was a progressive decrease in species number in all

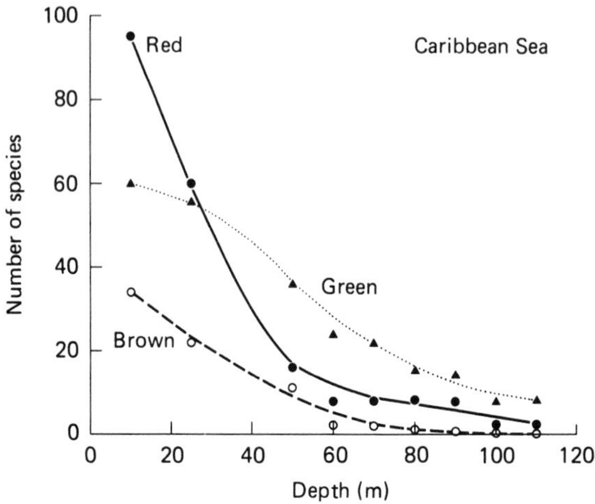

Fig. 12.3. Variation of taxonomic composition of benthic algal flora with depth in a tropical water. Plotted from data of Taylor (1959).

three groups with depth (Fig. 12.3). Proportionately, this decrease was least for the green algae and in the lower 75% of the sublittoral zone there were more green species than either red or brown. Similar distributions have been observed in the Pacific. In Hawaiian coastal waters, green algal species, although overall less numerous, penetrated as deeply as red algal species, and more deeply than the brown.[194] In Eniwetok atoll lagoon, Gilmartin (1960) found that although the numbers of green and red algal species were comparable at depths down to 65 m (both greatly exceeding the number of brown species), the greens appeared (on the basis of visual observation) to be dominant in terms of biomass at all stations down to this depth, at which irradiance was 2–4% of that at the surface.

Algal distributions in the Mediterranean show some similarities. On a Corsican headland, Molinier (1960) found that green algae penetrated as deeply as 80 m, being replaced below this by red algae. A particularly valuable quantitative study, of the depth distribution of algal biomass on vertical rock faces off Malta, was carried out by Larkum, Drew & Crossett (1967). They measured the dry mass of algal biomass within the three taxonomic groups per unit area of the cliff face, as a function of depth: the results are shown in Fig. 12.4. Down to about 10 m depth brown algae were dominant but their contribution to total biomass decreased sharply below this. Green algae became significant at about 15 m and were the major component of the community from 20 to 60 m (the lowest depth studied).

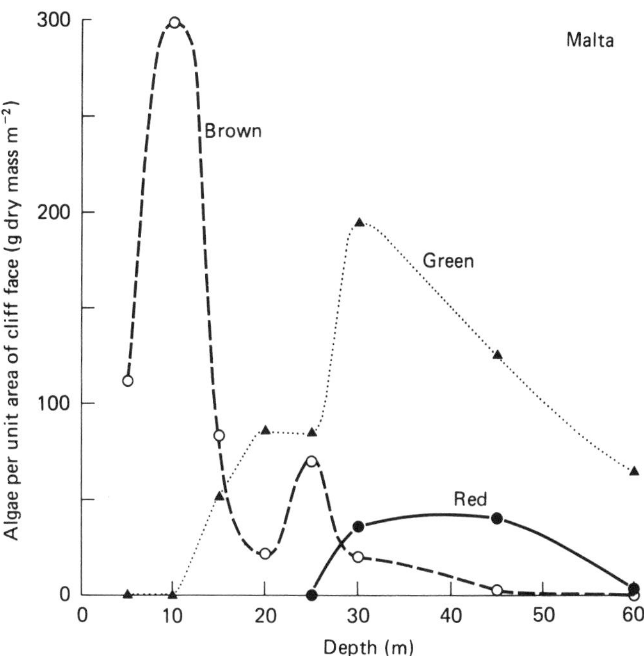

Fig. 12.4. Variation of biomass per unit area of cliff face of green, brown and red algae with depth on vertical rock faces in the Mediterranean Sea (Malta, 36° N). (Plotted from data of Crossett, Drew & Larkum, 1965.)

Red algae became significant components only at 30 m and decreased again below 45 m, proportionately as well as in absolute terms.

Littler *et al.* (1985, 1986) used a submersible to carry out a detailed survey of the depth distribution of algae on the San Salvador Seamount, the top of which forms a ~1 km^2 flat plateau about 80 m below the ocean surface, 6.5 km north of San Salvador Island (Bahamas). The plateau region and the sides of the seamount down to 90 m depth are dominated (% cover) by the brown alga *Lobophora variegata*, although a wide variety of green brown and red algae are present. From 90 m down to about 130 m, green and red algae are present but green dominate, the community consisting mainly of four species of the calcareous genus *Halimeda*, especially *H. copiosa*. From 130 down to 189 m the assemblage is dominated by a crustose red alga, *Peyssonelia*, but two frondose chlorophyte species are also abundant (at least above 157 m). From 189 down to 268 m the dominant species is a crustose coralline red alga, but upwards of 210 m small amounts of the green alga *Ostreobium* appear.

It is clear from these various studies that the most significant difference between the colourless and the slightly yellow coastal waters, so far as algal distribution is concerned, is the very much greater success of green algae in the former. This is just what would be expected on the basis of chromatic adaptation. As we have seen, the underwater light field in such waters becomes particularly rich in that blue waveband in which the green algae carry out most of their light harvesting. Except at the greatest depths there is a high proportion of green as well as blue light. Some species of green algae, in virtue of their containing the carotenoid siphonaxanthin (which when bound to protein *in vivo* absorbs in the 500–550 nm region), have an enhanced ability to absorb green light. Yokohama *et al.* (1977) and Yokohama (1981), in a survey of green algal species in Japanese waters, found that in three orders (Ulvales, Cladophorales and Siphonocladales) siphonaxanthin is present in deep-water species but absent from those growing in shallow water: a plausible example of chromatic adaptation. In certain other orders (Codiales, Derbesiales and Caulerpales) siphonaxanthin is present in all species, even in those from shallow waters: in the latter, Yokohama suggests that siphonaxanthin may be an evolutionary relic, from deep-water ancestors. Amongst 14 species of the marine Chaetophoraceae (Chlorophyta), O'Kelly (1982) found five with lutein (the most abundant higher plant/green algal xanthophyll) and no siphonaxanthin, four with siphonaxanthin and no lutein, and five with both. The species having just lutein were found only in the mid- to upper-intertidal habitats, those possessing only siphonaxanthin were confined to the subtidal region, while those containing both pigments occupied intermediate, wide-ranging habitats.

Direct experimental evidence supporting the chromatic adaptation theory has been obtained by Levring (1966, 1968). He measured the photosynthetic rates of samples of green, brown and red algae suspended in bottles at a series of depths in turbid nearshore water (highest transmittance in the green) off the Swedish and North Carolina coasts, and in clear oceanic water (transmittance highest in the blue) in the Gulf Stream. To compare the ways in which photosynthesis and irradiance varied with depth he used a parameter, q, which may be regarded as the ratio of the vertical attenuation coefficient for photosynthetic rate to the vertical attenuation coefficient for irradiance. If a particular type of alga becomes increasingly ill-adapted to the spectral distribution of the light with increasing depth, then q will be greater than 1 (i.e. photosynthesis will decrease faster than irradiance); if the alga is better adapted to the spectral distribution found at great depth, then q will be less than 1. Below 10 m, for green algae, q was 1.2–1.3 in the turbid/coloured water and ~0.8 in the colourless water, indicating that they were better adapted for photosynthesis at depth in the less-coloured water. In the case of red algae, q was ~0.8 in the turbid/coloured water, indicating improved adaptation with depth, and ~1.0 in the colourless water, indicating little change in adaptation with depth. For the brown alga, *Fucus*, q was ~1.0 in the coloured/turbid water, but variable (above and below 1.0) in locations with colourless water.

In the case of phytoplankton also there is some, at least suggestive, evidence for phylogenetic chromatic adaptation.[679,820] In stratified oligotrophic blue ocean waters, chlorophyll *b*, which is indicative of the presence of green chlorophytes and prochlorophytes, is concentrated near the bottom of the euphotic zone where the light is predominantly blue-green. In near-surface, nitrate-rich waters, diatoms predominate: their major accessory pigment, fucoxanthin, efficiently harvests the green light which is abundantly present at these lesser depths. Pick (1991) studied the distribution of different pigment types of picocyanobacteria in 38 lakes of varying optical and trophic status. Some cyanobacterial picoplankton strains contain only phycocyanin and allophycocyanin biliprotein pigments, with absorption peaks at ~620 and 650 nm, respectively, while other strains also contain phycoerythrin, absorbing in the green at ~550 nm. Pick found that as light attenuation amongst the lakes increased, a trend which would be accompanied by a shift in the underwater spectral distribution from the green towards the red, so the percentage of phycoerythrin-containing picocyanobacteria significantly decreased.

Evidence against chromatic adaptation

On the basis of the relation between water optical type and algal distribution outlined above, we would predict that in any clear, colourless coastal water, with maximum penetration in the blue waveband, green algae should be certainly a major, and probably a dominant, component of the algal biomass throughout much of the middle and lower sublittoral. The water bathing the coast of South Australia, in the region of the Great Australian Bight and the Gulf of St. Vincent, is of a clear, colourless oceanic type: there is little river runoff in this dry region. We might thus expect the depth distribution of benthic algae to be similar to that described above for Caribbean, central Pacific or Mediterranean locations. In fact, a series of thorough studies by Shepherd & Womersley (1970, 1971, 1976) and Shepherd & Sprigg (1976) has shown this not to be so: the distribution is in reality much more similar to that found on northern European coasts, with

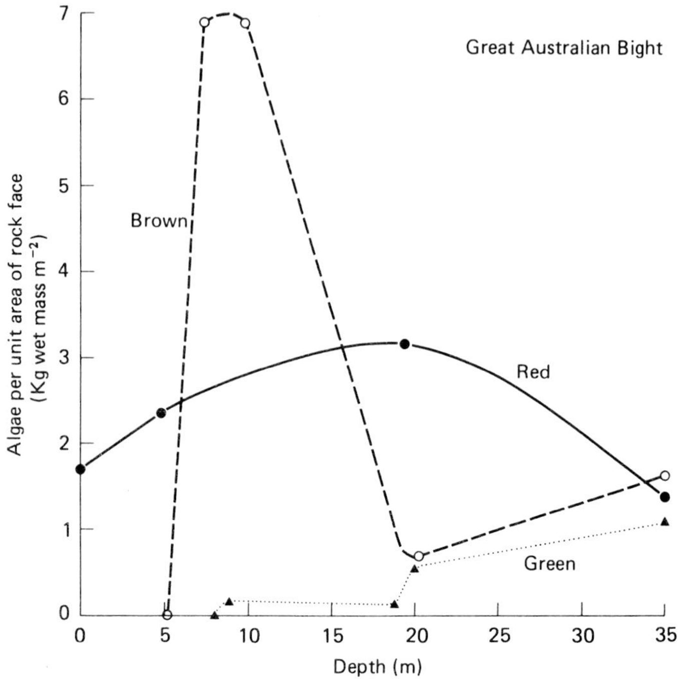

Fig. 12.5. Variation of biomass per unit area of rock face of green, brown and red algae with depth in a clear, colourless southern hemisphere temperate water (Pearson Island, Great Australian Bight, 34° S). Plotted from data of Shepherd & Womersley (1971).

the mid-sublittoral zone dominated by large brown algal species, giving way in the lower sublittoral to a dense cover of red algae. In calmer waters the brown algae dominate the upper sublittoral too, whereas on rough-water coasts they are replaced in this zone by a short turf of the (presumably) surge-resistant coralline red alga *Corallina*.[1003] In terms of biomass, green algae are generally a minor component of the algal community at all depths. At one site (Pearson Island, Great Australian Bight), however, where the water was particularly clear and colourless (oceanic water type IA), the Chlorophyta (mainly *Caulerpa* sp.) constituted a significant fraction of the community from 20 to 35 m (Fig. 12.5) but even here never attained the dominant position that green algae occupy in the underwater flora of Malta or Eniwetok atoll. The poor showing of the green algae in these waters cannot be explained on optical grounds.

An alternative possibility is temperature, an environmental parameter which is known to be a major determinant of seaweed geographical distribution. According to Lüning (1990), in his book on *Seaweeds: their environment, biogeography and ecophysiology*',

> '... the worldwide distribution patterns of seaweeds are mainly determined by global temperature gradients. Deeply imprinted temperature demands, evolved to mirror the geological cycle of cooling and heating of the earth at higher latitudes, keep species of algae apart.'

The South Australian waters are comparatively cool, attaining temperatures of 18–20 °C at the surface in summer. The temperature of the surface water off Malta at the time of the study by Larkum *et al.* was about 27 °C.[166] It may be that the siphonaxanthin-containing green algae, being perhaps mainly of tropical origin, require, as a group, higher temperatures for growth, than the brown and red algae.

Our consideration of chromatic adaptation has so far been couched in qualitative terms; in terms, for example, of whether a particular pigment absorption band is located within or outside the predominant waveband of light penetrating to a given depth. Dring (1981) has attempted to carry out a quantitative test of the theory by calculating how the amount of photosynthesis per unit irradiance might be expected to change with depth in different kinds of algae in waters of various optical types. The calculations for each alga were carried out using a measured photosynthetic action spectrum for the alga and a series of spectral distributions of irradiance at increasing depth computed for a given water type using the spectral transmission data of Jerlov (1976). In effect, what is being calculated is the extent to which the matching of the active pigment composition to the spectral distribution of available light, i.e. chromatic adaptation, should result in increased or diminished efficiency of utilization of available light with increasing depth, and therefore increased or diminished competitive ability, of one algal type compared to another. The results of such calculations can then be compared with observed algal depth distribution data to see if the predictions of the theory correspond to observation.

The findings are somewhat mixed. Some of the predictions accord reasonably well with observation. For example, for coastal waters types 3 and 9 (both of which would have sufficient yellow colour to cause rapid attenuation of the blue waveband) calculations predict that with increasing depth there should be a decrease in photosynthetic effectiveness of green algae such as *Ulva* sp., little change for kelp (brown algae) such as *Laminaria saccharina*, and an increase for red algal species having phycoerythrin as the main biliprotein. This fits in quite well with the algal distributions in such waters on northern European coasts where the maximum depths to which the algae penetrate are generally found to be in the order red > brown > green.

On the other hand, these calculations and earlier ones by Larkum *et al.* (1967) indicate that in colourless oceanic waters, at the deepest limits of algal growth, green algae and thin-bladed brown algae should perform as well as or better than the phycoerythrin-containing red algae. We have in fact already noted that at least in warm oceanic waters green algae do indeed penetrate deeply and are sometimes dominant throughout most of the sublittoral zone. Nevertheless, at extreme depth in such waters they are eventually replaced by red algae. A problem with these calculations is that the validity of the results is highly dependent on the accuracy of the action spectrum attributed to the alga. Dring used action spectra data measured with just one wavelength at a time. As we saw in a previous chapter (§10.3), such action spectra can be misleading because some wavelengths of light when presented singly are incapable of exciting both photosystems to the same extent. The errors arising in this way are not great in the case of the green and brown algae but can be substantial in the case of the red algae in which photosystems I and II have rather different action spectra. If red-algal action spectra are measured with supplementary green light (to ensure that photosystem II is always functioning) then substantially higher relative activity in the blue region is obtained.[260] It may be that if the calculations were carried out again with a corrected red-algal action spectrum, the increased activity in the blue might be sufficient to give the red algae a significant advantage over the green algae in the blue-green light field at the bottom of the euphotic zone in oceanic water.

Another objection to the chromatic adaptation theory is based on the fact that if any particular pigment mixture, whether it be that of the green, the brown or the red algae, is raised to a high enough concentration per unit area of thallus, then there is eventually almost complete light absorption at all wavelengths and the algae become virtually black. Absorption by chlorophyll, for example, in the green waveband, although certainly low, is not infinitesimal and at high chlorophyll concentrations becomes substantial. A mature ivy leaf containing about 60 $\mu g\ cm^{-2}$ chlorophyll absorbs not only, as we might expect, ~100% of the red light incident upon it, but also ~70% of the green (550 nm) light it receives.[496] Ramus *et al.* (1976) argue that if a seaweed is optically thick as are, for example, *Codium fragile* (green) and *Chondrus crispus* (red), then it does not matter what colour it is, and they go on to conclude that the red algae are phylogenetically no better adapted to utilize the ambient light at great depth than are their green counterparts. That optically thick algae have roughly the same light-harvesting capacity (approaching 100% at all wavelengths) whatever the nature of their pigments is certainly true. What must be borne in mind, however, is that if an alga is faced with the problem of absorbing light from an ambient field which is rich in the green waveband, then it is much more efficient in terms of the biochemical economy of the cell to achieve this with a pigment such as phycoerythrin which has its absorption peaks in that waveband than to do it with chlorophyll which absorbs only weakly. The protein cost (remembering that chlorophyll has to be part of a pigment–protein) of achieving high absorption in the green with chlorophyll will be much greater than the protein cost of achieving it with phycoerythrin. It is true that within each of the three algal groups there are some species which have adopted the strategy of having a thick, deeply coloured, thallus and thus absorbing most of the incident light at all wavelengths, and it would be reasonable to say that algae in this category do not make use of chromatic adaptation: as Ramus *et al.* point out, it does not matter which set of pigments they contain. In most marine algal species, however, absorption is far from complete in parts of, or throughout, the spectrum, and so the degree to which their spectral absorption matches the spectral quality of the field is of great significance for the efficient utilization of the incident radiation.

Significance of phylogenetic chromatic adaptation

The evidence taken together seems to me to lead to the conclusion that chromatic adaptation is a major factor influencing the depth distribution of the three types of benthic marine algae, but that it is not the only one, and in some instances other factors prevail. The fact that all algae which grow at depths where the light is predominantly green or blue-green have pigments, whether specialized carotenoids such as fucoxanthin or siphonaxanthin, or biliproteins such as R-phycoerythrin, which enhance their absorption in that spectral region, cannot in my view be plausibly regarded merely as coincidence. The algae evolved in the sea; this means that they evolved in an environment throughout most of the illuminated part of which the light field had a predominantly blue-green character. Given the enormous importance of absorption spectra in determining the efficiency of collection of light energy, it is hard to believe that the types of pigment system which evolved would not be influenced by, and thus correspond to, the spectral character of the prevailing light. The argument that the specialized algal pigment systems are merely adaptations to low irradiance I do not find acceptable. That they are adaptations to limiting light levels seems unquestionable, but if the low-intensity light field has a markedly non-uniform distribution across the spectrum, then for efficient utilization of cellular resources it is better to make pigments whose absorption bands are well-placed to harvest the dim light. It is in principle possible to achieve any required level of light collection at any depth with, say, just the green pigment mixture found in surface-dwelling Chlorophyta provided the pigment concentration per unit area is made high enough: the fact that the algae have in the main not chosen that, biochemically expensive, form of adaptation to low irradiance, but have evolved pigments absorbing in the spectral regions where the light actually occurs, indicates that they are adapted not just to low light, but to low light of a particular spectral character.

Thus, the predominance of red algae in the deepest part of the euphotic zone can be plausibly attributed to their possession of phycoerythrin – the most efficient way (in terms of quanta collected per unit of protein invested) of harvesting the dim blue-green light prevailing at those depths. The failure of green algal species lacking siphonaxanthin to penetrate to great depths we may consider to be a consequence of their comparatively poor absorption capacity for the prevailing underwater light field. The greater depth distribution of the siphonaxanthin-containing green species, and the brown algae we may reasonably regard as being due to the enhanced capacity for absorption in the plentiful 500–550 nm waveband, conferred by their specialized carotenoids.

Chromatic adaptation is not, of course, the whole story. We noted early on that there is no reason why it should have anything to do with the relative success of the different algal groups near the surface. The presence of non-siphonaxanthin green species near the surface is not an example of chromatic adaptation, but their failure to penetrate the depths is such an example. The predominance of coralline reds near the surface in rough-water locations in South Australia has everything to do with their resistance to wave movement and nothing to do with their pigment composition. The presence of significant levels of green algae in the 20–35 m zone in the clear water of Pearson Island, South Australia (Fig. 12.5), we may regard as an example of chromatic adaptation, but their much poorer performance overall in South Australian than in the optically similar Mediterranean or central Pacific waters indicates that other factors such as, perhaps, temperature can prevail over chromatic adaptation in determining algal depth distribution. Within the deep-water red-alga zone, the transition at the lowest depths from fleshy macroscopic red algae to encrusting coralline red algae[805] has no obvious explanation in terms of pigment differences. In the very dim light available at those depths, production rates can only be very low, in which case the rates of loss by grazing and respiration become very important. It may be that the encrusting coralline reds are less susceptible to grazing than the fleshy species and/or have a lower respiration rate. The predominance of sea grasses on sandy bottoms down to great depths in clear colourless water, e.g. to 35 m in the St Vincent Gulf, South Australia[807] or 50 m in the Red Sea,[564] is not due to their being chromatically better adapted than algae (although they should be reasonably well adapted to the bluish light prevailing in such habitats) but to their ability, by means of their roots and rhizomes, to colonize these unstable substrates.

In short, phylogenetic chromatic adaptation is by no means the only factor responsible for the depth distribution of the different plant groups under water: we should, however, continue to regard it as one important factor in that set of interacting factors which finally determines that distribution.

12.2 Ontogenetic adaptation – intensity

Adaptation of the photosynthetic apparatus to changes in light quality (intensity and/or spectral distribution) within a given species can take the form of changes in the total amount of pigment per cell or in the ratio of different pigments, or both.

Chlorophylls and other pigments

It is generally found (with a few exceptions) amongst the algae, both unicellular and multicellular, that as the light intensity during growth decreases (even without alteration of spectral composition), the content of their photosynthetic pigments increases: two- to five-fold increases are commonly observed. Many studies of this have now been carried out, in a wide taxonomic range of algal species: reviews may be found in Richardson, Beardall & Raven (1983) and Falkowski & LaRoche (1991). A convenient parameter in terms of which to express this phenomenon is the carbon: chlorophyll *a* ratio of the cells (C:chl *a*). On the basis of an analysis of literature data for eight diatoms, two green algae, one euglenid, and two cyanobacteria, Geider (1987) concluded that C:chl *a* increases linearly (i.e. decreasing pigment content) with increased light level at constant temperature, but decreased exponentially (i.e. increasing pigment) with increased temperature at constant light level.

Accessory photosynthetic pigments also increase as the growth light intensity is decreased, and indeed generally do so to a greater extent than chlorophyll *a*.[202] In the higher plants and green algae, the ratio of chlorophyll *b* to *a* increases with diminishing light intensity. For example in the green flagellate *Dunaliella tertiolecta*, simultaneously with a 2.6-fold rise in chlorophyll *a* content, the *a*/*b* ratio fell from 5.6 to 2.3 when the growth irradiance was reduced from 400 to 20 μeinsteins m^{-2} s^{-1}.[245] In the marine dinoflagellate *Glenodinium*, as the light intensity during growth was lowered over the range 30 to 2.5 W m^{-2}, the chlorophyll *a* content per cell rose progressively, by about 80%, but the cellular concentration of the light-harvesting peridinin/chl *a* protein rose seven-fold.[721] It has been shown with cultures of the cryptomonads *Chroomonas*,[248] and *Cryptomonas*,[913] the cyanobacteria *Anacystis*,[956] *Oscillatoria*,[264] and *Synechococcus*,[455] the unicellular rhodophyte *Porphyridium*,[103,192,552] and the red macroalga *Griffithsia*,[965] that the ratio of biliprotein pigment to chlorophyll *a* increases with diminishing light intensity: the increases can be several-fold.

Corresponding changes have been observed in the field. In plants of *Chondrus crispus* growing in sunlit sites at 3–4 m depth, the ratio of phycoerythrin to chlorophyll remained high during the winter but underwent a 60% fall in late spring/early summer: in plants in shaded sites at the same depth the ratio remained high during the summer.[752] When the sunlit plants became shaded due to a dense growth of an epiphytic diatom in August they regained much of their phycoerythrin. In the sublittoral red alga *Gracilaria compressa* in the Adriatic Sea, the distal portions of the fronds which received direct illumination were yellow-green in colour and contained 0.065% (dry mass) chlorophyll and only traces of phycoerythrin. The proximal portions which were shaded were purplish-red, and contained 0.085% chlorophyll and 0.82% phycoerythin.[125]

An apparent exception to the rule that accessory pigments increase more than chlorophyll *a* as growth light levels decrease is fucoxanthin. In the brown algae *Sphacelaria* and *Laminaria*, and in the diatoms *Nitzschia* and *Phaeodactylum*, fucoxanthin was observed to increase somewhat less than chlorophyll *a* as light intensity diminished.[107, 201, 813]

The radiant intensity in full sunlight is so high that, quite apart from causing photoinhibition, it can be quite lethal, in part because of the inability of some plants safely to handle the very high rates of energy absorption by chlorophyll and other pigments, resulting in photooxidation of cell material. Carotenoids, however, can in various ways exert a protective effect against such photooxidation,[150] and another adaptive response of algae to high light intensity is, as well as reducing the levels of photosynthetic light-harvesting pigments in the manner we have discussed, to increase the cellular concentration of photoprotective carotenoids. The halophilic unicellular chlorophyte *Dunaliella salina*, which occurs in salt ponds, when grown in full sunlight makes so much photoprotective β-carotene that the cells turn red. This extra β-carotene is not coupled in to the photosynthetic system, and in fact acts as a colour filter, with the consequence that the cells show greatly diminished photosynthetic activity in the blue spectral region.[569]

Paerl, Tucker & Bland (1983) found that the ratio of carotenoid to chlorophyll in surface blooms of the blue-green alga *Microcystis aeruginosa* rose progressively during the summer, to a high value, and attributed this to the carotenoids having a protective role. The carotenoid which reached the highest concentration was zeaxanthin, a xanthophyll which is in fact now thought to exert an important photoprotective function in plants generally, but by mechanisms other than simple interception of the light.[143] In the marine cyanobacterium *Synechococcus*, Kana *et al.* (1988) found that as the light intensity during growth varied from 30 to 2000 μeinsteins $m^{-2}\ s^{-1}$, so the cellular content of β-carotene and chlorophyll *a* diminished several-fold in parallel, but the zeaxanthin concentration remained the same. These authors suggested that in this case β-carotene is entirely part of the photosynthetic system, whereas zeaxanthin by contrast has a wholly photoprotective function.

Photosynthetic units

The increases in photosynthetic pigment content that occur in algae as the light intensity at which they are grown is lowered, can be due to an increase (per cell, or per unit biomass) in the number of photosynthetic units, or in the average size (as absorption cross-section) of the photosynthetic unit, or both.[243] In most green plants – the algae as well as the angiosperms – it appears that the increase in chlorophyll content during shade adaptation is largely due to an increase in the number of photosynthetic units. This has been shown for higher plants,[75] for the unicellular chlorophytes *Scenedesmus obliquus*[256] and *Dunaliella tertiolecta*,[245] and for the multicellular green species *Ulva lactuca*.[613] In *Chlorella pyrenoidosa* the five-fold increase in chlorophyll during shade adaptation was mainly due to an increased number of photosynthetic units, but there was also a 50% increase in the number of chlorophyll molecules per unit.[652] In *Chlamydomonas reinhardtii*, Neale & Melis (1986) found an actual change in the proportion of photosystem I and photosystem II reaction centres as light intensity during growth was altered. The high-light cells, with half the chlorophyll content of the low-light cells, contained slightly less than half as many Photosystem I centres, but almost as many photosystem II centres, as the low-light cells. The photosystem II/photosystem I ratio shifted from near unity in the low-light cells to greater than two in high-light cells.

In the diatoms *Skeletonema costatum*[245] and *Chaetoceros danicus*[697] it appears that the increase in cellular chlorophyll during shade adaptation is mainly due to an increase in the number of chlorophyll *a* molecules per photosynthetic unit, and the same appears to be true of the chrysophyte *Isochrysis galbana*.[697] The diatom *Phaeodactylum tricornutum*, on the other hand, responds to low light by increasing the number of photosynthetic units per cell, without increasing the unit size.[268]

In the marine dinoflagellate *Glenodinium* it does seem likely that the great increase (seven-fold) in the cellular level of the peridinin-chlorophyll *a* protein resulting from a 12-fold decrease in growth irradiance,[721] especially when compared with the comparatively modest increase in chlorophyll (80%), is associated with a substantial increase in the number of these pigment–protein molecules per photosynthetic unit. Shade adaptation in the symbiotic dinoflagellates (zooxanthellae) of coral is seemingly due to an increase in the size, but not the number, of photosynthetic units per cell.[240, 242] In the estuarine dinoflagellate *Prorocentrum mariae-lebouriae*, however, shade adaptation appears to involve increases in both the size and the number of photosynthetic units.[155]

In the unicellular red alga *Porphyridium cruentum*, Levy & Gantt (1988) found that acclimation to low light intensity was accompanied by slightly more than a doubling of biliprotein content, but little change in the amount of chlorophyll or number of photosynthetic units. They concluded that adaptation of this alga to varying light levels involved changes in size of the photosystem II antenna, with little effect on photosystem I. In cases such as *Gracilaria* in the Adriatic Sea referred to above, where shading of macrophytic red algae brings about massive increases in phycoerythrin content, this must be accompanied by increases in the average light-harvesting capacity of the photosynthetic unit. The same is probably true in low-light-grown cultures of the cryptophyte *Cryptomonas*, which show a six-fold increase in phycoerythrin (relative to high-light cultures) compared with only two-fold increases in chlorophylls *a* and *c*.[913]

Shade adaptation in the cyanobacterium *Anacystis nidulans*, comparing cells grown at 10 μeinsteins m^{-2} s^{-1} to those grown at 100 μeinsteins m^{-2} s^{-1}, was accompanied by a doubling in the number of photosynthetic units per cell, but with no change in the number of chlorophyll molecules per photosynthetic unit: the number of phycocyanin molecules per cell, however, tripled.[956] In cultures of the very common bloom-forming cyanobacterium *Microcystis aeruginosa*, shade adaptation led to a 2½-fold increase in the number of photosynthetic units per cell but with little change in the number of chlorophyll molecules per unit.[743] Raps *et al.* (1985) found the low-light (40 μeinsteins m^{-2} s^{-1}) cells to have a 2.6-fold greater concentration of phycobilisomes than the high-light (270 μeinsteins m^{-1} s^{-1}) cells: phycobilisome structure and composition (phycocyanin/allophycocyanin) were the same at both light intensities. In contrast, with a marine *Synechococcus* strain, Kana & Glibert (1987*a*) found considerable changes in phycobilisome composition as growth irradiance varied. Between 700 and 30 μeinsteins m^{-2} s^{-1}, the phycoerythrin/phycocyanin ratio rose from 3 to 14. The phycoerythrin content of the cells increased 20-fold over this range, while chlorophyll content only doubled, suggesting a major increase in the average absorption cross-section of the photosynthetic units.

Changes in electron carriers and carboxylase

In higher plant species which are able to adapt to low light conditions, the increase in pigment during shade adaptation is not accompanied by an increase in the content of the photosynthetic electron transfer components – the cytochromes, ferredoxin and plastoquinone. The levels of these per unit mass may remain about the same or may decrease somewhat.[75,997] The level per unit mass of leaf is markedly lowered in shade-adapted plants.[607] These changes are of adaptive value since at low light intensities the plant will not be able to carry out high rates of electron transfer and carboxylation anyway and so can achieve biosynthetic economies by refraining from increasing, or preferably decreasing, its content of electron transfer components and carboxylase at the same time as it increases its pigment content and photosynthetic unit number. Thus, in shade-adapted higher plants there is an increase in the ratio of pigment assemblies to the pools of electron carriers, and an even bigger increase in the ratio of pigment assemblies to carboxylase.

The same sort of changes have been shown to accompany shade adaptation in some unicellular algae. In the chlorophyte *Scenedesmus obliquus*, while the chlorophyll content per g fresh weight increased by 64% (in the low-light-grown, relative to the high-light-grown cells), the cytochrome *f* content decreased by 33%, and the carboxylase activity also decreased.[256, 806] In shade-adapting *Chlamydomonas reinhardtii*, while the cellular chlorophyll content doubled, the cytochrome *f* level remained about the same.[656] In another chlorophyte *Tetraedron minimum*, reduction of growth irradiance from 500 to 50 μeinsteins m^{-2} s^{-1} led to a five-fold increase in cellular chlorophyll and photosynthetic unit number, but the amount of Rubisco per cell remained the same.[255] In partial contrast to some of the above observations, in the marine chlorophyte *Dunaliella tertiolecta*, the cellular cytochrome *f* concentration increased several-fold in parallel with chlorophyll and photosynthetic unit number during shade adaptation, but in this case also Rubisco content per cell remained about the same.[888] In the marine diatom *Phaeodactylum tricornutum*, grown in continuous culture, while the chlorophyll and carotenoid content rose progressively by about 100% as light intensity was lowered from 12 to 0.5 klux, the Rubisco activity per cell fell to less than 25% of the value present in high-light-grown cells.[50] Thus, in algae the preferred strategy for coping with low light levels appears to be an increase in the number or size (or both) of the pigment assemblies (photosynthetic units) with no increase, or an actual decrease, in the synthesis of carboxylase.

Photosynthetic consequences of light/shade adaptation

The physiological consequences of these biochemical changes are manifested as changes in the dependence of photosynthetic performance on light intensity. If photosynthetic rate per unit chlorophyll is measured as a function of irradiance then it is in some cases found that rates exhibited by low- and high-light-adapted cells or tissues are much the same at low irradiance, but the low-light-adapted plants level off and reach light saturation first. It is generally the case that high-light-adapted cells require a higher light intensity to reach saturation and achieve a higher light-saturated photosynthetic rate, as is shown for *S. obliquus* in Fig. 12.6: the E_k values were about 40 and 110 Wm m^{-2} for the low- and high-light-adapted cells, respectively, in this green alga. Comparable increases in E_c during high-light-adaptation have been observed in a wide range of algal types.

In the case of *S. obliquus*, the high-light-adapted cells also achieve a higher light-saturated photosynthetic rate per unit of cellular biomass (packed cell volume), presumably reflecting the higher cellular content of electron transfer components and carboxylase. This is not observed in all species. In some, such as the dinoflagellate *Glenodinium*,[721] and the chlorophyte, *Chlamydomonas reinhardtii*[656] the high-light-adapted forms achieve a higher light-saturated rate per mg chlorophyll but only about the same light-saturated rate per unit cellular biomass, as the low-light-adapted forms. It seems likely that in such species the carboxylase content does not change during light adaptation.

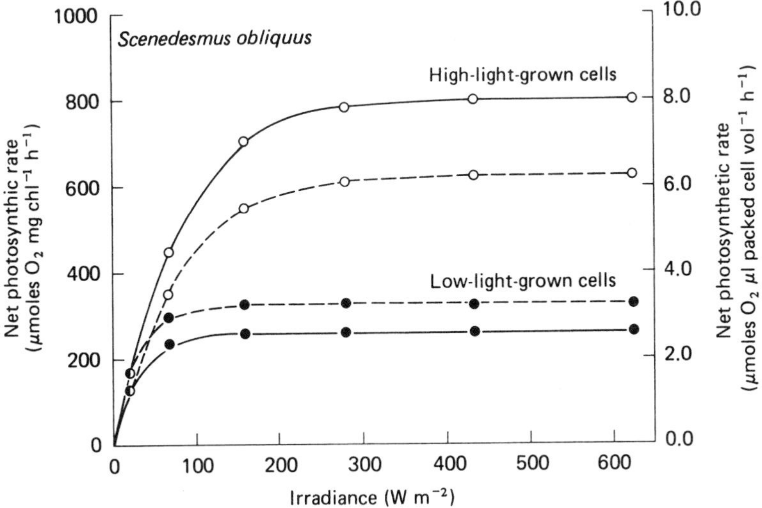

Fig. 12.6. Photosynthetic characteristics (P versus E_d) of cells of *Scenedesmus obliquus* (Chlorophyceae) grown under high (○) or low (●) light intensity. The specific photosynthetic rate is expressed either per mg chlorophyll (——) or per unit cellular biomass (packed cell volume, -----). The cells were grown in continuous culture under high (28 W m^{-2}) or low (5 W m^{-2}) irradiance, and contained 7.8 and 12.8 mg chlorophyll ml^{-1} packed cell volume, respectively. Plotted from data of Senger & Fleischhacker (1978).

We have noted that both high- and low-light-adapted forms of *S. obliquus* and certain other algae have about the same photosynthetic rate per mg chlorophyll at low light intensities: the P versus E_d curve has about the same slope for both types of cells in the light-limited region. This is because at such intensities, electron transfer and carboxylation capacity are present in excess. Photosynthetic rate is determined entirely by the rate of photon capture, and so is determined by the amount of pigment present. Strictly speaking, it is the rate of photosynthesis per unit absorptance rather than per unit chlorophyll which we would expect to be the same for both types of cell in low light, and the ratio of absorptance to chlorophyll can change. The slope, α, of the P versus E_d curve in the light-limited region is equal to $\bar{a}_c\phi_m$ where $\bar{a}_c$ is the specific absorption coefficient (per mg chlorophyll) for PAR, and ϕ_m is the maximum quantum yield, for the algal cells (§10.2). We do not expect ϕ_m to change during shade adaptation: however, we saw earlier (§9.5) that $\bar{a}_c$ is a function of the intracellular pigment concentration as well as of cell size and shape. The more concentrated the pigments within the cell, the less efficiently they collect light, and so the lower the value of $\bar{a}_c$. Thus the increase in pigment content that takes place during shade adaptation is likely (assuming no change in the ratio of pigments present) to lead to some reduction in $\bar{a}_c$, and therefore in α: moderate decreases (7–34%) in α have in fact been observed in certain diatom species during shade adaptation.[697] In an alga in which there was a substantial increase in the ratio of some other light-harvesting pigment to chlorophyll during shade adaptation, and therefore a diminution in the proportionate contribution of chlorophyll to light absorption, we might expect the shade-adapted cells to show a higher rate of photosynthesis per unit chlorophyll at low irradiance than cells grown at high light intensity. Such an effect is in fact found in, for example, those cyanobacterial and red-algal species which increase their biliprotein/chlorophyll ratio during shade adaptation.[263,264,743,456,552]

When shade-adapted cells achieve about the same photosynthetic rate per unit pigment at low irradiance as high-light-grown cells, but contain more pigment per unit cellular biomass, they consequently have a higher photosynthetic rate per unit biomass than the high-light-grown cells under these conditions. This can be seen for *Scenedesmus obliquus* in Fig. 12.6. The difference becomes more pronounced in accordance with the extent to which pigment content increases and is particularly marked for example in the case of *Chlorella vulgaris*.[874] It is at these low light intensities, when the supply of excitation energy is the limiting factor, that the higher concentrations of light-harvesting pigment within the low-light-grown cells gives them the advantage over the high-light-grown cells.

Shade adaptation in aquatic angiosperms

All the adaptive changes we have considered so far are in the biochemical composition and consequent functioning of the photosynthetic apparatus. There are also other forms of ontogenetic adaptation, particularly in multicellular species, which can enable plants to cope with variation in the intensity of the light field. In higher plants, both aquatic and terrestrial, adapting to shade, at the same time as chlorophyll content (% of dry mass) increases, the leaves increase in area so as to intercept more light and also become thinner. Because of the increase in specific leaf area (area per unit dry mass of leaf), the chlorophyll content per unit area increases less than content per unit mass during shade adaptation, or may remain the same, or may decrease. Chlorophyll content (per g fresh weight) has been shown to increase not only with shade, but also with growth temperature, in certain submersed freshwater angiosperm species.[45]

Spence & Chrystal (1970) studied light-intensity adaptation in certain species of the fresh-water angiosperm genus *Potamogeton* from Scottish lochs. In *P. polygonifolius*, a shallow-water species (occurring from the waters edge to 0.6 m depth), shade plants (grown at 6% of full sunlight) had a specific leaf area three times that of plants grown in full sunlight: the shaded leaves were 0.04 mm and the sunlit leaves 0.12 mm thick. The chlorophyll content in the shaded leaves was about one third higher than that in the sunlit leaves on a dry mass basis. Because of the great increase in area, however, the shade leaves contained only about half as much chlorophyll per unit area as the sunlit leaves. The dark-respiration rate per unit area was 27% lower in the shaded leaves – a greater reduction might have been expected given that leaf thickness was reduced by two-thirds. Presumably as a consequence of the diminution in respiration rate, the light compensation point in the shaded leaves was lowered by about the same amount.

P. obtusifolius, a deeper-water species (depth range 0.5–3.0 m) appeared to lack the ability to change its specific leaf area in accordance with light intensity, there being little difference in this respect between plants grown at 100% and 6% of full sunlight. There was also little change in the chlorophyll content per unit mass or leaf area. However, the specific leaf area of *P. obtusifolius*, (~ 2 cm^2 mg^{-1}) is already greater than that of *P. polygonifolius* (sunlit leaves, 0.48 cm^2 mg^{-1}, shaded leaves, 1.43 cm^2 mg^{-1}), and so we may suppose that *P. obtusifolius* has already taken this step during evolution, as part of its phylogenetic adaptation to the deeper water, and it may be that further increases in specific leaf area would not be possible. The respiration rate of *P. obtusifolius* per unit area of leaf was, even in the sunlit leaves, as low as one third, and the light compensation point about half that of the shaded leaves of *P. polygonifolius*, further evidence for the superior phylogenetic adaptation of *P. obtusifolius* to low light. This species still retained some capacity for ontogenetic adaptation, however, since its shaded leaves had a very much lower respiration rate than its sunlit leaves, so much so that its light compensation point was reduced by $\sim 90\%$.

The submersed macrophyte *Potamogeton perfoliatus*, a species inhabiting turbid brackish tidal waters in Chesapeake Bay, responds to a diminution in ambient light by increasing its chlorophyll concentration. Goldsborough & Kemp (1988) found that in plants transferred to a light intensity 11% of ambient, Chl *a* cm^{-2} leaf area increased by 20% in 3 days and 50% after 17 days: on returning the plants to normal light intensities, chlorophyll content reverted to normal in about 3 days. The increased pigment content led to a marked increase in the photosynthetic rate per g dry weight of shoot at low light intensity, with a consequent reduction in the light compensation point: P_{max} in saturating light did not change during the shade treatment. Shading was accompanied by very marked stem elongation, as well as an increase in specific leaf area.

While *P. perfoliatus* seems to have the capacity to adapt to a reduction in available light, many submersed macrophytes cannot cope so well. Seagrasses for example, which typically occur in clear, relatively colourless waters, are particularly sensitive to increased turbidity in the water, resulting from human activity such as dredging or effluent disposal.[606,660]

12.3 Ontogenetic adaptation – spectral quality

Within the underwater environment the spectral composition of the light varies greatly, both with depth within a given water body and from one water body to another, in accordance with the absorption spectrum of the water. As we saw in the previous section, laboratory studies show that variation in intensity alone can bring about marked changes in pigment composition, photosynthetic characteristics and morphology of aquatic plants. Laboratory studies have also shown that intensity is not the whole story and that the spectral quality of the incident light has, for some plants, a specific role to play.

Chromatic adaptation within the blue-green algae

The most clear-cut examples are to be found amongst the blue-green algae (Cyanobacteria), and it was indeed in a blue-green algal species that ontogenetic chromatic adaptation was first described. Gaidukov (1902) observed that *Oscillatoria rubescens* was red in colour when grown in green light, and blue-green when grown in orange light: he attributed these colour changes to the synthesis of different kinds of pigment. Boresch (1921) showed that the colour changes are due to shifts in the types of biliproteins synthesized by the algae: the red cells contain predominantly phycoerythrin, the blue-green cells mainly phycocyanin. This phenomenon was considered by Engelmann & Gaidukov (1902) to be an example of *complementary chromatic adaptation*, the pigment induced by a specific waveband being one which absorbs that waveband (phycoerythrin and phycocyanin have their absorption peaks in the green (~565 nm) and the red (~620 nm), respectively): 'complementary' because the pigment has the complementary colour to that of the light which induces it.

Not all blue-green algal species show chromatic adaptation, and amongst those that do there are variations in the form that adaptation takes. Tandeau de Marsac (1977) separated cyanobacteria into three groups. In group I strains the biliprotein composition is not affected by the spectral quality of the light in which they are grown, i.e. they lack chromatic adaptation. In group II strains, only the phycoerythrin content of the cells is affected by light quality, being very low in red light and high in green light; phycocyanin content remains high in cells grown under either kind of light. In group III strains the synthesis of both the major biliproteins is affected by the spectral composition of the light: like the previous group their phycoerythrin content is high in green light and low in red, but their phycocyanin content, while still substantial in green light, is 1.6- to 3.7-fold higher in red light. Of 69 strains examined, 25 were in group I, 13 in group II and 31 in group III.[109,908]

The phycocyanin formed in group I and group II strains has two polypeptide subunits, α and β. Bryant (1981) found that in 24 of the 31 group III strains, the phycocyanin formed in green light also contained just two subunits, but the phycocyanin from cells grown in red light contained four subunits, α_1, α_2, β_1, and β_2. The particular subunits present in phycocyanin in the green-light-grown cells were α_2 and β_1. These, being always present, Bryant referred to as constitutive: the other two, α_1 and β_2, formed in red light, he referred to as inducible. Thus, in these 24 strains it appeared that the increased phycocyanin formation induced in red light consisted of synthesis of different kinds of phycocyanin subunit. It was not possible to determine from the data whether the new phycocyanin subunits associated to form a distinct phycocyanin species, $(\alpha_1\beta_2)_n$, or whether they aggregated with the two constitutive subunits, α_2 and β_1, to form a hybrid phycocyanin. In the remaining seven group III strains, no α_1 and β_2 phycocyanin subunits were found in red-light-grown cells: the extra phycocyanin in these cases may have consisted just of additional α_2 and β_1 subunits. In *Tolypothrix tenuis*, Ohki *et al.* (1985) found that chromatic adaptation in green light resulted in a one for one substitution of phycoerythrin for phycocyanin, so that phycobilisome size remained constant.

The specific nature of the light treatments required to bring about changes in biliprotein composition have been studied in one of the group III strains, *Calothrix* 7101 (formerly *Tolypothrix tenuis*), by Fujita & Hattori (1960, 1962*a*, *b*) and Diakoff & Scheibe (1973). It has been found convenient to study the controlling influence of light in *Calothrix* 7101 by first depleting the cells of biliproteins by exposure to intense light for 24 h in the absence of nitrogen, then giving a brief light treatment of the desired

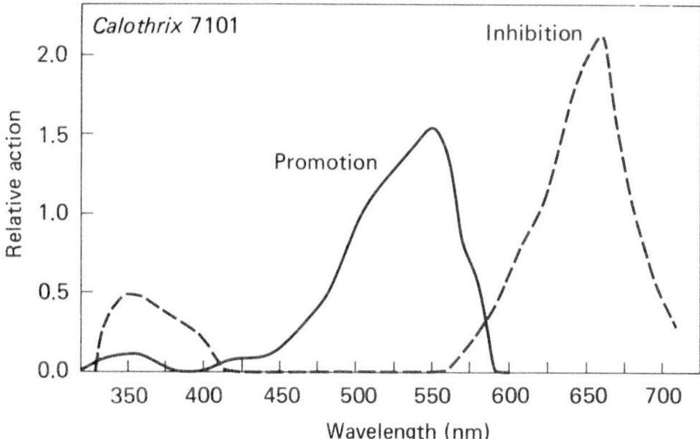

Fig. 12.7. Action spectra for the promotion and inhibition of phycoerythrin synthesis in the blue-green alga *Calothrix* 7101 (*Tolypothrix tenuis*) (after Diakoff & Scheibe, 1973). The reciprocal of the amount of radiant energy (J cm^{-2}) required to produce a 25% increase (promotion) or decrease (inhibition) in the amount of phycoerythrin (as a proportion of total biliprotein) synthesized in the dark after the light treatment is (after normalization to 500 or 680 nm) plotted against wavelength.

spectral quality, and then placing the cells in the dark with a nitrogen source, and following biliprotein synthesis. After a saturating dose of red light the cells synthesize phycocyanin but not phycoerythrin during the subsequent dark incubation: after a saturating dose of green light they continue to make phycocyanin but now also make phycoerythrin. If red and green light treatments are given alternately, the subsequent pattern of biliprotein synthesis is determined by the colour of the last light treatment, i.e. green or red light will each reverse the effect of the other.[276]

The action spectrum for the promotion of phycoerythrin synthesis has a peak in the green at about 550 nm with an additional small peak in the UV at about 350 nm (Fig. 12.7): the action spectrum for inhibition of phycoerythrin synthesis has a peak in the red at about 660 nm, and also has a subsidiary peak in the UV at about 350 nm.[190,275] The shape of these action spectra are consistent with the photoreceptive pigment system being itself biliprotein in nature. Bogorad (1975) has suggested the term 'adaptochrome' for any photoreceptive pigment involved in the regulation of biliprotein synthesis.

Scheibe (1972) has proposed that this controlling pigment is analogous to phytochrome and has two interconvertible forms: one (P_G) would have its absorption maximum in the green and on irradiation with green light undergo photoconversion to another form (P_R) with an absorption maximum in the red region. On irradiation with red light P_R would be photoconverted to P_G

$$P_G \underset{\text{red } h\nu}{\overset{\text{green } h\nu}{\rightleftarrows}} P_R$$

Scheibe points out that both forms of the pigment could be biologically active, P_G promoting phycocyanin and P_R promoting phycoerythrin formation, or else just one form could be active, its presence or absence determining which of two possible differentiative pathways is to be followed. Oelmüller *et al.* (1988) have shown that in *Fremyella diplosiphon* green light induces the formation of the phycoerythrin messenger RNA and red light induces formation of the phycocyanin mRNA.

Although in light-grown cells the spectral quality of the light influences the pattern of synthesis of biliproteins, light is not essential for the synthesis of these pigments to occur: facultatively heterotrophic cyanobacteria continue to make biliproteins when grown on carbohydrate medium in the dark. When such strains in group III are grown in the dark their biliprotein composition is similar to that of cells grown in red light: high in phycocyanin and low in phycoerythrin.[908] Bryant (1981) found that the phycocyanin synthesized by *Calothrix* strains 7101 and 7601 in the dark included the inducible subunits, α_1, and β_2, found in red-light-grown but not green-light-grown cells. This was the case whether the inocula used were derived from cultures grown in red or green light. Bryant concluded that it is not exposure to red light, but the absence of green light which turns on the synthesis of the inducible phycocyanin subunits. If there is indeed a photoreversible pigment involved in the control of biliprotein synthesis, then it is presumably the P_G form which is synthesized by cells growing in the dark.

Chromatic adaptation in eukaryotic algae

Red algae do show changes in their pigment composition in response to changes in the spectral quality of the light field in which they are grown. The direction of the pigment change, however, quite apart from its quantitative extent, seems to depend on the intensity of the light. Brody & Emerson (1959) determined the ratio of phycoerythrin to chlorophyll in the unicellular red alga *Porphyridium cruentum* grown in green light (546 nm – absorbed mainly by phycoerythrin) or blue light (436 nm – absorbed mainly by chlorophyll), at low (~ 0.1 W m^{-2}) or high (25–62 W m^{-2}) intensity. At low irradiance the cells grown in green light had a phycoerythrin/chlorophyll

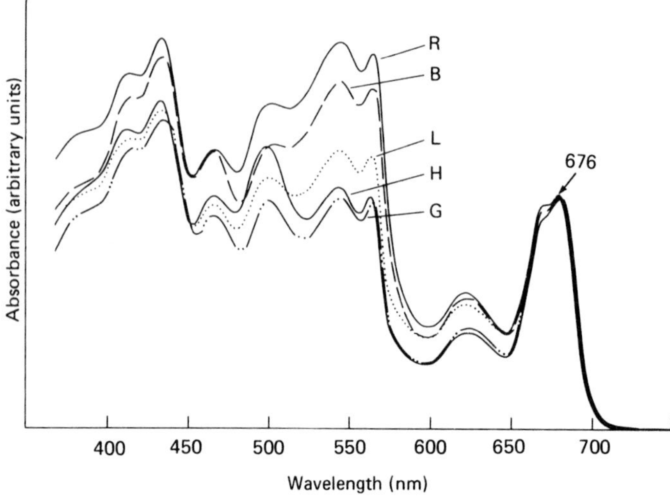

Fig. 12.8. Effect of spectral quality of light field during growth on the photosynthetic pigment composition of the unicellular red alga *Porphyridium cruentum* (by permission, from Ley & Butler (1980), *Plant Physiology*, **65**, 714–22). The spectra are normalized to the same absorbance value at 676 nm. The letter labelling each spectrum indicates the light field in which the cells were grown: R = red, B = blue, L = low-intensity white, H = high-intensity white, G = green.

ratio more than twice that of the cells grown in blue light. In a *Cryptomonas* species isolated from the deep chlorophyll layer in the western Pacific Ocean, Kamiya & Miyachi (1984) found the phycoerythrin/chlorophyll ratio to be much higher in cells grown in green light than in cells grown in blue or red light, the same low irradiance (0.8 W m^{-2}) being used in each case. We may regard these pigment changes as complementary chromatic adaptation: the cells increased the proportion of that pigment which best absorbed the light to which they were exposed.

In the *Porphyridium* cells grown at high monochromatic irradiance, however, the position was reversed: the phycoerythrin/chlorophyll ratio of cells grown in green light was less than 50% of that in cells grown in blue light. Similarly, Ley & Butler (1980) found that cells of *P. cruentum* grown in high-intensity red (112 μeinsteins m^{-2} s^{-1}) or blue (50 μeinsteins m^{-2} s^{-1}) light have about twice the phycoerythrin/chlorophyll ratio of cells grown in high-intensity green (99 μeinsteins m^{-2} s^{-1}) light (Fig. 12.8). In the multicellular red alga *Porphyra*, Yocum & Blinks (1958) found that plants exposed to high-intensity blue light (436 nm, 24 W m^{-2}) for 10 days contained more phycoerythrin and less chlorophyll than plants exposed for the same time to high-intensity green light (546 nm, 17.5 W m^{-2}).

These changes in pigment composition induced by high-intensity monochromatic illumination are accompanied by changes in the photosynthetic characteristics of the plants. Yocum & Blinks (1958) found that marine red algae which had just been collected or had been kept under green light for 10 days showed low photosynthetic efficiency in the region of the chlorophyll red absorption band, the action spectrum falling well below the absorption spectrum between 650 and 700 nm. Plants which had been kept for 10 days under blue light, however, were highly efficient in the red region (as well as showing some increase in the blue region) the action and absorption spectra now approximately coinciding between 650 and 700 nm. Clearly, what had resulted from the exposure to blue light (itself absorbed by chlorophyll) was an increase in the efficiency of utilization of light absorbed by chlorophyll.

The nature of adaptive changes that take place within the photosynthetic system under these various moderate-to-high intensity, spectrally selective light regimes has been clarified by the detailed study of Ley & Butler (1980) on *Porphyridium cruentum*. By careful analysis of the absorption spectra and the fluorescence behaviour they have been able to arrive at conclusions concerning the absorption characteristics of, and energy transfer between, the two photosystems in cells grown in light of various spectral compositions. In red algae, chlorophyll *a* is the main light-harvesting pigment for photosystem I and phycoerythrin is the main pigment in photosystem II. Cells growing in red or blue light will be receiving much more excitation energy in chlorophyll than in phycoerythrin, which, given a photosynthetic system initially adapted to a broad spread of wavelengths, will lead to a much greater input of energy into photosystem I than photosystem II. To ensure a balanced operation of the photosynthetic system, the cells must increase the absorption cross-section of photosystem II relative to that of photosystem I. This they achieve partly by increasing the ratio of phycoerythrin to chlorophyll, but also by putting a much larger proportion of their chlorophyll into photosystem II. In addition, Ley & Butler found that these red- or blue-light-grown cells have a diminished probability of energy transfer from photosystem II to photosystem I, which helps to keep excitation energy in photosystem II. The increased effectiveness of red light in *Porphyra*, observed by Yocum & Blinks after prolonged exposure of the alga to blue light, we may now plausibly explain in terms of an increased proportion of chlorophyll in photosystem II, and therefore a more balanced functioning of the photosystems in red light. Cells growing in green light will, if they are initially adapted to a mixture of wavelengths, receive far more excitation energy in photosystem II than I. Their adaptive response is to lower the absorption cross-section of photosystem II relative to that of photosystem I by reducing the phycoerythrin/chlorophyll ratio, and including nearly all their chlorophyll in photosystem I: PSI chlorophyll/PSII chlorophyll is $\sim$20 in green-light-grown, compared to $\sim$1.5 in red- or blue-light-grown cells.[559] In addition, the green-light-grown cells show a greater probability of energy transfer from photosystem II to I than cells grown in red or blue light.

The red- or blue-light-grown cells have a higher phycoerythrin content than the green-light-grown cells on a per cell basis, as well as in proportion to chlorophyll. It might be argued that although it makes sense to put more chlorophyll into photosystem II, to make more phycoerythrin in red light is not a useful adaptive response since this biliprotein has little absorption above 600 nm and so will not collect the red photons. However, it may be that what the cell 'detects' is not specifically the spectral nature of the incident light, but simply the fact that photosystem I is receiving more energy than photosystem II, and so it responds with a general increase in the pigment complement of photosystem II. Red-dominated light fields do not occur naturally in the marine environment and so we should not expect red algae to show specific adaptation to them.

From the work of Ley & Butler it seems that the general principle on which chromatic adaptation to relatively high intensity, spectrally confined, light fields takes place is adjustment of the composition and properties of the photosystems in such a way as to lead to their being excited at about the same rate, and thus to efficient photosynthesis. The complementary chromatic adaptation observed by Brody & Emerson (1959) in *P. cruentum* exposed to low-intensity blue and green light, since it operates in a contrary fashion so far as pigment changes are concerned, must have a different basis. At low light intensity the total rate of supply of excitation energy rather than imbalance between the photosystems becomes the major constraint on photosynthesis. It may be that the best strategy for the algae growing in dim green or blue-green light is to make whatever pigments will best capture the available light, to incorporate these predominantly into photosystem II (biliproteins are in this photosystem, anyway) and transfer a proportion of the absorbed energy to photosystem I.

The fact that the chromatic adaptation in cyanobacteria, involving changes in the ratio of phycoerythrin to phycocyanin, is entirely of the complementary type is not surprising. Both the pigments involved are biliproteins and both feed their excitation energy to photosystem II, so the problem of imbalance between the photosystems does not arise. Growth of the blue-green alga *Anacystis nidulans* in red ($\lambda > 650$ nm) light does lead to a 75% fall in chlorophyll content with little change in phycocyanin content.[440] We may now interpret this as an attempt by the cells to reduce excitation of photosystem I to the point at which it is in balance with photosystem II. In the cyanobacterium *Synechococcus* 6301, the phycocyanin/chlorophyll ratio increases in red light (absorbed by chlorophyll *a*, mainly associated with photosystem I) and decreases in yellow light (absorbed by phycocyanin, associated exclusively with photosystem II). Manodori & Melis (1986) interpreted these changes as representing the achievement of a balanced excitation of the two photosystems by adjustment of photosystem stoichiometry: the actual antenna size of the individual photosystems apparently did not alter.

The blue-light effect

The underwater light field in oceanic waters becomes, as we have seen earlier, predominantly blue-green and eventually blue with increasing depth. There is evidence for the existence of specific effects of blue light on the development of the photosynthetic apparatus of certain phytoplankton species. Wallen & Geen (1971*a*, *b*) found that cells of the marine diatom *Cyclotella nana* grown in blue light (0.8 W m^{-2}) contained 20% more chlorophyll *a* and had a 70% higher light-saturated photosynthetic rate per cell than cells grown in white light at the same intensity. Jeffrey & Vesk (1977, 1978) observed that in cells of the marine diatom *Stephanopyxis turris* grown in blue-green light (4 W m^{-2}), the cellular content of all the chloroplast pigments (chlorophylls *a* and *c*, fucoxanthin and other carotenoids) was about twice that in cells grown under the same intensity of white light (Fig. 12.9). Both the number of chloroplasts per cell and the number of three-thylakoid compound lamellae within each chloroplast were higher in the cells grown in blue-green light (Fig. 12.10). Furthermore, the blue/green-light-grown cells were better able to utilize low intensity (the same irradiance as during growth) blue-green light for photosynthesis than the white-light-grown cells, achieving a 42% higher rate. Vesk & Jeffrey (1977) examined a number of other marine phytoplankton species. Substantial increases in pigment content (55–146%) in blue-green-light-grown cells compared to white-light-grown (both at 4 W m^{-2}) cells were observed in five diatom, one dinoflagellate and one cryptomonad species. Minor increases (17–39%) were found in two diatom, two dinoflagellate, one prymnesiophyte, one chrysophyte and one chlorophyte species. There was no chlorophyll increase in two diatom and one prymnesiophyte species. Thus, the ability to increase pigment content in a blue-green light field is common, but not universal amongst marine phytoplankton. There was

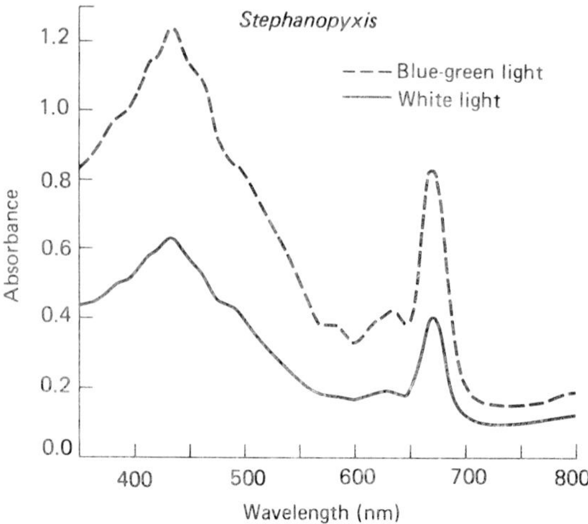

Fig. 12.9. Absorbance spectra of cells of the marine diatom *Stephanopyxis turris* grown in white (——) or blue-green (------) light (4 W m^{-2}, in each case). (By permission, from Vesk & Jeffrey (1977), *Journal of Phycology*, **13**, 280–8.) The cells are suspended at a concentration of 200 000 cells ml^{-1}

little alteration in the proportions of the different photosynthetic pigments, and so these blue-light-induced changes are not examples of complementary chromatic adaptation. Evidence that the increases in pigment content were accompanied by increases in the number of thylakoids per chloroplast was obtained in a dinoflagellate and a cryptomonad, as well as certain diatom species: in some diatom species the number of chloroplasts also increased.[414]

12.4 Ontogenetic adaptation – depth

We have seen that aquatic plants undergo adaptive changes in response to changes in the intensity and spectral quality of the ambient light. It is not surprising, therefore, that in species which in nature occur over a range of depths, adaptive changes with depth can often be observed. In this section we shall examine some of the changes that have been observed in aquatic plants accompanying increases in depth, and consider to what extent the lowered intensity or the altered spectral composition are responsible.

Depth variation of pigment composition in multicellular plants

Ramus *et al.* (1976) suspended samples of two green- (*Ulva lactuca* and *Codium fragile*) and two red-algal species (*Chondrus crispus* and *Porphyra*

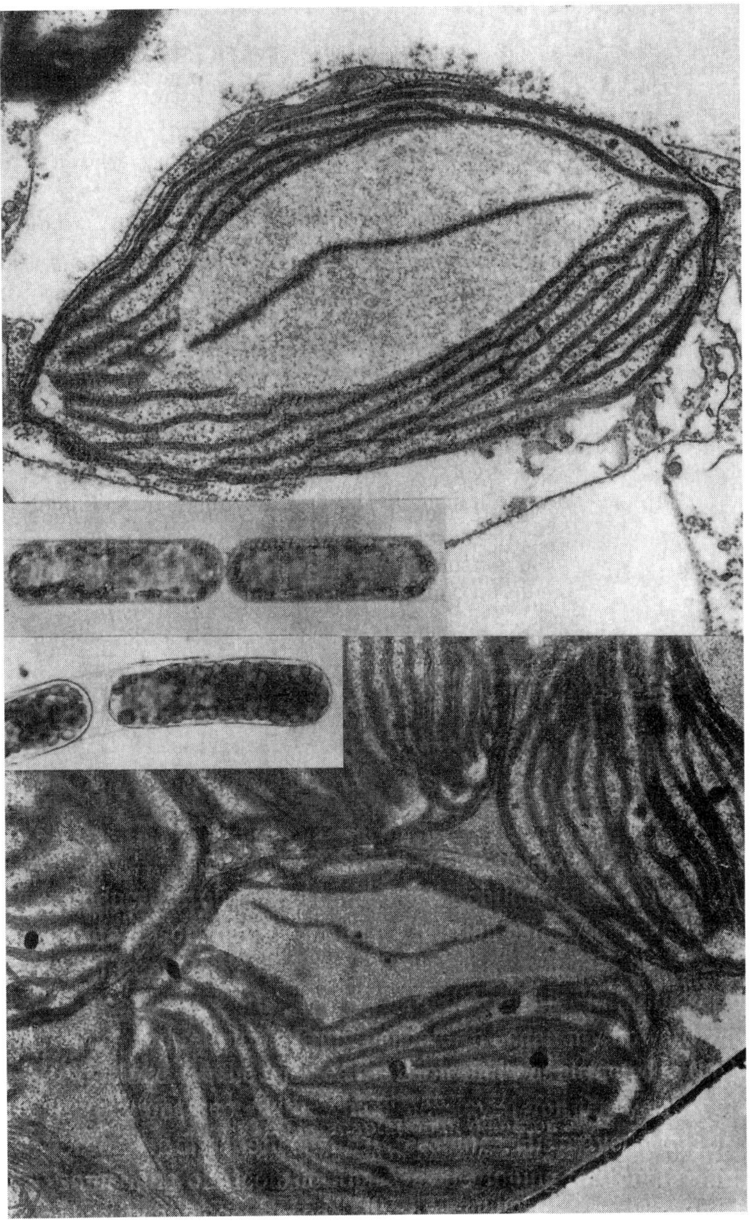

Fig. 12.10. Chloroplast number and structure in cells of the marine diatom *Stephanopyxis turris* grown in white or blue-green light (by permission, from Jeffrey & Vesk (1977), *Journal of Phycology*, **13**, 271–9). Upper: electron micrograph of chloroplast in cell grown in white (4 W m^{-2}) light. Inset is light micrograph of whole cells. Lower: electron micrograph of chloroplasts in cell grown in blue-green (4 W m^{-2}) light showing increased number of three-thylakoid compound lamellae. Inset is light micrograph of whole cells showing increased number of chloroplasts.

umbilicalis) for seven-day periods at depths of 1 and 10 m in harbour water at Woods Hole, Massachusetts, USA. Pigment compositions were determined. The samples were then reversed in position – the 1-m samples being lowered to 10 m, and *vice versa* – and after a further seven-day period, further pigment analyses were carried out. In all four species the level of photosynthetic pigments increased in the plants held at 10 m, relative to those held at 1 m, the increase for chlorophyll *a* being about 1.4-fold in the red algae and 2.3 to 3.4-fold in the green. In *U. lactuca*, chlorophyll *b* increased about 5-fold, so that the *b*/*a* ratio rose by 50%, but in *C. fragile* the *b*/*a* ratio remained the same. In the red-algal species the increase in phycoerythrin in the 10 m relative to the 1 m samples was greater than the increase in chlorophyll *a*, so that the phycoerythrin/chlorophyll ratio rose by 50–60% with increased depth.

The light field at 10 m depth at this site, being predominantly yellow-green, differed greatly from that at 1 m depth in spectral composition as well as intensity. In an attempt to determine the effects of diminished intensity alone, Ramus *et al.* analysed the pigments from algae taken from sunny and shaded sites in the intertidal region.[739] Shade increased pigment levels to about the same extent as submersion to 10 m depth. In *U. lactuca* and *P. umbilicalis*, the chlorophyll *b*/*a* and the phycoerythrin/chlorophyll ratios, respectively, did increase in the shade but to a lesser extent than that induced by increased depth. Ramus *et al.* concluded that the greater shifts in pigment ratio associated with increased depth were indeed a response to the changed spectral composition of the light, not just to the lowered intensity.

Wheeler (1980*a*) carried out transplantation experiments with the sporophytes of the giant kelp *Macrocystis pyrifera* (Phaeophyta). When plants growing at 12 m depth were moved up to 1 m depth, then over a 10-day period the pigment content per unit area of thallus diminished markedly, the decreases being 52% for chlorophyll *a*, 35% for chlorophyll *c* and 68% for fucoxanthin. When the plants were replaced at 12-m depth pigment levels increased again to their original values in 18 days. Readaptation to increased depth was accompanied by an increase in the molar ratio of fucoxanthin to chlorophyll *a* from 0.50 to 0.77 but by a decrease in the chlorophyll *c*/*a* ratio from 0.50 to 0.31. When samples of the intertidal brown algae *Ascophyllum nodosum* and *Fucus vesiculosus* were suspended at 4 m depth for seven days, concentrations of fucoxanthin and chlorophylls *a* and *c* per g fresh mass rose 1.5- to three-fold.[740] However, in these species fucoxanthin increased somewhat less than chlorophyll *a* so that the fucoxanthin/chlorophyll *a* ratio diminished by about 20%: the chlorophyll *c*/*a* ratio remained the same during depth adaptation.

Interesting though transplantation experiments are, the alternative approach of studying plants of a given species growing naturally at different depths is of greater ecological relevance. Plants of the siphonous green alga *Halimeda tuna* growing in the Adriatic Sea at 5 m had a total chlorophyll content per g dry mass which was 50 % greater than that in plants growing at 2 m depth:[847] the chlorophyll *a*/*b* ratio fell only slightly with increased depth, from 2.03 to 1.86. The chlorophyll content of the green benthic alga *Udotea petiolata* growing in Spanish coastal water increased with depth by 50% over the range 5 to 20 m, most of the increase taking place between 5 and 10 m:[694] the *a*/*b* ratio did not change with depth.

Wiginton & McMillan (1979) studied a number of seagrass species growing at various depths off the Virgin Islands and the coast of Texas, USA. Three of the species showed no significant increase in chlorophyll content with increasing depth down to their maximum depths of 12 to 18 m. In a fourth species, *Halophila decipiens*, however, which penetrated to 42 m, the chlorophyll content changed only slightly from 7 to 18 m, but approximately doubled between 18 and 42 m. The chlorophyll *a*/*b* ratio fell from 1.72 at 18 m to 1.49 at 42 m. Plants of this species from a shallow but shaded site had the same chlorophyll content and *a*/*b* ratio as plants from 42 m depth: this was taken to indicate that the pigment changes associated with increased depth were due to the decrease in total intensity rather than to the change in spectral quality. Plants of the isoetid species *Littorella uniflora* growing at 2.3 m depth (20% of subsurface irradiance) in a Danish lake contained about 65% more chlorophyll (dry weight basis), and had a lower chlorophyll *a*/*b* ratio (2.6 compared to 3.2) in their leaves than plants growing at 0.2 m depth (70% of subsurface irradiance).[849]

In the brown alga *Dictyota dichotoma* growing in Spanish coastal water, Perez-Bermudez, Garcia-Carrascosa, Cornejo & Segura (1981) found that the chlorophyll *a* and *c* contents (on a mass basis) increased by 17 and 53%, respectively, between 0 and 20 m depth, the *a*/*c* ratio decreasing from 2.35 to 1.80: at 10 m depth the values were intermediate. The fucoxanthin content, on the other hand, decreased by 42% between 0 and 10 m depth, and then increased again to its original value between 10 and 20 m depth: the possible functional significance of this is discussed in a later section (§12.5). Plants growing in a shaded site at the surface contained 35% more chlorophyll *a*, 63% more chlorophyll *c* and 20% more fucoxanthin than plants growing in a sunny site. In a study on the red alga *Chondrus crispus* growing from 0 to 20 m depth off Massachusetts, Rhee & Briggs (1977) found that the phycoerythrin/chlorophyll ratio in midsummer remained approximately constant down to about 10 m depth, but approximately doubled between 10 and 13 m, remaining at this value at greater depth.

On the basis of the studies that have so far been carried out, it appears that many benthic aquatic species can increase the content and/or alter the proportions of their photosynthetic pigments with increasing depth.

Depth variation of pigment composition in unicellular algae

The majority of phytoplankton cells are subjected to continually varying irradiance as they circulate through the mixed layer and do not have the time to change their pigment composition in accordance with the prevailing light field. When as in some of the clearer, less-coloured waters, light penetrates so well that the euphotic zone is deeper than the mixed layer, however, then in the stable water below the thermocline the phytoplankton can remain at approximately the same depth for long periods and there is evidence that they do indeed adapt to the light conditions prevailing at such depths. We noted earlier (§11.1) that phytoplankton isolated from the deep chlorophyll maximum in the oceans have about twice as much chlorophyll per unit biomass as phytoplankton from the surface layer.[468] This is not necessarily adaptation within a species: whether members of the same species are found both in a highly pigmented form in the oceanic deep chlorophyll maximum and in a less-pigmented form near the surface, or whether the taxonomic composition of the phytoplankton population is entirely different at the two depths, is not known.

In inland waters with their usually more rapid attenuation of light with depth, it is usually the case that the depth of the mixed layer is greater than or equal to that of the euphotic zone, so there is little opportunity for depth adaptation by the phytoplankton. There are, however, some lakes such as Crater Lake, Oregon, USA, and Lake Tahoe, California–Nevada, USA, with water so clear and colourless that the euphotic zone is deeper than the mixed layer and a deep chlorophyll maximum develops at 75–100 m depth in the growing season.[539, 924] Even in lakes in which light does not penetrate deeply, if there is for some reason particularly stable thermal stratification so that circulation is restricted to a very shallow layer, then the euphotic depth can exceed the mixed depth. This appears to be the case in Lake Lovojärvi in Finland: in the summer this was found to have an epilimnion about 2.5 m deep but with photosynthesis extending down to about 3.5 m.[442] A chlorophyll maximum, consisting largely of cells of the blue-green alga *Lyngbya limnetica* with a particularly high chlorophyll content, developed at about 3.5 m.

In Lake Tahoe, Tilzer & Goldman (1978) showed that in June and September (particularly the latter) the chlorophyll content of the phytoplankton (mainly diatoms) per unit biomass increased with depth. In September, for example, the chlorophyll *a* content (μg per mg wet mass) was 0.61, 1.8 and 3.06 at 0, 50 and 105 m depth, respectively. Since one diatom species, *Fragilaria vaucheriae*, constituted about 58% of the total biomass at all three depths, it seems likely that the increase in chlorophyll content of the total biomass represented true adaptation within this, and perhaps other, species. In June also, when three other species constituted most of the biomass, a doubling of chlorophyll *a* content between 20 and 50 m was accompanied only by comparatively small changes in the population composition, suggesting that chlorophyll content had increased within species. In Lake Lovojärvi, in contrast, there was a marked change in population composition between the mixed layer and the chlorophyll maximum in the hypolimnion, with *Lyngbya* being unimportant in the former and dominant in the latter: the chlorophyll maximum in this lake we may regard as being due to phylogenetic rather than ontogenetic adaptation. Similarly, in clear-water lakes in northwestern Ontario, Canada,[250] the deep chlorophyll maximum (4–10 m depth) was in each case dominated by a single large, colonial chrysophycean flagellate species (the particular species varying from one lake to another), presumably adapted to this particular niche.

Given the demonstrated *in vitro* pigment adaptation of marine phytoplankton species to blue light, and taking into account also the data for Lake Tahoe (optically similar to ocean water), we may reasonably suspect that the deep chlorophyll maximum in oceanic waters is caused at least in part by increased pigment synthesis promoted by the low intensity and/or the predominantly blue character of the prevailing light, within species which occur over a range of depths. It seems likely that there are present, in addition, specialized, highly pigmented species for which the bottom of the euphotic zone is the preferred niche. Studies of the taxonomic composition of the phytoplankton population as a function of depth in stratified oceanic waters could provide information on the relative importance of phylogenetic and ontogenetic adaptation in the establishment of the deep chlorophyll maximum.

Not all unicellular algae are planktonic. Any species which normally occur attached to, or within, fixed structures are thereby freed from the continually varying character of the light field experienced by the plankton and so have the opportunity to adapt to the light field prevailing at the depths where they are located. Little seems to be known about pigment and photosynthetic adaptation within those unicellular species which occur on the surface of sediments or rocks, or epiphytically upon larger algae.

Studies have, however, been carried out on the symbiotic dinoflagellates (zooxanthellae) which occur within corals. There is a general tendency, but to varying extents in different coral species, for the pigment content of the zooxanthellae to increase with depth. Leletkin, Zvalinskii & Titlyanov (1981) compared the zooxanthellae of the coral *Pocillopora verrucosa* growing at 20 and 45 m depth in the Timor Sea. Those from coral at 45 m contained about 1.5 times as much chlorophylls *a* and *c*, β-carotene and diadinoxanthin, and 2.4 times as much peridinin per cell as zooxanthellae from coral at 20 m. The number of chlorophyll molecules per photosynthetic unit was 42% higher in the cells from 45 m than in those from 20 m, but the number of photosynthetic units per cell was about the same. Thus the pigment adaptation to increased depth appears largely to consist of an increase in the pigment complement per photosynthetic unit.[242]

Depth variation of photosynthetic characteristics in multicellular algae

Plants of the subtidal red alga *Ptilota serrata* collected from 24 m depth off northeastern America showed light saturation of photosynthesis at about 116 μeinsteins m^{-2} s^{-1}, whereas plants from 6 m saturated at about 182 μeinsteins m^{-2} s^{-1}.[600] At very low light intensities (7–14 μeinsteins m^{-2} s^{-1}) both kinds of plant achieved the same photosynthetic rate per g dry mass but at higher intensity the deep-water plants levelled off first and achieved a maximum photosynthetic rate only half that of the shallow-water plants (Fig. 12.11). It seems likely that the shallow-water plants contained a more active carboxylation system.

A similar adaptive response has been observed by Glenn & Doty (1981) in the tropical red alga *Eucheuma striatum*. They attached thalli of this alga (grown in shallow water) at a series of points to a line extending from 1 to 9.5 m depth on a Hawaiian reef: irradiance diminished by about a factor of 10 over this range. After a month the thalli were recovered for photosynthetic measurements. A progressive decrease in photosynthetic capacity (light-saturated photosynthetic rate per g dry mass) with depth was observed, the value in thalli from 9.5 m being about half that in material from 1 m. Thus, *Eucheuma* seems to divest itself of surplus carboxylating capacity as depth increases: a response which should be of benefit to the alga.

In the case of the giant kelp *Macrocystis pyrifera*, a single frond can extend from the holdfast on the sea floor up through as many as 30 m of water column and thus through a very considerable light gradient. Wheeler (1980*a*) found that in immature fronds, 12–14 m long, growing in 12 m of

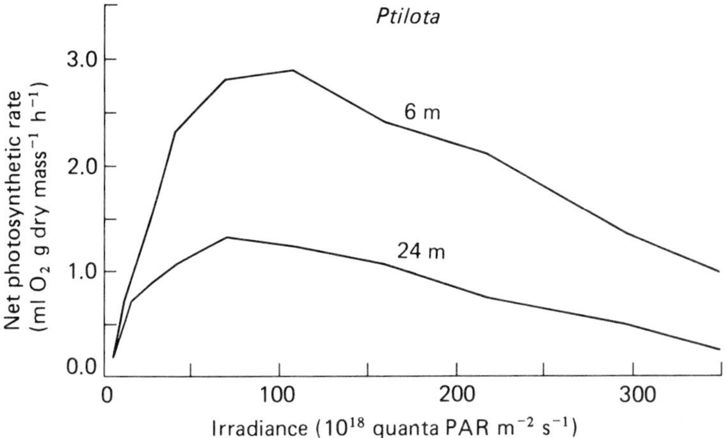

Fig. 12.11. Photosynthesis as a function of irradiance in plants of the subtidal red alga *Ptilota serrata* collected from 6 m and 24 m depth (after Mathieson & Norall, 1975).

water the pigment content rose by about 50% in the first 2–4 m above the holdfast and then remained approximately constant up to about 12 m. The light-saturated photosynthetic rate per unit area rose progressively with distance from the holdfast, 1.5- to three-fold, reaching a maximum at about 10–12 m. Thus, over this part of the frond the photosynthetic capacity per unit area rose with diminishing depth, which we may regard as an adaptive response since it is only at the lesser depths that there is enough light to support a high photosynthetic rate per unit area.

Depth variation of photosynthetic characteristics in phytoplankton

As we have noted earlier, depth adaptation of the photosynthetic system in phytoplankton is only to be expected in waters in which density stratification permits some of the cells to remain for long periods at depths where circulation is minimal but there is sufficient light for photosynthesis to take place. This situation exists throughout most of the ocean in the lower part of the euphotic zone. By what criteria can we recognize true depth adaptation of the phytoplankton photosynthetic system? The primary criterion is that cells from deep water should achieve a higher photosynthetic rate per unit cellular biomass than cells from the surface, in the light field prevailing in deep water. The deep-water cells would generally achieve this by investing an increased proportion of their cellular substance in light-harvesting pigments, possibly accompanied by lower respiration. A secondary criterion is that the deep-water cells should have a lower photosynthetic rate per unit cellular biomass than the surface cells at saturating light intensity: this would be a consequence of the deep-water cells, in the interests of biochemical economy, synthesizing only as much carboxylase as they can make use of at the low irradiance values existing at the greater depths.

On its own, a reduction in the light saturation parameter, E_k, is not an entirely satisfactory indication of adaptation to low irradiance.[1008] Since $E_k = P_m/\alpha$ (see §10.1), any environmental factor which reduces the light-saturated photosynthetic rate per mg chlorophyll to a greater extent than it affects the slope of the initial part of the P (per mg chlorophyll) versus E_d curve, will reduce E_k. Yentsch & Lee (1966) showed, by analysing a large amount of published data for both natural and cultured phytoplankton populations, that E_k tends to be linearly related to P_m and they emphasize the desirability of presenting P_m as well as E_k values in studies on light adaptation by phytoplankton. Nevertheless if, when measured under standardized conditions, one phytoplankton population is found to have a lower E_k value (or light saturation value) than another, then while we cannot on this ground alone conclude that its light-harvesting capability has increased, we may reasonably suppose that its ratio of light-harvesting capacity to carboxylation rate has increased, i.e. it has either increased its pigment content or divested itself of some of its surplus carboxylation capacity, or both, all of which can be regarded as forms of shade adaptation. It is generally found in the oceans that once thermal stratification has been established, phytoplankton isolated from the lower part of the euphotic zone have a lower E_k value than phytoplankton from near the surface.[874] Phytoplankton from the surface, 50 m and 100 m depth (100%, 10% and 1% of subsurface irradiance) in the Sargasso Sea in October, had E_k values of about 600, 300 and 60 μeinsteins $m^{-2} s^{-1}$ (Fig. 12.12*a*).[778] In the winter when the thermocline breaks down, the algae are circulated rapidly enough to prevent their becoming adapted to the irradiance value at any depth, and it is found that the cells from all depths show the same degree of light adaptation and indeed are rather similar in their properties to summer surface phytoplankton (Fig. 12.12*b*).[778, 874] It can be seen from the curves in Fig. 12.12*a* that the phytoplankton from the lower part of the euphotic zone not only have a lower E_k, but are much more susceptible to photoinhibition by higher light intensity than are the surface phytoplankton.

Shimura & Ichimura (1973) found that phytoplankton from near the bottom of the euphotic zone in the ocean had a higher photosynthetic activity than surface phytoplankton when both were placed deep in the euphotic zone. In addition the phytoplankton from the deep layer showed a higher ratio of photosynthetic activity in green light to that in red light than surface layer phytoplankton. Thus the phytoplankton growing deep in the euphotic zone of the ocean appeared to be the better adapted to the dim, predominantly blue-green, light prevailing in that region. Similarly, Neori *et al.* (1984) found, for temperate and polar oceanic stations, that both absorption and chlorophyll *a* fluorescence excitation spectra of phytoplankton samples showed enhancement in the blue-to-green part of the spectrum (470–560 nm) relative to that at 440 nm, with increasing depth. They attributed this change to an increase in the concentration of photosynthetic accessory pigments, relative to chlorophyll *a*.

In nature in thermally stratified waters, the cells below the thermocline will be at a lower temperature than those near the surface and as we saw in an earlier section (§11.3), lowering the temperature automatically lowers E_k (by diminishing P_m without affecting α), quite apart from any alterations in the biochemical composition of the depth-adapted cells. That there are nevertheless inherent changes in the light relations of the cells is readily shown (as in the experiments in Fig. 12.12) by comparing the deep- and shallow-water phytoplankton at the same temperature. The *in situ* lowering of the E_k value of the deep phytoplankton due to lowered temperature is in fact of little ecological significance since the light intensity at that depth will be well below saturation in any case.

To what extent the shade adaptation of the deep phytoplankton in the ocean represents true ontogenetic adaptation (i.e. physiological adaptation within species occurring throughout the illuminated water column) or is phylogenetic adaptation (the predominance in deeper water of particular species genetically adapted to low irradiance values) is unknown. However, Neori *et al.* 1984) believe that the changes in absorption and fluorescence properties with depth that they observed (see above), were predominantly due to shade adaptation within existing species: the depth distribution of species did not seem to account for the spectral changes, and furthermore, similar changes could be detected in phytoplankton samples within one day of transferring them to low light levels.

In the clear, colourless water of Lake Tahoe in which, as in the ocean, the illuminated region extends below the thermocline, Tilzer & Goldman (1978) found that the increase in chlorophyll content of the deep-water cells at the time of maximum thermal stratification (September) was accompanied by a diminution in the light saturation onset parameter: the E_k values were about 80, 50 and 15 W (PAR) m^{-2}, for cells from 0, 50 and 105 m depth, respectively. The phytoplankton in this lake satisfied the primary criterion for true depth adaptation of the photosynthetic system. In the light field prevailing in deep water, the deep-water cells achieved a substantially higher photosynthetic rate per unit biomass than cells from the surface (Fig. 12.13). Their light-saturated rate of photosynthesis per unit biomass was not, however, significantly lower than that of the shallow-water cells, suggesting that the deep-water cells in this lake had not adopted the additional shade adaptation strategy of reducing their carboxylase content.

The less turbulent the upper mixed layer of a water body is, the more time the phytoplankton cells have to adapt as they undergo vertical circulation. This suggests that even within a mixing layer, photosynthetic differences between the cells at the top and the bottom might develop as turbulence diminishes. Lewis *et al.* (1984) obtained evidence that this can indeed happen. In Bedford Basin (Nova Scotia, Canada) they measured ΔP_m, the difference between P_m (per unit chl) at the surface and at the base of the mixing region, and also measured the rate of turbulent kinetic energy (TKE) dissipation, ϵ ($m^2\ s^{-3}$), on 19 occasions over the period of a month. While on many occasions, especially at high rates of TKE dissipation, ΔP_m did not differ significantly from zero, on other occasions, as ϵ diminished to low values, ΔP_m increased markedly, indicating that photoadaptation had in fact occurred.

Morphogenetic adaptation to depth

We have noted earlier that the adaptation of angiosperms to increased shade includes a morphogenetic response: the leaves increase in area while simultaneously becoming thinner. Spence, Campbell & Chrystal (1973) studied five submerged species of *Potamogeton* in several Scottish lochs. They found that specific leaf area (SLA, $cm^2\ mg^{-1}$ leaf dry mass) increased linearly with water depth. The rate of increase with depth appeared to be typical of the loch (and therefore, presumably, of the water type) rather than of the species, the more rapid rates of increase being observed in the more highly coloured and therefore more attenuating waters. For example, the specific leaf area of *P. perfoliatus* increased from about 0.6 to 1.3 $cm^2\ mg^{-1}$ from near the surface to 4 m depth in the brown water of L. Uanagan, but increased only from about 0.7 to 1.1 $cm^2\ mg^{-1}$ from the surface down to 6 m depth in the comparatively colourless waters of L. Croispol. It seemed that a given specific leaf area could be achieved in light fields of markedly different spectral quality, and it was thought likely that the total irradiance (400–750 nm) was of more importance in determining the SLA than any aspect of the spectral quality such as the blue/red or red/far-red ratios.

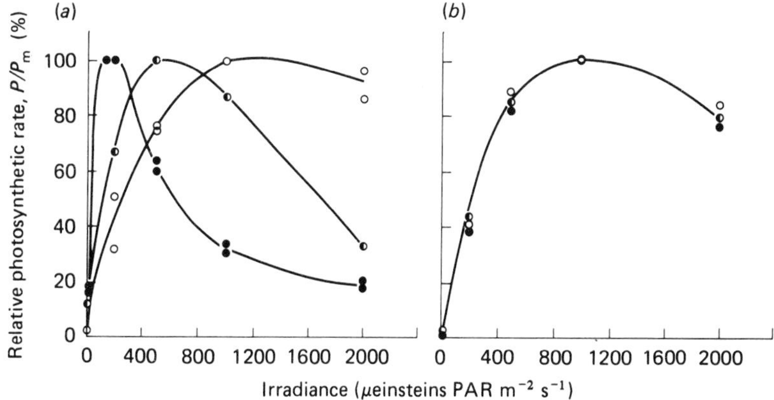

Fig. 12.12. Depth adaptation by marine phytoplankton (after Ryther & Menzel, 1959). These P versus E_d curves for Sargasso Sea phytoplankton were determined by the ^{14}C method at the temperature prevailing at the sea surface. Samples taken from 0 m (○), 50 m (◐) and 100 m (●) depth. (*a*) October; water stratified with thermocline at 25–50 m. (*b*) November; water isothermal down to > 150 m depth.

Lipkin (1979) found that in communities of the seagrass *Halophila stipulacea* growing in the northern Red Sea, leaf area increased about 2.5-fold with depth from the surface down to 30 m (Fig. 12.14). Although this was at first considered to be an ontogenetic adaptation controlled by light intensity, culture experiments indicated that the differences were genetically controlled and Lipkin concluded that true ecotypes of this seagrass were present in the same locality within short distances, i.e. adaptation was apparently phylogenetic rather than ontogenetic.

The siphonous green alga *Halimeda tuna* growing in the Adriatic Sea shows depth-dependent changes in thallus morphology. Mariani Colombo *et al.* (1976) observed an increase in the total surface number of branches and number of segments with increasing depth over the range of 7 to 16 m. The overall effect is to increase the photosynthetic surface exposed to light.

12.5 Significance of ontogenetic adaptation of the photosynthetic system

From our consideration of this topic so far, it is clear that the ability to adapt the characteristics of their photosynthetic systems to the prevailing light field is widespread amongst aquatic plants. To what extent is such adaptation of ecological significance?

In broad terms, the significance of ontogenetic adaptation is that it enables a species to exploit a wider range of habitats. Suitability of habitat

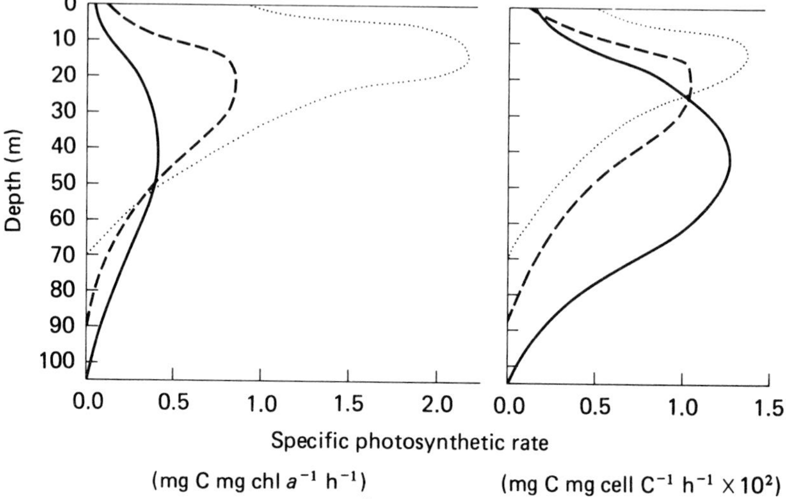

Fig. 12.13. Variation with depth of specific photosynthetic rates of phytoplankton originating from different depths, in Lake Tahoe (California–Nevada, USA) (by permission, from Tilzer & Goldman (1978), *Ecology*, **59**, 810–21. Copyright © 1978, the Ecological Society of America). The phytoplankton were collected from the surface (.....), 50 m (– – – –) and 105 m (——), and incubated with $^{14}CO_2$ at a series of depths between the surface and 105 m. (*a*) Specific photosynthetic rate per unit chlorophyll. (*b*) Specific rate per unit cellular biomass, expressed as carbon. The lower maximum photosynthetic rate per unit chlorophyll of the deeper-water cells is due to their higher pigment content.

for aquatic plant photosynthesis varies with depth; with water type, particularly with regard to optical character; with time, as the seasonal cycle progresses; and in the intensity of competition between plants. The significance of ontogenetic adaptation of the photosynthetic apparatus might therefore be assessed in terms of the extent to which it enables particular species to increase the range of depth and of water types that they can exploit, the proportion of the year in which they can grow, and the success with which they withstand competition from other species.

Depth and shade adaptation

Given the demonstrated ability of so many phytoplankton species under laboratory conditions to increase their light-harvesting capability in dim light, especially in dim blue-green light, and given also the observed shade adaptation of natural phytoplankton populations taken from deep in the

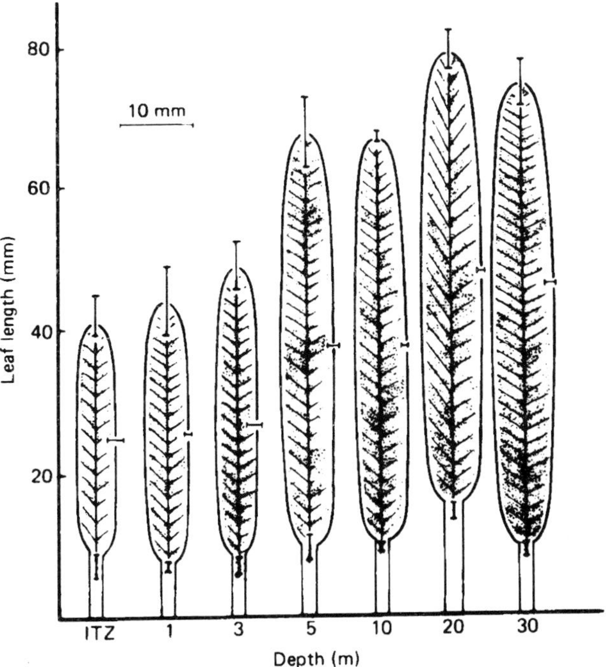

Fig. 12.14. Change in length and area of leaves of the seagrass *Halophila stipulacea* (Sinai, northern Red Sea) with depth (by permission, from Lipkin (1979), *Aquatic Botany*, **7**, 119–28). ITZ = intertidal zone.

euphotic zone of stratified waters, we may reasonably conclude that ontogenetic adaptation of the phytoplankton photosynthetic apparatus to the low-intensity light field in these deep layers is a real phenomenon, although it may well coexist with phylogenetic adaptation, certain species being perhaps permanently adapted to the greater depths. The shade-adapted phytoplankton growing at the bottom of the euphotic zone in the ocean contribute only a small proportion of the integral photosynthesis of the water column. It may therefore be concluded that their shade adaptation, while highly relevant to our understanding of how that ecological niche comes to be filled, is not of much significance for the total primary production of the oceans. Primary production in the region of the deep chlorophyll maximum, although a small proportion of the total is not a trivial one (see, for example, Fig. 11.1). Furthermore, there is evidence for a significant enhancement of zooplankton biomass around the deep chlorophyll maximum; Ortner, Wiebe & Cox (1980) consider that this is a depth zone of particularly intense trophic activity. Thus, when all trophic levels are taken into account these shade-adapted phytoplankton may be of more significance for the ecology of the ocean than their contribution to primary production would indicate.

So far as total primary production is concerned, it is true that most of it takes place in the upper region of the euphotic zone and so, as Steemann Nielsen (1975) points out, the most relevant form of adaptation is that of the phytoplankton in the surface layer to the high irradiances prevailing there; for example, their greater ability to withstand photoinhibition. There is evidence that one of the roles of carotenoids, specifically those xanthophylls not involved in light harvesting, is to protect plants against damage by excessive light. There is also evidence that in red algae, for example, in which carotenoids contribute little to photosynthesis, the ratio of carotenoid to chlorophyll increases if the cells are grown at high light intensity.[192,559] It is thus a plausible supposition that the acquisition of resistance to photoinhibition by phytoplankton involves increased synthesis of protective carotenoids.

In the case of the multicellular benthic algae and angiosperms, while there is evidence for an increase in light-harvesting capacity as growth irradiance is decreased, it is less well documented than in phytoplankton species, perhaps because of the greater difficulty in culturing these plants under controlled laboratory conditions. Although more studies are required, particularly in the field, we may reasonably conclude on the basis of the information already available that in some species which grow over a range of depth, adaptation in the deeper-growing individuals consists in part of an increase in pigment content per unit biomass. An alternative or additional strategy is a reduction in respiration rate; morphological changes which increase light interception by the plants may also occur.

Any adaptive ability which enables an aquatic plant species to cope with the variation in light regime over a range of depth should in principle also enable it to cope with the variation in light quality that will be found within a range of optical water types. The extent to which the ability of particular species to grow in a range of water types is due to ontogenetic adaptation of the photosynthetic apparatus has not been the subject of systematic study.

Given that a common ontogenetic response of aquatic plants to variations in light regime is to vary the light-harvesting capacity of the cells, how important is chromatic adaptation as a component of this? How important, that is, are changes in the shape of the absorption spectra (due to changes in the proportions of the pigments) of the cells, as opposed to changes in absolute absorption across the whole spectrum? In aquatic plants which do not contain biliproteins, i.e. higher plants and algae other than the

Rhodophyta, Cryptophyta and Cyanophyta, changes in pigment ratios in response to changes in the ambient light are relatively small, where they occur at all. In certain green species (algae and higher plants) there is some increase in the ratio of chlorophyll *b* to *a*, during shade adaptation: this should somewhat increase the absorption in the 450–480 nm waveband relative to other parts of the spectrum. In the dinoflagellate *Glenodinium*, the increase in concentration of the peridinin/chlorophyll *a*-protein associated with growth at low irradiance causes absorption to rise more in the 500–550 nm region than elsewhere in the spectrum.[724] These comparatively subtle changes in the shape of the absorption spectrum will indeed have the effect of matching the absorption spectrum of the cells better to the spectral distribution of the blue-green light field which prevails at increased depth in many marine waters and so can legitimately be regarded as examples of ecologically useful chromatic adaptation. Nevertheless, the effects of such changes are quantitatively less important for total light capture than the increase in absorption throughout the spectrum resulting from the general increase in pigment concentration. For clear-cut instances of chromatic adaptation we must turn to the biliprotein-containing groups, the Cyanophyta and Rhodophyta.

The changes in the levels of phycoerythrin and phycocyanin in blue-green algae in response to growth in light of differing spectral quality are such as to lead to drastic alterations in the absorption spectra. As shown in Fig. 12.15, the phycocyanin-dominated cells absorb much more strongly in the 600–650 nm region than the phycoerythrin-dominated cells, which in turn absorb more strongly in the 500–575 nm region. There is no doubt that the former cells would be better equipped to collect light from an underwater light field of an orange/red character, while the latter cells would harvest more photons from a greenish light field. Underwater light fields with a predominantly orange/red character would almost never be found in marine systems, but do exist in inland waters strongly coloured by soluble and particulate humic materials (see Fig. 6.4*d*). Green light fields are, of course, common in natural waters. Thus, light climates to which these two kinds of cell would be well matched do occur in aquatic ecosystems. Unfortunately there appear to be no field data on the extent to which adaptation of this type actually happens in nature.

The ratio of biliprotein to chlorophyll can also, as we have seen, change in accordance with the prevailing light. As light intensity diminishes, even when the spectral composition is unaltered, the biliprotein/chlorophyll ratio increases to an extent varying from one blue-green- or red-algal species to another. Since this change can be brought about by variation in

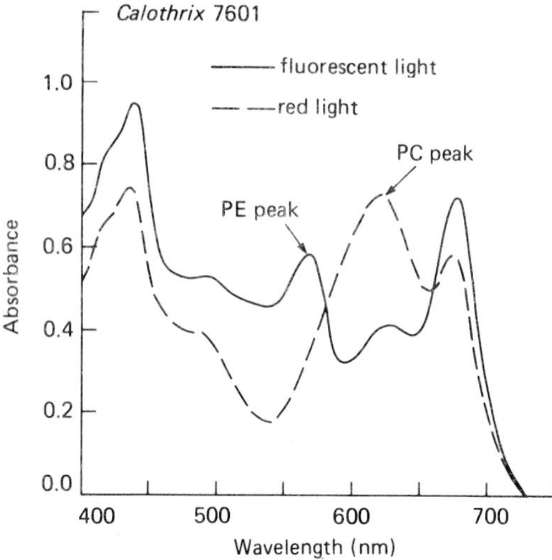

Fig. 12.15. Absorption spectrum of blue-green algae as influenced by spectral quality of light field during growth (after Bennett & Bogorad, 1973). Cultures of *Calothrix* 7601 (formerly *Fremyella diplosiphon*) were grown in red light or in fluorescent light (enriched in the green-yellow waveband), and the absorbance spectra of the whole cells were determined on suspensions containing 0.68 mg dry mass ml^{-1}. The presence of high levels of phycoerythrin (PE) in the fluorescent-light-grown cells (——) and of phycocyanin (PC) in the red-light-grown cells (-----) is apparent from the spectra.

intensity alone, it could be argued that this is simply 'intensity adaptation' and is not true chromatic adaptation since it does not require a change in spectral composition to bring it about. On the other hand, diminution of light intensity in the underwater environment, as opposed to the laboratory, is most commonly associated with increase in depth in which case it is associated with a change in spectral quality, in particular with a removal of red light. In the course of evolution, therefore, the common association between low irradiance and the virtual absence of red light might well lead to the development of regulatory mechanisms which respond automatically to lowered intensity by increasing the synthesis of pigments, such as phycoerythrin, which absorb in spectral regions other than the red waveband. The plant does not necessarily have to possess a chromatically specific detection system in order to undergo chromatic adaptation. The increase in biliprotein/chlorophyll ratio as light intensity diminishes might reasonably be regarded as both intensity and chromatic adaptation: while it is triggered off by falling intensity, at the same time it makes the cells chromatically better adapted to the particular spectral character that dim underwater light fields usually have.

Little information is presently available on the extent to which this kind of adaptation is of ecological significance. Limited field studies indicating an increase in biliprotein/chlorophyll ratio with depth have been carried out on two red-algal species (§12.4).

The non-complementary kind of chromatic adaptation that occurs at relatively high light intensities in red algae, in which the adaptation seems to be directed towards achieving a balanced excitation of the two photosystems (§12.3) is accompanied by substantial changes in the absorption spectrum (Fig. 12.8). This kind of adaptation has not yet been described in the field. It is most likely to be found amongst algae growing at relatively shallow depths. On the basis of the laboratory studies and theoretical explanations[103,559,1012] one might predict that with increasing depth and therefore increasing green/red ratio in the underwater light, the phycoerythrin/chloropyll ratio would at first decrease to ensure equal excitation of the photosystems and then increase again as the need to capture all available quanta became paramount. Although this has not yet been described for a red alga, what could be an analogous phenomenon has been reported for a brown alga. As we noted earlier, in *Dictyota dichotoma* at a Spanish coastal site, the fucoxanthin/chlorophyll ratio decreased by 42% between 0 and 10 m depth, and then increased again to its original value between 10 and 20 m depth.[694] If, as is likely to be the case, fucoxanthin transfers its excitation energy primarily to photosystem II, then the increasingly blue-green character of the light with increasing depth might lead to greater excitation of photosystem II than photosystem I, and a decrease in the fucoxanthin/chlorophyll ratio could rectify this. At greater depths still, the need to harvest all available quanta becomes of dominating importance and so the fucoxanthin content is increased again.

In the more densely vegetated parts of the aquatic environment, the ability to compete with other plants for light and/or to tolerate shading by other plants is of great importance. On *a priori* grounds it seems likely that plants which find themselves in dim light, not because of great depth but because of overshadowing by other plants, will undergo similar kinds of shade adaptation to those associated with increased depth, but the topic has received little experimental attention in the case of aquatic ecosystems. An example of shade tolerance enabling a plant to withstand competition for light is the previously mentioned increase in phycoerythrin content in *Chondrus crispus* plants overgrown by epiphytic diatoms.[752]

A special case of shade adaptation is that of the microalgae living in and under the annual sea ice in North and South polar marine environments. In the Antarctic pack-ice these algae are present at a concentration in the region of 100 mg chl *a* m^{-2}, and since the area covered by the annual expansion and contraction of the ice is about 15 million km^2, and the algae are released into the water during the summer thaw, it seems likely that these ice algae make a substantial contribution to the productivity of the polar marine ecosystem.[119,934] The sea ice algal community is dominated by pennate diatoms, and its photosynthesis saturates at low incident irradiance values, indicating a high level of shade adaptation.[160,686,763] While there is some indication of a modest capacity for ontogenetic shade adaptation within this community,[161,686] it appears likely that the shade adaptation it exhibits is mainly phylogenetic in nature, i.e. the species present are likely to be genetically adapted to low irradiance.[160]

Seasonal adaptation – multicellular benthic algae

Away from the tropics the average irradiance during daylight hours changes cyclically with the seasons during the year and we may reasonably suppose that it would be of advantage to the aquatic flora to adapt their photosynthetic systems to the seasonally varying light climate. This should especially be the case for the perennial benthic algae which survive throughout the year and therefore must photosynthesize as best they can at all seasons. However, the seasonal variation in temperature which accompanies the variation in irradiance does itself impose certain changes on the photosynthetic behaviour of aquatic plants. Light compensation and saturation points increase with temperature since the rates of respiration and carboxylation both increase with temperature whereas the rate of the light reactions is relatively unaffected. King & Schramm (1976) in their studies on a large number of benthic algal species from the Baltic found that there was a marked lowering of the compensation point in the winter compared to the summer or autumn. Since their measurements were carried out at the water temperature prevailing at the time, it was impossible to say whether there was any adaptation or whether the lower winter temperatures accounted entirely for the lower compensation points. In most of the species there was no marked seasonal change in the light saturation point but in two *Laminaria* (Phaeophyta) species the irradiance required to achieve saturation was much lower in winter than in autumn. In one of these species, *L. saccharina*, the light-saturated photosynthetic rate per g dry mass was about the same in both seasons suggesting that the reduction

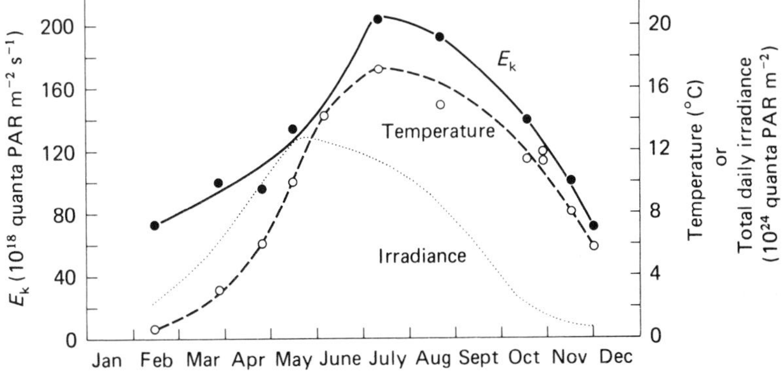

Fig. 12.16. Seasonal variation of saturation onset parameter (E_k) of phytoplankton, temperature and total daily irradiance at a Danish coastal station. Plotted from data of Steemann Nielsen & Hansen (1961). The temperature (– – ○ – –) and E_k (— ● —) values are for the surface layer and E_k values were measured at the prevailing temperature. Irradiance values have been converted from lux to quanta m^{-2} s^{-1} using an approximate factor. The values for total daily irradiance (······) are for days of average cloudiness throughout the year.

in light saturation point in the winter was not merely due to a fall in carboxylation rate caused by the lower temperature.

In the giant kelps *Macrocystis integrifolia* and *Nereocystis luetkeana*, growing off Vancouver Island (B.C., Canada), Smith, Wheeler & Srivastava (1983) and Wheeler, Smith & Srivastava (1984) found that the chlorophyll *a* and fucoxanthin levels (per unit area of blade) were high in winter, declined in spring–summer, and then rose in autumn once again to the winter level. The seasonal changes in pigment were substantial (two- to three-fold). Although a case might be made that the changes in pigment levels were an adaptive response to seasonally changing insolation, they appeared in fact to correlate best with the nitrate concentration in the water, i.e. perhaps these algae simply make more photosynthetic pigments when they have the nitrogen to do so.

In a number of red-algal species it has been shown that the ratio of phycoerythrin to chlorophyll is higher in the winter than in the summer.[124, 628, 752] It seems likely in such cases that the irradiance required to achieve saturation would, quite apart from temperature effects, be lower in the winter, but experimental confirmation is required.

In the intertidal red alga *Bostrychia binderi*, growing in a subtropical location (Florida, USA), Davis & Dawes (1981) found little seasonal variation in chlorophyll content, but the light-saturated photosynthetic rate was at its highest in the late summer and autumn and at its lowest in the winter, the summer/autumn values being two- to four-fold greater than the winter values. Since all the measurements were carried out at 30 °C, this probably represents a true seasonal variation in carboxylation capacity.

Seasonal adaptation – phytoplankton

Seasonal temperature changes make interpretation difficult in the case of the phytoplankton also. Steemann Nielsen & Hansen (1961) measured the light saturation onset parameter, E_k, of phytoplankton in the surface layer in Danish coastal waters at the prevailing temperature at intervals throughout a year. It varied about three-fold being highest in midsummer and lowest in midwinter. A comparison of the variation in E_k, daily insolation and water temperature throughout the year (Fig. 12.16) suggests that most of the seasonal change in the light saturation parameter can be attributed to the change in temperature: however, the value of E_k in the spring seems to be somewhat greater than would be expected on the basis of temperature alone and it may be that the rise in E_k at this time of year is in part due to adaptation of the photosynthetic system to the higher irradiance.

The values of α and P_m^B for the phytoplankton of a Nova Scotia coastal water were found by Platt & Jassby (1976) to vary about five-fold over a period of 2.75 yr. Measurements were carried out at the prevailing water temperature. Values tended to be highest in the summer and early autumn but subsidiary peaks at other times of the year were also sometimes observed. Statistical analysis indicated that photosynthetic capacity (P_m^B, light-saturated photosynthetic rate per mg chlorophyll) was strongly correlated with temperature but that α (the initial slope of the P versus E_d curve) was not. These findings are in accordance with the known insensitivity of the photochemical processes, but sensitivity of carboxylation, to temperature.

The E_k value of the phytoplankton in a Japanese pond was found to vary about three-fold during the year, rising to a maximum in the summer and falling to a minimum in the winter:[19] however, these measurements were carried out at the prevailing water temperature.

Even when the effects of temperature are allowed for, there is the problem of determining whether any apparent seasonal change in the photosynthetic behaviour of the population is true ontogenetic adaptation within species, or merely represents changes in which species are present, since there are marked successional changes in the taxonomic composition of the phytoplankton population through the year, both in the sea[748] and in lakes.[750] A study by Durbin, Krawiec & Smayda (1975) has shown how the relative importance of different types of phytoplankton (separated in terms of size), both as components of total biomass and as contributors to primary production, varied in a northeast American coastal water throughout most of one year (Fig. 12.17). Larger cells (from 20 μm to > 100 μm), mainly diatoms, dominated the winter-spring bloom, while small cells (< 20 μm), mainly flagellates, dominated the summer population. Even within given size classes the species composition changed with time: for example, different diatom species were dominant in each of the three abundance peaks during the February to May period.

The photosynthetic capacity of the phytoplankton population in a Canadian Pacific coastal inlet was found by Hobson (1981) to rise rapidly many-fold from a very low value in March–April up to a peak in the June to August period, and then to decrease rapidly in the mid-August to September period, down to a very low value again in October. Although the measurements were carried out at the prevailing water temperature, and although P_m^B for any given species tends to increase with temperature (§11.3), it seemed that only a small part of the variation (~ 17%) was accounted for by the changes in temperature. The major changes in P_m^B in early April and August/September coincided with changes in the taxonomic composition of the phytoplankton, this being dominated by unidentified nanoflagellates at times when photosynthetic capacity was low, and by diatoms (*Chaetoceros*, *Thalassiosira*) or dinoflagellates (*Gymnodinium*, *Peridinium*) when capacity was high. Hobson suggested that the nanoflagellates which dominated the phytoplankton in the winter were genetically adapted to conditions of short days and low irradiance.

Situations like these, with a continually varying species composition of the phytoplankton population, are probably the rule rather than the exception in fresh as well as marine waters and so any apparent seasonal adaptation of phytoplankton photosynthetic characteristics (after allowing for temperature effects) is in most cases attributable to changes in the algal species present, i.e. is phylogenetic rather than ontogenetic adaptation. It is possible, of course, that species such as *Skeletonema costatum*, which in many coastal waters are present in significant amounts throughout the year, do indeed adapt to the changing irradiance values, as they are known to do under laboratory conditions, but this is hard to demonstrate in the presence of all the other species. A further complication is that even within the population of a given phytoplankton species at a given geographic location there can be marked genetic heterogeneity, and changes in genetic

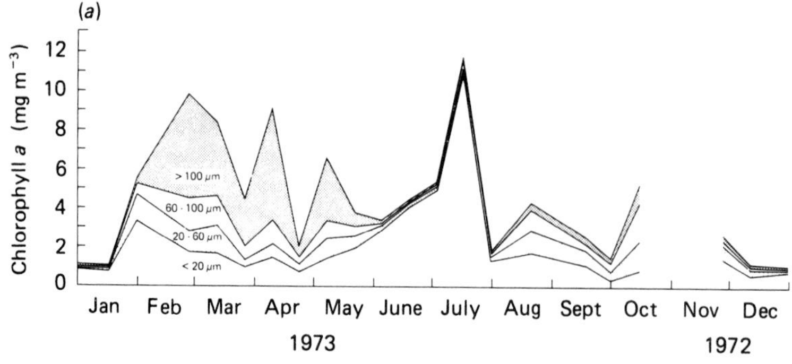

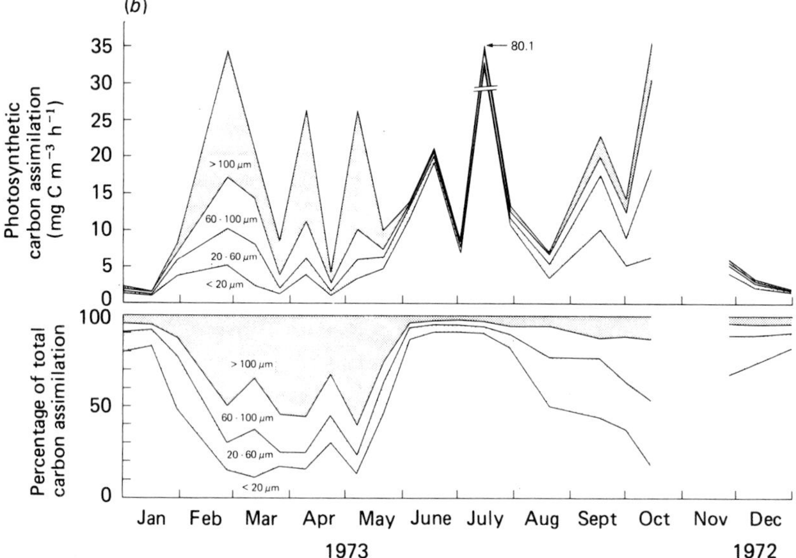

Fig. 12.17. Seasonal variation in the contribution of different size fractions of the phytoplankton to total biomass and to primary production in a north temperate coastal water (by permission, from Durbin, Krawiec & Smayda (1975), *Marine Biology*, **32**, 271–87). The data are for Narragansett Bay, Rhode Island, USA. (*a*) Cumulative graph of chlorophyll *a* content of the different size fractions. (*b*) Upper: cumulative graph of photosynthetic carbon assimilation (per unit volume) of the different phytoplankton size fractions. Lower: photosynthetic carbon assimilation by the different size fractions expressed as a percentage of assimilation by the total population.

composition through the seasons. By analysis of isoenzyme patterns and physiological characteristics of individual clones of *S. costatum* isolated from Narragansett Bay (RI, USA), Gallagher (1980, 1982) was able to show that the winter bloom populations of the diatom were on average genetically different from the summer bloom populations, although neither were genetically homogeneous.

In high-latitude and high-altitude lakes, the phytoplankton in the dim light under the ice in winter have a higher cellular chlorophyll content and a lower light-saturation intensity than the phytoplankton present in the summer.[443,449,926] In one such case, Lake Pääjärvi, Finland, different algal species were dominant at the different times of the year.[443] In an Austrian alpine lake, however, the species composition did not vary greatly through the year, being mainly dominated by the dinoflagellate species *Gymnodinium uberrimum*,[920] and Tilzer & Schwarz (1976) attributed the much higher (two- to four-fold) chlorophyll content and lower E_k values, or cells under the ice to actual adaptation within the species.

Evidence for comparatively short-term photosynthetic adaptation of phytoplankton, over periods of a few days, has come from a number of field studies.[441,707,794] Platt & Jassby (1976) found the value of α for phytoplankton in Nova Scotia coastal waters to be correlated with the average solar irradiance over the previous three days. In Lough Neagh (N. Ireland) Jones (1978) found that the light saturation onset parameter, E_k, for the dominating blue-green-algal population was positively correlated with the average daily irradiance over the previous five days, i.e. the higher the average irradiance in the previous five days, the higher the light intensity required to achieve saturation. The changes in E_k appeared to be due more to changes in α (the initial slope of the P versus E_d curve) rather than in the light-saturated rate (P_m) per mg chlorophyll ($\alpha = P_m/E_k$). Since $\alpha = \bar{a}_c \phi_m$, these changes can plausibly be attributed to an increase in the specific average absorption coefficient per unit chlorophyll, associated with a decrease in the cellular chlorophyll content at higher light intensity (§12.2). In view of the short time period during which the changes occurred, we may reasonably assume that they correspond to true ontogenetic adaptation within the species rather than to changes in the population structure.

Thus it may be that while there are substantial seasonal changes in the photosynthetic properties of the phytoplankton population as a whole, due to the succession of species, there are also, on a shorter time scale, adaptive changes in the photosynthetic properties of individual species, during the few weeks that they persist as significant members of the population.

12.6 Rapid adaptation of the photosynthetic system

Quite apart from the seasonal cycle in solar irradiance, there are of course extreme changes in the intensity of PAR every day: not only those due to the regular diurnal change in solar altitude, but also the rapid fluctuations in irradiance resulting from variations in cloud cover (Fig. 2.7).

Flagellate migration

Within the illuminated water column, there is at a given time for any particular species in the phytoplankton population a particular depth at which the light intensity is optimal for photosynthesis: high enough to give the maximum rate without causing photoinhibition. This optimal depth will vary with solar altitude during the day. The ability to move within the water column to whatever depth suits it best would clearly be of advantage to an alga; many groups have this capability.

The possession of flagella is common in all the algal divisions other than the Rhodophyta, Phaeophyta, Bacillariophyceae and Cyanophyta. Vertical migration by flagellated cells has been studied most amongst the dinoflagellates. In some cases the behaviour pattern is for the cells to migrate downwards away from the surface as the light becomes more intense, towards noon, and then to ascend again during the remainder of the day and the night.[73,920] Migration appears to be in part phototactic, but can be positive or negative within any given species, the direction of migration being determined by the light intensity at the water surface. For example, Blasco (1978) found that off the coast of Baja California, USA, the marine dinoflagellate *Ceratium furca* migrated towards the sea surface at 07.00 h when the solar irradiance was 280 W m^{-2}, but away from the sea surface at 12.00 h when the irradiance was 900 W m^{-2}.

Not all dinoflagellate species show a regular up-and-down diurnal migration but instead appear to use their powers of locomotion to move themselves to and remain at a particular depth. Heaney & Talling (1980) found that in Esthwaite Water, England, in the spring and summer, *Ceratium hirundinella* showed surface avoidance but not vertical migration. The dinoflagellate cells were found in maximum density at the depth, 3–4 m, at which irradiance was reduced to about 10% of that penetrating the surface.

Patterns of dinoflagellate movement can be influenced by the availability of nitrate in the water. Cullen & Horrigan (1981) studied the movement of the marine species *Gymnodinium splendens* in a 2 m-deep laboratory tank under a 12 h dark/12 h light regime. When nitrate was present throughout the container, the cells spent most of the day near the surface and most of the night at greater depth: they started swimming upwards before the end of the dark period and downwards before the end of the light period. When the nitrate in the water became depleted, the light-saturated photosynthetic rate per unit biomass decreased, and the cells formed a layer at a depth corresponding to the light level at which photosynthesis was saturated. The behaviour pattern of this species was also observed in the sea off the California coast. During the night (01.10 h) the cells were at 18 m depth, in the region of the nitrocline (the layer in which nitrate concentration is rising from the low value in the upper waters to the high value of deep water). By 10.50 h in the morning the cells had moved up into nitrate-depleted water at 14 m, a depth at which the light intensity was sufficient to saturate photosynthesis. It therefore seems that *G. splendens* moves down at night to a depth where it can accumulate nitrogen and then up during the day to a depth where, although nitrate is lacking, it can photosynthesize at its maximum rate. Its behaviour may be chemotactic as well as phototactic.

In lakes, a common (but not universal) flagellate behaviour pattern is for the cells to congregate in the surface water by day and migrate below the thermocline at night. Such behaviour has been described, for example, in the cryptomonad flagellate *Cryptomonas marssonii* in a Finnish brown-water lake,[782] and in the colonial green flagellate *Volvox* in Lake Cahora Bassa (Mozambique).[848] In both cases the adaptive significance was considered to be nutrient retrieval (in this case phosphate – generally the limiting nutrient in inland waters) from deep water at night, followed by a return to the illuminated, but nutrient-depleted, upper layer to photosynthesize during the day.

Vertical movement by blue-green algae

Blue-green algae have a completely different mechanism for moving up and down in the water column, involving the formation and collapse of gas-filled vacuoles within the cells. A comprehensive account of these structures and their likely ecological significance may be found in the monograph on the blue-green algae by Fogg *et al.* (1973) and the reviews by Walsby (1975) and Reynolds & Walsby (1975).

The gas vacuoles (Fig. 12.18) are cylindrical structures about 0.07 μm in diameter and of variable length, 0.3–0.4 μm being common but up to 2 μm being recorded. There are many in each cell. The membrane by which they are bounded is composed of protein, just one protein species, molecular

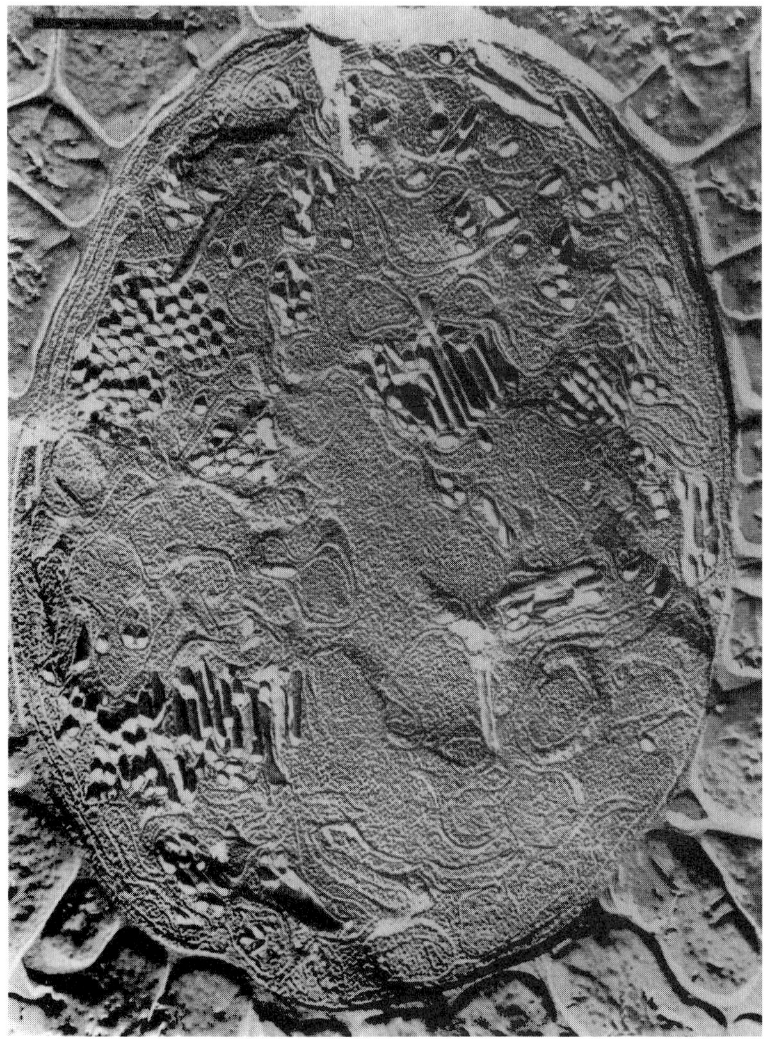

Fig. 12.18. Freeze-fractured cell of *Anabaena flos-aquae* showing cylindrical gas vacuoles longitudinally and in cross-section (by courtesy of Professor D. Branton). The bar corresponds to 1.0 μm.

weight 20.6 kDa, being present[967] The gas mixture within the vacuole is the same as that in the surrounding solution and at the same partial pressures. When the turgor pressure within the surrounding cytoplasm rises above a certain critical value, the gas vacuoles collapse. New vacuoles are formed by *de novo* synthesis rather than by re-inflation of collapsed vacuoles: small vesicles are formed first which then increase in length. Formation of gas vacuoles increases the buoyancy of the blue-green-algal cells, making them float towards the surface: collapse of the vacuoles makes the cells sink. It is the gas-vacuole volume/cell volume ratio that determines whether, and how fast, the cells rise or sink, or remain at the same depth.

Vacuole formation is favoured by dim light, by a lack of inorganic carbon (CO_2, HCO_3^-) for photosynthesis, and by a sufficiency of inorganic nitrogen.[259,506,684,973] Under these circumstances photosynthesis proceeds rather slowly and the availability of nitrogen both facilitates the synthesis of vacuole membrane and diverts the products of photosynthesis towards the manufacture of cell materials rather than sugar accumulation. Having formed vacuoles the cells float up towards the surface. With increased light intensity and CO_2 availability (resulting from diffusion of atmospheric CO_2), the rate of photosynthesis increases. If, as will commonly be the case, mineral nitrogen concentration decreases towards the surface, photosynthesis leads increasingly to sugar accumulation rather than cell growth, turgor pressure within the cell increases and, eventually, the vacuoles begin to collapse. Depending on how far this process goes the cells will either simply slow down in their upward motion and eventually come to a halt at a particular depth, or will begin to actually sink down again towards the deeper layers where nitrogen concentrations are higher. Thus gas-vacuole buoyancy control can be used to enable the blue-green-algal population to find a particular depth that suits it and this is what seems to happen to forms such as *Oscillatoria*: these occur as small-diameter single filaments with high resistance to movement through the water, respond slowly to buoyancy changes, and so normally remain in a layer at a particular depth, usually several metres. Species with large colonies, on the other hand, can sink or rise rapidly through the water and since marked buoyancy changes can take place within time periods as short as an hour or so, can respond rapidly to changing conditions during the day. It is in the forms such as *Microcystis* and *Anabaena*, with large colonies, that the algae will frequently rise all the way up to the surface under still conditions, forming the surface scums so commonly seen on eutrophic lakes in the summer. They are redistributed again from the surface either by collapse of the gas vacuoles or by wind action.

While it appears likely that in most cases the advantage conferred on blue-green algae by buoyancy control resides in the ability of the cells to locate themselves at a depth where light intensity is optimal, there is some evidence that the more rapidly moving forms, such as *Microcystis* and *Anabaena*, can indeed periodically sink down through the thermocline where, like the flagellates referred to earlier, they can take up phosphate before rising up to photosynthesize again in the illuminated, but nutrient-depleted, surface layer.[285] In hypertrophic Hartbeespoort Dam (South Africa), nutrients are always present in excess, but wind speeds and water turbulence are low, and enormous populations of *M. aeruginosa* develop. Zohary & Robarts (1989) attribute the success of this alga to its ability, by means of its strong buoyancy, to maintain itself in the shallow diurnal mixed layer which forms each day under the intense sunlight incident on this subtropical impoundment. Other, non-buoyant, species sediment through the density gradient and are lost. Modelling studies also indicate that the formation of diurnal mixed layers greatly favours positively buoyant algae such as *Microcystis*.[397] Being trapped near the surface in intense sunlight is, of course, a considerable hazard to phytoplankton. The Hartbeespoort *Microcystis*, however, appeared to be well adapted to intense light: it had a low cellular chlorophyll content and a high photosynthetic saturation irradiance (E_k up to 1230 μeinsteins m^{-2} s^{-1}).[1022]

Epipelic algal movements

Planktonic diatoms are unable to move within the water, although they of course undergo passive movement determined by sinking and turbulence within the water. On surfaces, however, diatoms can move by a mechanism involving secretion of mucilage. The algal community which inhabits sediments is known as the *epipelon* and diatoms are often the most conspicuous members.[771] Diatoms in sediments both in fresh and marine waters start to move up to the surface of the sediments before dawn.[126,772] In the afternoon the cells begin to migrate down into the sediments again, this movement continuing during the first few hours of darkness. In the case of benthic diatoms in a fresh-water lake it was found that they would not leave the surface of the sediments for as long as these were illuminated at 750 lux, but migrated downwards again at light intensities of 75 lux or lower.[348] The advantage of moving up onto the illuminated surface during daylight is clear enough: what is less obvious is the benefit the cells derive from burrowing down into the sediments during the hours of darkness – the greater availability of mineral nutrients is a possible explanation.

In turbid estuarine waters, the arrival of the flood tide causes a marked reduction in light intensity. Diatoms in tidal mud flats where the water is turbid can show a tidal rhythm superimposed on the diurnal one. In the River Avon estuary, England, 1–2 h before tidal flooding of the sediment at any particular point, the diatoms burrow beneath the surface again: the disappearance of the green or brown film of algae on the surface of the mud moves as a wave along the banks in advance of the tide.[773] In contrast, in the clear water of the River Eden estuary, Scotland, the movement of the benthic diatoms was found to be strictly diurnal.[695]

Diurnal movements onto the surface during the day and down into the sediments during darkness occur with benthic euglena and other flagellates, and blue-green-algal species, as well as diatoms. A full account of the epipelon is given in Round's (1981) book on *The ecology of algae*.

Chloroplast movements

The rate of light collection by chloroplasts depends not only on their pigment complement but also on their position and orientation within the cell. In some aquatic plants this can vary with light intensity. Amongst the higher plants[960] the general pattern of behaviour is that in low-intensity light the chloroplasts move to a position in the cell such that light absorption is maximized – they spread themselves out adjoining, and parallel to, those cell walls which face the incident light. In high-intensity light (saturating for photosynthesis), the chloroplasts move to a position such that light absorption is minimized, perhaps to reduce photoinhibition: they move away from that cell wall which is directly exposed to the light and align themselves adjoining, and parallel to, those cell walls which are most shaded, typically the side walls so that not only are they edge-on rather than face-on to the light, but also some are shaded by others. In the case of the aquatic angiosperm *Potamogeton crispus*, growing in a Scottish loch, leaves collected from 0.25 m depth had their chloroplasts around the side walls of the cells, whereas leaves from 2.5 m depth had their chloroplasts distributed over the surface facing the light.[854] An alternative way of reducing light collection in high-intensity light is for the chloroplasts to assemble in a cluster around the nucleus: this has been observed in leaves of the seagrass *Halophila stipulacea* growing at 0.5 m depth in the Mediterranean.[199]

The action spectrum for chloroplast movement in higher plants has a peak in the blue region at about 450 nm: the photoreceptor may be a flavin or a carotenoid.[359] The mechanism may be related to that of cytoplasmic streaming: the chloroplasts appear to move together with the cytoplasm rather than through it.

Nultsch & Pfau (1979) studied chloroplast movements in a large number of littoral and sublittoral marine algae. In most of the brown algal species, the chloroplasts moved to the cell walls facing the light at low irradiance and to the side walls parallel to the light direction at high irradiance (Fig. 12.19): the change from one position to the other took 1–2 h. No clear-cut light-induced movements of the chloroplasts were observed in the green- or

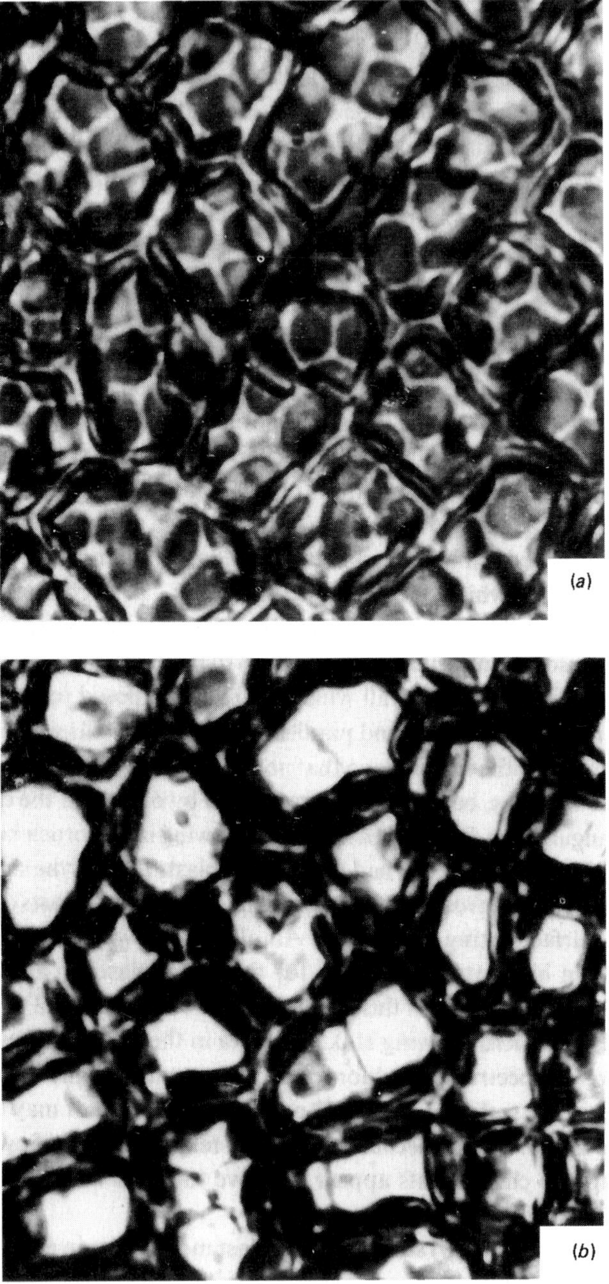

Fig. 12.19. Light-induced movement of the chloroplasts in the brown alga *Laminaria saccharina* (by permission from Nultsch & Pfau (1979), *Marine Biology*, **51**, 77–82). Arrangement of the chloroplasts in (*a*) low intensity (1000 lux) and (*b*) high intensity (10000 lux) light. Magnification ~4000×.

red-algal species studied. In the siphonaceous green alga *Caulerpa racemosa*, which grows in tropical shallow water reef areas, Horstmann (1983) observed that in bright sunlight the chloroplasts are retracted from the fronds into the stolon.

In the filamentous green algae *Mougeotia* and *Mesotaenium*, which have a single, flat centrally located rectangular chloroplast in each cell, the characteristic movement of the chloroplast in response to a change in the light regime is to turn on its longitudinal axis, rather than to move around the wall. It turns face-on to moderate light, and edge-on to intense light. Haupt (1973) and co-workers have shown that the photoreceptor controlling this movement is phytochrome. In another filamentous, coenocytic alga, *Vaucheria sessilis* (Xanthophyceae), if the filament is illuminated at one point with low-intensity blue light, the chloroplasts and other organelles which are normally carried along by the streaming cytoplasm are caused to aggregate in the illuminated part of the cell.[74] The photosynthetic implications of this are not yet clear.

In the case of floating, planktonic algae, oriented at random with respect to the light, movement alone to one part of the cell or another is not likely – if the cell contains only one or a few chloroplasts – to make much difference to the rate of light collection. In some cases the chloroplast can reduce its absorption cross-section in bright light by shrinking: this has been observed in a dinoflagellate[893] and in diatoms.[353,467] In the marine centric diatom *Lauderia borealis*, which has about 50 chloroplasts per cell, Kiefer (1973) observed that in the first 2 min of exposure to intense light (244 W m^{-2}) the chloroplasts contracted in size (Fig. 12.20*a*). During the following 30–60 min the chloroplasts, which in low light were distributed evenly around the periphery, moved to the valvar ends of the cell forming two aggregates of equal size (Fig. 12.20*b*). The changes in size and position of the chloroplasts were accompanied by a decrease of about 40% in the absorbance of the suspension at 440 nm (probably an underestimate since scattering always contributes some spurious absorbance): thus by a combination of shrinkage and aggregation this diatom is able to substantially reduce its rate of energy collection in bright light.

Rhythms in the photosynthetic system

Diurnal rhythms in photosynthetic capacity (photosynthetic rate per unit biomass at light saturation) have been described for phytoplankton[195,346,726,458,588] (reviewed by Sournia, 1974, and more recently by Prézelin, 1992) and for multicellular benthic algae.[102,304,447] Photosynthetic capacity rises, often several-fold, during the early part of the day to a maximum

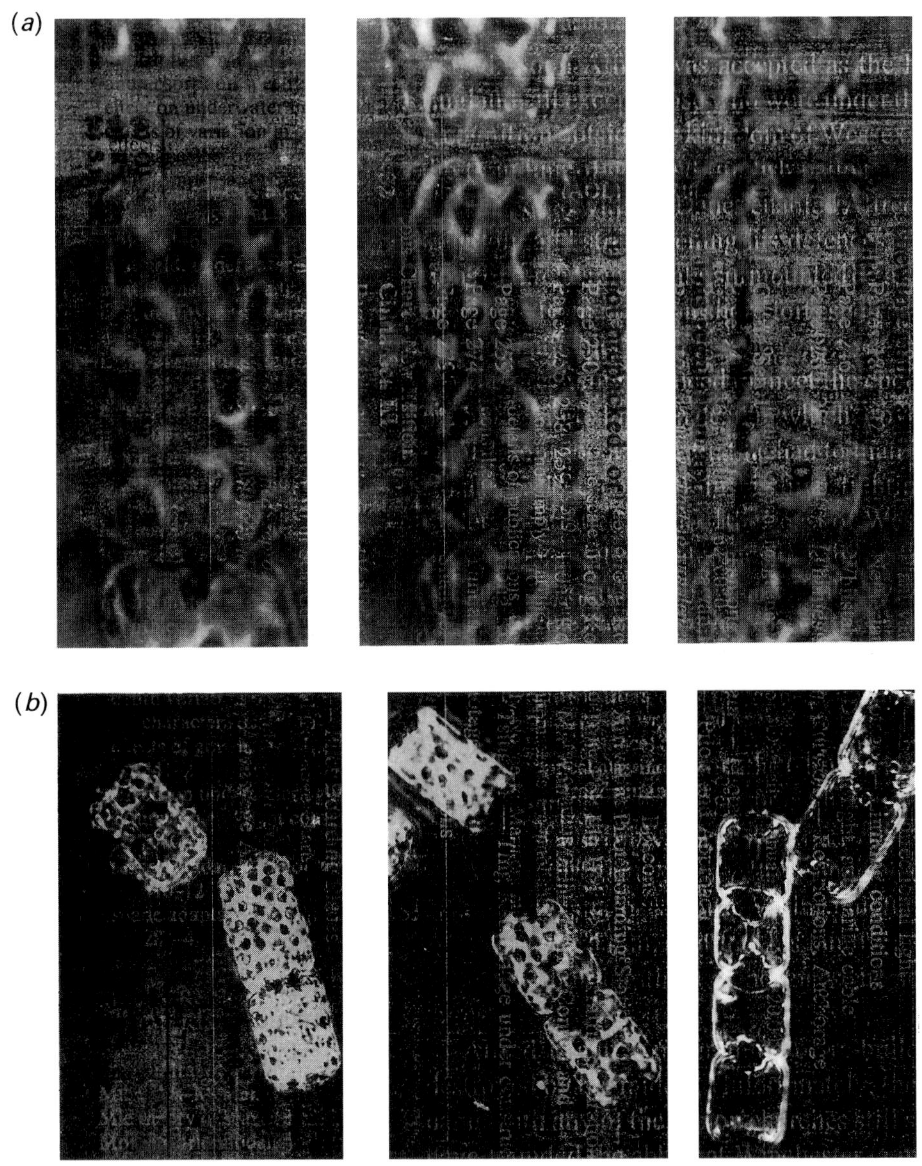

Fig. 12.20. Chloroplast shrinkage and aggregation induced by high light intensity in the marine diatom *Lauderia borealis* (by permission, from Kiefer (1973), *Marine Biology*, **23**, 39–46). In each case the series of micrographs shows increasing light-induced change from left to right. (*a*) Contraction of chloroplasts. (*b*) Movement of chloroplasts to valvar ends of the cell, followed by aggregation.

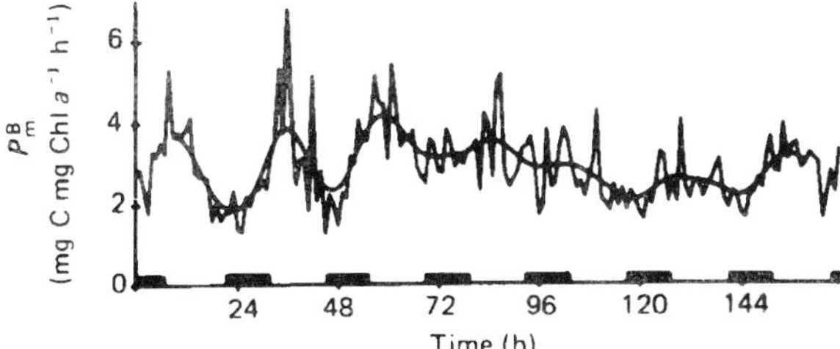

Fig. 12.21. Diurnal rhythm in the photosynthetic capacity (P_m) of phytoplankton in the St Lawrence estuary, Canada, over a seven-day period (by permission, from Demers & Legendre (1981), *Marine Biology*, **64**, 243–50).

which can be reached any time from about mid-morning to mid-afternoon and then declines during the rest of the day and remains low during the night. Fig. 12.21 shows the time course of variation in P_m for the phytoplankton in the St Lawrence estuary during the course of a week.[186] In some species, both benthic and planktonic, it has been shown that this diurnal variation is not directly controlled by environmental parameters but is a true circadian rhythm which continues for several cycles in cells kept in the dark.

With cultures, a daily periodicity in P_m has been shown to be present in some, but not all, marine diatom, dinoflagellate and chrysophyte species:[345] the species that showed no day/night differences in P_m were small, rapidly dividing types, whereas larger, slower growing cells almost uniformly showed marked changes. Detailed studies of this rhythm in marine dinoflagellates have been carried out by Sweeney and co-workers. The photosynthetic rate per cell in cultured *Glenodinium* and *Ceratium* varies two- to six-fold during the cycle, when measured under light-limiting as well as light-saturating conditions.[725,727] The fact that the rate-variation occurs under light-limited conditions led Prézelin & Sweeney (1977) to conclude that it is variation in the light reactions rather than the dark reactions of photosynthesis which are primarily responsible for the diurnal rhythm: it had previously been shown that carboxylase activity does not vary during the cycle.[122] There is no significant change in the pigment content or the absorption spectrum of the cells during the cycle, but the initial slope (α) of the photosynthesis *versus* irradiance curve varies in parallel with the photosynthetic capacity. This implies that the quantum yield of photosynthesis – the efficiency with which the cells use the quanta that have been absorbed – changes in a cyclical manner. For natural phytoplankton populations also, off the Southern California coast, Harding *et al.* (1982*a*) found that α and P_m varied (three- to nine-fold) in parallel during the day: this applied to both diatom- and dinoflagellate-dominated assemblages. Changes in chlorophyll content were much smaller and did not correlate with the changes in photosynthetic activity (which was, in any event, expressed per unit chlorophyll). Prézelin & Sweeney suggest that there is a cyclical variation in the proportion of the total photosynthetic units present which are functional. In the multicellular brown alga *Spatoglossum pacificum* the mechanism may be different. There is evidence that the approximately two-fold daily variation in photosynthetic capacity in this alga is due to variation in the activity of the CO_2-fixing enzyme system.[1004]

What, if any, advantage algae derive from cyclical variations in their photosynthetic capability remains unknown. Conceivably, after the main photosynthetic period is finished each day, certain proteins of the photosynthetic apparatus might be broken down so that their amino acids can be used for other cellular purposes: the proteins would then be resynthesized each morning. These daily variations in photosynthetic properties may affect the amount of primary production that is actually carried out. The effect of variation in light-saturated photosynthetic rate may be, at least in part, offset by the fact that solar irradiance itself varies roughly in parallel. The rate of light-limited photosynthesis also shows diurnal variation, however, and this may reduce the amount of primary production that the cells achieve early and late in the day. Harding *et al.* (1982*b*) calculated, for Californian coastal phytoplankton, that as a result of the daily periodicity in the parameters of the photosynthesis *versus* irradiance curves, the primary production achieved would be 19–39% less than if the values of P_m and α remained constant all day at their maximum values.

In calculations of integral daily photosynthesis, although the diurnal variation of solar irradiance is taken into account, it is commonly assumed that the values of α and P_m observed at mid-day, apply throughout the day. Making this assumption for the Californian phytoplankton yielded values for daily integral photosynthesis ranging from 15% greater to 20% less than values calculated taking account of periodicities in α and P_m. The variation in size and direction of the discrepancy is caused by variation in the timing of the photosynthetic activity maximum: when this is not at mid-day, the mid-day values of α and P_m come closer to the mean values, and so the errors may not be so great. For phytoplankton in the St Lawrence estuary (Canada), Vandevelde *et al.* (1989) found that calculations of integral daily photosynthesis made using the mid-day values of α and P_m gave production estimates 15–43% greater than those which took into account the variation in these photosynthetic parameters during the day.

Ramus & Rosenberg (1980) measured the actual photosynthetic rate of five intertidal species (two green, two brown, one red) at hourly intervals during the day in sea-water tanks (such that E_d was 70% of the subsurface value) exposed to ambient sunlight. The most common behaviour pattern on sunny days was for the photosynthetic rate (per unit chlorophyll *a*) to reach its peak in the morning, while solar irradiance was still well below its daily maximum, to decline somewhat during the middle part of the day under the most intense Sun, and then sometimes to show a partial recovery late in the afternoon, before declining again to zero, with falling light levels, towards sunset. Ramus & Rosenberg concluded that an afternoon depression in photosynthesis must be of normal occurrence (on sunny days) in these five species in their intertidal habitat. On cloudy days, when light saturation does not occur, the curve of photosynthetic rate is approximately symmetrical about the time of the Sun's zenith, and in the case of the most light-sensitive species (the rhodophyte *Gracilaria foliifera*) it seemed that total daily production would actually be higher than on sunny days. The characteristic sunny-day diurnal pattern observed in this work is likely to result from the simultaneous operation of a number of independent processes, particularly photoinhibition coming into operation when solar intensities are at their highest, together with the underlying circadian rhythm in photosynthetic capacity with its pre-mid-day morning peak.[102,447]

Although, as we have seen, no diurnal fluctuation in pigment content was observed in marine dinoflagellates under laboratory conditions, Yentsch & Scagel (1958) have described such a fluctuation in a natural phytoplankton population dominated by diatoms. In a north American coastal (Pacific) water they found that the chlorophyll content per cell fell by 50–60% during late morning, remained approximately constant during the afternoon, and then rose again during the evening. Auclair *et al.* (1982) studied fluctuations in chlorophyll content of estuarine phytoplankton under semi-natural conditions. Water from the St Lawrence estuary was transferred to a 1200 l tank, exposed to sunlight and the chlorophyll content per litre of water was followed over the next 45 h. Cell numbers showed little change. Chlorophyll content fluctuated up and down in a cyclical manner, through an approximately two-fold range with maxima in chlorophyll content occurring about every 6 h. Auclair *et al.* were able to correlate the variation in pigment content with tidal movements. Assuming the rhythm in the tank was a continuation of that in the estuary, it seemed that the times of maximum chlorophyll content corresponded to the periods of maximum current speed and turbulence and therefore probably, as a result of enforced circulation, to periods of lower average irradiance on the cells. The times of minimum chlorophyll content corresponded to periods of slack water in which a stability gradient was established resulting in a higher irradiance on the cells. It makes sense that the cells should increase their light-harvesting capacity during periods of low light intensity: what is noteworthy about this system is that they do it so rapidly, doubling their chlorophyll content in 2–3 h. Regular cyclical changes in phytoplankton pigment content may be more common in nature than we realize. The patchiness of the horizontal distribution of phytoplankton and the additional complicating factor of water movement make it difficult to unequivocally demonstrate such changes in large water bodies.

12.7 Highly productive aquatic ecosystems

To achieve a high rate of primary production, an aquatic plant community must achieve a high rate of collection of light energy and an efficient utilization of this absorbed energy by its photosynthetic system, followed by conversion of photosynthate to new cell material. In this section we are concerned with the ways in which some natural and man-made ecosystems manage to do this.

The most effective strategy which aquatic plants can follow to minimize the extent to which their photosynthesis is limited by lack of light is to attach themselves to surfaces at depths sufficiently shallow to ensure that for most of the day sufficient light is available to permit active photosynthesis. By 'choosing' to be benthic rather than planktonic, plants avoid the problem of being carried by water circulation down to depths where the light is insufficient for growth. Furthermore, the attached benthic flora can, as a consequence of water movement (tides, currents, wind-induced circulation), exploit the nutrients in a much larger volume than that in its immediate vicinity at any moment. The non-motile members of the phytoplankton, by contrast, are carried around in the moving mass of water and so have less opportunity to find new nutrient supplies. The rooted members of the benthic flora – seagrasses in marine waters and the various kinds of aquatic angiosperms in fresh waters – have the additional advantage that they can derive nutrients from the sediments in which they are growing.

It is thus not surprising that the most productive natural aquatic plant communities are benthic. In marine waters the most productive systems are brown algal beds, seagrass beds and coral reefs. The kelps, brown algae of the genera *Laminaria* and *Macrocystis*, form dense forests in the sublittoral zone of rocky coasts in cool waters. According to Mann & Chapman

(1975), these plants achieve annual net production rates in the range 1000–2000 g C m^{-2}. The intertidal brown seaweeds such as *Fucus* and *Ascophyllum* in temperate and subarctic latitudes have annual net production rates of 500–1000 g C m^{-2}. Tropical sublittoral seagrass beds dominated by the genus *Thalassia* achieve annual net production rates of 500–1500 g C m^{-2} and in temperate waters beds dominated by genera such as *Zostera* fix in the range of 100–1500 g C m^{-2}.

In coral reefs primary production is carried out by multicellular algae and seagrasses, as well as by the symbiotic zooxanthellae living within the cells of the coral. Total annual primary production on coral reefs is typically in the range of 300–5000 g C m^{-2}.[554, 650]

If the immediate products of photosynthesis, which are carbohydrate in nature, are to be used for cell growth and multiplication, proteins, nucleic acids and other cell constituents must be synthesized, for which mineral nutrients are required. A problem faced by aquatic plants is that at the time of year when high solar altitude favours photosynthesis in any one species, it favours it in all so that the period of maximum potential primary production is also a period of low nutrient levels in the water. Certain productive brown algae in the genus *Laminaria* circumvent this problem by separating the period of maximum growth from that of maximum photosynthesis.[448, 572, 591] In the summer, the plants accumulate photosynthate in the form of mannitol and laminaran, but grow only slowly. In the winter, although the light intensity is low, the nutrient levels in the water are at their highest and this is when the kelps grow rapidly at the expense of their stored carbohydrate.

Taking the world ocean as a whole, most of the primary production is carried out by phytoplankton. Within coastal embayments and estuaries, however, it is commonly the case that the benthic macrophytes account for most of the primary production. For St Margaret's Bay, Nova Scotia, Canada, Mann (1972) found that seaweed (mainly brown algae) productivity averaged over the whole 138 km^2 of the bay was 603 g C m^{-2} yr^{-1}, compared to 190 g C m^{-2} yr^{-1} for phytoplankton. In the Newport River estuary, North Carolina, USA, the contribution of seagrasses to annual primary production was estimated to be 2.5 times that of phytoplankton.[912]

Beds of submerged macrophytes in inland waters have productivities comparable to those of marine macrophytes. In the temperate zone, values range from below 10 to about 500 g C m^{-2} yr^{-1}, whereas in tropical regions annual production may exceed 1000 g C m^{-2}.[986]

The major factor responsible for the lower productivity of the phytoplankton, compared to the benthic plants, is, as we have noted, the lower average irradiance they receive as a consequence of vertical circulation. One solution to this problem, adopted by the blue-green algae (§12.6), is the evolution of a flotation mechanism enabling the algae to move vertically within the water to a depth which suits them, without (as in the dinoflagellates) the need for continual expense of energy to drive flagella. This may in part account for the fact that the natural water bodies with the highest phytoplankton productivity are dominated by blue-green algae: annual net yields from such water bodies are commonly in the region of 300–1000 g C m^{-2} but values in the region of 2000 g C m^{-2} have been reported.[338]

The circulation problem is also in some measure solved if the water body is very shallow, so that the cells never get very far away from the light. This is the solution generally adopted in man-made high-yield aquatic ecosystems such as sewage oxidation ponds or algal mass culture systems[906] in which a high rate of phytoplankton primary production is the aim: depths of 10–90 cm are commonly used. It is generally arranged in such systems that mineral nutrients are available in excess, so that the rate of primary production is limited by the rate of supply of PAR, and sometimes also of CO_2, to the system. Bannister (1974*a*, *b*, 1979) has developed a theoretical treatment of phytoplankton growth under mineral nutrient- and CO_2-saturated conditions by means of which the steady-state growth rate can be expressed as a function of the incident irradiance and the photosynthetic and respiratory characteristics of the cells. Goldman (1979*a*, *b*) has comprehensively reviewed the topic of outdoor mass culture of algae. He concludes that yields in excess of 30–40 g dry matter m^{-2} day^{-1} (equivalent to ~5000–7000 g C m^{-2} yr^{-1}) are unlikely to be exceeded: to date, such rates have been achieved only for short periods (less than a month), and over the long term, yields of 10–20 g m^{-2} day^{-1} are more typical. A crucial limiting factor in such systems is light saturation of photosynthesis. Even in the very dense algal suspensions which develop, the cells near the surface are exposed to irradiances well above their saturation points. Of the three requirements for high productivity listed at the beginning of this section, such cells will have a high rate of collection of light energy, and (given the plentiful nutrient supply) rapid conversion of photosynthate to new cell material: what they will lack is efficient utilization of the absorbed energy in photosynthesis, since their carboxylation system will be unable to keep pace with the high rate of arrival of excitation energy. The inability of algae to increase their photosynthetic rate linearly with irradiance all the way up to full sunlight is likely to remain an insuperable obstacle in the path of increasing yields from outdoor mass culture. To develop, by mutation or genetic engineering, algal types with increased carboxylase and lowered pigment content would not be a solution, because such cells would be poorly adapted to the lower light intensities existing at greater depths. It is essential that the algal suspension should be sufficiently deep and dense to absorb nearly all the light incident upon it. This means that a gradient of irradiance, from full sunlight to near darkness, must exist within the suspension, and no one type of alga can be optimally adapted to all the intensities present. In principle, a laminar arrangement with a range of algal types stacked one above the other in thin layers separated by transparent boundaries, with the most Sun-adapted type at the top and shade-adapted type at the bottom, could increase the yields, but the technical problems and capital cost would be considerable.

References and author index

The numbers in italic following each item are pages where the author's work is mentioned

1. Abel, K. M. (1984). Inorganic carbon source for photosynthesis in the seagrass *Thalassia hemprichii* (Ehrenb.) Aschers. *Plant Physiol.*, **76**, 776–81. *327*
2. Abiodun, A. A. & Adeniji, H. A. (1978). Movement of water columns in Lake Kainji. *Remote Sens, Env.*, **7**, 227–34. *185, 207*
3. Ackleson, S. & Spinrad, R. W. (1988). Size and refractive index of individual marine particulates: a flow cytometric approach. *Appl. Opt.*, **27**, 1270–7. *111*
4. Adams, M. S., Guilizzoni, P. & Adams, S. (1978). Relationship of dissolved inorganic carbon to macrophyte photosynthesis in some Italian lakes. *Limnol. Oceanogr.*, **23**, 912–19. *328–9*
5. Aiken, G. R. & Malcolm, R. L. (1987). Molecular weight of aquatic fulvic acids by vapor pressure osmometry. *Geochim. Cosmochim. Acta*, **51**, 2177–84. *62*
6. Aiken, J. (1981). The Undulating Oceanographic Recorder. *J. Plankton Res.*, **3**, 551–60. *118*
7. Aiken, J. (1985). The Undulating Oceanographic Recorder Mark 2. In A. Zirino (ed.), *Mapping strategies in chemical oceanography* (pp. 315–32). *118*
8. Aiken, J. & Bellan, I. (1990). Optical oceanography: an assessment of a towed method. In P. J. Herring, A. K. Campbell, M. Whitfield & L. Maddock (eds), *Light and life in the sea* (pp. 39–58). Cambridge: Cambridge University Press. *118*
9. Alberte, R. S., Friedman, A. L., Gustafson, D. L., Rudnick, M. S. & Lyman, H. (1981). Light-harvesting systems of brown algae and diatoms. *Biochim. Biophys. Acta*, **635**, 304–16. *230*
10. Alberte, R. S., Wood, A. M., Kursar, T. A. & Guillard, R. R. L. (1984). Novel phycoerythrins in marine *Synechococcus* spp. *Plant Physiol.*, **75**, 732–9. *240*
11. Allen, E. D. & Spence, D. H. N. (1981). The differential ability of aquatic plants to utilize the inorganic carbon supply in fresh waters. *New Phytol.*, **87**, 269–83. *325, 327*
12. Amos, C. L. & Alfoldi, T. T. (1979). The determination of suspended sediment concentration in a macrotidal system using Landsat data. *J. Sedim. Petrol.*, **49**, 159–74. *207, 194*

13. Amos, C. L. & Topliss, B. J. (1985). Discrimination of suspended particulate matter in the Bay of Fundy using the Nimbus 7 Coastal Zone Colour Scanner. *Canad. J. Remote Sens.*, 11, 85–92. *194*
14. Andersen, R. A. (1987). Synurophyceae classis nov., a new class of algae. *Amer. J. Bot.*, **74**, 337–53. *228*
15. Anderson, G. C. (1969). Subsurface chlorophyll maximum in the Northeast Pacific Ocean. *Limnol. Oceanogr.*, **14**, 386–91. *319, 320*
16. Anderson, J. M. & Barrett, J. (1986). Light-harvesting pigment–protein complexes of algae. *Encycl. Plant Physiol. (n.s.)*, **19**, 269–85. *235*
17. Antia, N. J. (1977). A critical appraisal of Lewin's Prochlorophyta. *Br. phycol. J.*, **12**, 271–6. *223*
18. Arnold, K. E. & Murray, S. N. (1980). Relationships between irradiance and photosynthesis for marine benthic green algae (Chlorophyta) of differing morphologies. *J. Exp. Mar. Biol. Ecol.*, **43**, 183–92. *221*
19. Aruga, Y. (1965). Ecological studies of photosynthesis and matter production of phytoplankton. I. Seasonal changes in photosynthesis of natural phytoplankton. *Biol. Mag. (Tokyo)*, **78**, 280–8. *412*
20. Arvesen, J. C., Millard, J. P. & Weaver, E. C. (1973). Remote sensing of chlorophyll and temperature in marine and fresh waters. *Astronaut. Acta*, **18**, 229–39. *176, 209*
21. Ashley, L. E. & Cobb, C. M. (1958). *J. Opt. Soc. Amer.*, **48**, 261–8. Quoted by Hodkinson & Greenleaves (1963). *87*
22. Asmus, R. M. & Asmus, H. (1991). Mussel beds: limiting or promoting phytoplankton? *J. Exp. Mar. Biol. Ecol.*, **148**, 215–32. *340*
23. Atlas, D. & Banister, T. T. (1980). Dependence of mean spectral extinction coefficient of phytoplankton on depth, water colour and species. *Limnol. Oceanogr.*, **25**, 157–9. *270*
24. Auclair, J. C., Demers, S., Frechette, M., Legendre, L. & Trump, C. L. (1982). High frequency endogenous periodicities of chlorophyll synthesis in estuarine phytoplankton. *Limnol. Oceanogr.*, **27**, 348–52. *427–8*
25. Aughey, W. H. & Baum, F. J. (1954). Angular-dependence light scattering – a high resolution recording instrument for the angular range 0.05–140°. *J. Opt. Soc. Amer*, **44**, 833–7. *95*
26. Austin, R. W. (1974*a*). The remote sensing of spectral radiance from below the ocean surface. In N. G. Jerlov & E. S. Nielsen (eds), *Optical aspects of oceanography* (pp. 317–44). London: Academic Press. *42, 43, 172, 183, 190, 290*
27. Austin, R. W. (1974*b*). Ocean colour analysis, part 2. *San Diego: Scripps Inst. Oceanogr.* *147*
28. Austin, R. W. (1980). Gulf of Mexico, ocean-colour surface-truth measurements. *Boundary-layer Meteorol.*, **18**, 269–85. *183*
29. Austin, R. W. (1981). Remote sensing of the diffuse attenuation coefficient of ocean water. In *Special topics in optical propagation*, 300 (pp. 18–1 to 18–9). Neuilly-sur-Seine: AGARD (NATO). *204*
30. Austin, R. W. & Petzold, T. J. (1977). Considerations in the design and evaluation of oceanographic transmissometers. In J. E. Tyler (ed.), *Light in the sea* (pp. 104–20). Stroudsbury: Dowden Hutchinson Ross. *92*
31. Austin, R. W. & Petzold, T. J. (1981). The determination of the diffuse

attenuation coefficient of sea water using the Coastal Zone Colour Scanner. In J. F. R. Gower (ed.), *Oceanography from space* (pp. 239–56). New York: Plenum. *204–5*

32. Bader, H. (1970). The hyperbolic distribution of particle sizes. *J. Geophys. Res.*, **75**, 2822–30. *86*
33. Badger, M. R. & Andrews, T. J. (1982). Photosynthesis and inorganic carbon usage by the marine cyanobacterium *Synechococcus* sp. *Plant Physiol.*, **70**, 517–23. *327, 332*
34. Badger, M. R., Kaplan, A. & Berry, J. A. (1980). Internal inorganic carbon pool of *Chlamydomonas reinhardtii. Plant Physiol.*, **66**, 407–13. *332*
35. Baker, K. S. & Frouin, R. (1987). Relation between photosynthetically available radiation and total insolation at the ocean surface under clear skies. *Limnol. Oceanogr.*, **32**, 1370–7. *31*
36. Baker, K. S. & Smith, R. C. (1979). Quasi-inherent characteristics of the diffuse attenuation coefficient for irradiance. *Soc. Photo-opt. Instrum. Eng.*, **208**, 60–3. *19, 133*
37. Balch, W. M., Abbott, M. R. & Eppley, R. W. (1989). Remote sensing of primary production I. A comparison of empirical and semi-analytical algorithms. *Deep-Sea Res.*, **36**, 281–95. *356*
38. Balch, W. M., Eppley, R. W. & Abbott, M. R. (1989). Remote sensing of primary production II. A semi-analytical algorithm based on pigments, temperature and light. *Deep-Sea Res.*, **36**, 1201–17. *355–6*
39. Balch, W., Evans, E., Brown, J., Feldman, G., McClain, C. & Esaias, W. (1992). The remote sensing of ocean primary productivity: use of a new data compilation to test satellite algorithms. *J. Geophys. Res.*, **97**, 2279–93. *355–6*
40. Balch, W. M., Holligan, P. M., Ackleson, S. G. & Voss, K. J. (1991). Biological and optical properties of mesoscale coccolithophore blooms in the Gulf of Maine. *Limnol. Oceanogr.*, **36**, 629–43. *150*
41. Bannister, T. T. (1974*a*). Production equation in terms of chlorophyll concentration, quantum yield, and upper limit to production. *Limnol. Oceanogr.*, **19**, 1–12. *430*
42. Bannister, T. T. (1974*b*). A general theory of steady state phytoplankton growth in a nutrient-saturated mixed layer. *Limnol. Oceanogr.*, **19**, 13–30. *430*
43. Bannister, T. T. (1979). Quantitative description of steady state, nutrient-saturated algal growth, including adaptation. *Limnol. Oceanogr.*, **24**, 76–96. *430*
44. Bannister, T. T. & Weidemann, A. D. (1984). The maximum quantum yield of phytoplankton photosynthesis *in situ. J. Plankton Res.*, **6**, 275–94. *301*
45. Barko, J. W. & Filbin, G. J. (1983). Influences of light and temperature on chlorophyll composition in submersed freshwater macrophytes. *Aquat. Bot.*, **15**, 249–55. *382*
46. Barrett, J. & Anderson, J. M. (1980). The P-700-chlorophyll *a* protein complex and two major light-harvesting complexes of *Acrocarpia paniculata* and other brown seaweeds. *Biochim. Biophys. Acta*, **590**, 309–23. *238*
47. Bauer, D., Brun-Cottan, J. C. & Saliot, A. (1971). Principe d'une mésure directe dans l'eau de mer du coefficient d'absorption de la lumière. *Cah. Oceanogr.*, **23**, 841–58. *53*

48. Bauer, D. & Ivanoff, A. (1970). Spectroirradiance-metre. *Cah. Oceanogr.*, **22**, 477–82. *123*

49. Bauer, D., & Morel, A. (1967). Etude aux petits angles de l'indicatrice de diffusion de la lumière par les eaux de mer. *Ann. Geophys.*, **23**, 109–23. *94*

50. Beardall, J. & Morris, I. (1976). The concept of light intensity adaptation in marine phytoplankton: some experiments with *Phaeodactylum tricornutum. Mar. Biol.*, **37**, 3777–87. *380*

51. Beardsley, G. F. (1968). Mueller scattering matrix of sea water. *J. Opt. Soc. Amer.*, **58**, 52–7. *94*

52. Beer, S. & Eshel, A. (1983). Photosynthesis of *Ulva* sp. II. Utilization of CO_2 and HCO_3^- when submerged. *J. Exp. Mar. Biol. Ecol.*, **70**, 99–106. *325*

53. Beer, S., Eshel, A. & Waisel, Y. (1977). Carbon metabolism in seagrasses I. The utilization of exogenous inorganic carbon species in photosynthesis. *J. Exp. Bot.*, **28**, 1180–9. *327*

54. Beer, S., Israel, A., Drechsler, Z. & Cohen, Y. (1990). Evidence for an inorganic carbon concentrating system, and ribulose-1, 5-bisphosphate carboxylase/oxygenase CO_2 kinetics. *Plant Physiol.*, **94**, 1542–6. *332*

55. Beer, S. & Wetzel, R. G. (1982). Photosynthetic carbon fixation pathways in *Zostera marina* and three Florida seagrasses. *Aquat. Bot.*, **13**, 141–6. *333*

56. Belay, A. (1981). An experimental investigation of inhibition of phytoplankton photosynthesis at lake surfaces. *New Phytol.*, **89**, 61–74. *273, 274, 284, 285*

57. Belay, A. & Fogg, G. E. (1978). Photoinhibition of photosynthesis in *Asterionella formosa* (Bacillariophyceae). *J. Phycol.*, **14**, 341–7. *285*

58. Bennett, A. & Bogorad, L. (1973). Complementary chromatic adaptation in a filamentous blue-green alga. *J. Cell Biol.*, **58**, 419–35. *407–8*

59. Berdalet, E. (1992). Effects of turbulence on the marine dinoflagellate *Gymnodinium nelsonii. J. Phycol.*, **28**, 267–72. *341*

60. Berger, W. H., Smetacek, V. S. & Wefer, G. (eds). (1989). *Productivity of the ocean: present and past.* Chichester: Wiley. *358*

61. Berner, T., Dubinsky, Z., Wyman, K. & Falkowski, P. G. (1989). Photoadaptation and the 'package' effect in *Dunaliella tertiolecta* (Chlorophyceae). *J. Phycol.*, **25**, 70–8. *259*

62. Berthold, G. (1882). Uber die Verteilung der Algen im Golf von Neapel nebst einem Verzeichnis der bisher daselbst beobachten Arten. *Mitt. Zool. Sta Neopol.*, **3**, 393–536. *361*

63. Bezrukov, L. B., Budnev, N. M., Galperin, M. D., Dzhilkibayev, Z. M., Lanin, O. Y. & Taraschanskiy, B. A. (1990). Measurement of the light attenuation coefficient of water in Lake Baikal. *Oceanology*, **30**, 756–9. *53*

64. Bidigare, R. R., Prézelin, B. B. & Smith, R. C. (1992). Bio-optical models and the problems of scaling. In P. G. Falkowski & A. D. Woodhead (eds), *Primary productivity and biogeochemical cycles in the sea* (pp. 175–212). New York: Plenum. *299, 356*

65. Bidigare, R. R., Smith, R. C., Baker, K. S. & Marra, J. (1987). Oceanic primary production estimates from measurements of spectral

irradiance and pigment concentrations. *Global Biogeochemical Cycles*, **1**, 171–86. *76, 297*

66. Bienfang, P. K., Szyper, J. P., Okamoto, M. Y. & Noda, E. K. (1984). Temporal and spatial variability of phytoplankton in a subtropical ecosystem. *Limnol. Oceanogr.*, **29**, 527–39. *138*
67. Bindloss, M. E. (1974). Primary productivity of phytoplankton in Loch Leven, Kinross. *Proc. Roy. Soc. Edin. (B)*, **74**, 157–81. *269, 337*
68. Biospherical Instruments Inc. (San Diego, Calif.). MER 1000, MER 1032, Spectroradiometers. *123*
69. Bird, D. F. & Kalff, J. (1989). Phagotrophic sustenance of a metalimnetic phytoplankton peak. *Limnol. Oceanogr.*, **34**, 155–62. *320*
70. Biscaye, P. E. & Eittreim, S. L. (1973). Variations in benthic boundary layer phenomena: nepheloid layer in the North American basin. In R. J. Gibbs (ed.), *Suspended solids in water* (pp. 227–60). New York: Plenum. *109*
71. Bishop, J. K. B. & Rossow, W. B. (1991). Spatial and temporal variability of global surface solar irradiance. *J. Geophys. Res.*, **96**, 16839–58. *33*
72. Bjornland, T. & Aguilar-Martinez, M. (1976). Carotenoids in red algae. *Phytochem.*, **15**, 291–6. *230*
73. Blasco, D. (1978). Observations on the diel migration of marine dinoflagellates off the Baja California coast. *Mar. Biol.*, **46**, 41–7. *416*
74. Blatt, M. R. & Briggs, W. R. (1980). Blue-light-induced cortical fibre reticulation concomitant with chloroplast aggregation in the alga, *Vaucheria sessilis*. *Planta*, **147**, 355–62. *423*
75. Boardman, N. K. (1977). Comparative photosynthesis of sun and shade plants. *Ann. Rev. Plant Physiol.*, **28**, 355–77. *377, 379*
76. Bochkov, B. F., Kopelevich, O. V. & Kriman, B. A. (1980). A spectrophotometer for investigating the light of sea water in the visible and ultraviolet regions of the spectrum. *Oceanology*, **20**, 101–4. *92*
77. Boczar, B. A. & Palmisano, A. C. (1990). Photosynthetic pigments and pigment–proteins in natural populations of Antarctic sea-ice diatoms. *Phycologia*, **29**, 470–7. *238*
78. Bogorad, L. (1975). Phycobiliproteins and complementary chromatic adaptation. *Ann. Rev. Plant Physiol.*, **26**, 369–401. *239, 386*
79. Boivin, L. P., Davidson, W. F., Storey, R. S., Sinclair, D. & Earle, E. D. (1986). Determination of the attenuation coefficients of visible and ultraviolet radiation in heavy water. *Appl. Opt.*, **25**, 877–82. *55, 56*
80. Bold, H. C. & Wynne, M. J. (1978). *Introduction to the algae*. Englewood Cliffs, N.J.: Prentice-Hall. *218*
81. Booth, C. R. (1976). The design and evaluation of a measurement system for photosynthetically active quantum scalar irradiance. *Limnol. Oceanogr.*, **21**, 326–36. *121*
82. Boresch, K. (1921). Die komplementäre chromatische Adaptation. *Arch. Protistenkd.*, **44**, 1–70. *384*
83. Borgerson, M. J., Bartz, R., Zaneveld, J. R. V. & Kitchen, J. C. (1990). A modern spectral transmissometer. *Proc. Soc. Photo-Opt. Instrum. Eng., Ocean Optics X*, **1302**, 373–85. *92*

84. Borowitzka, M. A. & Larkum, A. W. D. (1976). Calcification in the green alga *Halimeda*. *J. Exp. Bot.*, **27**, 879–93. *324*
85. Borstad, A., Brown, R. M. & Gower, J. F. R. (1980). Airborne remote sensing of sea surface chlorophyll and temperature along the outer British Columbia coast. *Proc. 6th Canadian Symp. Rem. Sens.*, pp. 541–9. *198, 210*
86. Boston, H. L. (1986). A discussion of the adaptations for carbon acquisition in relation to the growth strategy of aquatic isoetids. *Aquat. Bot.*, **26**, 259–70. *330, 333*
87. Bowes, G. (1987). Aquatic plant photosynthesis: strategies that enhance carbon gain. In R. M. M. Crawford (ed.), *Plant life in aquatic and amphibious habitats* (pp. 79–98). Oxford: Blackwell. *250, 332*
88. Bowker, D. E. & LeCroy, S. R. (1985). Bright spot analysis of ocean-dump plumes using Landsat MSS. *Int. J. Remote Sens.*, **6**, 759–71. *207*
89. Bowling, L. C. (1988). Optical properties, nutrients and phytoplankton of freshwater coastal dune lakes in south-east Queensland. *Aust J. Mar. Freshwater Res.*, **39**, 805–15. *69, 104, 142*
90. Bowling, L. C., Steane, M. S. & Tyler, P. A. (1986). The spectral distribution and attenuation of underwater irradiance in Tasmanian inland waters. *Freshwater Biol.*, **16**, 313–35. *69, 103, 104, 143, 150*
91. Boynton, W. R., Kemp, W. M. & Keefe, C. W. (1982). A comparative analysis of nutrients and other factors influencing estuarine phytoplankton productivity. In V. S. Kennedy (ed.), *Estuarine comparisons*. New York: Academic. *359*
92. Braarud, T. & Klem, A. (1931). *Hvalradets Skrift*, **1**, 1–88. Quoted by Steemann-Nielsen (1974). *315*
93. Brakel, W. H. (1984). Seasonal dynamics of suspended-sediment plumes from the Tana and Sabaki Rivers, Kenya: analysis of Landsat imagery. *Remote Sens. Environ.*, **16**, 165–73. *207, 191*
94. Breen, P. A. & Mann, K. H. (1976). Changing lobster abundance and the destruction of kelp beds by sea urchins. *Mar. Biol.*, **34**, 137–42. *342*
95. Bricaud, A. & Morel, A. (1986). Light attenuation and scattering by phytoplanktonic cells: a theoretical model. *Appl. Opt.*, **25**, 571–80. *110, 212*
96. Bricaud, A. & Morel, A. (1987). Atmospheric corrections and interpretation of marine radiances in CZCS imagery: use of a reflectance model. *Oceanologica Acta, Proc. Spatial Oceanography Symp. (Brest, Nov. 1985)*, pp. 33–50. *188*
97. Bricaud, A., Morel, A. & Prieur, L. (1981). Absorption by dissolved organic matter of the sea (yellow substance) in the UV and visible domains. *Limnol. Oceanogr.*, **26**, 43–53. *57, 62, 63, 64, 66*
98. Bricaud, A., Morel, A. & Prieur, L. (1983). Optical efficiency factors of some phytoplankters. *Limnol. Oceanogr.*, **28**, 816–32. *110, 111*
99. Bricaud, A. & Stramski, D. (1990). Spectral absorption coefficients of living phytoplankton and nonalgal biogenous matter: a comparison between the Peru upwelling and the Sargasso Sea. *Limnol. Oceanogr.*, **35**, 562–82. *73, 76*
100. Bristow, M., Nielsen, D., Bundy, D. & Furtek, R. (1981). Use of water Raman emission to correct airborne laser fluorosensor data for

effects of water optical attenuation. *Appl. Opt.*, **20**, 2889–906. *201–3*

101. Bristow, M. P. F., Bundy, D. H., Edmonds, C. M., Ponto, P. E., Frey, B. E. & Small, L. F. (1985). Airborne laser fluorosensor survey of the Columbia and Snake rivers: simultaneous measurements of chlorophyll, dissolved organics and optical attenuation. *Int. J. Remote Sens.*, **6**, 1707–34. *204, 210*
102. Britz, S. J. & Briggs, W. R. (1976). Circadian rhythms of chloroplast orientation and photosynthetic capacity in *Ulva. Plant Physiol.*, **58**, 22–7. *109, 423, 427*
103. Brody, M. & Emerson, R. (1959). The effect of wavelength and intensity of light on the proportion of pigments in *Porphyridium cruentum. Amer. J. Bot.*, **46**, 433–40. *376, 387–8, 390, 409*
104. Brown, J. & Simpson, J. H. (1990). The radiometric determination of total pigment and seston and its potential use in shelf seas. *Estuar. Coast. Shelf Sci.*, **31**, 1–9. *198*
105. Brown, J. S. (1987). Functional organization of chlorophyll *a* and carotenoids in the alga *Nannochloropsis salina. Plant Physiol.*, **83**, 434–7. *237*
106. Brown, O. B., Evans, R. H., Brown, J. W., Gordon, H. R., Smith, R. C. & Baker, K. S. (1985). Phytoplankton blooming off the U.S. East coast: a satellite description. *Science*, **229**, 163–7. *212*
107. Brown, T. E. & Richardson, F. L. (1968). The effect of growth environment on the physiology of algae: light intensity. *J. Phycol.*, **4**, 38–54. *376*
108. Bruning, K., Lingeman, R. & Ringelberg, J. (1992). Estimating the impact of fungal parasites on phytoplankton populations. *Limnol. Oceanogr.*, **37**, 252–60. *341*
109. Bryant, D. A. (1981). The photoregulated expression of multiple phycocyanin genes. *Eur. J. Biochem*, **119**, 425–9. *385, 387*
110. Brylinsky, M. (1980). Estimating the productivity of lakes and reservoirs. In E. D. Le Cren & R. H. Lowe-McConnell (eds), *The functioning of freshwater ecosystems.* Cambridge: Cambridge University Press. *303, 359*
111. Buchwald, M. & Jencks, W. P. (1968). Properties of the crustacyanins and the yellow lobster shell pigments. *Biochemistry*, **7**, 844–59. *234*
112. Bukata, R. P. & Bruton, J. E. (1974). ERTS-1 digital classifications of the water regimes comprising Lake Ontario. *Proc. 2nd Canad. Symp. Rem. Sens.*, pp. 627–34. *195, 207, 211*
113. Bukata, R. P., Harris, G. P. & Bruton, J. E. (1974). The detection of suspended solids and chlorophyll *a* utilizing multispectral ERTS-1 data. *Proc. 2nd Canad. Symp. Rem. Sens.*, pp. 551–64. *191, 195*
114. Bukata, R. P., Jerome, J. H. & Bruton, J. E. (1988). Particulate concentrations in Lake St Clair as recorded by a shipborne multispectral optical monitoring system. *Remote Sens. Environ.*, **25**, 201–29. *175*
115. Bukata, R. P., Jerome, J. H., Bruton, J. E. & Jain, S. C. (1979). Determination of inherent optical properties of Lake Ontario coastal waters. *Appl. Opt.*, **18**, 3926–32. *103*
116. Bukata, R. P., Jerome, J. H., Bruton, J. E. & Jain, S. C. (1980). Nonzero subsurface irradiance reflectance at 670 nm from Lake Ontario water masses. *Appl. Opt.*, **19**, 2487–8. *186*

117. Bukaveckas, P. A. & Driscoll, C. T. (1991). Effects of whole-lake base addition on the optical properties of three clearwater acidic lakes. *Can. J. Fish. Aquat. Sci.*, **48**, 1030–40. *103*
118. Bullerjahn, G. S., Matthijs, H. C. P., Mur, L. R. & Sherman, L. A. (1987). Chlorophyll–protein composition of the thylakoid membrane from *Prochlorothrix hollandica*, a prokaryote containing chlorophyll b. *Eur. J. Biochem.*, **168**, 295–300. *238*
119. Bunt, J. S. (1963). Diatoms of Antarctic sea ice as agents of primary production. *Nature*, **199**, 1255–8. *410*
120. Burger-Wiersma, T., Veenhuis, M., Korthals, H. J., Van de Wiel, C. C. M. & Mur, L. R. (1986). A new prokaryote containing chlorophylls a and b. *Nature*, **320**, 262–4. *223, 228, 231*
121. Burt, W. V. (1958). Selective transmission of light in tropical Pacific waters. *Deep-Sea Res.*, **5**, 51–61. *64*
122. Bush, K. J. & Sweeney, B. M. (1972). The activity of ribulose diphosphate carboxylase in extracts of *Gonyaulax polyedra* in the day and night phases of the circadian rhythm of photosynthesis. *Plant Physiol.*, **50**, 446–51. *425*
123. Butler, W. L. (1978). Energy distribution in the photochemical apparatus of photosynthesis. *Ann. Rev. Plant Physiol.*, **29**, 345–78. *247, 310*
124. Calabrese, G. (1972). Research on red algal pigments. 2. Pigments of *Petroglossum nicaeense* (Duby) Schotter (Rhodophyceae, Gigartinales) and their seasonal variations at different light intensities. *Phycologia*, **11**, 141–6. *411*
125. Calabrese, G. & Felicini, G. P. (1973). Research on red algal pigments. 5. The effect of white and green light on the rate of photosynthesis and its relationship to pigment components in *Gracilaria compressa. Phycologia*, **12**, 195–9. *376*
126. Callame, B. & Debyser, J. (1954). Observations sur les mouvements des diatomées à la surface des sédiments marins de la zone intercotidale. *Vie milieu*, **5**, 242–9. *420*
127. Cambridge, M. L., Chiffings, A. W., Britton, C., Moore, L. & McComb, A. J. (1986). The loss of seagrass in Cockburn Sound, Western Australia II. Possible causes of seagrass decline. *Aquat. Bot.*, **24**, 269–85. *341*
128. Campbell, E. E. & Bate, G. C. (1987). Factors influencing the magnitude of phytoplankton primary production in a high-energy surf zone. *Estuar. Coast. Shelf Sci.*, **24**, 741–50. *280*
129. Campbell, J. W. & Esaias, W. E. (1983). Basis for spectral curvature algorithms in remote sensing of chlorophyll. *Appl. Opt.*, **22**, 1084–92. *176, 194, 199*
130. Campbell, J. W. & O'Reilly, J. E. (1988). Role of satellites in estimating primary productivity on the northwest Atlantic continental shelf. *Continental Shelf Res.*, **8**, 179–204. *354*
131. Carder, K. L., Hawes, S. K., Baker, K. A., Smith, R. C., Steward, R. G. & Mitchell, B. G. (1991). Reflectance model for quantifying chlorophyll a in the presence of productivity degradation products. *J. Geophys. Res.*, **96**, 20599–611. *198*
132. Carder, K. L., Stewart, R. G., Harvey, G. R. & Ortner, P. B. (1989). Marine humic and fulvic acids: their effects on remote sensing of ocean chlorophyll. *Limnol. Oceanogr.*, **34**, 68–81. *61, 63, 64, 65*
133. Carder, K. L., Tomlinson, R. D. & Beardsley, G. F. (1972). A

technique for the estimation of indices of refraction of marine phytoplankton. *Limnol. Oceanogr.*, **17**, 833–9. *111*

134. Caron, L., Dubacq, J. P., Berkaloff, C. & Jupin, H. (1985). Subchloroplast fractions from the brown alga *Fucus serratus*: phosphatidylglycerol contents. *Plant Cell Physiol.*, **26**, 131–9. *238*
135. Caron, L., Remy, R. & Berkaloff, C. (1988). Polypeptide composition of light-harvesting complexes from some brown algae and diatoms. *FEBS Lett.*, **229**, 11–15. *238*
136. Carpenter, D. J. & Carpenter, S. M. (1983). Modelling inland water quality using Landsat data. *Remote Sens. Environ.*, **13**, 345–52. *206, 207*
137. Carpenter, E. J. & Romans, K. (1991). Major role of the cyanobacterium *Trichodesmium* in nutrients cycling in the North Atlantic Ocean. *Science*, **254**, 1356–8. *339*
138. Carpenter, R. C. (1985). Relationship between primary production and irradiance in coral reef algal communities. *Limnol. Oceanogr.*, **30**, 784–93. *282*
139. Chen, Z., Curran, P. J. & Hanson, J. D. (1992). Derivative reflectance spectroscopy to estimate suspended sediment concentration. *Remote Sens. Environ.*, **40**, 67–77. *194*
140. Chisholm, S. W., Armbrust, E. V. & Olson, R. J. (1986). The individual cell in phytoplankton ecology: cell cycles and applications of flow cytometry. In T. Platt & W. K. W. Li (eds), *Photosynthetic picoplankton* (pp. 343–69). *Can. Bull. Fish Aquat. Sci.* **214**. *111, 226*
141. Chisholm, S. W. & Morel, F. M. M. (eds) (1991). What controls phytoplankton production in nutrient-rich areas of the open sea? *Limnol. Oceanogr. (special issue)*, **36** (No. 8). *341*
142. Chisholm, S. W., Olson, R. J., Zettler, E. R., Goericke, R., Waterbury, J. B. & Welschmeyer, N. A. (1988). A novel free-living prochlorophyte abundant in the oceanic euphotic zone. *Nature*, **334**, 340–3. *223, 227, 228, 231*
143. Chow, W. S. (1993). Photoprotection and photoinhibitory damage. In J. Barber (ed.), *Molecular processes of photosynthesis. Adv. Molec, Cell Biol., Vol. 7*, Greenwich, Conn: JAI Press. *285, 377*
144. Chrystal, J. & Larkum, A. W. D. (1987). Pigment–protein complexes and light harvesting in Eustigmatophyte algae. In J. Biggins (ed.), *Progress in photosynthesis research* (pp. 189–92). Dordrecht: Martinus Nijhoff. *237*
145. Chu, Z.-X. & Anderson, J. M. (1985). Isolation and characterization of a siphonaxanthin–chlorophyll a/b protein complex of Photosystem I from a *Codium* species (Siphonales). *Biochim. Biophys. Acta*, **806**, 154–60. *235, 237*
146. Clarke, G. L. (1939). The utilization of solar energy by aquatic organisms. In F. R. Poulton (ed.), *Problems of lake biology* (pp. 27–38). American Association for the Advancement of Science. *290*
147. Clarke, G. L. & Ewing, G. C. (1974). Remote spectroscopy of the sea for biological production studies. In N. G. Jerlov & E. S. Nielsen (eds), *Optical aspects of oceanography* (pp. 389–413). London: Academic Press. *209*
148. Clarke, G. L., Ewing, G. C. & Lorenzen, C. J. (1970). Spectra of

backscattered light from the sea obtained from aircraft as a measure of chlorophyll concentration. *Science*, **167**, 1119–21. *172, 175, 196–7*

149. Clarke, G. L. & James, H. R. (1939). Laboratory analysis of the selective absorption of light by sea water. *J. Opt. Soc. Am.*, **29**, 43–55. *54*
150. Clayton, R. K. (1980). *Photosynthesis: physical mechanisms and chemical patterns*. Cambridge: Cambridge University Press. *376*
151. Cleveland, J. S., Perry, M. J., Kiefer, D. A. & Talbot, M. C. (1989). Maximal yield of photosynthesis in the northwestern Sargasso Sea. *J. Mar. Res.*, **47**, 869–86. *297, 302, 356*
152. Cloern, J. E. (1987). Turbidity as a control on phytoplankton biomass and productivity in estuaries. *Continental Shelf Res.*, **7**, 1367–81. *139, 321*
153. Cloern, J. E. (1991). Tidal stirring and phytoplankton bloom dynamics in an estuary. *J. Mar. Res.*, **49**, 203–21. *318, 345*
154. Cloern, J. E., Alpine, A. E., Cole, B. E., Wong, R. L. J., Arthur, J. F. & Ball, M. D. (1983). River discharge controls phytoplankton dynamics in the Northern San Francisco Bay estuary. *Estuar. Coast. Shelf Sci.*, **16**, 415–29. *315*
155. Coats, D. W. & Harding, L. W. (1988). Effect of light history on the ultrastructure and physiology of *Prorocentrum mariae-lebouriae* (Dinophyceae). *J. Phycol.*, **24**, 67–77. *378*
156. Cole, B. E. & Cloern, J. E. (1984). Significance of biomass and light availability to phytoplankton productivity in San Francisco Bay. *Mar. Ecol. Prog. Ser.*, **17**, 15–24. *321–2*
157. Cole, B. E. & Cloern, J. E. (1987). An empirical model for estimating phytoplankton productivity in estuaries. *Mar. Ecol. Prog. Ser.*, **36**, 299–305. *321–2*
158. Colijn, F., Admiraal, W., Baretta, J. W. & Ruardij, P. (1987). Primary production in a turbid estuary, the Ems–Dollard: field and model studies. *Continental Shelf Res.*, **7**, 1405–9. *138, 139*
159. Colman, B. & Gehl, K. A. (1983). Physiological characteristics of photosynthesis in *Porphyridium cruentum*: evidence for bicarbonate transport in a unicellular red alga. *J. Phycol.*, **19**, 216–19. *327*
160. Cota, G. F. (1985). Photoadaptation of high Arctic ice algae. *Nature*, **315**, 219–22. *280, 410*
161. Cota, G. F. & Sullivan, C. W. (1990). Photoadaptation, growth and production of bottom ice algae in the Antarctic. *J. Phycol.*, **26**, 399–411. *410*
162. Cox, C. & Munk, W. (1954). Measurement of the roughness of the sea surface from photographs of the sun's glitter. *J. Opt. Soc. Amer.*, **44**, 838–50. *45, 127*
163. Craigie, J. S. (1974). Storage Products. In W. D. P. Stewart (ed.), *Algal physiology and biochemistry* (pp. 206–35). Oxford: Blackwell. *250*
164. Critchley, C. (1981). Studies on the mechanism of photoinhibition in higher plants. *Plant Physiol.*, **67**, 1161–5. *285*
165. Critchley, C. (1988). The molecular mechanism of photoinhibition – facts and fiction. *Aust. J. Plant Physiol.*, **15**, 27–41. *285*
166. Crossett, R. N., Drew, E. A. & Larkum, A. W. D. (1965). Chromatic adaptation in benthic marine algae. *Nature*, **207**, 547–8. *367, 371*
167. Cullen, J. J. (1982). The deep chlorophyll maximum: comparing vertical

profiles of chlorophyll a. *Can J. Fish, Aquat. Sci.*, **39**, 791–803. *320*

168. Cullen, J. J. & Horrigan, S. G. (1981). Effects of nitrate on the diurnal vertical migration, carbon to nitrogen ratio, and the photosynthetic capacity of the dinoflagellate, *Gymnodinium splendens. Mar. Biol.*, **62**, 81–9. *416–7*
169. Cullen, J. J., Lewis, M. R., Daviss, C. O. & Barber, R. T. (1992). Photosynthetic characteristics and estimated growth rates indicate grazing is the proximate control of primary production in the equatorial Pacific. *J. Geophys. Res.*, **97**, 639–54. *340*
170. Cunningham, F. X. & Schiff, J. A. (1986). Chlorophyll–protein complexes from *Euglena gracilis* and mutants deficient in chlorophyll b. *Plant Physiol.*, **80**, 223–30. *235, 237*
171. Curran, R. J. (1972). Ocean colour determination through a scattering atmosphere. *Appl. Opt.*, **11**, 1857–66. *186*
172. Cushing, D. H. (1988). The flow of energy in marine ecosystems, with special reference to the continental shelf. In H. Postma & J. J. Zijlstra (eds), *Continental Shelves* (pp. 203–30). Amsterdam: Elsevier. *358*
173. Dale, H. M. (1986). Temperature and light: the determining factors in maximum depth distribution of aquatic macrophytes in Ontario, Canada. *Hydrobiologia*, **133**, 73–7. *338*
174. Davies-Colley, R. J. (1987). Optical properties of the Waikato River, New Zealand. *Mitt. Geol.-Paläont. Inst. Univ. Hamburg, SCOPE/UNEP Sonderbd.*, **64**, 443–60. *70, 73, 98, 104*
175. Davies-Colley, R. J. (1992). Yellow substance in coastal and marine waters round the South Island, New Zealand. *N.Z. J. Mar. Freshwater Res.*, **26**, 311–22. *67*
176. Davies-Colley, R. J., Pridmore, R. D. & Hewitt, J. E. (1986). Optical properties of some freshwater phytoplanktonic algae. *Hydrobiologia*, **133**, 165–78. *110, 259*
177. Davies-Colley, R. J. & Vant, W. N. (1987). Absorption of light by yellow substance in freshwater lakes. *Limnol. Oceanogr.*, **32**, 416–25. *63, 70*
178. Davies-Colley, R. J. & Vant, W. N. (1988). Estimates of optical properties of water from Secchi disk depths. *Water Res. Bull.*, **24**, 1329–35. *120, 143*
179. Davies-Colley, R. J., Vant, W. N. & Latimer, G. J. (1984). Optical characterization of natural waters by PAR measurement under changeable light conditions. *N.Z. J. Mar. Freshwater Res.*, **18**, 455–60. *118, 143*
180. Davies-Colley, R. J., Vant, W. N. & Smith, D. G. (1994). *Colour and clarity of natural waters*. Chichester: Ellis Horwood. (In press.) *150*
181. Davies-Colley, R. J., Vant, W. N. & Wilcock, R. J. (1988). Lake water colour: comparison of direct observations with underwater spectral irradiance. *Water Res. Bull.*, **24**, 11–18. *149, 150*
182. Davis, M. A. & Dawes, C. J. (1981). Seasonal photosynthetic and respiratory responses of the intertidal red alga, *Bostrychia binderi* Harvey (Rhodophyta, Ceramiales) from a mangrove swamp and a salt marsh. *Phycologia*, **20**, 165–73. *411–2*
183. Day, H. R. & Felbeck, G. J. (1974). Production and analysis of a

humic-acid-like exudate from the aquatic fungus *Aureobasidium pullulans*. *J. Amer. Water Works Ass.*, **66**, 484–8. *58*

184. Degens, E. T., Guillard, R. R. L., Sackett, W. M. & Hellebust, J. A. (1968). Metabolic fractionation of carbon isotopes in marine plankton – I. Temperature and respiration experiments. *Deep-Sea Res.*, **15**, 1–9. *327*
185. Dekker, A. G., Malthus, T. J. & Seyhan, E. (1991). Quantitative modelling of inland water quality for high-resolution MSS systems. *IEEE Trans. Geosci. Remote Sens.*, **29**, 89–95. *195, 210*
186. Demers, S. & Legendre, L. (1981). Mélange vertical et capacité photosynthétique du phytoplancton estuarien (estuaire du Saint Laurent). *Mar. Biol.*, **64**, 243–50. *425*
187. Dera, J. & Gordon, H. R. (1968). Light field fluctuations in the photic zone. *Limnol. Oceanogr.*, **13**, 697–9. *117*
188. Desa, E. S. & Desa, B. A. E. (1991). The design of an in-water spectrograph for irradiance measurements in the ocean. *NIO (Goa, India). Unpublished*, 25p. *123*
189. Di Toro, D. M. (1978). Optics of turbid estuarine water: approximations and applications. *Water Res.*, **12**, 1059–68. *98*
190. Diakoff, S. & Scheibe, J. (1973). Action spectra for chromatic adaptation in *Tolypothrix tenuis*. *Plant Physiol.*, **51**, 382–5. *385–6*
191. Dixon, G. K. & Merrett, M. J. (1988). Bicarbonate utilization by the marine diatom *Phaeodactylum tricornutum* Bohlin. *New Phytol.*, **109**, 47–51. *327*
192. Döhler, G., Bürstell, H. & Jilg-Winter, G. (1976). Pigment composition and photosynthetic CO_2 fixation of *Cyanidium caldarium* and *Porphyridium aerugineum*. *Biochem. Physiol. Pflanzen*, **170**, 103–10. *376, 406*
193. Dokulil, M. (1979). Optical properties, colour and turbidity. In H. Loffler (ed.), *Neusiedlersee: the limnology of a shallow lake in Central Europe* (pp. 151–67). The Hague: Junk. *67*
194. Doty, M. S., Gilbert, W. J. & Abbott, I. A. (1974). Hawaiian marine algae from seaward of the algal ridge. *Phycologia*, **13**, 345–57. *366*
195. Doty, M. S. & Oguri, M. (1957). Evidence for a photosynthetic daily periodicity. *Limnol. Oceanogr.*, **2**, 37–40. *423*
196. Dragesund, O. & Gjøsaeter, J. (1988). The Barents Sea. In H. Postma & J. J. Zijlstra (eds), *Continental shelves* (pp. 339–61). Amsterdam: Elsevier. *359*
197. Drechsler, Z. & Beer, S. (1991). Utilization of inorganic carbon by *Ulva lactuca*. *Plant Physiol.*, **97**, 1439–44. *325*
198. Drew, E. A. (1978). Factors affecting photosynthesis and its seasonal variation in the seagrasses *Cymodocea nodosa* (Ucria) Aschers., and *Posidonia oceanica* (L.) Delile in the Mediterranean. *J. Exp. Mar. Biol. Ecol.*, **31**, 173–94. *350*
199. Drew, E. A. (1979). Physiological aspects of primary production in seagrasses. *Aquat. Bot.*, **7**, 139–50. *280, 350, 421*
200. Dring, M. J. (1981). Chromatic adaptation of photosynthesis in benthic marine algae: an examination of its ecological significance using a theoretical model. *Limnol. Oceanogr.*, **26**, 271–84. *371–2*
201. Dring, M. J. (1986). Pigment composition and photosynthetic action spectra of sporophytes of *Laminaria* (Phaeophyta) grown in

different light qualities and irradiances. *Br. Phycol. J.*, **21**, 199–207. *376*

202. Dring, M. J. (1990). Light harvesting and pigment composition in marine phytoplankton and macroalgae. In P. J. Herring, A. K. Campbell, M. Whitfield & L. Maddock (eds), *Light and life in the sea* (pp. 89–103). Cambridge: Cambridge University Press. *376*
203. Duarte, C. M. (1991). Seagrass depth limits. *Aquat. Bot.*, **40**, 363–77. *316–7*
204. Dubinsky, Z. & Berman, T. (1976). Light utilization efficiencies of phytoplankton in Lake Kinneret (Sea of Galilee). *Limnol. Oceanogr.*, **21**, 226–30. *300*
205. Dubinsky, Z. & Berman, T. (1979). Seasonal changes in the spectral composition of downwelling irradiance in Lake Kinneret (Israel). *Limnol. Oceanogr.*, **24**, 652–63. *141, 269*
206. Dubinsky, Z. & Berman, T. (1981). Light utilization by phytoplankton in Lake Kinneret (Israel). *Limnol. Oceanogr.*, **26**, 660–70. *292, 300, 303, 305*
207. Duchrow, R. M. & Everhart, W. H. (1971). Turbidity measurement. *Amer. Fish. Soc. Trans.*, **100**, 682–90. *109*
208. Dugdale, R. C. & Goering, J. T. (1967). Uptake of new and regenerated forms of nitrogen in primary productivity. *Limnol. Oceanogr.*, **12**, 196–206. *357*
209. Dugdale, R. C. & Wilkerson, F. P. (1991). Low specific nitrate uptake rate: a common feature of high-nutrient low-chlorophyll marine ecosystems. *Limnol. Oceanogr.*, **36**, 1678–88. *340*
210. Duntley, S. Q. (1963). Light in the sea. *J. Opt. Soc. Amer.*, **53**, 214–33. *154*
211. Duntley, S. Q., Wilson, W. H. & Edgerton, C. F. (1974). Ocean colour analysis, part 1. In (pp. Ref. 74–10, 41pp). San Diego: Scripps Inst. Oceanogr. *146*
212. Dunton, K. H. & Jodwalis, C. M. (1988). Photosynthetic performance of *Laminaria solidungula* measured *in situ* in the Alaskan High Arctic. *Mar. Biol.*, **98**, 277–85. *281*
213. Durbin, E. G., Krawiec, R. W. & Smayda, T. J. (1975). Seasonal studies on the relative importance of different size fractions of phytoplankton in Narragansett Bay (U.S.A.). *Mar. Biol.*, **32**, 271–87. *413–4*
214. Duysens, L. N. M. (1956). The flattening of the absorption spectrum of suspensions as compared to that of solutions. *Biochim. Biophys. Acta*, **19**, 1–12. *255, 257, 261*
215. Dwivedi, R. M. & Narain, A. (1987). Remote sensing of phytoplankton. An attempt from the Landsat Thematic Mapper. *Int. J. Remote Sens.*, **8**, 1563–9. *198, 211*
216. Effler, S. W., Brooks, C. M., Perkins, M. G., Meyer, M. & Field, F. D. (1987). Aspects of the underwater light field of eight central New York lakes. *Water Res. Bull.*, **23**, 1193–201. *140*
217. Effler, S. W., Roop, R. & Perkins, M. G. (1988). A simple technique for estimating absorption and scattering coefficients. *Water Res. Bull.*, **24**, 397–404. *103*
218. Effler, S. W., Wodka, M. C. & Field, S. D. (1984). Scattering and absorption of light in Onondaga Lake. *J. Environ. Eng.*, **110**, 1134–45. *103*

219. Egle, K. (1960). Menge und Verhältnis der Pigmente. In W. Ruhland (ed.), *Encyclopaedia of plant physiology* (pp. 444–96). Berlin: Springer-Verlag. *227*

220. Ekstrand, S. (1992). Landsat TM based quantification of chlorophyll a during algae blooms in coastal waters. *Int. J. Remote Sens.*, **13**, 1913–26. *198*

221. Elser, J. J. & Kimmel, B. L. (1985). Photoinhibition of temperate lake phytoplankton by near-surface irradiance: evidence from vertical profiles and field experiments. *J. Phycol.*, **21**, 419–27. *288*

222. Elterman, P. (1970). Integrating cavity spectroscopy. *Appl. Opt.*, **9**, 2140–2. *54*

223. Emerson, R. (1958). Yield of photosynthesis from simultaneous illumination with pairs of wavelengths. *Science*, **127**, 1059–60. *310*

224. Emerson, R. & Lewis, C. M. (1942). The photosynthetic efficiency of phycocyanin in *Chroococcus* and the problem of carotenoid participation in photosynthesis. *J. gen. Physiol.*, *25*, 579–95. *307, 309*

225. Emerson, R. & Lewis, C. M. (1943). The dependence of the quantum yield of *Chlorella* photosynthesis on wave length of light. *Amer. J. Bot.*, **30**, 165–78. *307, 308–9*

226. Engelmann, T. W. (1883). Farbe und Assimilation. *Bot. Zeit.*, **41**, 1–13, 17–29. *361*

227. Engelmann, T. W. (1884). Untersuchungen über die quantitativen Beziehungen zwischen Absorption des Lichtes und Assimilation in Pflanzenzellen. *Bot. Zeit.*, **42**, 81–93, 97–108. *307*

228. Engelmann, T. W. & Gaidukov, N. I. (1902). Ueber experimentelle Erzeugung zweckmässiger Aenderungen der Färbung pflanzlicher Chromophylle durch farbiges Licht. *Arch. Anat. Physiol. Lpz.: Physiol. Abt*, 333–5. Quoted by Rabinowitch (1945). *384*

229. EOSAT/NASA (1987). *System concept for wide-field-of-view observations of ocean phenomena from space*. *180*

230. Eppley, R. W. (1968). An incubation method for estimating the carbon content of phytoplankton in natural samples. *Limnol. Oceanogr.*, **13**, 547–82. *228*

231. Eppley, R. W. & Peterson, B. J. (1979). Particulate organic matter flux and planktonic new production in the ocean. *Nature*, **282**, 677–80. *357*

232. Ertel, J. R., Hedges, J. I., Devel, A. H., Richey, J. E. & Ribeiro, M. N. G. (1986). Dissolved humic substances of the Amazon River system. *Limnol. Oceanogr.*, **31**, 739–54. *61*

233. Ertel, J. R., Hedges, J. I. & Perdue, E. M. (1984). Lignin signature of aquatic humic substances. *Science*, **223**, 485–7. *59*

234. ESA (1992). *Medium resolution imaging spectrometer (MERIS). Draft brochure*. Paris: European Space Agency. *180*

235. Evans, G. T. & Taylor, F. R. J. (1980). Phytoplankton accumulation in Langmuir cells. *Limnol. Oceanogr.*, **25**, 840–5. *287*

236. Exton, R. J., Houghton, W. M., Esaias, W., Harriss, R. C., Farmer, F. H. & White, H. H. (1983). Laboratory analysis of techniques for remote sensing of estuarine parameters using laser excitation. *Appl. Opt.*, **22**, 54–65. *204*

237. Fahnenstiel, G. L. & Scavia, D. (1987). Dynamics of Lake Michigan

phytoplankton: the deep chlorophyll layer. *J. Great Lakes Res.*, **13**, 285–95. *321*

238. Fahnenstiel, G. L., Schelske, C. L. & Moll, R. A. (1984). *In situ* quantum efficiency of Lake Superior phytoplankton. *J. Great Lakes Res.*, **10**, 399–406. *297, 300*
239. Falkowski, P. G. (1981). Light-shade adaptation and assimilation numbers. *J. Plankton Res.*, **3**, 203–16. *353, 354*
240. Falkowski, P. G. & Dubinsky, Z. (1981). Light-shade adaptation of *Stylophora pistillata*, a hermatypic coral from the Gulf of Eilat. *Nature*, **289**, 172–4. *278, 378*
241. Falkowski, P. G., Dubinsky, Z. & Wyman, K. (1985). Growth-irradiance relationships in phytoplankton. *Limnol. Oceanogr.*, **30**, 311–21. *259*
242. Falkowski, P. G., Jokiel, P. L. & Kinzie, R. A. (1990). Irradiance and corals. In Z. Dubinsky (ed.), *Coral reefs* (pp. 89–107). Amsterdam: Elsevier. *378, 398*
243. Falkowski, P. G. & LaRoche, J. (1991). Acclimation to spectral irradiance in algae. *J. Phycol.*, **27**, 8–14. *375, 377*
244. Falkowski, P. G. & Owens, T. G. (1978). Effects of light intensity on photosynthesis and dark respiration in six species of marine phytoplankton. *Mar. Biol.*, **45**, 289–95. *277*
245. Falkowski, P. G. & Owens, T. G. (1980). Light-shade adaptation: two strategies in marine phytoplankton. *Plant Physiol.*, **66**, 592–5. *376, 377, 378*
246. Falkowski, P. G. & Woodhead, A. D. (eds) (1992). *Primary productivity and biogeochemical cycles in the sea*. New York: Plenum. *314*
247. Faller, A. J. (1978). Experiments with controlled Langmuir circulations. *Science*, **201**, 618–20. *287*
248. Faust, M. A. & Gantt, E. (1973). Effect of light intensity and glycerol on the growth, pigment composition and ultrastructure of *Chroomonas* sp. *J. Phycol.*, **9**, 489–95. *376*
249. Fee, E. J. (1969). A numerical model for the estimation of photosynthetic production, integrated over time and depth, in natural waters. *Limnol. Oceanogr.*, **14**, 906–11. *352*
250. Fee, E. J. (1976). The vertical and seasonal distribution of chlorophyll in lakes of the Experimental Lakes Area, northwestern Ontario: implication for primary production estimates. *Limnol. Oceanogr.*, **21**, 767–83. *320, 397*
251. Ferrari, G. M. & Tassan, S. (1991). On the accuracy of determining light absorption by 'yellow substance' through measurements of induced fluorescence. *Limnol. Oceanogr.*, **36**, 777–86. *66*
252. Fiedler, P. C. & Laurs, R. M. (1990). Variability of the Columbia River plume observed in visible and infrared satellite imagery. *Int. J. Remote Sens.*, **11**, 999–1010. *213*
253. Fischer, J. & Kronfeld, U. (1990). Sun-stimulated fluorescence. 1: Influence of oceanic properties. *Int. J. Remote Sens.*, **11**, 2125–47. *201*
254. Fischer, J. & Schlüssel, P. (1990). Sun-stimulated chlorophyll fluorescence. 2: impact of atmospheric properties. *Int. J. Remote Sens.*, **11**, 2149–62. *201*
255. Fisher, T., Shurtz-Swirski, R., Gepstein, S. & Dubinsky, Z. (1989). Changes in the levels of ribulose-1, 5-bisphosphate carboxylase/

oxygenase (Rubisco) in *Tetraedron minimum* (Chlorophyta) during light and shade adaptation. *Plant Cell Physiol.*, **30**, 221–8. *380*

256. Fleischhacker, P. & Senger, H. (1978). Adaptation of the photosynthetic apparatus of *Scenedesmus obliquus* to strong and weak light conditions. *Physiol. Plant*, **43**, 43–51. *377, 379*
257. Fogg, G. E. (1991). The phytoplanktonic way of life. *New Phytol.*, **118**, 191–232. *314*
258. Fogg, G. E. & Thake, B. (1987). *Algal cultures and phytoplankton ecology*. Madison: University of Wisconsin Press. *314*
259. Fogg, G. E., Stewart, W. D. P., Fay, P. & Walsby, A. E. (1973). *The blue-green algae*. London: Academic Press. *417, 419*
260. Fork, D. C. (1963). Observations of the function of chlorophyll a and accessory pigments in photosynthesis. In *Photosynthesis mechanisms in green plants* (pp. 352–61). Washington: National Academy of Science–National Research Council. *311, 312, 372*
261. Foss, P., Guillard, R. R. L. & Liaaen-Jensen, S. (1984). Prasinoxanthin – a chemosystematic marker for algae. *Phytochem.*, **23**, 1629–33. *231*
262. Fox, L. E. (1983). The removal of dissolved humic acid during estuarine mixing. *Estuar. Coast. Shelf Sci.*, **16**, 431–40. *61*
263. Foy, R. H. & Gibson, C. E. (1982*a*). Photosynthetic characteristics of planktonic blue-green algae: the response of twenty strains grown under high and low light. *Br. Phycol. J.*, **17**, 169–82. *382*
264. Foy, R. H. & Gibson, C. E. (1982*b*). Photosynthetic characteristics of planktonic blue-green algae: changes in photosynthetic capacity and pigmentation of *Oscillatoria redekei* Van Goor under high and low light. *Br. Phycol. J.*, **17**, 183–93. *376, 382*
265. French, C. S. (1960). The chlorophylls *in vivo* and *in vitro*. In W. Ruhland (ed.), *Encyclopaedia of Plant Physiology* (pp. 252–97). Berlin: Springer-Verlag. *229, 261*
266. French, C. S., Brown, J. S. & Lawrence, M. C. (1972). Four universal forms of chlorophyll *a*. *Plant Physiol.*, **49**, 421–9. *234*
267. Friedman, A. L. & Alberte, R. S. (1984). A diatom light-harvesting pigment–protein complex. *Plant Physiol.*, **76**, 483–9. *238*
268. Friedman, A. L. & Alberte, R. S. (1986). Biogenesis and light regulation of the major light-harvesting chlorophyll-protein of diatoms. *Plant Physiol.*, **80**, 43–51. *378*
269. Friedman, E., Poole, L., Cherdak, A. & Houghton, W. (1980). Absorption coefficient instrument for turbid natural waters. *Appl. Opt.*, **19**, 1688–93. *53*
270. Fritsch, F. E. (1948). *The structure and reproduction of the algae*. Cambridge: Cambridge University Press. *218*
271. Frost, B. W. (1980). Grazing. In I. Morris (ed.), *The physiological ecology of phytoplankton* (pp. 465–91). Oxford: Blackwell. *340*
272. Fry, E. S. & Kattawar, G. W. (1988). Measurement of the absorption coefficient of ocean water using isotropic illumination. *Proc. Soc. Photo-Opt. Instrum. Eng., Ocean Optics IX*, **925**, 142–8. *54*
273. Fry, E. S., Kattawar, G. W. & Pope, R. M. (1992). Integrating cavity absorption meter. *Appl. Opt.*, **31**, 2055–65. *54*
274. Fujita, Y. & Hattori, A. (1960). Effect of chromatic lights on phycobilin formation in a blue-green alga, *Tolypothrix tenuis*. *Plant Cell Physiol.*, **1**, 293–303. *385–6*

275. Fujita, Y. & Hattori, A. (1962*a*). Photochemical interconversion between precursors of phycobilin chromoproteids in *Tolypothrix tenuis*. *Plant Cell Physiol.*, **3**, 209–20. *385–6*
276. Fujita, Y. & Hattori, A. (1962*b*). Changes in composition of cellular material during formation of phycobilin chromoproteids in a blue-green alga, *Tolypothrix tenuis*. *J. Biochem.*, **52**, 38–42. *385–6*
277. Furnas, M. J. & Mitchell, A. W. (1988). Photosynthetic characteristics of Coral Sea picoplankton (<2 μm size fraction). *Biol. Oceanogr.*, **5**, 163–82. *279*
278. Gagliardini, D. A., Karszenbaum, H., Legeckis, R. & Klemas, V. (1984). Application of Landsat MSS, NOAA/TIROS AVHRR, and Nimbus CZCS to study the La Plata River and its interaction with the ocean. *Remote Sens. Environ.*, **15**, 21–36. *194*
279. Gaidukov, N. (1902). Ueber den Einfluss farbigen Lichts auf die Faerbung levender Oscillarien. *Abh. Preuss Akad. Wiss. Berlin, No. 5*. Quoted by Bogorad (1975). *384*
280. Gallagher, J. C. (1980). Population genetics of *Skeletonema costatum* (Bacillariophyceae) in Narragansett Bay. *J. Phycol.*, **16**, 464–74. *415*
281. Gallagher, J. C. (1982). Physiological variation and electrophoretic banding patterns of genetically different seasonal populations of *Skeletonema costatum* (Bacillariophyceae). *J. Phycol.*, **18**, 148–62. *415*
282. Gallegos, C. L., Correll, D. L. & Pierce, J. W. (1990). Modelling spectral diffuse attenuation, absorption, and scattering coefficients in a turbid estuary. *Limnol. Oceanogr.*, **35**, 1486–502. *65, 101, 139*
283. Ganf, G. G. (1974). Incident solar irradiance and underwater light penetration as factors controlling the chlorophyll *a* content of a shallow equatorial lake (Lake George, Uganda). *J. Ecol.*, **62**, 593–609. *269*
284. Ganf, G. G. (1975). Photosynthetic production and irradiance–photosynthesis relationships of the phytoplankton from a shallow equatorial lake (Lake George, Uganda). *Oecologia*, **18**, 165–83. *278, 318, 346*
285. Ganf, G. G. & Oliver, R. L. (1982). Vertical separation of light and available nutrients as a factor causing replacement of green algae by blue-green algae in the plankton of a stratified lake. *J. Ecol.*, **70**, 829–44. *420*
286. Ganf, G. G., Oliver, R. L. & Walsby, A. E. (1989). Optical properties of gas-vacuolate cells and colonies of *Microcystis* in relation to light attenuation in a turbid, stratified reservoir (Mount Bold Reservoir, South Australia). *Aust J. Mar. Freshwater Res.*, **40**, 595–611. *70, 98, 104, 110*
287. Gantt, E. (1975). Phycobilisomes: light-harvesting pigment complexes. *BioScience*, **25**, 781–8. *239*
288. Gantt, E. (1977). Recent contributions in phycobiliproteins and phycobilisomes. *Photochem. Photobiol.*, **26**, 685–9. *239*
289. Gantt, E. (1986). Phycobilisomes. *Encycl. Plant Physiol. (n.s.)*, **19**, 260–8. *243*
290. Gates, D. M. (1962). *Energy exchange in the biosphere*. New York: Harper & Row. *35*
291. Geider, R. J. (1987). Light and temperature dependence of the carbon

to chlorophyll a ratio in microalgae and cyanobacteria: implications for physiology and growth of phytoplankton. *New Phytol.*, **106**, 1–34. *375*

292. Geider, R. J. & Osborne, B. A. (1987). Light absorption by a marine diatom: experimental observations and theoretical calculations of the package effect in a small *Thalassiosira* species. *Mar. Biol.*, **96**, 299–308. *261*
293. Gelin, C. (1975). Nutrients, biomass and primary productivity of nanoplankton in eutrophic Lake Vombsjön, Sweden. *Oikos*, **26**, 121–39. *269, 337*
294. Gerard, V. A. (1986). Photosynthetic characteristics of giant kelp (*Macrocystis pyrifera*) determined *in situ*. *Mar. Biol.*, **90**, 473–82. *281*
295. Gershun, A. (1936). O fotometrii mutnykk sredin. *Tr. Gos. Okeanogr. Inst.*, **11**, 99. Quoted by Jerlov (1976). *24*
297. Gessner, F. (1937). Untersuchungen uber Assimilation und Atmung submerser Wasserpflanzen. *Jahrb. Wiss. Bot.*, **85**, 267–328. *331*
298. Gilmartin, M. (1960). The ecological distribution of the deep water algae of Eniwetok atoll. *Ecology*, **41**, 210–21. *366*
299. Gitelson, A. A. & Kondratyev, K. Y. (1991). Optical models of mesotrophic and eutrophic water bodies. *Int. J. Remote Sens.*, **12**, 375–85. *175*
300. Glazer, A. N. (1981). Photosynthetic accessory proteins with bilin prosthetic groups. In M. D. Hatch & N. K. Boardman (eds), *The biochemistry of plants* (pp. 51–96). New York: Academic Press. *239, 240*
301. Glazer, A. N. (1985). Light harvesting by phycobilisomes. *Ann. Rev. Biophys. Biophys. Chem*, **14**, 47–77. *222, 239, 240, 243*
302. Glazer, A. N. & Cohen-Bazire, G. (1975). A comparison of Cryptophytan phycocyanins. *Arch. Mikrobiol.*, **104**, 29–32. *240*
303. Glazer, A. N. & Melis, A. (1987). Photochemical reaction centres: structure, organization and function. *Ann. Rev. Plant Physiol.*, **38**, 11–45. *243*
304. Glenn, E. P. & Doty, M. S. (1981). Photosynthesis and respiration of the tropical red seaweeds, *Eucheuma striatum* (Tambalang and Elkhorn varieties) and *E. denticulatum*. *Aquat. Biol.*, **10**, 353–64. *398, 423*
305. Glibert, P. M., Dennett, M. R. & Goldman, J. C. (1985). Inorganic carbon uptake by phytoplankton in Vineyard Sound, Massachusetts I. Measurements of the photosynthesis-irradiance response of winter and early-spring assemblages'. *J. Exp. Mar. Biol. Ecol.*, **85**, 21–36. *288*
306. Gliwicz, M. Z. (1986). Suspended clay concentration controlled by filter-feeding zooplankton in a tropical reservoir. *Nature*, **323**, 330–2. *105*
307. Goedheer, J. C. (1969). Energy transfer from carotenoids to chlorophyll in blue-green, red and green algae and greening bean leaves. *Biochim. Biophys. Acta*, **172**, 252–65. *312*
308. Goericke R. & Repeta, D. J. (1992). The pigments of *Prochlorococcus marinus*: the presence of divinyl chlorophyll a and b in a marine prokaryote. *Limnol. Oceanogr.*, **37**, 425–33. *226, 227, 231*
309. Goldman, C. R., Mason, D. T. & Wood, B. J. B. (1963). Light injury

and inhibition in Antarctic freshwater plankton. *Limnol. Oceanogr.*, **8**, 313–22. *285*

310. Goldman, J. C. (1979*a*). Outdoor algal mass cultures. I. Applications. *Water Res.*, **13**, 1–19. *430*
311. Goldman, J. C. (1979*b*). Outdoor algal mass cultures. II. Photosynthetic yield limitations. *Water Res.*, **13**, 119–36. *430*
312. Goldsborough, W. J. & Kemp, W. M. (1988). Light responses of a submersed macrophyte: implications for survival in turbid tidal waters. *Ecology*, **69**, 1775–86. *383–4*
313. Gordon, H. R. (1978). Removal of atmospheric effects from satellite imagery of the oceans. *Appl. Opt.*, **17**, 1631–6. *186*
314. Gordon, H. R. (1985). Ship perturbation of irradiance measurements at sea. *Appl. Opt.*, **24**, 4172–82. *118*
315. Gordon, H. R. (1989*a*). Can the Lambert–Beer law be applied to the diffuse attenuation coefficient of ocean water? *Limnol. Oceanogr.*, **34**, 1389–409. *126*
316. Gordon, H. R. (1989*b*). Dependence of the diffuse reflectance of natural waters on the sun angle. *Limnol. Oceanogr.*, **34**, 1484–9. *126, 146*
317. Gordon, H. R. (1991). Absorption and scattering estimates from irradiance measurements: Monte Carlo simulations. *Limnol. Oceanogr.*, **36**, 769–77. *92, 126*
318. Gordon, H. R. & Brown, O. B. (1973). Irradiance reflectivity of a flat ocean as a function of its optical properties. *Appl. Opt*, **12**, 1549–51. *126*
319. Gordon, H. R., Brown, O. B. & Jacobs, M. M. (1975). Computed relationships between the inherent and apparent optical properties of a flat, homogeneous ocean. *Appl. Opt.*, **14**, 417–27. *126, 146, 147*
320. Gordon, H. R. & Clark, D. K. (1980). Atmospheric effects in the remote sensing of phytoplankton pigments. *Boundary-layer Meteorol.*, **18**, 299–313. *172*
321. Gordon, H. R. & Clark, D. K. (1981). Clear water radiances for atmospheric correction of coastal zone scanner imagery. *Appl. Opt.*, **20**, 4175–80. *186*
322. Gordon, H. R., Clark, D. K., Brown, J. W., Brown, O. B., Evans, R. H. & Broenkow, W. W. (1983). Phytoplankton pigment concentrations in the Middle Atlantic Bight: comparison of ship determinations and CZCS estimates. *Appl. Opt.*, **22**, 20–36. *186–8, 197–8, 212*
323. Gordon, H. R., Clark, D. K., Mueller, J. L. & Hovis, W. A. (1980). Phytoplankton pigments from the Nimbus-7 Coastal Zone Scanner: comparison with surface measurements. *Science*, **210**, 63–6. *198, 211*
324. Gordon, H. R. & Ding, K. (1992). Self-shading of in-water optical instruments. *Limnol. Oceanogr.*, **37**, 491–500. *118*
325. Gordon, H. R. & McLuney, W. R. (1975). Estimation of the depth of sunlight penetration in the sea for remote sensing. *Appl. Opt.*, **14**, 413–16. *170*
326. Gordon, H. R. & Morel, A. Y. (1983). *Remote assessment of ocean colour for interpretation of satellite visible imagery. A review.* New York: Springer. *81, 198*
327. Gordon, J. I. (1969). San Diego: Scripps Inst. Oceanogr. Ref 67–27. Quoted by Austin (1974a). *42, 43*

328. Gorham, E. (1957). The chemical composition of lake waters in Halifax County, Nova Scotia. *Limnol. Oceanogr.*, **2**, 12–21. *68*

329. Gorham, E., Dean, W. E. & Sanger, J. E. (1983). The chemical composition of lakes in the north-central United States. *Limnol. Oceanogr.*, **28**, 287–301. *72*

330. Gower, J. F. R. (1980). Observations of fluorescence of *in situ* chlorophyll *a* in Saanich Inlet. *Boundary-layer Meteorol.*, **18**, 235–45. *201*

331. Grew, G. W. (1981). *Real-time test of MOCS algorithm during Superflux 1980*. NASA-Publ. CP-2188. 301 pp. Quoted by Campbell & Esaias (1983). *194, 199*

331*a*. Grobbelaar, J. U. (1989). The contribution of phytoplankton productivity in turbid freshwaters to their trophic status. *Hydrobiologia*, **173**, 127–33. *315*

332. Guard-Friar, D. & MacColl, R. (1986). Subunit separation (α, α′, β) of cryptomonad biliproteins. *Photochem. Photobiol.*, **43**, 81–5. *243*

333. Gugliemelli, L. A., Dutton, H. J., Jursinic, P. A. & Siegelman, H. W. (1981). Energy transfer in a light-harvesting carotenoid-chlorophyll *c*-chlorophyll *a*-protein of *Phaeodactylum tricornutum*. *Photochem. Photobiol.*, **33**, 903–7. *230*

334. Gurfink, A. M. (1976). Light field in the surface layers of the sea. *Oceanology.*, **15**, 295–9. *45*

335. Haardt, H. & Maske, H. (1987). Specific *in vivo* absorption coefficient of chlorophyll a at 675 nm. *Limnol. Oceanogr.*, **32**, 608–19. *76, 259*

336. Halldal, P. (1974). Light and photosynthesis of different marine algal groups. In N. G. Jerlov & E. S. Nielsen (eds), *Optical aspects of oceanography* (pp. 345–60). London: Academic Press. *133, 265*

337. Hamilton, M. K., Daviss, C. O., Pilorz, S. H., Rhea, W. J. & Carder, K. L. (1992). Examination of chlorophyll distribution in Lake Tahoe, using the Airborne Visible and Infrared Imaging Spectrometer (AVIRIS). *Proceedings of Third AVIRIS Workshop (in press)*. *210*

338. Hammer, U. T. (1980). Primary production: geographical variations. In E. D. Le Cren & R. H. Lowe-McConnell (eds), *The functioning of freshwater ecosystems* (pp. 235–46). Cambridge: Cambridge University Press. *359, 430*

339. Hanisak, M. D. (1979). Growth patterns of *Codium fragile* ssp. *tomentosoides* in response to temperature, irradiance, salinity and nitrogen source. *Mar. Biol.*, **50**, 319–22. *348*

340. Harder, R. (1921). *Jahrb. Wiss. Botan.*, **60**, 531. Quoted by Rabinowitch (1971). *325*

341. Harder, R. (1923). Uber die Bedeutung von Lichtintensität und Wellenlänge fur die Assimilation Färbiger Algen. *Z. Bot.*, **15**, 305–55. *361*

342. Harding, L. W. & Coats, D. W. (1988). Photosynthetic physiology of *Prorocentrum mariae-lebouriae* (Dinophyceae) during its subpycnocline transport in Chesapeake Bay. *J. Phycol.*, **24**, 77–89. *280*

343. Harding, L. W., Itsweire, E. C. & Esaias, W. E. (1992). Determination of phytoplankton chlorophyll concentrations in the Chesapeake Bay with aircraft remote sensing. *Remote Sens. Environ.*, **40**, 79–100. *176, 210*

344. Harding, L. W., Meeson, B. W. & Fisher, T. R. (1986). Phytoplankton production in two East coast estuaries: photosynthesis-light functions and patterns of carbon assimilation in Chesapeake and Delaware Bays. *Estuar. Coast. Shelf Sci.*, **23**, 773–806. *321*

345. Harding, L. W., Meeson, B. W., Prézelin, B. B. & Sweeney, B. M. (1981). Diel periodicity of photosynthesis in marine phytoplankton. *Mar. Biol.*, **61**, 95–105. *425*

346. Harding, L. W., Prézelin, B. B., Sweeney, B. M. & Cox, J. L. (1982*a*). Diel oscillations of the photosynthesis–irradiance (*P–I*) relationship in natural assemblages of phytoplankton. *Mar. Biol.*, **67**, 167–78. *423, 425–6*

347. Harding, L. W., Prézelin, B. B., Sweeney, B. M. & Cox, J. L. (1982*b*). Primary production as influenced by diel periodicity of phytoplankton photosynthesis. *Mar. Biol.*, **67**, 179–86. *426*

348. Harper, M. A. (1969). Movement and migration of diatoms on sand grains. *Br. Phycol. J.*, **4**, 97–103. *420*

349. Harrington, J. A., Schiebe, F. R. & Nix, J. F. (1992). Remote sensing of Lake Chicot, Arkansas: monitoring suspended sediments, turbidity, and Secchi depth with Landsat data. *Remote Sens. Environ.*, **39**, 15–27. *206*

350. Harris, G. P. (1980). The measurement of photosynthesis in natural populations of phytoplankton. In I. Morris (ed.), *The physiological ecology of phytoplankton* (pp. 129–87). Oxford: Blackwell. *343*

351. Harris, G. P. (1986). *Phytoplankton ecology: structure, function and fluctuation*. London: Chapman & Hall. *314*

352. Harris, G. P., Heaney, S. I. & Talling, J. F. (1979). Physiological and environmental constraints in the ecology of the planktonic dinoflagellate *Ceratium hirundinella. Freshwater Biol.*, **9**, 413–28. *141*

353. Harris, G. P. & Piccinin, B. B. (1977). Photosynthesis by natural phytoplankton populations. *Arch. Hydrobiol.*, **80**, 405–57. *285, 286, 288, 423*

354. Harrison, P. J., Fulton, J. D., Taylor, F. J. R. & Parsons, T. R. (1983). Review of the biological oceanography of the Strait of Georgia: pelagic environment. *Can. J. Fish. Aquat. Sci.*, **40**, 1064–94. *139*

355. Harron, J. W., Hollinger, A. B., Jain, S. C., Kemenade, C. V. & Buxton, R. A. H. (1983). Presentation. *Annual Meeting, Optical Society of America. 91*

356. Harvey, G. R., Boran, D. A., Chesal, L. A. & Tokar, J. M. (1983). The structure of marine fulvic and humic acids. *Marine Chem.*, **12**, 119–32. *59*

357. Harvey, G. R., Boran, D. A., Piotrowicz, S. R. & Weisel, C. P. (1984). Synthesis of marine humic substances from unsaturated lipids. *Nature*, **309**, 244–6. *59*

358. Hatcher, B. G., Chapman, A. R. O. & Mann, K. H. (1977). An annual carbon budget for the kelp *Laminaria longicruris. Mar. Biol.*, **44**, 85–96. *347, 349*

359. Haupt, W. (1973). Role of light in chloroplast movement. *BioScience*, **23**, 289–96. *421, 423*

360. Haxo, F. T. (1985). Photosynthetic action spectrum of the coccolithophorid, *Emiliania huxleyi* (Haptophyceae): 19′ hexanoyloxyfucoxanthin as antenna pigment. *J. Phycol.*, **21**, 282–7. *313*

361. Haxo, F. T. & Blinks, L. R. (1950). Photosynthetic action spectra of marine algae. *J. Gen. Physiol.*, **33**, 389–422. *307, 308*

362. Haxo, F. T., Kycia, J. H., Somers, G. F., Bennett, A. & Siegelman, H. W. (1976). Peridinin–chlorophyll *a* proteins of the dinoflagellate *Amphidinium carterae* (Plymouth 450). *Plant Physiol.*, **57**, 297–303. *238*

363. Heaney, S. I. & Talling, J. F. (1980). Dynamic aspects of dinoflagellate distribution patterns in a small productive lake. *J. Ecol.*, **68**, 75–94. *416*

364. Heckey, R. E. & Fee, E. J. (1981). Primary production and rates of algal growth in Lake Tanganyika. *Limnol. Oceanogr.*, **26**, 532–47. *141*

365. Hellström, T. (1991). The effect of resuspension on algal production in a shallow lake. *Hydrobiologia*, **213**, 183–90. *144*

366. Hickey, J. R., Alton, B. M., Griffith, F. J., Jacobowitz, H., Pellegrino, P., Maschhoff, R. H., Smith, E. A. & van den Haar, T. H. (1982). Extraterrestrial solar irradiance variability. Two and one-half years of measurement from Nimbus 7. *Solar Energy*, **29**, 125–7. *27*

367. Hiller, R. G., Anderson, J. M. & Larkum, A. W. D. (1991). The chlorophyll–protein complexes of algae. In H. Scheer (ed.), *Chlorophylls* (pp. 529–47). Boca Raton: CRC Press. *235*

368. Hiller, R. G. & Larkum, A. W. D. (1985). The chlorophyll–protein complexes of *Prochloron* sp. (Prochlorophyta). *Biochim. Biophys. Acta*, **806**, 107–15. *238*

369. Hiller, R. G., Larkum, A. W. D. & Wrench, P. W. (1988). Chlorophyll proteins of the prymnesiophyte *Pavlova lutherii* (Droop) comb. nov.: identification of the major light-harvesting complex. *Biochim. Biophys. Acta*, **932**, 223–31. *238*

370. Hiller, R. G., Wrench, P. M., Gooley, A. P., Shoebridge, G. & Breton, J. (1993). The major intrinsic light-harvesting protein of *Amphidinium*: characterization and relation to other light-harvesting proteins. *Photochem. Photobiol.*, **57** (in press). *238*

371. Hobson, L. A. (1981). Seasonal variations in maximum photosynthetic rates of phytoplankton in Saanich Inlet, Vancouver Island, British Columbia. *J. Exp. Mar. Biol. Ecol.*, **52**, 1–13. *413*

372. Hodkinson, J. R. & Greenleaves, J. I. (1963). Computations of light-scattering and extinction by spheres according to diffraction and geometrical optics and some comparisons with the Mie theory. *J. Opt. Soc. Amer.*, **53**, 577–88. *87, 89*

373. Hoge, F. E., Berry, R. E. & Swift, R. N. (1986). Active-passive airborne colour measurement. 1: Instrumentation. *Appl. Opt.*, **25**, 39–47. *202–4*

374. Hoge, F. E. & Swift, R. N. (1981). Airborne simultaneous spectroscopic detection of laser-induced water Raman backscatter and fluorescence from chlorophyll *a* and other naturally occurring pigments. *Appl. Opt.*, **20**, 3197–205. *201–2, 210*

375. Hoge, F. E. & Swift, R. N. (1983). Airborne dual laser excitation and mapping of phytoplankton photopigments in a Gulf Stream warm core ring. *Appl. Opt.*, **22**, 2272–81. *210*

376. Hoge, F. E. & Swift, R. N. (1986). Active-passive correlation spectroscopy: a new technique for identifying ocean colour algorithm spectral regions. *Appl. Opt.*, **25**, 2571–83. *199–200, 201*

377. Hoge, F. E., Swift, R. N. & Yungel, J. K. (1986). Active-passive

airborne ocean colour measurement. 2: Applications. *Appl. Opt.*, **25**, 48–57. *210*

378. Hoge, F. E., Wright, C. W. & Swift, R. N. (1987). Radiance-ratio algorithm wavelengths for remote oceanic chlorophyll determination. *Appl. Opt.*, **26**, 2082–94. *200, 201, 202–4*

379. Højerslev, N. & Lundgren, B. (1977). Inherent and apparent optical properties of Icelandic waters. 'Bjarni Saemundsson Overflow 73'. *Univ. Copenhagen, Inst. Phys. Oceanogr. Rep.*, **33**, 63pp. *101*

380. Højerslev, N. K. (1973). Inherent and apparent optical properties of the western Mediterranean and the Hardangerfjord. *Univ. Copenhagen, Inst. Phys. Oceanogr. Rep.*, **21**, 70pp. *102*

381. Højerslev, N. K. (1975). A spectral light absorption meter for measurements in the sea. *Limnol. Oceanogr.*, **20**, 1024–34. *54, 120, 121, 151*

382. Højerslev, N. K. (1977). Inherent and apparent optical properties of the North Sea. Fladen Ground experiment – Flex 75. *Univ. Copenhagen, Inst. Phys. Oceanogr. Rep.*, **32**, 68pp. *102*

383. Højerslev, N. K. (1979). On the origin of yellow substance in the marine environment. In *17th General Assembly of I.A.P.S.O. (Canberra, 1979)* (pp. Abstracts, 71). *61*

384. Højerslev, N. K. (1981). Assessment of some suggested algorithms on sea colour and surface chlorophyll. In J. F. R. Gower (ed.), *Oceanography from space* (pp. 347–53). New York: Plenum. *205*

385. Højerslev, N. K. (1986). Variability of the sea with special reference to the Secchi disc. *Proc. Soc. Photo-Opt. Instrum. Eng., Ocean Optics VIII*, **637**, 294–305. *120*

386. Højerslev, N. K. (1988). Natural occurrences and optical effects of gelbstoff. *Univ. Copenhagen, Inst. Phys. Oceanogr. Rep.*, **50**, 30pp. *64, 65, 67*

386a. Højerslev, N. K. & Trabjerg, I. (1990). A new perspective for remote sensing of plankton pigments and water quality. *Univ Copenhagen, Inst. Phys. Oceanogr. Rep.*, **51**, 10pp. *57*

387. Holligan, P. M., Aarup, T. & Groom, S. B. (1989). The North Sea: satellite colour atlas. *Continental Shelf Res.*, **9**, 667–765. *209, 213*

388. Holligan, P. M., Viollier, M., Harbour, D. S., Camus, P. & Champagne-Philippe, M. (1983). Satellite and ship studies of coccolithophore production along a continental shelf edge. *Nature*, **304**, 339–42. *111*

389. Holm-Hansen, O. & Mitchell, B. G. (1991). Spatial and temporal distribution of phytoplankton and primary production in the western Bransfield Strait region. *Deep-Sea Res.*, **38**, 961–80. *279*

390. Holmes, R. W. (1970). The Secchi disk in turbid coastal zones. *Limnol. Oceanogr.*, **15**, 688–94. *119*

391. Holyer, R. J. (1978). Toward universal multispectral suspended sediment algorithms. *Remote Sens. Environ.*, **7**, 323–38. *192*

392. Horstmann, U. (1983). Cultivation of the green alga *Caulerpa racemosa*, in tropical waters and some aspects of its physiological ecology. *Aquaculture*, **32**, 361–71. *423*

393. Houghton, W. M., Exton, R. J. & Gregory, R. W. (1983). Field investigation of techniques for remote laser sensing of oceanographic parameters. *Remote Sens. Environ.*, **13**, 17–32. *204*

394. Hovis, W. (1978). The Coastal Zone Colour Scanner (CZCS)

experiment. In C. R. Madrid (ed.), *The Nimbus 7 users' guide*. Beltsville: Management & Technical Services Co. *179*

395. Hovis, W. A., Clark, D. K., Anderson, F., Austin, R. W., Wilson, W. H., Baker, E. T., Ball, D., Gordon, H. R., Mueller, J. L., El-Sayed, S. Z., Sturm, B., Wrigley, R. C. & Yentsch, C. S. (1980). Nimbus-7 Coastal Zone Colour Scanner: system description and initial imagery. *Science*, **210**, 60–3. *179, 211*
396. Hovis, W. A. & Leung, K. C. (1977). Remote sensing of ocean colour. *Opt. Eng.*, **16**, 158–66. *176, 179*
397. Humphries, S. E. & Lyne, V. D. (1988). Cyanophyte blooms: the role of cell buoyancy. *Limnol. Oceanogr.*, **33**, 79–91. *420*
398. Hutchinson, G. E. (1967). *A treatise on limnology, Vol. 2*. New York: Wiley. *340, 341, 345*
399. Hutchinson, G. E. (1975). *A treatise on limnology, Vol. 3*. New York: Wiley-Interscience. *218, 327, 329, 331*
400. Ikusima, I. (1970). Ecological studies on the productivity of aquatic plant communities. *Bot. Mag. (Tokyo)*, **83**, 330–41. *279*
401. Ingram, K. & Hiller, R. G. (1983). Isolation and characterization of a major chlorophyll a/c_2 light-harvesting protein from a *Chroomonas* species (Cryptophyceae). *Biochim. Biophys. Acta*, **772**, 310–19. *238*
402. International Association for the Physical Sciences of the Ocean (1979). *Sun Report (Report of the working group on symbols, units and nomenclature in physical oceanography)*. Paris. *6, 80*
403. Iturriaga, R., Mitchell, B. G. & Kiefer, D. A. (1988). Microphotometric analysis of individual particle absorption spectra. *Limnol. Oceanogr.*, **33**, 128–35. *259*
404. Iturriaga, R. & Siegel, D. A. (1989). Microphotometric characterization of phytoplankton and detrital absorption properties in the Sargasso Sea. *Limnol. Oceanogr.*, **34**, 1706–26. *52, 73, 75, 76*
405. Ivanoff, A., Jerlov, N. & Waterman, T. H. (1961). A comparative study of irradiance, beam transmittance and scattering in the sea near Bermuda. *Limnol. Oceanogr.*, **6**, 129–48. *64*
406. Iverson, R. L. & Curl, H. (1973). Action spectrum of photosynthesis for *Skeletonema costatum* obtained with carbon-14. *Physiol. Plant.*, **28**, 498–502. *307, 308–9*
407. Jamart, B. M., Winter, D. F., Banse, K., Anderson, G. C. & Lam, R. K. (1977). A theoretical study of phytoplankton growth and nutrient distribution in the Pacific Ocean off the northwestern U.S. coast. *Deep-Sea Res.*, **24**, 753–73. *320*
408. James, H. R. & Birge, E. A. (1938). A laboratory study of the absorption of light by lake waters. *Trans. Wisc. Acad. Sci. Arts Lett.*, **31**, 1–154. *59, 67, 68*
409. Jassby, A. T. & Platt, T. (1976). Mathematical formulation of the relationship between photosynthesis and light for phytoplankton. *Limnol. Oceanogr.*, **21**, 540–7. *277, 283*
410. Jaworski, G. H. M., Talling, J. F. & Heaney, S. I. (1981). The influence of carbon dioxide depletion on growth and sinking rate of two planktonic diatoms in culture. *Br. phycol. J.*, **16**, 395–410. *331*
411. Jeffrey, S. W. (1961). Paper chromatographic separation of chlorophylls and carotenoids in marine algae. *Biochem. J.*, **80**, 336–42. *228*
412. Jeffrey, S. W. (1972). Preparation and some properties of crystalline

chlorophyll c_1 and c_2 from marine algae. *Biochim. Biophys. Acta*, **279**, 15–33. *228*

413. Jeffrey, S. W. (1976). The occurrence of chlorophyll c_1 and c_2 in algae. *J. Phycol.*, **12**, 349–54. *228*
414. Jeffrey, S. W. (1984). Responses of unicellular marine plants to natural blue-green light environments. In H. Senger (ed.), *Blue light effects in biological systems* (pp. 497–508). Berlin: Springer. *392*
415. Jeffrey, S. W., Sielicki, M. & Haxo, F. T. (1975). Chloroplast pigment patterns in dinoflagellates. *J. Phycol.*, **11**, 374–84. *227, 228, 230*
416. Jeffrey, S. W. & Vesk, M. (1977). Effect of blue-green light on photosynthetic pigments and chloroplast structure in the marine diatom *Stephanopyxis turris*. *J. Phycol.*, **13**, 271–9. *391–3*
417. Jeffrey, S. W. & Vesk, M. (1978). Chloroplast structural changes induced by white light in the marine diatom *Stephanopyxis turris*. *J. Phycol.*, **14**, 238–40. *391–3*
418. Jeffrey, S. W. & Wright, S. W. (1987). A new spectrally distinct component in preparations of chlorophyll c from the microalga *Emiliania huxleyi* (Prymnesiophyceae). *Biochim. Biophys. Acta*, **894**, 180–8. *227*
419. Jerlov, N. G. (1951). Optical studies of ocean water. *Rep. Swedish Deep-Sea Exped.*, **3**, 1–59. *78*
420. Jerlov, N. G. (1961). Optical measurements in the eastern North Atlantic. *Medd. Oceanogr. Inst. Goteborg, Ser. B*, **8**, 40 pp. *94*
421. Jerlov, N. G. (1974). Significant relationships between optical properties of the sea. In N. G. Jerlov & E. S. Nielsen (eds), *Optical aspects of oceanography* (pp. 77–94). London: Academic Press. *174*
422. Jerlov, N. G. (1976). *Marine Optics*. Amsterdam: Elsevier. *44, 61, 62, 64, 65, 78, 81, 86, 95, 101, 102, 111, 124, 133, 145, 150, 154, 365, 371*
423. Jerlov, N. G. & Nygard, K. (1969). A quanta and energy meter for photosynthetic studies. *Univ. Copenhagen Inst. Phys. Oceanogr. Rep.*, **10**, 29 pp. *113, 116, 139*
424. Jerlov (Johnson), N. G. & Liljequist, G. (1938). On the angular distribution of submarine daylight and the total submarine illumination. *Sven. Hydrogr.-Biol. Komm. Skr., Ny Ser. Hydrogr.*, **14**, 15 pp. Quoted by Jerlov (1976). *154*
425. Jerome, J. H., Bukata, R. P. & Bruton, J. E. (1983). Spectral attenuation and irradiance in the Laurentian Great Lakes. *J. Great Lakes Res.*, **9**, 60–8. *140*
426. Jerome, J. H., Bukata, R. P. & Bruton, J. E. (1988). Utilizing the components of vector irradiance to estimate the scalar irradiance in natural waters. *Appl. Opt.*, **27**, 4012–18. *146*
427. Jewson, D. H. (1976). The interactions of components controlling net phytoplankton photosynthesis in a well-mixed lake (Lough Neagh, Northern Ireland). *Freshwater Biol.*, **6**, 551–76. *202, 278, 318, 337, 338*
428. Jewson, D. H. (1977). Light penetration in relation to phytoplankton content of the euphotic zone of Lough Neagh, N. Ireland. *Oikos*, **28**, 74–83. *268, 269, 292*
429. Jewson, D. H. & Taylor, J. A. (1978). The influence of turbidity on net phytoplankton photosynthesis in some Irish lakes. *Freshwater Biol.*, **8**, 573–84. *67*

430. Jitts, H. R. (1963). The simulated *in situ* measurement of oceanic primary production. *Aust. J. Mar. Freshwater Res.*, **14**, 139–47. *272*
431. Jitts, H. R., Morel, A. & Saijo, Y. (1976). The relation of oceanic primary production to available photosynthetic irradiance. *Aust. J. Mar. Freshwater Res.*, **27**, 441–54. *279*
432. Johnson, R. W. (1978). Mapping of chlorophyll *a* distributions in coastal zones. *Photogramm. Eng. Rem. Sens.*, **44**, 617–24. *210*
433. Johnson, R. W. (1980). Remote sensing and spectral analysis of plumes from ocean dumping in the New York Bight Apex. *Remote Sens. Env.*, **9**, 197–209. *209*
434. Johnson, R. W. & Harriss, R. C. (1980). Remote sensing for water quality and biological measurements in coastal waters. *Photogramm. Eng. Rem. Sens.*, **46**, 77–85. *176*
435. Joint, I. R. (1986). Physiological ecology of picoplankton in various oceanographic provinces. In T. Platt & W. K. W. Li (eds), *Photosynthetic picoplankton* (pp. 287–309). *202*
436. Joint, I. R. & Pomroy, A. J. (1981). Primary production in a turbid estuary. *Estuar. Coast. Shelf Sci.*, **13**, 303–16. *321*
437. Jones, D. & Wills, M. S. (1956). The attenuation of light in sea and estuarine waters in relation to the concentration of suspended solid matter. *J. Mar. Biol. Ass. U.K.*, **35**, 431–44. *108*
438. Jones, L. W. & Kok, B. (1966). Photoinhibition of chloroplast reactions. *Plant Physiol.*, **41**, 1037–43. *285, 286*
439. Jones, L. W. & Myers, J. (1964). Enhancement in the blue-green alga, *Anacystis nidulans. Plant Physiol.*, **39**, 938–46. *312*
440. Jones, L. W. & Myers, J. (1965). Pigment variations in *Anacystis nidulans* induced by light of selected wavelengths. *J. Phycol.*, **1**, 6–13. *390–1*
441. Jones, R. I. (1978). Adaptations to fluctuating irradiance by natural phytoplankton communities. *Limnol. Oceanogr.*, **23**, 920–6. *415*
442. Jones, R. I. & Ilmavirta, V. (1978*a*). A diurnal study of the phytoplankton in the eutrophic lake Lovojärvi, Southern Finland. *Arch. Hydrobiol.*, **83**, 494–514. *343, 396–7*
443. Jones, R. I. & Ilmavirta, V. (1978*b*). Vertical and seasonal variation of phytoplankton photosynthesis in a brown-water lake with winter ice cover. *Freshwater Biol.*, **8**, 561–72. *415*
444. Jones, R. I. & Arvola, L. (1984). Light penetration and some related characteristics in small forest lakes in Southern Finland. *Verh. Internat. Verein. Limnol.*, **22**, 811–16. *141*
445. Jørgensen, E. G. (1968). The adaptation of plankton algae. II. Aspects of the temperature adaptation of *Skeletonema costatum. Physiol. Plant.*, **21**, 423–7. *336*
446. Jupp, B. P. & Spence, D. H. N. (1977). Limitations on macrophytes in a eutrophic lake, Loch Leven. *J. Ecol.*, **65**, 175–86, 431–6. *341, 342*
447. Kageyama, A., Yokohama, Y. & Nisizawa, K. (1979). Diurnal rhythms of apparent photosynthesis of a brown alga, *Spatoglossum pacificum. Botanica Marina*, **22**, 199–201. *423, 427*
448. Kain, J. M. (1979). A view of the genus *Laminaria. Oceanogr. Mar. Biol. Ann. Rev.*, **17**, 101–61. *342, 429*
449. Kalff, J., Welch, H. E. & Holmgren, S. K. (1972). Pigment cycles in two high-arctic Canadian lakes. *Ver. internat. Verein Limnol.*, **18**, 250–6. *415*

450. Kalle, K. (1937). *Annln. Hydrogr. Berl.*, **65**, 276–82. Quoted by Kalle (1966). *61*
451. Kalle, K. (1961). What do we know about the 'Gelbstoff'? *Union Geod. Geophys. int. Monogr.*, **10**, 59–62. Quoted by Jerlov (1976). *65*
452. Kalle, K. (1966). The problem of the gelbstoff in the sea. *Oceanogr. Mar. Biol. Ann. Rev.*, **4**, 91–104. *61, 62*
453. Kamiya, A. & Miyachi, S. (1984). Blue-green and green light adaptations on photosynthetic activity in some algae collected from the sub-surface layer in the western Pacific Ocean. In H. Senger (ed.), *Blue light effects in biological systems* (pp. 517–28). Berlin: Springer. *388*
454. Kan, K. S. & Thornber, J. P. (1976). The light-harvesting chlorophyll a/b-protein complex of *Chlamydomonas reinhardtii*. *Plant Physiol.*, **57**, 47–52. *237*
455. Kana, T. M. & Glibert, P. M. (1987*a*). Effect of irradiances up to 2000 μE m^{-2} s^{-1} on marine *Synechococcus* WH7803 – I. Growth, pigmentation and cell composition. *Deep-Sea Res.*, **34**, 479–95. *376, 379*
456. Kana, T. M. & Glibert, P. M. (1987*b*). Effect of irradiances up to 2000 μE m^{-2} s^{-1} on marine *Synechococcus* WH7803 – II. Photosynthetic responses and mechanisms. *Deep-Sea Res.*, **34**, 499–516. *382*
457. Kana, T. M., Glibert, P. M., Goericke, R. & Welschmeyer, N. A. (1988). Zeaxanthin and β-carotene in *Synechococcus* WH7803 respond differently to irradiance. *Limnol. Oceanogr.*, **33**, 1623–7. *377*
458. Kana, T. M., Watts, J. L. & Glibert, P. M. (1985). Diel periodicity in the photosynthetic capacity of coastal and offshore phytoplankton assemblages. *Mar. Ecol. Prog. Ser.*, **25**, 131–9. *423*
459. Karelin, A. K. & Pelevin, V. N. (1970). The FMPO-64 marine underwater irradiance meter and its application in hydro-optical studies. *Oceanology*, **10**, 282–5. *121*
460. Katoh, T. & Ehara, T. (1990). Supramolecular assembly of fucoxanthin–chlorophyll–protein complexes isolated from a brown alga, *Petalonia fascia*. Electron microscope studies. *Plant Cell Physiol.*, **31**, 439–47. *236, 239*
461. Keeley, J. E. (1990). Photosynthetic pathways in freshwater aquatic plants. *Trends Ecol. Evol.*, **5**, 330–3. *250, 333*
462. Kelly, M. G., Thyssen, N. & Moeslund, B. (1983). Light and the annual variation of oxygen- and carbon-based measurements of productivity in a macrophyte-dominated river. *Limnol. Oceanogr.*, **28**, 503–15. *350*
463. Khorram, S. (1981*a*). Use of ocean colour scanner data in water quality mapping. *Photogramm. Eng. Rem. Sens.*, **47**, 667–76. *210*
464. Khorram, S. (1981*b*). Water quality mapping from Landsat digital data. *Int. J. Rem. Sens.*, **2**, 145–53. *192, 206*
465. Khorram, S., Cheshire, H., Geraci, A. L. & Rosa, G. L. (1991). Water quality mapping of Augusta Bay, Italy from Landsat-TM data. *Int. J. Remote Sens.*, **12**, 803–8. *192, 206, 209*
466. Kieber, R. J., Xianling, Z. & Mopper, K. (1990). Formation of carbonyl compounds from UV-induced photodegradation of humic substances in natural waters: fate of riverine carbon in the sea. *Limnol. Oceanogr.*, **35**, 1503–15. *72*

467. Kiefer, D. A. (1973). Chlorophyll *a* and fluorescence in marine centric diatoms: responses of chloroplasts to light and nutrient stress. *Mar. Biol.*, **23**, 39–46. *423–4*

468. Kiefer, D. A., Olson, R. J. & Holm-Hansen, O. (1976). Another look at the nitrite and chlorophyll maxima in the central North Pacific. *Deep-Sea Res.*, **23**, 1199–208. *320, 396*

469. Kiefer, D. A. & SooHoo, J. B. (1982). Spectral absorption by marine particles of coastal waters of Baja California. *Limnol. Oceanogr.*, **27**, 492–9. *76*

470. Killilea, S. D., O'Carra, P. & Murphy, R. F. (1980). Structures and apoprotein linkages of phycoerythrobilin and phycocyanobilin. *Biochem. J.*, **187**, 311–20. *242*

471. Kim, H., Fraser, R. S., Thompson, L. L. & Bahethi, O. (1980). A design study for an advanced ocean colour scanner system. *Boundary-layer Meteorol.*, **18**, 315–27. *176*

472. Kim, H., McClain, C. R. & Hart, W. D. (1979). Chlorophyll gradient map from high-altitude ocean-colour-scanner data. *Appl. Opt.*, **18**, 3715–16. *189, 210*

473. Kim, H. H. & Linebaugh, G. (1985). Early evaluation of thematic mapper data for coastal process studies. *Adv. Space Res.*, **5**, 21–9. *198, 211*

474. King, R. J. & Schramm, W. (1976). Photosynthetic rates of benthic marine algae in relation to light intensity and seasonal variations. *Mar. Biol.*, **37**, 215–22. *274–5, 276, 277, 280, 281, 282, 285, 350, 410–1*

475. Kirk, J. T. O. (1975*a*). A theoretical analysis of the contribution of algal cells to the attenuation of light within natural waters. I. General treatment of suspensions of pigmented cells. *New Phytol.*, **75**, 11–20. *255, 256–7*

476. Kirk, J. T. O. (1975*b*). A theoretical analysis of the contributions of algal cells to the attenuation of light within natural waters. II. Spherical cells. *New Phytol.*, **75**, 21–36. *255, 257, 270*

477. Kirk, J. T. O. (1976*a*). A theoretical analysis of the contribution of algal cells to the attenuation of light within natural waters. III. Cylindrical and spheroidal cells. *New Phytol.*, **77**, 341–58. *255, 256–7, 259, 260, 268, 270*

478. Kirk, J. T. O. (1976*b*). Yellow substance (gelbstoff) and its contribution to the attenuation of photosynthetically active radiation in some inland and coastal southeastern Australian waters. *Aust. J. Mar. Freshwater Res.*, **27**, 61–71. *60, 63, 66, 68, 71, 83*

479. Kirk, J. T. O. (1977*a*). Use of a quanta meter to measure attenuation and underwater reflectance of photosynthetically active radiation in some inland and coastal southeastern Australian waters. *Aust. J. Mar. Freshwater Res.*, **28**, 9–21. *66, 68, 137, 139, 141, 142, 145, 146, 150*

480. Kirk, J. T. O. (1977*b*). Thermal dissociation of fucoxanthin-protein binding in pigment complexes from chloroplasts of *Hormosira* (Phaeophyta). *Plant Sci. Lett.*, **9**, 373–80. *230*

481. Kirk, J. T. O. (1979). Spectral distribution of photosynthetically active radiation in some south-eastern Australian waters. *Aust. J. Mar. Freshwater Res.*, **30**, 81–91. *118, 132, 135*

482. Kirk, J. T. O. (1980*a*). Relationship between nephelometric turbidity

and scattering coefficients in certain Australian waters. *Aust. J. Mar. Freshwater Res.*, **31**, 1–12. *99, 150, 167*

483. Kirk, J. T. O. (1980*b*). Spectral absorption properties of natural waters: contribution of the soluble and particulate fractions to light absorption in some inland waters of southeastern Australia. *Aust. J. Mar. Freshwater Res.*, **31**, 287–96. *52, 68, 72, 75, 76, 82, 83, 84*
484. Kirk, J. T. O. (1981*a*). A Monte Carlo study of the nature of the underwater light field in, and the relationships between optical properties of, turbid yellow waters. *Aust. J. Mar. Freshwater Res.*, **32**, 517–32. *79, 97, 126, 146, 147, 153, 157, 158, 160–3, 167*
485. Kirk, J. T. O. (1981*b*). Estimation of the scattering coefficient of natural waters using underwater irradiance measurements. *Aust. J. Mar. Freshwater Res.*, **32**, 533–9. *97, 98, 103*
486. Kirk, J. T. O. (1981*c*). *A Monte Carlo procedure for simulating the penetration of light into natural waters.* CSIRO Division of Plant Industry, Technical Paper No. 36, 18 pp. *126, 153, 157*
487. Kirk, J. T. O. (1982). Prediction of optical water quality. In E. M. O'Loughlin & P. Cullen (eds), *Prediction in water quality* (pp. 307–26). Canberra: Australian Academy of Science. *149*
488. Kirk, J. T. O. (1984). Dependence of relationship between inherent and apparent optical properties of water on solar altitude. *Limnol. Oceanogr.*, **29**, 350–6. *146, 164*
489. Kirk, J. T. O. (1985). Effect of suspensoids (turbidity) on penetration of solar radiation in aquatic ecosystems. *Hydrobiologia*, **125**, 195–208. *91*
490. Kirk, J. T. O. (1986). Optical properties of picoplankton suspensions. In T. Platt & W. K. W. Li (eds), *Photosynthetic picoplankton* (pp. 501–20). *Can. Bull. Fish Aquat. Sci.*, **214**. *269*
491. Kirk, J. T. O. (1988). Optical water quality – what does it mean and how should we measure it? *J. Water Pollut. Control Fed.*, **60**, 194–7. *127, 149, 164*
492. Kirk, J. T. O. (1989*a*). The upwelling light stream in natural waters. *Limnol. Oceanogr.*, **34**, 1410–25. *21, 99*
493. Kirk, J. T. O. (1989*b*). The assessment and prediction of optical water quality. In *13th Fed. Conv. Aust. Water Wastewater Assoc.*, 89/2 (pp. 504–7). Canberra: Institution of Engineers. *127, 149, 164*
494. Kirk, J. T. O. (1991). Volume scattering function, average cosines and the underwater light field. *Limnol. Oceanogr.*, **36**, 455–67. *11, 19, 146, 147, 164*
495. Kirk, J. T. O. (1992). Monte Carlo Modelling of the performance of a reflective tube absorption meter. *Appl. Opt.*, **31**, 6463–8. *54*

495*a*. Kirk, J. T. O. Unpublished data. *63, 66, 68, 69, 141, 142*

496. Kirk, J. T. O. & Goodchild, D. J. (1972). Relationship of photosynthetic effectiveness of different kinds of light to chlorophyll content and chloroplast structure in greening wheat and in ivy leaves. *Aust. J. biol. Sci.*, **25**, 215–41. *372*
497. Kirk, J. T. O. & Tilney-Bassett, R. A. E. (1978). *The plastids* (2nd edn). Amsterdam: Elsevier. *217, 227, 231*
498. Kirk, J. T. O. & Tyler, P. A. (1986). The spectral absorption and scattering properties of dissolved and particulate components in relation to the underwater light yield of some tropical Australian freshwaters. *Freshwater Biol.*, **16**, 573–83. *69, 84, 103, 142*

499. Kirkman, H. & Reid, D. D. (1979). A study of the seagrass *Posidonia australis* in the carbon budget of an estuary. *Aquat. Bot.*, **7**, 173–83. *347–8*

500. Kishino, M., Booth, C. R. & Okami, N. (1984). Underwater radiant energy absorbed by phytoplankton, detritus, dissolved organic matter, and pure water. *Limnol. Oceanogr.*, **29**, 340–9. *67, 70, 297*

501. Kishino, M. & Okami, N. (1984). Instrument for measuring downward and upward spectral irradiances in the sea. *La mer (Bull. Soc. franco-japon. d'oceanogr.)*, **22**, 37–40. *123*

502. Kishino, M., Okami, N., Takahashi, M. & Ichimura, S. (1986). Light utilization efficiency and quantum yield of phytoplankton in a thermally stratified sea. *Limnol. Oceanogr.*, **31**, 557–66. *266, 293, 297, 300, 302*

503. Kishino, M., Takahashi, M., Okami, N. & Ichimura, S. (1985). Estimation of the spectral absorption coefficients of phytoplankton in the sea. *Bull. Marine Sci.*, **37**, 634–42. *76*

504. Klemas, V., Bartlett, D., Philpot, W., Rogers, R. & Reed, L. (1974). Coastal and estuarine studies with ERTS-1 and Skylab. *Remote Sens. Env.*, **3**, 153–74. *207, 208*

505. Klemas, V., Borchardt, J. F. & Treasure, W. M. (1973). Suspended sediments observations from ERTS-1. *Remote Sens. Env.*, **2**, 205–21. *191*

506. Klemer, A. R., Feuillade, J. & Feuillade, M. (1982). Cyanobacterial blooms: carbon and nitrogen limitations have opposite effects on the buoyancy of *Oscillatoria. Science*, **215**, 1629–31. *419*

507. Kling, G. W. (1988). Comparative transparency, depth of mixing and stability of stratification in lakes of Cameroon, West Africa. *Limnol. Oceanogr.*, **33**, 27–40. *141*

508. Koblents-Mishke, O. I. (1965). Primary production in the Pacific. *Oceanology*, **5**, 104–16. *344, 359*

509. Koblents-Mishke, O. I. (1979). Photosynthesis of marine phytoplankton as a function of underwater irradiance. *Soviet Plant Physiol.*, **26**, 737–46. *286, 303*

510. Koenings, J. P. & Edmundson, J. A. (1991). Secchi disk and photometer estimates of light regimes in Alaskan lakes: effects of yellow colour and turbidity. *Limnol. Oceanogr.*, **36**, 91–105. *120, 140*

511. Koepke, P. (1984). Effective reflectance of oceanic whitecaps. *Appl. Opt.*, **23**, 1816–24. *43*

512. Kolber, Z., Wyman, K. D. & Falkowski, P. G. (1990). Natural variation in photosynthetic energy conversion efficiency: a field study in the Gulf of Maine. *Limnol. Oceanogr.*, **35**, 72–9. *302, 356*

513. Kondratyev, K. Y. (1954). *Radiant solar energy*. Leningrad. Quoted by Robinson (1966). *41*

514. Kondratyev, K. Y. & Pozdniakov, D. V. (1990). Passive and active optical remote sensing of the inland water phytoplankton. *ISPRS J. Photogramm. Remote Sens.*, **44**, 257–94. *210*

515. Kopelevich, O. V. (1982). The 'yellow substance' in the ocean according to optical data. *Oceanology*, **22**, 152–6. *63, 65*

516. Kopelevich, O. V. (1984). On the influence of river and eolian suspended matter on the optical properties of sea water. *Oceanology*, **24**, 331–4. *105*

517. Kopelevich, O. V. & Burenkov, V. I. (1971). The nephelometric method for determining the total scattering coefficient of light in sea water. *Izv. Atmos. Oceanic Phys.*, **7**, 835–40. *95, 101, 102, 103*
518. Kopelevich, O. V. & Burenkov, V. I. (1977). Relation between the spectral values of the light absorption coefficients of sea water, phytoplanktonic pigments, and the yellow substance. *Oceanology*, **17**, 278–82. *62, 64*
519. Kopelevich, O. V. & Mezhericher, E. M. (1979). Improvement of the method of 'inversion' of the spectral values of the luminance coefficient of the sea. *Oceanology*, **19**, 621–4. *190*
520. Kowalik, W. S., Marsh, S. E. & Lyon, R. J. P. (1982). A relation between Landsat digital numbers surface reflectance, and the cosine of the solar zenith angle. *Remote Sens. Env.*, **12**, 39–35. *185*
521. Kramer, C. J. M. (1979). Degradation by sunlight of dissolved fluorescing substances in the upper layers of the eastern Atlantic Ocean. *Neth J. Sea Res.*, **13**, 325–9. *72*
522. Kullenberg, G. (1968). Scattering of light by Sargasso Sea water. *Deep-Sea Res.*, **15**, 423–32. *93, 101, 108*
533. Kullenberg, G. (1984). Observations of light scattering functions in two oceanic areas. *Deep-Sea Res.*, **31**, 295–316. *94*
534. Kullenberg, G., Lundgren, B., Malmberg, S. A., Nygard, K. & Højerslev, N. K. (1970). Inherent optical properties of the Sargasso Sea. *Univ. Copenhagen Inst. Phys. Oceanogr. Rep.*, **11**, 18pp. Quoted by Jerlov (1976). *64*
535. Kuring, N., Lewis, M. R., Platt, T. & O'Reilly, J. E. (1990). Satellite-derived estimates of primary production on the northwest Atlantic continental shelf. *Continental Shelf Res.*, **10**, 461–84. *355*
536. Langmuir, I. (1938). Surface motion of water induced by wind. *Science*, **87**, 1119–23. *286*
537. Larkum, A. W. D. & Barrett, J. (1983). Light-harvesting processes in algae. *Adv. Botan. Res.*, **10**, 3–219. *235*
538. Larkum, A. W. D., Drew, E. A. & Crossett, R. N. (1967). The vertical distribution of attached marine algae in Malta. *J. Ecol.*, **55**, 361–71. *366–7, 372*
539. Larson, D. W. (1972). Temperature, transparency, and phytoplankton productivity in Crater Lake, Oregon. *Limnol. Oceanogr.*, **17**, 410–17. *321, 396*
540. Lathrop, R. G. & Lillesand, T. M. (1986). Use of Thematic Mapper data to assess water quality in Green Bay and central Lake Michigan. *Photogramm. Eng. Remote Sens.*, **52**, 671–80. *195, 206, 209*
541. Lathrop, R. G. & Lillesand, T. M. (1989). Monitoring water quality and river plume transport in Green Bay, Lake Michigan with SPOT-1 imagery. *Photogramm. Eng. Remote Sens.*, **55**, 349–54. *193, 209*
542. Latimer, P. & Rabinowitch, E. (1959). Selective scattering of light by pigments *in vivo*. *Arch. Biochem. Biophys.*, **84**, 428–41. *265*
543. Laws, E. A., Tullio, G. R. D., Carder, K. L., Betzer, P. R. & Hawes, S. (1990). Primary production in the deep blue sea. *Deep-Sea Res.*, **37**, 715–30. *272, 297*
544. Le Cren, E. D. & Lowe-McConnell, R. H. (eds). (1980). *The functioning*

of freshwater ecosystems. Cambridge: Cambridge University Press. *359*

545. Leletkin, V. A., Zvalinskii, V. I. & Titlyanov, E. A. (1981). Photosynthesis of zooxanthellae in corals of different depths. *Soviet Plant Physiol.*, **27**, 863–70. *398*
546. Lester, W. W., Adams, M. S. & Farmer, A. M. (1988). Effects of light and temperature on photosynthesis of the nuisance alga *Cladophora glomerata* (L.) Kutz from Green Bay, Michigan. *New Phytol.*, **109**, 53–8. *278*
547. Levavasseur, G., Edwards, G. E., Osmond, C. B. & J. Ramus (1991). Inorganic carbon limitation of photosynthesis in *Ulva rotundata* (Chlorophyta). *J. Phycol.*, **27**, 667–72. *327*
548. Lévêque, C. Quoted in Le Cren & Lowe-McConnell (1980). *346*
549. Levring, T. (1959). Submarine illumination and vertical distribution of algal vegetation. In *Proc. 9th Int. Bot. Congr.* (pp. 183–193). Canada: University of Toronto Press. *363*
550. Levring, T. (1966). Submarine light and algal shore zonation. In R. Bainbridge, G. C. Evans & O. Rackham (eds), *Light as an ecological factor* (pp. 305–18). Oxford: Blackwell. *368*
551. Levring, T. (1968). Photosynthesis of some marine algae in clear, tropical oceanic water. *Bot. Mar.*, **11**, 72–80. *368*
552. Levy, I. & Gantt, E. (1988). Light acclimation in *Porphyridium purpureum* (Rhodophyta): growth, photosynthesis and phycobilisomes. *J. Phycol.*, **24**, 452–8. *376, 378, 382*
553. Lewin, R. A. (1976). Prochlorophyta as a proposed new division of algae. *Nature.*, **261**, 697–8. *223*
554. Lewis, J. B. (1977). Processes of organic production on coral reefs. *Biol. Rev.*, **52**, 305–47. *429*
555. Lewis, M. R., Horne, E. P. W., Cullen, J. J., Oakey, N. S. & Platt, T. (1984). Turbulent motions may control phytoplankton photosynthesis in the upper ocean. *Nature*, **311**, 49–50. *402*
556. Lewis, M. R., Warnock, R. E., Irwin, B. & Platt, T. (1985). Measuring photosynthetic action spectra of natural phytoplankton populations. *J. Phycol.*, **21**, 310–15. *307*
557. Lewis, M. R., Warnock, R. E. & Platt, T. (1985). Absorption and photosynthetic action spectra for natural phytoplankton populations: implications for production in the open sea. *Limnol. Oceanogr.*, **30**, 794–806. *76*
558. Lewis, W. M. & Canfield, D. (1977). Dissolved organic carbon in some dark Venezuelan waters and a revised equation for spectrophotometric determination of dissolved organic carbon. *Arch. Hydrobiol.*, **79**, 441–5. *68, 71*
559. Ley, A. C. & Butler, W. L. (1980). Effects of chromatic adaptation on the photochemical apparatus of photosynthesis in *Porphyridium cruentum. Plant Physiol.*, **65**, 714–22. *388–90, 406, 409*
560. Li, W. K. W. (1986). Experimental approaches to field measurements: methods and interpretation. In T. Platt & W. K. W. Li (eds), *Photosynthetic picoplankton* (pp. 251–86). *Can. Bull. Fish Aquat. Sci.*, **214**. *111*
561. LI-COR Inc. (Lincoln, Nebraska). LI-1800 UW Spectroradiometer. *123*
562. Lillesand, T. M., Johnson, W. L., Deuell, R. L., Lindstrom, O. M. &

Meisner, D. E. (1983). Use of Landsat data to predict the trophic state of Minnesota lakes. *Photogramm. Eng. Remote Sens.*, **49**, 219–29. *206*

563. Lindell, L. T., Steinvall, O., Jonsson, M. & Claeson, T. (1985). Mapping of coastal-water turbidity using LANDSAT imagery. *Int. J. Remote Sens.*, **6**, 629–42. *206*
564. Lipkin, Y. (1977). Seagrass vegetation of Sinai and Israel. In C. P. McRoy & C. Helfferich (eds), *Seagrass ecosystems* (pp. 263–93). New York: Marcel Dekker. *375*
565. Lipkin, Y. (1979). Quantitative aspects of seagrass communities particularly of those dominated by *Halophila stipulacea*, in Sinai (northern Red Sea). *Aquat. Bot.*, **7**, 119–28. *380, 403, 405*
566. Littler, M. M., Littler, D. S., Blair, S. M. & Norris, J. N. (1985). Deepest known plant life discovered on an uncharted seamount. *Science*, **227**, 57–9. *367*
567. Littler, M. M., Littler, D. S., Blair, S. M. & Norris, J. N. (1986). Deep-water plant communities from an uncharted seamount off San Salvador Island, Bahamas: distribution, abundance, and primary productivity. *Deep-Sea Res.*, **33**, 881–92. *316, 367*
568. Lodge, D. M. (1991). Herbivory on freshwater macrophytes. *Aquat. Bot.*, **41**, 195–224. *342*
569. Loeblich, L. A. (1982). Photosynthesis and pigments influenced by light intensity and salinity in the halophile *Dunaliella salina* (Chlorophyta). *J. Mar. Biol. Ass. U.K.*, **62**, 493–508. *377*
570. Lorenzen, C. J. (1968). Carbon/chlorophyll relationships in an upwelling area. *Limnol. Oceanogr.*, **13**, 202–4. *228*
571. Love, R. J. R. & Robinson, G. G. C. (1977). The primary productivity of submerged macrophytes in West Blue Lake, Manitoba. *Can. J. Bot.*, **55**, 118–27. *351*
572. Lüning, K. (1971). Seasonal growth of *Laminaria hyperborea* under recorded underwater light conditions near Heligoland. In D. J. Crisp (ed.), *Fourth European Marine Biology Symposium* (pp. 347–61). Cambridge: Cambridge University Press. *429*
573. Lüning, K. (1990). *Seaweeds: their environment, biogeography & ecophysiology*. New York: Wiley. *371*
574. Lüning, K. & Dring, M. J. (1979). Continuous underwater light measurement near Helgoland (North Sea) and its significance for characteristic light limits in the sublittoral region. *Helgoländer Wiss. Meeresunters*, **32**, 403–24. *316, 349*
575. Lynch, M. & Shapiro, J. (1981). Predation, enrichment and phytoplankton community structure. *Limnol. Oceanogr.*, **26**, 86–102. *341*
576. Maberley, S. C. (1990). Exogenous sources of inorganic carbon for photosynthesis by marine macroalgae. *J. Phycol.*, **26**, 439–49. *327*
577. Maberley, S. C. & Spence, D. H. N. (1983). Photosynthetic inorganic carbon use by freshwater plants. *J. Ecol.*, **71**, 705–24. *327*
578. Macauley, J. M., Clark, J. R. & Price, W. A. (1988). Seasonal changes in the standing crop and chlorophyll content of *Thalassia testudinum* Banks ex König and its epiphytes in the northern Gulf of Mexico. *Aquat. Bot.*, **31**, 277–87. *348*
579. MacColl, R. & Guard-Friar, D. (1983*a*). Phycocyanin 612: a biochemical and photophysical study. *Biochemistry*, **22**, 5568–72. *240*

580. MacColl, R. & Guard-Friar, D. (1983*b*). Phycocyanin 645: the chromophore assay of phycocyanin 645 from the cryptomonad *Chroomonas* species. *J. Biol. Chem.*, **258**, 14327–9. *240, 243*

581. MacColl, R. & Guard-Friar, D. (1987). *Phycobiliproteins*. Boca Raton: CRC Press. *239*

582. MacColl, R., Guard-Friar, D. & Csatorday, K. (1983). Chromatographic and spectroscopic analysis of phycoerythrin 545 and its subunits. *Arch. Mikrobiol.*, **135**, 194–8. *240*

583. MacFarlane, N. & Robinson, I. S. (1984). Atmospheric correction of Landsat MSS data for a multidate suspended sediment algorithm. *Int. J. Remote Sens.*, **5**, 561–76. *185*

584. Madsen, T. V. & Sand-Jensen, K. (1991). Photosynthetic carbon assimilation in aquatic macrophytes. *Aquat. Bot.*, **41**, 5–40. *326, 327, 330*

585. Malcolm, R. I. (1990). The uniqueness of humic substances in each of soil, stream and marine environments. *Analyt. Chim. Acta*, **232**, 19–30. *59, 62*

586. Malone, T. C. (1977*a*). Environmental regulation of phytoplankton productivity in the lower Hudson estuary. *Estuar. Coast. Mar. Sci.*, **5**, 157–71. *321, 337*

587. Malone, T. C. (1977*b*). Light-saturated photosynthesis by phytoplankton size fractions in the New York Bight, U.S.A. *Mar. Biol.*, **42**, 281–92. *337*

588. Malone, T. C. (1982). Phytoplankton photosynthesis and carbon-specific growth: light-saturated rates in a nutrient-rich environment. *Limnol. Oceanogr.*, **27**, 226–35. *423*

589. Malone, T. C. & Neale, P. J. (1981). Parameters of light-dependent photosynthesis for phytoplankton size fractions in temperate estuarine and coastal environments. *Mar. Biol.*, **61**, 289–97. *302*

590. Mann, K. H. (1972). Ecological energetics of the seaweed zone in a marine bay on the Atlantic coast of Canada. II. Productivity of the seaweeds. *Mar. Biol.*, **14**, 199–209. *429*

591. Mann, K. H. & Chapman, A. R. O. (1975). Primary production of marine macrophytes. In J. P. Cooper (ed.), *Photosynthesis and productivity in different environments* (pp. 207–23). Cambridge: Cambridge University Press. *428–9*

592. Manodori, A. & Melis, A. (1986). Cyanobacterial acclimation to Photosystem I or Photosystem II light. *Plant Physiol.*, **82**, 185–9. *391*

593. Mariani Colombo, P., Orsenigo, M., Solazzi, A. & Tolomio, C. (1976). Sea depth effects on the algal photosynthetic apparatus. IV. Observations on the photosynthetic apparatus of *Halimeda tuna* (Siphonales) at sea depths between 7 and 16 m. *Mem. Biol. mar. Ocean*, **6**, 197–208. *403*

594. Marra, J. (1978). Phytoplankton photosynthetic response to vertical movement in a mixed layer. *Mar. Biol.*, **46**, 203–8. *277, 284, 286*

594*a*. Marshall, B. R. & Smith, R. C. (1990). Raman scattering and in-water ocean optical properties. *Appl. Opt.*, **29**, 71–84. *150*

595. Marshall, C. T. & Peters, R. H. (1989). General patterns in the seasonal development of chlorophyll a for temperate lakes. *Limnol. Oceanogr.*, **34**, 856–67. *345*

596. Martin, J. H., Gordon, R. M. & Fitzwater, S. E. (1991). The case for iron. *Limnol. Oceanogr.*, **36**, 1793–1802. *340*

597. Maske, H. & Haardt, H. (1987). Quantitative in vivo absorption spectra of phytoplankton: detrital absorption and comparison with fluorescence excitation spectra. *Limnol. Oceanogr.*, **32**, 620–33. *76*
598. Mathieson, A. C. (1979). Vertical distribution and longevity of subtidal seaweeds in northern New England, U.S.A. *Bot. Mar.*, **30**, 511–20. *364*
599. Mathieson, A. C. & Burns, R. L. (1971). Ecological studies of economic red algae. *J. Exp. Mar. Biol. Ecol.*, **7**, 197–206. *102, 335*
600. Mathieson, A. C. & Norall, T. L. (1975). Physiological studies of subtidal red algae. *J. Exp. Mar. Biol. Ecol.*, **20**, 237–47. *398–9*
601. Mathis, P. & Paillotin, G. (1981). Primary processes of photosynthesis. In M. D. Hatch & N. K. Boardman (eds), *The biochemistry of plants* (pp. 97–161). New York: Academic Press. *246*
602. McAllister, H. A., Norton, T. A. & Conway, E. (1967). A preliminary list of sublittoral marine algae from the west of Scotland. *Br. Phycol. Bull.*, **3**, 175–84. *364*
603. McKim, H. L., Merry, C. J. & Layman, R. W. (1984). Water quality monitoring using an airborne spectroradiometer. *Photogramm. Eng. Remote Sens.*, **50**, 353–60. *175*
604. McLachlan, A. J. & McLachlan, S. M. (1975). The physical environment and bottom fauna of a bog lake. *Arch. Hydrobiol.*, **76**, 198–217. *67*
605. McMahon, T. G., Raine, R. C. T., Fast, T., Kies, L. & Patching, J. W. (1992). Phytoplankton biomass, light attenuation and mixing in the Shannon estuary. *J. Mar. Biol. Ass. U.K.*, **72**, 709–20. *139*
606. McRoy, C. P. & Helfferich, C. (1980). Applied aspects of seagrasses. In R. C. Phillips & C. P. McRoy (eds), *Handbook of seagrass biology* (pp. 297–343). New York: Garland STPM Press. *324, 384*
607. Medina, E. (1971). Effect of nitrogen supply and light intensity during growth on the photosynthetic capacity and carboxydismutase activity of leaves of *Atriplex patula* spp. *hastata*. *Carnegie Inst. Wash. Yearbook*, **70**, 551–9. *379*
608. Megard, R. O., Combs, W. S., Smith, P. D. & Knoll, A. S. (1979). Attenuation of light and daily integral rates of photosynthesis attained by planktonic algae. *Limnol. Oceanogr.*, **24**, 1038–50. *140, 269, 292*
609. Melack, J. M. (1979). Photosynthesis and growth of *Spirulina platensis* (Cyanophyta) in an equatorial lake (Lake Simbi, Kenya). *Limnol. Oceanogr.*, **24**, 753–60. *141*
610. Melack, J. M. (1981). Photosynthetic activity of phytoplankton in tropical African soda lakes. *Hydrobiologia*, **81**, 71–85. *304*
611. Meyers-Schulte, K. J. & Hedges, J. I. (1986). Molecular evidence for a terrestrial component of organic matter dissolved in ocean water. *Nature*, **321**, 61–3. *59, 62*
612. Mie, G. (1908). Beiträge zur Optik trüber Medien, speziell kolloidalen Metall-lösungen. *Ann. Phys.*, **25**, 377–445. *86–89*
613. Mishkind, M. & Mauzerall, D. (1980). Kinetic evidence for a common photosynthetic step in diverse seaweeds. *Mar. Biol.*, **58**, 39–96. *377*
614. Mitchell, B. G., Brody, E. A., Holm-Hansen, O., McClain, C. & Bishop, J. (1991). Light limitation of phytoplankton biomass and macronutrient utilization in the Southern Ocean. *Limnol. Oceanogr.*, **36**, 1662–77. *340*

615. Mitchell, B. G. & Holm-Hansen, O. (1991*a*). Bio-optical properties of Antarctic Peninsula waters: differentiation from temperate ocean models. *Deep-Sea Res.*, **38**, 1009–28. *261*
616. Mitchell, B. G. & Holm-Hansen, O. (1991*b*). Observations and modelling of the Antarctic phytoplankton crop in relation to mixing depth. *Deep-Sea Res.*, **38**, 981–1007. *279, 318*
617. Mitchell, B. G. & Kiefer, D. A. (1988). Chlorophyll a specific absorption and fluorescence excitation spectra for light-limited phytoplankton. *Deep-Sea Res.*, **35**, 639–63. *76*
618. Mitchelson, E. G., Jacob, N. J. & Simpson, J. H. (1986). Ocean colour algorithms from the Case 2 waters of the Irish Sea in comparison to algorithms from Case 1 waters. *Continental Shelf Res.*, **5**, 403–15. *198*
619. Mittenzwey, K.-H., Ullrich, S., Gitelson, A. A. & Kondratiev, K. Y. (1992). Determination of chlorophyll a of inland waters on the basis of spectral reflectance. *Limnol. Oceanogr.*, **37**, 147–9. *195*
620. Mizusawa, M., Kageyama, A. & Yokohama, Y. (1978). Physiology of benthic algae in tide pools. *Jap. J. Phycol.*, **26**, 109–14. *333, 334*
621. Mobley, C. D. (1992). The optical properties of water. In *Handbook of optics, Second edition*, ed. M. Bass. New York: McGraw-Hill (in press). *6*
622. Molinier, R. (1960). Etudes des biocénoses marines du Cap Corse. *Vegetatio*, **9**, 121–92, 219–312. *366*
623. Monahan, E. C. & Pybus, M. J. (1978). Colour, ultraviolet absorbance and salinity of the surface waters off the west coast of Ireland. *Nature*, **274**, 782–4. *61*
624. Monteith, J. L. (1973). *Principles of environmental physics*. London: Edward Arnold. *30, 31, 33, 34, 35, 38*
625. Monteith, J. L. & Unsworth, M. H. (1990). *Principles of environmental physics*. London: Edward Arnold. *34*
626. Moon, P. (1940). *J. Franklin Inst.*, **230**, 583. Quoted by Monteith (1973). *31*
627. Moon, P. (1961). *The scientific basis of illuminating engineering*. New York: Dover. *28*
628. Moon, R. E. & Dawes, C. J. (1976). Pigment changes and photosynthetic rates under selected wavelengths in the growing tips of *Euchaema isiforme* (C. Agardh) J. Agardh var *denudatum* Cheney during vegetative growth. *Br. Phycol. J.*, **11**, 165–74. *411*
629. Mopper, K., Xianling, Z., Kieber, R. J., Kieber, D. J., Sikorski, R. J. & Jones, R. D. (1991). Photochemical degradation of dissolved organic carbon and its impact on the oceanic carbon cycle. *Nature*, **353**, 60–2. *72, 79*
630. Morel, A. (1966). Etude expérimentale de la diffusion de la lumière par l'eau, les solutions de chlorure de sodium et l'eau de mer optiquement pures. *J. Chim. Phys.*, **10**, 1359–66. *95*
631. Morel, A. (1973). Diffusion de la lumière par les eaux de mer. Résultats expérimentaux et approche théorique. In *Optics of the sea* (pp. 3.1–1 to 3.1–76). Neuilly-sur-Seine: NATO. *100, 102, 108*
632. Morel, A. (1974). Optical properties of pure water and pure seawater. In N. G. Jerlov & E. S. Nielsen (eds), *Optical aspects of oceanography* (pp. 1–24). London: Academic Press. *56, 100, 101, 105*
633. Morel, A. (1978). Available, usable and stored radiant energy in

relation to marine photosynthesis. *Deep-sea Res.*, **25**, 673–88. *266, 293, 297, 300, 303, 305*

634. Morel, A. (1980). In-water and remote measurement of ocean colour. *Boundary-layer Meteorol.*, **18**, 177–201. *190, 197*

635. Morel, A. (1982). Optical properties and radiant energy in the waters of the Guinea dome and of the Mauritanian upwelling area: relation to primary production. *Rapp. P. V. Reun. Cons int. explor. Mer.*, **180**, 94–107. *100, 101, 130, 132, 138*

636. Morel, A. (1987). Chlorophyll-specific scattering coefficient of phytoplankton. A simplified theoretical approach. *Deep-Sea Res.*, **34**, 1093–105. *110, 111*

637. Morel, A. (1991). Light and marine photosynthesis: a spectral model with geochemical and climatological implications. *Prog. Oceanog.*, **26**, 263–306. *293, 295, 354, 355–6*

638. Morel, A. & Berthon, J.-F. (1989). Surface pigments, algal biomass profiles, and potential production of the euphotic layer: relationships reinvestigated in view of remote-sensing applications. *Limnol. Oceanogr.*, **34**, 1545–62. *355*

639. Morel, A. & Bricaud, A. (1981). Theoretical results concerning light absorption in a discrete medium, and application to specific absorption of phytoplankton. *Deep-Sea Res.*, **28**, 1375–93. *255, 257, 259, 261*

640. Morel, A. & Bricaud, A. (1986). Inherent optical properties of algal cells including picoplankton: Theoretical and experimental results. In T. Platt & W. K. W. Li (eds), *Photosynthetic picoplankton* (pp. 521–59). *Can. Bull. Fish Aquat. Sci.*, **214**. *110*

641. Morel, A. & Gentili, B. (1991). Diffuse reflectance of oceanic waters: its dependence on sun angle as influenced by the molecular scattering contribution. *Appl. Opt.*, **30**, 4427–38. *146*

642. Morel, A., Lazzara, L. & Gostan, J. (1987). Growth rate and quantum yield time response for a diatom to changing irradiances (energy and colour). *Limnol. Oceanogr.*, **32**, 1066–84. *259*

643. Morel, A. & Prieur, L. (1975). *Analyse spectrale des coefficients d'attenuation diffuse, de reflection diffuse, d'absorption et de retrodiffusion pour diverses régions marines.* Rep. No. 17. Laboratoire d'Oceanographie Physique, Villefranche-sur-Mer. 157pp. Quoted by Jerlov (1976). *82*

644. Morel, A. & Prieur, L. (1977). Analysis of variations in ocean colour. *Limnol. Oceanogr.*, **22**, 709–22. *55, 56, 77, 78, 81, 108, 146, 147, 148, 197, 200*

645. Morel, A. & Smith, R. C. (1974). Relation between total quanta and total energy for aquatic photosynthesis. *Limnol. Oceanogr.*, **19**, 591–600. *5, 294*

646. Morris, I. & Farrell, K. (1971). Photosynthetic rates, gross patterns of carbon dioxide assimilation and activities of ribulose diphosphate carboxylase in marine algae grown at different temperatures. *Physiol. Plant*, **25**, 372–7. *336*

647. Morrow, J. H., Chamberlin, W. S. & Kiefer, D. A. (1989). A two-component description of spectral absorption by marine particles. *Limnol. Oceanogr.*, **34**, 1500–9. *73, 76*

648. Munday, J. C. & Alföldi, T. T. (1979). Landsat test of diffuse reflectance models for aquatic suspended solids measurement. *Remote Sens. Env.*, **8**, 169–83. *192*

649. Muñoz, J. & Merrett, M. J. (1988). Inorganic carbon uptake by a small-celled strain of *Stichococcus bacillaris*. *Planta*, **175**, 460–4. *185, 191, 325, 327, 332*

650. Muscatine, L. (1990). The role of symbiotic algae in carbon and energy flux in reef corals. In Z. Dubinsky (ed.), *Coral Reefs* (p. 75–87). Amsterdam: Elsevier. *429*

651. Myers, J. & Graham, J.-R. (1963). Enhancement in *Chlorella*. *Plant Physiol.*, **38**, 105–16. *310*

652. Myers, J. & Graham, J.-R. (1971). The photosynthetic unit in *Chlorella* measured by repetitive short flashes. *Plant Physiol.*, **48**, 282–6. *377*

653. Nakamura, K., Ogawa, T. & Shibata, K. (1976). Chlorophyll and peptide compositions in the two photosystems of marine green algae. *Biochim. Biophys. Acta*, **423**, 227–36. *228*

654. Nakayama, N., Itagaki, T. & Okada, M. (1986). Pigment composition of chlorophyll-protein complexes isolated from the green alga *Bryopsis maxima*. *Plant Cell Physiol.*, **27**, 311–17. *237*

654*a*. NASA (1987). *HIRIS Instrument Panel Report*. *173, 177*

655. Neale, P. J. (1987). Algal photoinhibition and photosynthesis in the aquatic environment. In D. J. Kyle, C. B. Osmond & C. J. Arntzen (eds), *Photoinhibition* (pp. 39–65). Amsterdam: Elsevier. *288*

656. Neale, P. J. & Melis, A. (1986). Algal photosynthetic membrane complexes and the photosynthesis-irradiance curve: a comparison of light-adaptation responses in *Chlamydomonas reinhardtii* (Chlorophyta). *J. Phycol.*, **22**, 531–8. *377, 379, 380*

657. Nelson, N. B. & Prézelin, B. B. (1990). Chromatic light effects and physiological modelling of absorption properties of *Heterocapsa pygmaea* (= *Glenodinium* sp.). *Mar. Ecol. Prog. Ser.*, **63**, 37–46. *259*

658. Neori, A., Holm-Hansen, O., Mitchell, B. G. & Kiefer, D. A. (1984). Photoadaptation in marine phytoplankton. *Plant Physiol.*, **76**, 518–24. *401*

659. Neuymin. G. G., Zemlyanya, L. A., Martynov, O. V. & Solov'yev, M. V. (1982). Estimation of the chloropyll concentration from measurements of the colour index in different regions of the ocean. *Oceanology*, **22**, 280–3. *174–5*

660. Neverauskas, V. P. (1988). Response of a *Posidonia* community to prolonged reduction in light. *Aquat. Bot.*, **31**, 361–6. *384*

661. Neville, R. A. & Gower, J. F. R. (1977). Passive remote sensing of phytoplankton via chlorophyll *a* fluorescence. *J. Geophys. Res.*, **82**, 3487–93. *175, 201*

662. Nolen, S. L., Wilhm, J. & Howick, G. (1985). Factors influencing inorganic turbidity in a great plains reservoir. *Hydrobiologia*, **123**, 109–17. *105*

663. Norton, T. A. (1968). Underwater observations on the vertical distribution of algae at St. Mary's, Isles of Scilly. *Br. Phycol. Bull.*, **3**, 585–8. *364*

664. Norton, T. A., Hiscock, K. & Kitching, J. A. (1977). The ecology of Lough Ine. XX. The Laminaria forest at Carrigathorna. *J. Ecol.*, **65**, 919–41. *363, 365*

665. Norton, T. A., McAllister, H. A., Conway, E. & Irvine, L. M. (1969). The marine algae of the Hebridean island of Colonsay. *Br. Phycol. J.*, **4**, 125–36. *364*

666. Novo, E. M. M., Hanson, J. D. & Curran, P. J. (1989). The effect of sediment type on the relationship between reflectance and suspended sediment concentration. *Int. J. Remote Sens.* **10**, 1283–9. *193*
667. Nultsch, W. & Pfau, J. (1979). Occurrence and biological role of light-induced chromatophore displacements in seaweeds. *Mar. Biol.*, **51**, 77–82. *421–3*
668. O'Carra, C. & O'h Eocha, C. (1976). Algal biliproteins and phycobilins. In T. W. Goodwin (ed.), *Chemistry and biochemistry of plant pigments* (pp. 328–76). London: Academic Press. *239, 240*
669. Oelmüller, R., Coxley, P. B., Federspiel, N., Briggs, W. R. & Grossman, A. R. (1988). Changes in accumulation and synthesis of transcripts encoding phycobilisome components during acclimation of *Fremyella diplosiphon* to different light qualities. *Plant Physiol.*, **88**, 1077–83. *387*
670. Oertel, G. F. & Dunstan, W. M. (1981). Suspended-sediment distribution and certain aspects of phytoplankton production off Georgia, U.S.A. *Marine Geol.*, **40**, 171–97. *139, 207, 332*
671. Ogura, N. & Hanya, T. (1966). Nature of ultraviolet absorption of sea water. *Nature*, **212**, 758. *57*
672. O'h Eocha, C. (1965). Phycobilins. In T. W. Goodwin (ed.), *Chemistry and biochemistry of plant pigments* (pp. 175–96). London: Academic Press. *241*
673. O'h Eocha, C. (1966). Biliproteins. In T. W. Goodwin (ed.), *Biochemistry of chloroplasts* (pp. 407–21). London: Academic Press. *240*
674. Ohki, K., Gantt, E., Lipschultz, C. A. & Ernst, M. C. (1985). Constant phycobilisome size in chromatically adapted cells of the cyanobacterium *Tolypothrix tenuis*, and variation in *Nostoc* sp. *Plant Physiol.*, **79**, 943–8. *385*
675. Oishi, T. (1990). Significant relationship between the backward scattering coefficient of sea water and the scatterance at 120°. *Appl. Opt.*, *29*, 4658–65. *95*
676. O'Kelly, C. J. (1982). Chloroplast pigments in selected marine Chaetophoraceae and Chaetosiphonaceae (Chlorophyta): the occurrence and significance of siphonaxanthin. *Bot. Mar.*, **25**, 133–7. *231, 368*
677. Oliver, R. L. (1990). Optical properties of waters in the Murray–Darling basin, South-eastern Australia. *Aust J. Mar. Freshwater Res.*, **41**, 581–601. *69, 98, 103, 110, 142*
678. Oltmanns, F. (1892). Uber die Kultur und Lebensbedingungen der Meeresalgen. *Jb. Wiss. Bot.*, **23**, 349–440. *361*
679. Ondrusek, M. E., Bidigare, R. R., Sweet, S. T., Defreitas, D. A. & Brooks, J. M. (1991). Distribution of phytoplankton pigments in the North Pacific Ocean in relation to physical and optical variability. *Deep-Sea Res.*, **38**, 243–66. *369*
680. Ong, L. J., Glazer, A. N. & Waterbury, J. B. (1984). An unusual phycoerythrin from a marine cyanobacterium. *Science*, **224**, 80–3. *240*
681. Ortner, P. B., Wiebe, P. H. & Cox, J. L. (1980). Relationships between oceanic epizooplankton distribution and the seasonal deep chlorophyll maximum in the northwestern Atlantic Ocean. *J. Mar. Res.*, **3**, 507–31. *320, 405*

682. Paasche, E. (1964). A tracer study of the inorganic carbon uptake during coccolith formation and photosynthesis in the coccolithophorid *Coccolithus huxleyi*. *Physiol. Plant.*, **3** (suppl.), 1–82. *327*

683. Paerl, H. W., Tucker, J. & Bland, P. T. (1983). Carotenoid enhancement and its role in maintaining blue-green algal (*Microcystis aeruginosa*) surface blooms. *Limnol. Oceanogr.*, **28**, 847–57. *377*

684. Paerl, H. W. & Ustach, J. F. (1982). Blue-green algal scums: an explanation for their occurrence during freshwater blooms. *Limnol. Oceanogr.*, **27**, 212–17. *419*

685. Palmer, K. F. & Williams, D. (1974). Optical properties of water in the near infrared. *J. Opt. Soc Amer.*, **64**, 1107–10. *57*

686. Palmisano, A. C., SooHoo, J. B. & Sullivan, C. W. (1985). Photosynthesis–irradiance relationships in sea ice microalgae from McMurdo Sound, Antarctica. *J. Phycol.*, **21**, 341–6. *280, 410*

687. Parson, W. W. (1991). Reaction Centres. In H. Scheer (ed.), *Chlorophylls* (pp. 1153–80). Boca Raton: CRC Press. *246*

688. Patel, B. N. & Merrett, M. J. (1986). Inorganic carbon uptake by the marine diatom *Phaeodactylum tricornutum*. *Planta*, **169**, 222–7. *332*

689. Pearse, J. F. & Hines, A. H. (1979). Expansion of a Central California kelp forest following the mass mortality of sea urchins. *Mar. Biol.*, **51**, 83–91. *342*

689*a*. Pegau, W. S. & Zaneveld, J. R. V. (1993). Temperature-dependent absorption of water in the red and near-infrared portions of the spectrum. *Limnol. Oceanogr.*, **38**, 188–92. *57*

690. Pelaez, J. & McGowan, J. A. (1986). Phytoplankton pigment patterns in the California Current as determined by satellite. *Limnol. Oceanogr.*, **31**, 927–50. *212*

691. Pelevin, V. N. (1978). Estimation of the concentration of suspension and chlorophyll in the sea from the spectrum of outgoing radiation measured from a helicopter. *Oceanology*, **18**, 278–82. *190*

692. Pelevin, V. N. & Rutkovskaya, V. A. (1977). On the optical classification of ocean waters from the spectral attenuation of solar radiation. *Oceanology*, **17**, 28–32. *79, 80*

693. Pennock, J. R. (1985). Chlorophyll distribution in the Delaware estuary: regulation by light limitation. *Estuar. Coast. Shelf Sci.*, **21**, 711–25. *318, 321*

694. Perez-Bermudez, P., Garcia-Carrascosa, M., Cornejo, M. J. & Segura, J. (1981). Water-depth effects in photosynthetic pigment content of the benthic algae *Dictyota dichotoma* and *Udotea petiolata*. *Aquat. Bot.*, **11**, 373–7. *395, 409*

695. Perkins, E. J. (1960). The diurnal rhythm of the littoral diatoms of the river Eden estuary, Fife. *J. Ecol.*, **48**, 725–8. *421*

696. Perry, M. J. & Porter, S. M. (1989). Determination of the cross-section absorption coefficient of individual phytoplankton cells by analytical flow cytometry. *Limnol. Oceanogr.*, **34**, 1727–38. *111*

697. Perry, M. J., Talbot, M. C. & Alberte, R. S. (1981). Photoadaptation in marine phytoplankton: response of the photosynthetic unit. *Mar. Biol.*, **62**, 91–101. *378, 382*

698. Peterson, D. H., Perry M. J., Bencala, K. E. & Talbot, M. C. (1987). Phytoplankton productivity in relation to light intensity: a simple

equation. *Estuar. Coast. Shelf Sci.*, **24**, 813–32. *283*

699. Petzold, T. J. (1972). *Volume scattering functions for selected ocean waters* No. Ref. 72–78, 79pp. Scripps Inst. Oceanogr. *93, 94, 97, 101, 106–7, 160*
700. Phillips, D. M. & Kirk, J. T. O. (1984). Study of the spectral variation of absorption and scattering in some Australian coastal waters. *Aust. J. Mar. Freshwater Res.*, **35**, 635–44. *92, 102*
701. Phillips, G. L., Eminson, D. & Moss, B. (1978). A mechanism to account for macrophyte decline in progressively eutrophicated freshwaters. *Aquat. Bot.*, **4**, 103–26. *341*
702. Pick, F. R. (1991). The abundance and composition of freshwater picocyanobacteria in relation to light penetration. *Limnol. Oceanogr.*, **36**, 1457–62. *369*
703. Plass, G. N. & Kattawar, G. W. (1972). Monte Carlo calculations of radiative transfer in the Earth's atmosphere–ocean system. I. Flux in the atmosphere and ocean. *J. Phys. Oceanogr.*, **2**, 139–45. *126*
704. Platt, T. (1969). The concept of energy efficiency in primary production. *Limnol. Oceanogr.*, **14**, 653–9. *304*
705. Platt, T. (1986). Primary production of the ocean water column as a function of surface light intensity: algorithms for remote sensing. *Deep-Sea Res.*, **33**, 149–63. *354*
706. Platt, T., Gallegos, C. L. & Harrison, W. G. (1980). The relationship between photosynthesis and light for natural assemblages of coastal marine phytoplankton. *J. Mar. Res.*, **38**, 687–701. *283*
707. Platt, T. & Jassby, A. D. (1976). The relationship between photosynthesis and light for natural assemblages of coastal marine phytoplankton. *J. Phycol.*, **12**, 421–30. *279, 302, 337, 412, 415*
708. Platt, T. & Li, W. K. W. (eds) (1986). Photosynthetic picoplankton. *Can. Bull. Fish. Aquat. Sci.*, **214**. *314*
709. Platt, T. & Sathyendranath, S. (1988). Oceanic primary production: estimation by remote sensing at local and regional scales. *Science*, **241**, 1613–20. *354, 358*
710. Platt, T., Sathyendranath, S. & Ravindran, P. (1990). Primary production by phytoplankton: analytic solutions for daily rates per unit area of water surface. *Proc. R. Soc. Lond. B*, **241**, 101–11. *356*
711. Platt, T., Sathyendranath, S., Ulloa, O., Harrison, W. G., Hoepffner, N. & Goes, J. (1992). Nutrient control of phytoplankton photosynthesis in the western North Atlantic. *Nature*, pp. 229–31. *354*
712. Platt, T. & Subba Rao, D. V. (1975). Primary production of marine microphytes. In J. P. Cooper (ed.), *Photosynthesis and productivity in different environments* (pp. 249–80). Cambridge: Cambridge University Press. *359*
713. Poole, H. H. (1945). The angular distribution of submarine daylight in deep water. *Sci. Proc. R. Dublin Soc.*, **24**, 29–42. *154*
714. Poole, H. H. & Atkins, W. R. G. (1929). Photoelectric measurements of submarine illumination throughout the year. *J. Mar. Biol. Ass. U.K.*, **16**, 297–324. *119*
715. Preisendorfer, R. W. (1957). Exact reflectance under a cardioidal luminance distribution. *Q. J. Roy. Meteorol. Soc.*, **83**, 540. Quoted by Jerlov (1976). *44*

716. Preisendorfer, R. W. (1959). Theoretical proof of the existence of characteristic diffuse light in natural waters. *J. Mar. Res.*, **18**, 1–9. *154, 155*

717. Preisendorfer, R. W. (1961). Application of radiative transfer theory to light measurements in the sea. *Union Geod. Geophys. Inst. Monogr.*, **10**, 11–30. *11, 12, 19, 21, 24, 98, 103, 152, 154, 166*

718. Preisendorfer, R. W. (1976). *Hydrologic optics*. Washington: U.S. Department of Commerce. *1, 6*

719. Preisendorfer, R. W. (1986*a*). *Eyeball optics of natural waters: Secchi disk science* (Tech. Memo. No. ERL PMEL-67, 90p). NOAA. *120*

720. Preisendorfer, R. W. (1986*b*). Secchi disk science: visual optics of natural waters. *Limnol. Oceanogr.*, **31**, 909–26. *120*

721. Prézelin, B. B. (1976). The role of peridin–chlorophyll *a* proteins in the photosynthetic light adaptation of the marine dinoflagellate, *Glenodinium* sp. *Planta*, **130**, 225–33. *376, 378, 380*

722. Prézelin, B. B. (1992). Diel periodicity in phytoplankton productivity. *Hydrobiologia*, **238**, 1–35. *423*

723. Prézelin, B. B., Bidigare, R. R., Matlick, H. A., Putt, M. & Hoven, B. V. (1987). Diurnal patterns of size-fractioned primary productivity across a coastal front. *Mar. Biol.*, **96**, 563–74. *302*

724. Prézelin, B. B., Ley, A. C. & Haxo, F. T. (1976). Effects of growth irradiance on the photosynthetic action spectra of the marine dinoflagellate, *Glenodinium* sp. *Planta*, **130**, 251–6. *407*

725. Prézelin, B. B., Meeson, B. W. & Sweeney, B. M. (1977). Characterization of photosynthetic rhythms in marine dinoflagellates. I. Pigmentation, photosynthetic capacity and respiration. *Plant Physiol.*, **60**, 384–7. *425*

726. Prézelin, B. B., Putt, M. & Glover, H. E. (1986). Diurnal patterns in photosynthetic capacity and depth-dependent photosynthesis-irradiance relationships in *Synechococcus* spp. and larger phytoplankton in three water masses in the Northwest Atlantic Ocean. *Mar. Biol.*, **91**, 205–17. *423*

727. Prézelin, B. B. & Sweeney, B. M. (1977). Characterization of photosynthetic rhythms in marine dinoflagellates. II. Photosynthesis-irradiance curves and *in vivo* chlorophyll fluorescence. *Plant Physiol.*, **60**, 388–92. *425–6*

728. Prézelin, B. B., Tilzer, M. M., Schofield, O. & Haese, C. (1991). The control of the production process of phytoplankton by the physical structure of the aquatic environment with special reference to its optical properties. *Aquatic Sciences*, **53**, 136–86. *302, 304*

729. Prieur, L. (1976). *Transfer radiatif dans les eaux de mer*. D.Sc., Univ. Pierre et Marie Curie, Paris. *146, 147*

730. Prieur, L. & Morel, A. (1971). Etude théorique du régime asymptotique: relations entre caractéristiques optiques et coefficient d'extinction relatif à la pénétration de la lumière du jour. *Cah. Oceanogr.* **23**, 35–47. *154*

731. Prieur, L. & Sathyendranath, S. (1981). An optical classification of coastal and oceanic waters based on the specific spectral absorption curves of phytoplankton pigments, dissolved organic matter, and other particulate materials. *Limnol. Oceanogr.*, **26**, 671–89. *82*

732. Prins, H. B. A. & Elzenga, J. T. M. (1989). Bicarbonate utilization: function and mechanism. *Aquat. Bot.*, **34**, 59–83. *326, 327*

733. Proctor, L. M. & Fuhrman, J. A. (1990). Viral mortality of marine bacteria and cyanobacteria. *Nature*, **343**, 60–2. *341*

734. Quickenden, T. I. & Irvin, J. A. (1980). The ultraviolet absorption spectrum of liquid water. *J. Chem. Phys.*, **72**, 4416–28. *55, 56*

735. Rabinowitch, E. I. (1945). *Photosynthesis, Vol I.* New York: Interscience. *361*

736. Rabinowitch, E. I. (1951). *Photosynthesis, Vol. II.1.* New York: Interscience. *326*

737. Radmer, R. & Kok, B. (1977). Photosynthesis: limited yields, unlimited dreams. *BioScience*, **27**, 599–605. *294*

738. Ramus, J., Beale, S. I. & Mauzerall, D. (1976). Correlation of changes in pigment content with photosynthetic capacity of seaweeds as a function of depth. *Mar. Biol.*, **37**, 231–8. *277, 281, 282, 372–3, 392–4*

739. Ramus, J., Beale, S. I., Mauzerall, D. & Howard, K. L. (1977). Changes in photosynthetic pigment concentration in seaweeds as a function of depth. *Mar. Biol.*, **37**, 223–9. *394*

740. Ramus, J., Lemons, F. & Zimmerman, C. (1977). Adaptation of light-harvesting pigments to downwelling light and the consequent photosynthetic performance of the eulittoral rockweeds *Ascophyllum nodosum* and *Fucus vesiculosus. Mar. Biol.*, **42**, 293–303. *394*

741. Ramus, J. & Rosenberg, G. (1980). Diurnal photosynthetic performance of seaweeds measured under natural conditions. *Mar. Biol.*, **56**, 21–8. *427*

742. Raps, S., Kycia, J. H., Ledbetter, M. C. & Siegelman, H. W. (1985). Light intensity adaptation and phycobilisome composition of *Microcystis aeruginosa. Plant Physiol.*, **79**, 983–7. *379*

743. Raps, S., Wyman, K., Siegelman, H. W. & Falkowski, P. G. (1983). Adaptation of the cyanobacterium *Microcystis aeruginosa* to light intensity. *Plant Physiol.*, **72**, 829–32. *379, 382*

744. Rattray, M. R., Howard-Williams, C. & Brown, J. M. A. (1991). The photosynthetic and growth rate responses of two freshwater angiosperms in lakes of different trophic status: responses to light and dissolved inorganic carbon. *Freshwater Biol.*, **25**, 399–407. *329*

745. Raven, J. A. (1970). Endogenous inorganic carbon sources in plant photosynthesis. *Biol. Rev.*, **45**, 167–221. *327, 331*

746. Raven, J. A., Osborne, B. A. & Johnston, A. M. (1985). Uptake of CO_2 by aquatic vegetation. *Plant Cell Environ.*, **8**, 417–25. *250, 326, 330, 332, 333*

747. Ravisankar, M., Reghunath, A. T., Sathiandanan, K. & Nampoori, V. P. N. (1988). Effect of dissolved NaCl, $MgCl_2$ and Na_2SO_4 in seawater on the optical attenuation in the region from 430 to 630 nm. *Appl. Opt.*, **27**, 3387–94. *57, 65*

748. Raymont, J. E. (1980). *Plankton and productivity in the oceans, vol 1. Phytoplankton* (2nd edn). Oxford: Pergamon. *218, 339, 340, 343, 358, 412*

749. Reiskind, J. B., Seamon, P. T. & Bowes, G. (1988). Alternative methods of photosynthetic carbon assimilation in marine macroalgae. *Plant Physiol.*, **87**, 686–92. *333*

750. Reynolds, C. S. (1984). *The ecology of freshwater phytoplankton*. Cambridge: Cambridge University Press. *314, 339, 340, 341, 345, 413*

751. Reynolds, C. S. & Walsby, A. E. (1975). Water blooms. *Biol. Rev.*, **50**, 437–81. *417*

752. Rhee, C. & Briggs, W. R. (1977). Some responses of *Chondrus crispus* to light. I. Pigmentation changes in the natural habitat. *Bot. Gaz.*, **138**, 123–8. *376, 395, 409, 411*

752*a*. Richardson, K., Beardall, J. & Raven, J. A. (1983). Adaptation of unicellular algae to irradiance: an analysis of strategies. *New Phytol.*, **93**, 157–91. *375*

753. Riebesell, U., Wolf-Gladrow, D. A. & Smetacek, V. (1993). Carbon dioxide limitation of marine phytoplankton growth rates. *Nature*, **361**, 249–51. *327, 330*

754. Ried, A., Hessenberg, B., Metzler, H. & Ziegler, R. (1977). Distribution of excitation energy amongst Photosystem I and Photosystem II in red algae. *Biochim. Biophys. Acta*, **459**, 175–86. *312*

755. Riegmann, F. & Colijn, F. (1991). Evaluation of measurements and calculation of primary production in the Dogger Bank area (North Sea) in summer 1988. *Mar. Ecol. Prog. Ser*, **69**, 125–32. *138*

756. Riley, G. A. (1942). The relationship of vertical turbulence and spring diatom flowerings. *J. Mar. Res.*, **5**, 67–87. *343*

757. Rimmer, J. C., Collins, M. B. & Pattiaratchi, C. B. (1987). Mapping of water quality in coastal waters using Airborne Thematic Mapper data. *Int. J. Remote Sens.*, **8**, 85–102. *209*

758. Ritchie, J. C. & Cooper, C. M. (1988). Comparison of measured suspended sediment concentrations with suspended sediment concentrations estimated from Landsat MSS data. *Int. J. Remote Sens.*, **9**, 379–87. *191, 192*

759. Ritchie, J. C. & Cooper, C. M. (1991). An algorithm for estimating surface suspended sediment concentrations with Landsat MSS digital data. *Water Res. Bull.*, **27**, 373–9. *192*

760. Ritchie, J. C., Cooper, C. M. & Schiebe, F. R. (1990). The relationship of MSS and TM digital data with suspended sediments, chlorophyll and temperature in Moon Lake, Mississippi. *Remote Sens. Environ.*, **33**, 137–48. *184, 193*

761. Ritchie, J. C., Cooper, C. M. & Yongqing, J. (1987). Using Landsat multispectral scanner data to estimate suspended sediments in Moon Lake, Mississippi. *Remote Sens. Environ.*, **23**, 65–81. *184, 191*

762. Ritchie, J. C., Schiebe, F. R. & McHenry, J. R. (1976). Remote sensing of suspended sediments in surface waters. *Photogramm. Eng. Rem. Sens.*, **42**, 1539–45. *191, 192*

763. Rivkin, R. B. & Putt, M. (1987). Photosynthesis and cell division by Antarctic microalgae: comparison of benthic, planktonic and ice algae. *J. Phycol.*, **23**, 233–9. *410*

764. Robarts, R. D. & Zohary, T. (1984). *Microcystis aeruginosa* and underwater light attenuation in a hypertrophic lake (Hartbeespoort Dam, South Africa). *J. Ecol.*, **72**, 1001–17. *259*

765. Robinson, N. (1966). *Solar radiation*. Amsterdam: Elsevier. *29*

766. Rochon, G. & Langham, E. J. (1974). Teledetection par satellite dans l'evaluation de la qualité de l'eau. *Verh. Int. Verein. Limnol.*, **19**, 189–96. *184*

767. Rodhe, W. (1969). Crystallization of eutrophication concepts in Northern Europe. In *Eutrophication: causes, consequences, correctives* (pp. 50–64). Washington: National Academy of Science. *359*

768. Roemer, S. C. & Hoagland, K. D. (1979). Seasonal attenuation of quantum irradiance (400–700 nm) in three Nebraska reservoirs. *Hydrobiologia*, **63**, 81–92. *140*

769. Roesler, C. S., Perry, M. J. & Carder, K. L. (1989). Modelling *in situ* phytoplankton absorption from total absorption spectra in productive inland marine waters. *Limnol. Oceanogr.*, **34**, 1510–23. *73*

770. Rotatore, C. & Colman, B. (1992). Active uptake of CO_2 by the diatom *Navicula pelliculosa*. *J. Exp. Bot.*, **43**, 571–6. *332*

771. Round, F. E. (1981). *The ecology of algae.* Cambridge: Cambridge University Press. *420, 421*

772. Round, F. E. & Happey, C. M. (1965). Persistent vertical-migration rhythms in benthic microflora. IV. A diurnal rhythm of the epipelic diatom association in non-tidal flowing water. *Br. Phycol. Bull.*, **2**, 465–71. *420*

773. Round, F. E. & Palmer, J. D. (1966). Persistent vertical-migration rhythms in benthic microflora. II. Field and laboratory studies on diatoms from the banks of the River Avon. *J. Mar. Biol. Ass. U.K.*, **46**, 191–214. *420–1*

774. Rouse, L. J. & Coleman, J. M. (1976). Circulation observations in the Louisiana Bight using Landsat imagery. *Remote Sens. Env.*, **5**, 55–66. *207*

775. Rowan, K. S. (1989). *Photosynthetic pigments of algae*. Cambridge: Cambridge University Press. *225, 231, 235, 239*

776. Ryther, J. H. (1956). Photosynthesis in the ocean as a function of light intensity. *Limnol. Oceanogr.*, **1**, 61–70. *343, 352–4*

777. Ryther, J. H. (1969). Photosynthesis and fish production in the sea. *Science*, **166**, 72–6. *340*

778. Ryther, J. H. & Menzel, D. (1959). Light adaptation by marine phytoplankton. *Limnol. Oceanogr.*, **4**, 492–7. *273, 274, 400, 403*

779. Ryther, J. H. & Yentsch, C. S. (1957). The estimation of phytoplankton production in the ocean from chlorophyll and light data. *Limnol. Oceanogr.*, **2**, 281–6. *353*

780. Sakamoto, M. (1966). Primary production by phytoplankton community in some Japanese lakes and its dependence on lake depth. *Arch. Hydrobiol.*, **62**, 1–28. *315*

781. Sakshaug, E. & Holm-Hansen, O. (1984). Factors governing pelagic production in polar seas. In O. Holm-Hansen, L. Bolis & R. Gilles (eds). *Marine phytoplankton and productivity* (p. 1–18). New York: Springer. *359*

782. Salonen, K., Jones, R. I. & Arvola, L. (1984). Hypolimnetic phosphorus retrieval by diel vertical migrations of lake phytoplankton. *Freshwater Biol.*, **14**, 431–8. *417*

783. Salvucci, M. E. & Bowes, G. (1982). Photosynthetic and photorespiratory responses of the aerial and submerged leaves of *Myriophyllum brasiliense*. *Aquat. Bot.*, **13**, 147–64. *279*

784. Salvucci, M. E. & Bowes, G. (1983). Two photosynthetic mechanisms mediating the low photorespiratory state in submersed aquatic angiosperms. *Plant Physiol.*, **73**, 488–96. *332*

785. Sand-Jensen, K. & Madsen, T. V. (1991). Minimum light requirements of submerged freshwater macrophytes in laboratory growth experiments. *J. Ecol.*, **79**, 749–64. *316*
786. San Pietro, A. (1971, 1972, 1980). *Methods in enzymology*. New York: Academic Press. *271*
787. Sasaki, T., Watanabe, S., Oshiba, G., Okami, N. & Kajihara, M. (1962). On the instrument for measuring angular distribution of underwater radiance. *Bull. Jap. Soc. Sci. Fish.*, **28**, 489–96. *124*
788. Sathyendranath, S., Gouveia, A. D., Shetye, S. R., Ravindran, P. & Platt, T. (1991). Biological control of surface temperature in the Arabian Sea. *Nature*, **349**, 54–6. *213*
789. Sathyendranath, S., Lazzara, L. & Prieur, L. (1987). Variations in the spectral values of specific absorption of phytoplankton. *Limnol. Oceanogr.*, **32**, 403–15. *259, 271*
790. Sathyendranath, S., Platt, T., Horne, E. P. W., Harrison, W. G., Ulloa, O., Outerbridge, R. & Hoepffner, N. (1991). Estimation of new production in the ocean by compound remote sensing. *Nature*, **353**, 129–33. *357*
791. Sathyendranath, S., Prieur, L. & Morel, A. (1989). A three-component model of ocean colour and its application to remote sensing of phytoplankton pigments in coastal waters. *Int. J. Remote Sens.*, **10**, 1373–94. *198*
792. Sauberer, F. (1945). Beiträge zur Kenntnis der optischen Eigenschaften der Kärntner Seen. *Arch. Hydrobiol.*, **41**, 259–314. *59*
793. Savastano, K. J., Faller, K. H. & Iverson, R. L. (1984). Estimating vegetation coverage in St. Joseph's Bay, Florida with an airborne multispectral scanner. *Photogramm. Eng. Remote Sens.*, **1984**, 1159–70. *211*
794. Savidge, G. (1988). Influence of inter- and intra-daily light-field variability on photosynthesis by coastal phytoplankton. *Mar. Biol.*, **100**, 127–33. *415*
795. Schanz, F. (1986). Depth distribution of phytoplankton and associated spectral changes in downward irradiance in Lake Zürich (1980/81). *Hydrobiologia*, **134**, 183–92. *141, 269*
796. Scheibe, J. (1972). Photoreversible pigment: occurrence in a blue-green alga. *Science*, **176**, 1037–9. *386–7*
797. Scherz, J. P., van Domelen, J. F., Holtje, K. & Johnson, W. (1974). Lake eutrophication as indicated by ERTS satellite imagery. In *Symposium on remote sensing and photo interpretation*, (pp. 247–58). Canada: International Society for Photogrammetry. *184, 207*
798. Schiebe, F. R., Harrington, J. A. & Ritchie, J. C. (1992). Remote sensing of suspended sediments: the Lake Chicot, Arkansas project. *Int. J. Remote Sens.*, **13**, 1487–509. *191*
799. Schmitz-Peiffer, A., Viehoff, T. & Grassl, H. (1990). Remote sensing of coastal waters by airborne lidar and satellite radiometer. Part 2: Measurements. *Int. J. Remote Sens.*, **11**, 2185–204. *204, 209*
800. Schnitzer, M. (1978). Humic substances: chemistry and reactions. In M. Schnitzer & S. U. Khan (eds), *Soil organic matter* (pp. 1–64). Amsterdam: Elsevier. *57*
801. Schofield, O., Prézelin, B. B., Smith, R. C., Stegmann, P. M., Nelson, N. B., Lewis, M. R. & Baker, K. S. (1991). Variability in spectral and nonspectral measurements of photosynthetic light utilization

efficiencies. *Mar. Ecol. Prog. Ser.*, **78**, 253–71. *297, 302*

802. Schulten, H.-R., Plage, B. & Schnitzer, M. (1991). A chemical structure for humic substances. *Naturwiss.*, **78**, 311–12. *58*

803. Scott, B. D. (1978). Phytoplankton distribution and light attenuation in Port Hacking estuary. *Aust. J. Mar. Freshwater Res.*, **29**, 31–44. *139*

804. Scribner, E. A. (1985). Unpublished data (personal communication). *139, 142*

805. Sears, J. R. & Cooper, R. A. (1978). Descriptive ecology of offshore, deep-water, benthic algae in the temperate western North Atlantic Ocean. *Mar. Biol.*, **44**, 309–14. *374*

806. Senger, H. & Fleischhacker, P. (1978). Adaptation of the photosynthetic apparatus of *Scenedesmus obliquus* to strong and weak light conditions. *Physiol. Plant.*, **43**, 35–42. *379, 381*

807. Shepherd, S. A. & Sprigg, R. C. (1976). Substrate sediments and subtidal ecology of Gulf St. Vincent and Investigator Strait. In C. R. Twidale, M. J. Tyler & B. P. Webb (eds), *Natural History of the Adelaide region* (pp. 161–74). Adelaide: Royal Society of South Australia. *369–70, 375*

808. Shepherd, S. A. & Womersley, H. B. S. (1970). The sublittoral ecology of West Island, South Australia: environmental features and algal ecology. *Trans. R. Soc. South Aust.*, **94**, 105–38. *369–70*

809. Shepherd, S. A. & Womersley, H. B. S. (1971). Pearson Island Expedition 1969. 7. The subtidal ecology of benthic algae. *Trans R. Soc. South Aust.*, **94**, 155–67. *369–70*

810. Shepherd, S. A. & Womersley H. B. S. (1976). The subtidal algal and seagrass ecology of St. Francis Island, South Australia. *Trans. R. Soc. South Aust.*, **100**, 177–91. *369–70*

811. Sherman, K., Grasslein, M., Mountain, D., Busch, D., O'Reilly, J. & Theroux, R. (1988). The continental shelf ecosystem off the northeast coast of the United States. In H. Postma & J. J. Zijlstra (eds), *Continental shelves* (pp. 279–337). Amsterdam: Elsevier. *359*

812. Shibata, K. (1959). Spectrophotometry of translucent biological materials – opal glass transmission method. *Meth. Biochem. Anal.*, **7**, 77–109. *51*

813. Shimura, S. & Fujita, Y. (1975). Changes in the activity of fucoxanthin-excited photosynthesis in the marine diatom *Phaeodactylum tricornutum* grown under different culture conditions. *Mar. Biol.*, **33**, 185–94. *376*

814. Shimura, S. & Ichimura, S. (1973). Selective transmission of light in the ocean waters and its relation to phytoplankton photosynthesis. *J. Oceanogr. Soc. Japan.*, **29**, 257–66. Quoted by Shimura & Fujita (1975). *400–1*

815. Short, N. M. (1976). *Mission to earth: Landsat views the world.* Washington: NASA. *177*

816. Shulenberger, E. (1978). The deep chlorophyll maximum and mesoscale environmental heterogeneity in the western half of the North Pacific central gyre. *Deep-Sea Res.*, **25**, 1193–208. *320*

817. Sieburth, J. M. & Jensen, A. (1968). Studies on algal substances in the sea. I. Gelbstoff (humic material) in terrestrial and marine waters. *J. Exp. Mar. Biol. Ecol.*, **2**, 174–89. *61*

818. Sieburth, J. M. & Jensen, A. (1969). Studies on algal substances in the sea. II. The formation of gelbstoff (humic material) by exudates of Phaeophyta. *J. Exp. Mar. Biol. Ecol.*, **3**, 279–89. *61*

819. Siegel, D. A. & Dickey, T. D. (1987). On the parameterization of irradiance for open sea photoprocesses. *J. Geophys. Res.*, **92**, 14648–62. *138*

820. Siegel, D. A., Iturriaga, R., Bidigare, R. R., Smith, R. C., Pak, H., Dickey, T. D., Marra, J. & Baker, K. S. (1990). Meridional variations of the springtime phytoplankton community in the Sargasso Sea. *J. Mar. Res.*, **48**, 379–412. *369*

821. Simenstad, C. A., Estes, J. A. & Kenyon, K. W. (1978). Aleuts, sea otters, and alternative stable-state communities. *Science*, **200**, 403–11. *342*

822. Slater, P. N. (1980). *Remote sensing: optics and optical systems.* Reading: Addison-Wesley. *172*

823. Smith, E. L. (1936). Photosynthesis in relation to light and carbon dioxide. *Proc. Natl. Acad. Sci. Wash.*, **22**, 504–11. *283, 353*

824. Smith, E. L. (1938). Limiting factors in photosynthesis: light and carbon dioxide. *J. Gen. Physiol.*, **22**, 21–35. *325–6*

825. Smith, F. A. & Walker, N. A. (1980). Photosynthesis by aquatic plants: effects of unstirred layers in relation to assimilation of CO_2 and HCO_3^- and to carbon isotopic discrimination. *New Phytol.*, **86**, 245–59. *331*

826. Smith, R. C. (1968). The optical characterization of natural waters by means of an 'extinction coefficient'. *Limnol. Oceanogr.*, **13**, 423–9. *116*

827. Smith, R. C. (1969). An underwater spectral irradiance collector. *J. Mar. Res.*, **27**, 341–51. *113*

828. Smith, R. C., Austin, R. W. & Tyler, J. E. (1970). An oceanographic radiance distribution camera system. *Appl. Opt.*, **9**, 2015–22. *125*

829. Smith, R. C. & Baker, K. S. (1978*a*). The bio-optical state of ocean waters and remote sensing. *Limnol. Oceanogr.*, **23**, 247–59. *269, 355*

830. Smith, R. C. & Baker, K. S. (1978*b*). Optical classification of natural waters. *Limnol. Oceanogr.*, **23**, 260–7. *80, 83*

831. Smith, R. C. & Baker, K. S. (1980). Biologically effective dose transmitted by culture bottles in ^{14}C productivity measurements. *Limnol. Oceanogr.*, **25**, 364–6. *286*

832. Smith, R. C. & Baker, K. S. (1981). Optical properties of the clearest natural waters (200–800 nm). *Appl. Opt.*, **20**, 177–84. *55, 56, 57*

833. Smith, R. C. & Baker, K. S. (1982). Oceanic chlorophyll concentrations as determined by satellite (Nimbus-7 Coastal Zone Colour Scanner). *Mar. Biol.*, **66**, 269–79. *211–2*

834. Smith, R. C. & Baker, K. S. (1984). The analysis of ocean optical data. *Proc. Soc. Photo-Opt. Instrum. Eng., Ocean Optics VII*, **489**, 119–26. *118*

835. Smith, R. C., Baker, K. S., Holm-Hansen, O. & Olson, R. (1980). Photoinhibition of photosynthesis in natural waters. *Photochem. Photobiol.*, **31**, 585–92. *286*

835*a*. Smith, R. C., Bidigare, R. R., Prézelin, B. B., Baker, K. S. & Brooks, J. M. (1987). Optical characterization of primary productivity across a coastal front. *Mar. Biol.*, **96**, 575–91. *303*

836. Smith, R. C., Booth, C. R. & Star, J. L. (1984). Oceanographic biooptical profiling system. *Appl. Opt.*, **23**, 2791–7. *118, 123*
837. Smith, R. C. Marra, J., Perry, M. J., Baker, K. S., Swift, E., Buskey, E. & Kiefer, D. A. (1989). Estimation of a photon budget for the upper ocean in the Sargasso Sea. *Limnol. Oceanogr.*, **34**, 1673–93. *76, 138*
838. Smith, R. C., Prézelin, B. B., Baker, K. S., Bidigare, R. R., Boucher, N. P., Coley, T., Karentz, D., MacIntyre, S., Matlick, H. A., Menzies, D., Ondrusek, M., Wan, Z. & Waters, K. J. (1992). Ozone depletion: ultraviolet radiation and phytoplankton biology in Antarctic waters. *Science*, **255**, 952–9. *124, 286*
839. Smith, R. C., Prézelin, B. B., Bidigare, R. R. & Baker, K. S. (1989). Bio-optical modelling of photosynthetic production in coastal waters. *Limnol. Oceanogr.*, **34**, 1524–44. *279, 356*
840. Smith, R. C. & Tyler, J. E. (1967). Optical properties of clear natural water. *J. Opt. Soc. Amer.*, **57**, 289–95. *82*
841. Smith, R. C. & Wilson, W. H. (1981). Ship and satellite bio-optical research in the California Bight. In J. F. R. Gower (ed.) *Oceanography from space* (pp. 281–94). New York: Plenum. *188*
842. Smith, R. G. & Bidwell, R. G. S. (1987). Carbonic anhydrase-dependent inorganic carbon uptake by the red macroalga *Chondrus crispus*. *Plant Physiol.*, **83**, 735–8. *327*
843. Smith, R. G. & Bidwell, R. G. S. (1989). Mechanism of photosynthetic carbon dioxide uptake by the red macroalga, *Chondrus crispus*. *Plant Physiol.*, **89**, 93–9. *327, 332*
844. Smith, R. G., Wheeler, W. N. & Srivastava, L. M. (1983). Seasonal photosynthetic performance of *Macrocystis integrifolia* (Phaeophyceae). *J. Phycol.*, **19**, 352–9. *281, 411*
845. Smith, W. O. (1977). The respiration of photosynthetic carbon in eutrophic areas of the ocean. *J. Mar. Res.*, **35**, 557–65. *277*
846. Smith, W. O. & Nelson, D. M. (1986). Importance of ice edge phytoplankton production in the Southern Ocean. *BioScience*, **36**, 251–7. *318*
847. Solazzi, A. & Tolomio, C. (1976). Sea depth effects on the algal photosynthetic apparatus. I. Chlorophyll pigments of *Halimeda tuna* Lam. (Chlorophyceae, Siphonales). *Mem. Biol. mar. Ocean*, **6**, 21–7. *395*
848. Sommer, U. & Gliwicz, Z. M. (1986). Long range vertical migration of *Volvox* in tropical Lake Cahora Bassa (Mozambique). *Limnol. Oceanogr.*, **31**, 650–3. *417*
849. Søndergaard, M. & Bonde, G. (1988). Photosynthetic characteristics and pigment content and composition in *Littorella uniflora* (L.) Aschers in a depth gradient. *Aquat. Bot.*, **32**, 307–19. *395*
850. Sosik, H. M., Chisholm, S. W. & Olson, R. J. (1989). Chlorophyll fluorescence from single cells: interpretation of flow cytometric signals. *Limnol. Oceanogr.*, **34**, 1749–61. *111, 259*
851. Sournia, A. (1974). Circadian periodicities in natural populations of marine phytoplankton. *Adv. Mar. Biol.*, **12**, 325–89. *343, 423*
852. Sournia, A. (1976). Primary production of sands in the lagoon of an atoll and the role of Foraminiferan symbionts. *Mar. Biol.*, **37**, 29–32. *316*
853. Spalding, M. H. (1989). Photosynthesis and photorespiration in

freshwater green algae. *Aquat. Bot.*, **34**, 181–209. *332*

854. Spence, D. H. N. (1976). Light and plant response in fresh water. In G. C. Evans, R. Bainbridge & O. Rackham (eds), *Light as an ecological factor* (pp. 93–133). Oxford: Blackwell. *316, 421*
855. Spence, D. H. N., Campbell, R. M. & Chrystal, J. (1971). Spectral intensity in some Scottish freshwater lochs. *Freshwater Biol.*, **1**, 321–37. *141*
856. Spence, D. H. N., Campbell, R. M. & Chrystal, J. (1973). Specific leaf areas and zonation of freshwater macrophytes. *J. Ecol.*, **61**, 317–28. *402*
857. Spence, D. H. N. & Chrystal, J. (1970). Photosynthesis and zonation of freshwater macrophytes. I. Depth distribution and shade tolerance. II. Adaptability of species of deep and shallow water. *New Phytol.*, **69**, 205–15, 217–27. *383*
858. Spencer, J. W. (1971). Fourier series representation of the position of the sun. *Search*, **2**, 172. *35*
859. Spilhaus, A. F. (1968). Observations of light scattering in seawater. *Limnol. Oceanogr.*, **13**, 418–22. *94*
860. Spillane, M. C. & Doyle, D. M. (1983). Final results for STREX and JASIN photoanalyses with preliminary search for whitecap algorithm. In *Whitecaps and the marine atmosphere, Report 5* (pp. 8–27). Galway, Ireland: University College. *42*
861. Spinrad, R. W. & Brown, J. W. (1986). Relative real refractive index of marine microorganisms: a technique for flow cytometric estimation. *Appl. Opt.*, **25**, 1930–4. *111*
862. Spinrad, R. W., Zaneveld, J. R. V. & Pak, H. (1978). Volume scattering function of suspended particulate matter at near-forward angles: a comparison of experimental and theoretical values. *Appl. Opt.*, **17**, 1125–30. *95*
863. Spitzer, D. & Wernand, M. R. (1979*a*). Photon scalar irradiance meter. *Appl. Opt.*, **18**, 1698–700. *121*
864. Spitzer, D. & Wernand, M. R. (1979*b*). Irradiance and absorption spectra measurements in the tropical East Atlantic. In *17th General Assembly of IAPSO (Canberra, 1979)* (pp. Abstracts p. 73). San Diego: International Association for the Physical Sciences of the Ocean. *54*
865. Stabeno, P. J. & Monahan, E. C. (1983). The influence of whitecaps on the albedo of the sea surface. In *Whitecaps and the marine atmosphere, Report 5* (pp. 78–93). Galway, Ireland: University College. *43*
866. Stadelmann, P., Moore, J. E. & Pickett, E. (1974). Primary production in relation to temperature structure, biomass concentration, and light conditions at an inshore and offshore station in Lake Ontario. *J. Fish, Res. Bd Canada*, **31**, 1215–32. *140*
867. Staehelin, L. A. (1986). Chloroplast structure and supramolecular organization of photosynthetic membranes. *Encycl. Plant Physiol. (n.s.)*, **19**, 1–84. *217*
868. Staehelin, L. A. & Arntzen, C. J. (1986). Photosynthesis III: photosynthetic membranes and light harvesting systems. *Encycl. Plant Physiol. (n.s.)*, **19**, 802p. *246*
869. Stauber, J. L. & Jeffrey, S. W. (1988). Photosynthetic pigments in fifty-one species of marine diatoms. *J. Phycol.*, **24**, 158–72. *227*

869*a*. Stavn, R. H. & Weidemann, A. D. (1988). Optical modelling of clear ocean light fields: Raman scattering effects. *Appl. Opt.*, **27**, 4002–11. *150*

870. Steele, J. H. & Baird, I. E. (1965). The chlorophyll *a* content of particulate organic matter in the northern North Sea. *Limnol. Oceanogr.*, **10**, 261–7. *227*

871. Steele, J. H. & Yentsch, C. S. (1960). The vertical distribution of chlorophyll. *J. Mar. Biol. Ass. U.K.*, **39**, 217–26. *320*

872. Steemann Nielsen, E. (1952). The use of radioactive carbon (^{14}C) for measuring organic production in the sea. *J. Cons. perm. int. Explor. Mer*, **18**, 117–40. *272*

873. Steemann Nielsen, E. (1974). Light and primary production. In N. G. Jerlov & E. S. Nielsen (eds), *Optical aspects of Oceanography* (pp. 361–88). London: Academic Press. *315*

874. Steemann Nielsen, E. (1975). *Marine photosynthesis*. Amsterdam: Elsevier. *277, 327, 335, 336, 337, 344, 382, 400, 406*

875. Steemann Nielsen, E. & Hansen, V. K. (1961). Influence of surface illumination on plankton photosynthesis in Danish waters (56° N) throughout the year. *Physiol. Plant.*, **14**, 595–613. *279, 411–2*

876. Steemann Nielsen, E. & Jørgensen, E. G. (1968). The adaptation of plankton algae. *Physiol Plant.*, **21**, 401–13. *335, 336*

877. Sternberg, R. W., Baker, E. T., McManus, D. A., Smith, S. & Morrison, D. R. (1974). An integrating nephelometer for measuring particle concentration in the deep sea. *Deep-Sea Res.*, **21**, 887–92. *109*

878. Stokes, A. N. (1975). Proof of a law for calculating absorption of light by cellular suspensions. *Arch. Biochem. Biophys.*, **167**, 393–4. *255*

879. Stramski, D. & Morel, A. (1990). Optical properties of photosynthetic picoplankton in different physiological states as affected by growth irradiance. *Deep-Sea Res.*, **37**, 245–66. *111*

880. Strong, A. E. (1974). Remote sensing of algal blooms by aircraft and satellite in Lake Erie and Utah Lake. *Remote Sens. Env.*, **3**, 99–107. *195, 211*

881. Strong, A. E. (1978). Chemical whitings and chlorophyll distributions in the Great Lakes as viewed by Landsat. *Remote Sens. Env.*, **7**, 61–72. *209, 211*

882. Stross, R. G. & Sokol, R. C. (1989). Runoff and flocculation modify underwater light environment of the Hudson River Estuary. *Estuar. Coast. Shelf Sci.*, **29**, 305–16. *139*

883. Stuermer, D. H. & Harvey, G. R. (1978). Structural studies on marine humus. *Mar. Chem.*, **6**, 55–70. *62*

884. Stuermer, D. H. & Payne, J. R. (1976). Investigation of seawater and terrestrial humic substances with carbon-13 and proton nuclear magnetic resonance. *Geochim. Cosmochim. Acta*, **40**, 1109–14. *62*

885. Stumpf, R. P. & Pennock, J. R. (1989). Calibration of a general optical equation for remote sensing of suspended sediments in a moderately turbid estuary. *J. Geophys. Res.*, **94**, 14363–71. *193*

886. Stumpf, R. P. & Pennock, J. R. (1991). Remote estimation of the diffuse attenuation coefficient in a moderately turbid estuary. *Remote Sens. Environ.*, **38**, 183–91. *139, 205*

887. Stumpf, R. P. & Tyler, M. A. (1988). Satellite detection of bloom and pigment distributions in estuaries. *Remote Sens. Environ.*, **24**, 385–404. *211*

887*a*. Sugihara, S., Kishino, M. & Okami, N. (1984). Contribution of Raman scattering to upward irradiance in the sea. *J. Oceanogr. Soc. Jap.*, **40**, 397–404. *150*

888. Sukenik, A., Bennett, J. & Falkowski, P. (1987). Light-saturated photosynthesis – limitation by electron transport or carbon fixation? *Biochim. Biophys. Acta*, **891**, 205–15. *380*

889. Sukenik, A., Livne, A., Neori, A., Yacobi, Y. Z. & Katcoff, D. (1992). Purification and characterization of a light-harvesting chlorophyll–protein complex from the marine eustigmatophyte *Nannochloropsis* sp. *Plant Cell Physiol.*, **33**, 1041–8. *237*

890. Surif, M. B. & Raven, J. A. (1989). Exogenous inorganic carbon sources for photosynthesis in seawater by members of the Fucales and Laminariales (Phaeophyta): ecological and taxonomic implications. *Oecologia*, **78**, 97–105. *332*

891. Suttle, C. A., Chan, A. M. & Cottrell, M. T. (1990). Infection of phytoplankton by viruses and reduction of primary productivity. *Nature*, **347**, 467–9. *341*

892. Sverdrup, H. U. (1953). On conditions for the vernal blooming of phytoplankton. *J. Cons. int. Explor. Mer*, **18**, 287–95. *318*

893. Swift, E. & Taylor, W. R. (1967). Bioluminescence and chloroplast movement in the dinoflagellate *Pyrocystis lunula*. *J. Phycol.*, **3**, 77–81. *423*

894. Sydor, M. (1980). Remote sensing of particulate concentrations in water. *Appl. Opt.*, **19**, 2794–800. *193*

895. Taguchi, S. (1976). Relationship between photosynthesis and cell size of marine diatoms. *J. Phycol.*, **12**, 185–9. *301*

896. Takahashi, M., Ichimura, S., Kishino, M. & Okami, N. (1989). Shade and chromatic adaptation of phytoplankton photosynthesis in a thermally stratified sea. *Mar. Biol.*, **100**, 401–9. *279*

897. Talling, J. F. (1957*a*). Photosynthetic characteristics of some freshwater plankton diatoms in relation to underwater radiation. *New Phytol.*, **56**, 29–50. *278, 317, 338*

898. Talling, J. F. (1957*b*). The phytoplankton population as a compound photosynthetic system. *New Phytol.*, **56**, 133–49. *268, 276, 317, 352–4*

899. Talling, J. F. (1960). Self-shading effects in natural populations of a planktonic diatom. *Wett. Leben*, **12**, 235–42. *269, 292*

900. Talling, J. F. (1970). Generalized and specialized features of phytoplankton as a form of photosynthetic cover. In *Prediction and measurement of photosynthetic productivity* (pp. 431–45). Wageningen: Pudoc. *67, 82*

901. Talling, J. F. (1971). The underwater light climate as a controlling factor in the production ecology of freshwater phytoplankton. *Mitt. int. Verein. Limnol.*, **19**, 214–43. *269, 317, 343*

902. Talling, J. F. (1976). The depletion of carbon dioxide from lake water by phytoplankton. *J. Ecol.*, **64**, 79–121. *327, 330*

903. Talling, J. F. (1979). Factor interactions and implications for the prediction of lake metabolism. *Arch. Hydrobiol. Beih. ergebn. Limnol*, **13**, 96–109. *330–1*

904. Talling, J. F., Wood, R. B., Prosser, M. V. & Baxter, R. M. (1973). The upper limit of photosynthetic productivity by phytoplankton: evidence from Ethiopian soda lakes. *Freshwater Biol.*, **3**, 53–76. *303*

905. Tam, C. K. N. & Patel, A. C. (1979). Optical absorption coefficients of water. *Nature*, **280**, 302–4. *56*
906. Tamiya, H. (1957). Mass culture of algae. *Ann. Rev. Plant Physiol.*, **8**, 309–34. *430*
907. Tanada, T. (1951). The photosynthetic efficiency of carotenoid pigments in *Navicula minima*. *Amer. J. Bot.*, **38**, 276–83. *309*
908. Tandeau de Marsac, N. (1977). Occurrence and nature of chromatic adaptation in cyanobacteria. *J. Bacteriol.*, **130**, 82–91. *385, 387*
909. Tassan, S. & Sturm, B. (1986). An algorithm for the retrieval of sediment content in turbid coastal waters from CZCS data. *Int. J. Remote Sens.*, **7**, 643–55. *190, 194*
910. Taylor, W. R. (1959). Distribution in depth of marine algae in the Caribbean and adjacent seas. In *Proc. 9th Int. Botan. Congr.* (pp. 193–7). Canada: University of Toronto Press. *365–6*
911. Taylor, W. R. (1964). Light and photosynthesis in intertidal benthic diatoms. *Helgoländer Wiss. Meeresunters*, **10**, 29–37. *280*
912. Thayer, G. W., Wolfe, D. A. & Williams, R. B. (1975). The impact of man on seagrass ecosystems. *Amer. Scientist*, **63**, 288–95. *429*
913. Thinh, L.-V. (1983). Effect of irradiance on the physiology and ultrastructure of the marine cryptomonad, *Cryptomonas* strain Lis (Cryptophyceae). *Phycologia*, **22**, 7–11. *376, 378*
914. Thom, R. M. & Albright, R. G. (1990). Dynamics of benthic vegetation standing-stock, irradiance and water properties in central Puget Sound. *Mar. Biol.*, **104**, 129–41. *349*
915. Thomas, W. H. & Gibson, C. H. (1990). Quantified small-scale turbulence inhibits a red tide dinoflagellate *Gonyaulax polyedra* Stein. *Deep-Sea Res.*, **37**, 1583–93. *341*
916. Thornber, J. P. (1986) . Biochemical characterization and structure of pigment–proteins of photosynthetic organisms. *Encycl. Plant Physiol. (n.s.)*, **19**, 98–142. *235*
917. Thornber, J. P., Morishige, D. T., Anandan, S. & Peter, G. F. (1991). Chlorophyll–Carotenoid proteins of higher plant thylakoids. In H. Scheer (ed.), *Chlorophylls* (pp. 549–85). Boca Raton: CRC Press. *235*
918. Thorne, S. W., Newcomb, E. H. & Osmond, C. B. (1977). Identification of chlorophyll *b* in extracts of prokaryotic algae by fluorescence spectroscopy. *Proc. Natl. Acad. Sci. Wash.*, **74**, 575–8. *223, 228*
919. Thurman, E. M. Quoted by Josephson, J. (1982). In Humic substances. *Environ. Sci. Technol.*, **16**, 20A–4A. *58*
920. Tilzer, M. M. (1973). Diurnal periodicity in the phytoplankton assemblage of a high mountain lake. *Limnol. Oceanogr.*, **18**, 15–30. *343, 415, 416*
921. Tilzer, M. M. (1983). The importance of fractional light absorption by photosynthetic pigments for phytoplankton productivity in Lake Constance. *Limnol. Oceanogr.*, **28**, 833–46. *269, 292, 346*
922. Tilzer, M. M. (1984). Seasonal and diurnal shifts of photosynthetic quantum yields in the phytoplankton of Lake Constance. *Verh. Internat. Verein. Limnol.*, **22**, 958–62. *302*
923. Tilzer, M. M. (1987). Prediction of productivity changes in Lake Tahoe at increasing phytoplankton biomass. *Int. Ver. Theor. Angew. Limnol. Verh.*, **20**, 407–13. *269*
924. Tilzer, M. M. & Goldman, C. R. (1978). Importance of mixing, thermal

stratification and light adaptation for phytoplankton productivity in Lake Tahoe (California–Nevada). *Ecology*, **59**, 810–21. *321, 396–7, 401–2, 404*

925. Tilzer, M. M., Goldman, C. R. & Amezaga, E. D. (1975). The efficiency of photosynthetic light energy utilization by lake phytoplankton. *Verh. int. verein. Limnol.*, **19**, 800–7. *304*
926. Tilzer, M. M. & Schwarz, K. (1976). Seasonal and vertical patterns of phytoplankton light adaptation in a high mountain lake. *Arch. Hydrobiol.*, **77**, 488–504. *415*
927. Timofeeva, V. A. (1971). Optical characteristics of turbid media of the sea-water type. *Izv. atmos. oceanic Phys.*, **7**, 863–5. *107*
928. Timofeeva, V. A. (1974). Optics of turbid water. In N. G. Jerlov & E. S. Nielsen (eds), *Optical aspects of oceanography* (pp. 177–219). London: Academic Press. *154*
929. Timofeeva, V. A. & Gorobets, F. I. (1967). On the relationship between the attenuation coefficients of collimated and diffuse light fluxes. *Izv. atmos. oceanic Phys.*, **3**, 166–9. *92*
930. Titus, J. E. & Adams, M. S. (1979). Coexistence and the comparative light relations of the submersed macrophytes *Myriophyllum spicatum* L. and *Vallisneria americana* Michx. *Oecologia*, **40**, 273–86. *279*
931. Topliss, B. J., Amos, C. L. & Hill, P. R. (1990). Algorithms for remote sensing of high concentration, inorganic suspended sediment. *Int. J. Remote Sens.*, **11**, 947–66. *192*
932. Townsend, D. W., Keller, M. D., Sieracki, M. E. & Ackleson, S. G. (1992). Spring phytoplankton blooms in the absence of vertical water column stratification. *Nature*, **360**, 59–62. *343*
933. Townsend, D. W. & Spinrad, R. W. (1986). Early spring phytoplankton blooms in the Gulf of Maine. *Continental Shelf Res.*, **6**, 515–29. *318*
934. Tranter, D. J. (1982). Interlinking of physical and biological processes in the Antarctic Ocean. *Oceanogr. Mar. Biol. Ann. Rev.*, **20**, 11–35. *410*
935. Tsuzuki, M. (1983). Mode of HCO_3^- utilization by the cells of *Chlamydomonas reinhardtii* grown under ordinary air. *Z. Pflanzenphysiol.*, **110**, 29–37. *327*
936. Tsuzuki, M. & Miyachi, S. (1990). Transport and fixation of inorganic carbon in photosynthesis of cyanobacteria and green algae. *Bot. Mag. (Tokyo) Special Issue*, **2**, 43–52. *332*
937. Tyler, J. E. (1960). Radiance distribution as a function of depth in an underwater environment. *Bull. Scripps Inst. Oceanogr.*, **7**, 363–411. *124, 147, 152, 154–5, 159*
938. Tyler, J. E. (1968). The Secchi disc. *Limnol. Oceanogr.*, **13**, 1–6. *119*
939. Tyler, J. E. (1975). The *in situ* quantum efficiency of natural phytoplankton populations. *Limnol. Oceanogr.*, **20**, 976–80. *138*
940. Tyler, J. E. (1978). Optical properties of water. In W. G. Driscoll & W. Vaughan (eds), *Handbook of optics* (pp. 15–1 to 15–38). New York: McGraw-Hill. *158*
941. Tyler, J. E. & Richardson, W. H. (1958). Nephelometer for the measurement of volume scattering function *in situ*. *J. Opt. Soc. Amer.*, **48**, 354–7. *94*
942. Tyler, J. E. & Smith, R. C. (1967). Spectroradiometric characteristics of

natural light under water. *J. Opt. Soc, Amer.*, **57**, 595–601. *200*

943. Tyler, J. E. & Smith, R. C. (1970). *Measurements of spectral irradiance underwater*. New York: Gordon & Breach. *32, 123, 131, 132, 135, 138, 139, 140, 148, 150*
944. Vadas, R. L. & Steneck, R. S. (1988). Zonation of deep water benthic algae in the Gulf of Maine. *J. Phycol.*, **24**, 338–46. *363*
945. van de Hulst, H. C. (1957). *Light scattering by small particles*. New York: Wiley. *89–90*
946. Van, T. K., Haller, W. T. & Bowes, G. (1976). Comparison of the photosynthetic characteristics of three submersed aquatic plants. *Plant Physiol.*, **58**, 761–8. *274–5, 276, 278, 279, 325*
947. Van Wijk, W. R. & Ubing, D. W. S. (1963). In W. R. Van Wijk (ed.), *Physics of plant environment*. Amsterdam: North Holland. Quoted by Monteith (1973). *38*
948. Vandevelde, T., Legendre, L., Demers, S. & Therriault, J. C. (1989). Circadian variations in photosynthetic assimilation and estimation of daily phytoplankton production. *Mar. Biol.*, **100**, 525–31. *426*
949. Vant, W. N. (1990). Causes of light attenuation in nine New Zealand estuaries. *Estuar. Coast. Shelf Sci.*, **31**, 125–37. *67, 98, 102, 140*
950. Vant, W. N. & Davies-Colley, R. J. (1984). Factors affecting clarity of New Zealand lakes. *N.Z. J. Mar. Freshwater Res.*, **18**, 363–77. *98, 104, 119*
951. Vant, W. N. & Davies-Colley, R. J. (1988). Water appearance and recreational use of 10 lakes of the North Island (New Zealand). *Verh. Internat. Verein. Limnol.*, **23**, 611–15. *119*
952. Vant, W. N., Davies-Colley, R. J., Clayton, J. S. & Coffey, B. T. (1986). Macrophyte depth limits on North Island (New Zealand) lakes of differing clarity. *Hydrobiologia*, **137**, 55–60. *316*
953. Venrick, E. L. (1984). Winter mixing and the vertical stratification of phytoplankton – another look. *Limnol. Oceanogr.*, **29**, 636–40. *320*
954. Verdin, J. P. (1985). Monitoring water quality conditions in a large Western reservoir with Landsat imagery. *Photogramm. Eng. Remote Sens.*, **51**, 343–53. *185, 206*
955. Vesk, M. & Jeffrey, S. W. (1977). Effect of blue-green light on photosynthetic pigments and chloroplast structure in unicellular marine algae from six classes. *J. Phycol.*, **13**, 280–8. *391–2*
956. Vierling, E. & Alberte, R. S. (1980). Functional organization and plasticity of the photosynthetic unit of the cyanobacterium *Anacystis nidulans*. *Physiol. Plant.*, **50**, 93–8. *376, 379*
957. Vincent, W. F., Neale, P. J. & Richerson, P. J. (1984). Photoinhibition: algal responses to bright light during diel stratification and mixing in a tropical alpine lake. *J. Phycol.*, **20**, 201–11. *288*
958. Viollier, M., Deschamps, P. Y. & Lecomte, P. (1978). Airborne remote sensing of chlorophyll content under cloudy skies as applied to the tropical waters in the Gulf of Guinea. *Remote Sens. Env.*, **7**, 235–48. *209*
959. Viollier, M., Tanré, D. & Deschamps, P. Y. (1980). An algorithm for remote sensing of water colour from space. *Boundary-layer Meteorol.*, **18**, 247–67. *188, 190, 199*
960. Virgin, H. I. (1964). Some effects of light on chloroplasts and plant protoplasm. In A. C. Giese (ed.), *Photophysiology* (pp. 273–303). New York: Academic Press. *421*

961. Visser, S. A. (1984). Seasonal changes in the concentration and colour of humic substances in some aquatic environments. *Freshwater Biol.*, **14**, 79–87. *59*
962. Vollenweider, R. A. (1969) (ed.) *A manual on methods for measuring primary production in aquatic environments.* Oxford: Blackwell. *271*
963. Vollenweider, R. A. (1970). Models for calculating integral photosynthesis and some implications regarding structural properties of the community metabolism of aquatic systems. In *Prediction and measurement of photosynthetic productivity* (pp. 455–72). Wageningen: Pudoc. *352*
964. Voss, K. J. (1989). Electro-optic camera system for measurement of the underwater radiance distribution. *Optical Engineering*, **28**, 241–7. *125*
965. Waaland, J. R., Waaland, S. D. & Bates, G. (1974). Chloroplast structure and pigment composition in the red alga *Griffithsia pacifica*: regulation by light intensity. *J. Phycol.*, **10**, 193–9. *376*
966. Walker, G. A. H., Buchholz, V. L., Camp, D., Isherwood, B., Glaspey, J., Coutts, R., Condal, A. & Gower, J. (1974). A compact multichannel spectrometer for field use. *Rev. sci. Instrum.*, **45**, 1349–52. *175*
967. Walker, J. E. & Walsby, A. E. (1983). Molecular weight of gas-vesicle protein from the planktonic cyanobacterium *Anabaena flos-aquae* and implications for structure of the vesicle. *Biochem J.*, **209**, 809–15. *418*
968. Wallen, D. G. & Geen, G. H. (1971*a*). Light quality in relation to growth, photosynthetic rates and carbon metabolism in two species of marine plankton algae. *Mar. Biol.*, **10**, 34–43. *391*
969. Wallen, D. G. & Geen, G. H. (1971*b*). Light quality and concentrations of protein, RNA, DNA and photosynthetic pigments in two species of marine plankton algae. *Mar. Biol.*, **10**, 44–51. *391*
970. Wallentinus, I. (1978). Productivity studies on Baltic macroalgae. *Bot. Mar.*, **21**, 365–80. *281, 282*
971. Walmsley, R. D. & Bruwer, C. A. (1980). Water transparency characteristics of South African impoundments. *J. Limnol. Soc. Sth Afr.*, **6**, 69–76. *141*
972. Walsby, A. E. (1975). Gas vesicles. *Ann. Rev. Plant Physiol.*, **26**, 427–39. *417*
973. Walsby, A. E. & Booker, M. J. (1976). The physiology of water-bloom formation by planktonic blue-green algae. *Br. Phycol. J.*, **11**, 200. *419*
974. Walsh, J. J. (1976). Herbivory as a factor in patterns of nutrient utilization in the sea. *Limnol. Oceanogr.*, **21**, 1–13. *340*
975. Walsh, J. J. (1981). A carbon budget for overfishing off Peru. *Nature*, **290**, 300–4. *359*
976. Waterbury, J. B., Watson, S. W., Valois, F. W. & Franks, D. G. (1986). Biological and ecological characterization of the marine unicellular cyanobacterium *Synechococcus*. In T. Platt & W. K. W. Li (eds), *Photosynthetic picoplankton* (pp. 71–120). *Can. Bull. Fish Aquat. Sci.*, **214**. *240, 339*
977. Waters, K. J., Smith, R. C. & Lewis, M. R. (1990). Avoiding ship-induced light-field perturbation in the determination of oceanic optical properties. *Oceanography*, **3**, 18–21. *118*

978. Webb, W. L., Newton, M. & Starr, D. (1974). Carbon dioxide exchange of *Alnus rubra*: a mathematical model. *Oecologia*, **17**, 281–91. *283*

979. Weeks, A. & Simpson, J. H. (1991). The measurement of suspended particulate concentrations from remotely-sensed data. *Int. J. Remote Sens.*, **12**, 725–37. *209*

980. Weidemann, A. D. & Bannister, T. T. (1986). Absorption and scattering coefficients in Irondequoit Bay. *Limnol. Oceanogr.*, **31**, 567–83. *68, 98, 103, 110, 140, 269*

981. Weidemann, A. D., Bannister, T. T., Effler, S. W. & Johnson, D. L. (1985). Particulate and optical properties during $CaCO_3$ precipitation in Otisco Lake. *Limnol. Oceanogr.*, **30**, 1078–83. *68, 98, 103*

982. Weinberg, S. (1975). Ecologie des Octocoralliaires communs du substrat dur sans la région de Banyuls-sur-Mer. *Bijdr. Dierk*, **45**, 50–70. *349*

983. Weller, R. A., Dean, J. P., Marra, J., Price, J. F., Francis, E. A. & Boardman, D. C. (1985). Three-dimensional flow in the upper ocean. *Science*, **227**, 1552–6. *287–8*

984. Welschmeyer, N. A. & Lorenzen, C. J. (1981). Chlorophyll-specific photosynthesis and quantum efficiency at subsaturating light intensities. *J. Phycol*, **17**, 283–93. *301*

984*a*. West, G. S. (1904). *The British freshwater algae*. Cambridge: Cambridge University Press. *219*

985. Westlake, D. F. (1967). Some effects of low-velocity currents on the metabolism of aquatic macrophytes. *J. exp. Bot.*, **18**, 187–205. *331*

986. Westlake, D. F. (1980*a*). Photosynthesis: macrophytes. In E. D. Le Cren & R. H. Lowe-McConnell (eds), *The functioning of freshwater ecosystems* (pp. 177–82). Cambridge: Cambridge University Press. *351, 429*

987. Westlake, D. F. (1980*b*). Biomass changes: macrophytes. In E. D. Le Cren & R. H. Lowe-McConnell (eds), *The functioning of freshwater ecosystems* (pp. 203–6). Cambridge: Cambridge University Press. *342, 350*

988. Westlake, D. F. (1980*c*). Effects of macrophytes. In E. D. Le Cren & R. H. Lowe-McConnell (eds), *The functioning of freshwater ecosystems* (pp. 161–2). Cambridge: Cambridge University Press. *270*

989. Whatley, J. M. (1977). The fine structure of *Prochloron. New Phytol.*, **79**, 309–13. *223*

990. Wheeler, J. R. (1976). Fractionation by molecular weight of organic substances in Georgia coastal water. *Limnol. Oceanogr.*, **21**, 846–52. *65*

991. Wheeler, W. N. (1980*a*). Pigment content and photosynthetic rate of the fronds of *Macrocystis pyrifera. Mar. Biol.*, **56**, 97–102. *394, 398–9*

992. Wheeler, W. N. (1980*b*). Effect of boundary layer transport on the fixation of carbon by the giant kelp *Macrocystis pyrifera. Mar. Biol.*, **56**, 103–10. *331*

993. Wheeler, W. N., Smith, R. G. & Srivastava, L. M. (1984). Seasonal photosynthetic performance of *Nereocystis luetkeana. Can. J. Bot.*, **62**, 664–70. *281, 282, 411*

994. Whitlock, C. H., Bartlett, D. S. & Gurganus, E. A. (1982). Sea foam reflectance and influence on optimum wavelength for remote sensing of ocean aerosols. *Geophys. Res. Lett.*, **9**, 719–22. *43*

995. Whitney, L. V. (1941). The angular distribution of characteristic diffuse light in natural waters. *J. Mar. Res.*, **4**, 122–31. *154*

996. Wiginton, J. R. & McMillan, C. (1979). Chlorophyll composition under controlled light conditions as related to the distribution of seagrasses in Texas and the U.S. Virgin Islands. *Aquat. Bot.*, **6**, 171–84. *395*

997. Wild, A., Ke, B. & Shaw, E. R. (1973). The effect of light intensity during growth of *Sinapis alba* on the electron-transport components. *Z. Pflanzenphysiol.*, **69**, 344–50. *379*

998. Wilhelm, C. (1990). The biochemistry and physiology of light-harvesting processes in chlorophyll b- and chlorophyll c-containing algae. *Plant Physiol. Biochem.*, **28**, 293–306. *235, 237*

999. Williams, D. F. (1984). Overview of the NERC airborne thematic mapper campaign of September 1982. *Int. J. Remote Sens.*, **5**, 631–4. *177*

1000. Williams, R. B. & Murdoch, M. B. (1966). Phytoplankton production and chlorophyll concentration in the Beaufort Channel, North Carolina. *Limnol. Oceanogr.*, **11**, 73–82. *338, 345*

1001. Wofar, M. V. M., Corre, P. L. & Birrien, J. L. (1983). Nutrients and primary production in permanently well-mixed temperate coastal waters. *Estuar. Coast. Shelf Sci.*, **17**, 431–46. *318*

1002. Wollman, F.-A. (1986). Photosystem I proteins. *Encycl. Plant Physiol. (n.s.)*, **19**, 487–95. *235*

1003. Womersley, H. B. S. (1981). Marine ecology and zonation of temperate coasts. In M. N. Clayton & R. J. King (eds), *Marine botany: an Australian perspective* (pp. 211–40). Melbourne: Longman Cheshire. *370*

1004. Yamada, T., Ikawa, T. & Nisizawa, K. (1979). Circadian rhythm of the enzymes participating in the CO_2-photoassimilation of a brown alga, *Spatoglossum pacificum*. *Bot. Mar.*, **22**, 203–9. *426*

1005. Yentsch, C. M. and others (1983). Flow cytometry and cell sorting: a technique for analysis and sorting of aquatic particles. *Limnol. Oceanogr.*, **28**, 1275–80. *111*

1006. Yentsch, C. S. (1960). The influence of phytoplankton pigments on the colour of sea water. *Deep-Sea Res.*, **7**, 1–9. *52*

1007. Yentsch, C. S. (1974). Some aspects of the environmental physiology of marine phytoplankton: a second look. *Oceanogr. mar. Biol. Ann. Rev.*, **12**, 41–75. *336, 337*

1008. Yentsch, C. S. & Lee, R. W. (1966). A study of photosynthetic light reactions, and a new interpretation of sun and shade phytoplankton. *J. mar. Res.*, **24**, 319–37. *400*

1009. Yentsch, C. S. & Phinney, D. A. (1989). A bridge between ocean optics and microbial ecology. *Limnol. Oceanogr.*, **34**, 1694–705. *76*

1010. Yentsch, C. S. & Scagel, R. F. (1958). Diurnal study of phytoplankton pigments. An *in situ* study in East Sound. Washington. *J. Mar. Res.*, **17**, 567–83. *427*

1011. Yeoh, H.-H., Badger, M. R. & Watson, L. (1981). Variation in kinetic properties of ribulose-1, 5-bisphosphate carboxylases among plants. *Plant Physiol.*, **67**, 1151–5. *324*

1012. Yocum, C. S. & Blinks, L. R. (1958). Light-induced efficiency and pigment alterations in red algae. *J. Gen. Physiol.*, **41**, 1113–17. *388–9, 409*

1013. Yoder, J. A., Atkinson, L. P., Lee, T. N., Kim, H. H. & McLain, C. R. (1981). Role of Gulf Stream frontal eddies in forming phytoplankton patches on the outer southeastern shelf. *Limnol. Oceanogr.*, **26**, 1103–10. *210*

1014. Yokohama, Y. (1973). A comparative study on photosynthesis-temperature relationships and their seasonal changes in marine benthic algae. *Int. Rev. Ges. Hydrobiol.*, **58**, 463–72. *333*

1015. Yokohama, Y. (1981). Distribution of the green light-absorbing pigments siphonaxanthin and siphonein in marine green algae. *Botanica Marina*, **24**, 637–40. *368*

1016. Yokohama, Y., Kageyama, A., Ikawa, T. & Shimura, S. (1977). A carotenoid characteristic of Chlorophycean seaweeds living in deep coastal waters. *Bot. Mar.*, **20**, 433–6. *231, 235, 368*

1017. Yokohama, Y. & Misonou, T. (1980). Chlorophyll *a:b* ratios in marine benthic green algae. *Jap. J. Phycol.*, **28**, 219–23. *228*

1018. Yu, M., Glazer, A. N., Spencer, K. G. & West, J. A. (1981). Phycoerythrins of the red alga *Callithamnion. Plant Physiol.*, **68**, 482–8. *241*

1019. Zaneveld, J. R. V., Bartz, R. & Kitchen, J. C. (1990). A reflective-tube absorption meter. *Proc. Soc. Photo-Opt. Instrum. Eng., Ocean Optics X*, **1302**, 124–36. *54*

1020. Zibordi, G., Parmiggiani, F. & Albertanza, L. (1990). Application of aircraft multispectral scanner data to algae mapping over the Venice Lagoon. *Remote Sens. Environ.*, **34**, 49–54. *211*

1021. Zijlstra, J. J. (1988). The North Sea ecosystem. In H. Postma & J. J. Zijlstra (eds), *Continental shelves* (pp. 231–77). Amsterdam: Elsevier. *359*

1022. Zohary, T. & Robarts, R. D. (1989). Diurnal mixed layers and the long-term dominance of *Microcystis aeruginosa. J. Plankton Res.*, **11**, 25–48. *420*

Index to symbols

Listed below are the numbers of the pages where each of the more commonly used symbols is first introduced and defined.

Index to organisms

Index to water bodies

Marine and estuarine waters

Fresh waters
(L. = Lake, Lough (Ireland) or Loch (Scotland); R. = River)

Subject Index

Page numbers in italic type refer to figures and tables.